气候变化教育研究丛书
上海终身教育研究院　总主编

气候变化教育的中国行动

汉英双语

第一辑

李家成　杨志平　顾惠芬　主编

上海交通大学出版社

内容提要

本书围绕"教育能为气候变化做什么"展开。气候变化教育涉及气象学、地理学、物理学、化学、社会学、经济学、政治学等多学科内容,其实施需要跨学科思维。以学生为中心的体验式教学、探究式学习对于气候变化教育最为有效,为此本书收集了来自全国各类学校的 35 个实践案例,围绕气候适应性导向下的校园建设、基于学科教学的气候变化教育、基于项目化学习的气候变化教育、基于综合活动的气候变化教育、学校家庭社区协同开展气候变化教育展开。这也是对国际气候变化议题的回应。因此,本书设计为汉英双语,作为第六届全球学习型城市大会的宣传材料,以此体现中国担当,传承中国智慧,贡献中国方案。

本书适合对环保教育、探究性活动教学感兴趣的中小学教师、学校管理者、教育研究者阅读。

图书在版编目(CIP)数据

气候变化教育的中国行动. 第一辑 : 汉、英 / 李家成,杨志平,顾惠芬主编. -- 上海 : 上海交通大学出版社, 2024.11. -- ISBN 978-7-313-31814-5

Ⅰ.P467-4

中国国家版本馆 CIP 数据核字第 2024N0A481 号

气候变化教育的中国行动(第一辑):汉英双语
QIHOU BIANHUA JIAOYU DE ZHONGGUO XINGDONG (DI-YI JI):HAN-YING SHUANGYU

主　　编:	李家成　杨志平　顾惠芬		
出版发行:	上海交通大学出版社	地　　址:	上海市番禺路 951 号
邮政编码:	200030	电　　话:	021-64071208
印　　刷:	上海锦佳印刷有限公司	经　　销:	全国新华书店
开　　本:	787mm×1092mm 1/16	印　　张:	38.25
字　　数:	942 千字		
版　　次:	2024 年 11 月第 1 版	印　　次:	2024 年 11 月第 1 次印刷
书　　号:	ISBN 978-7-313-31814-5		
定　　价:	188.00 元(全两册)		

版权所有　侵权必究
告 读 者:如发现本书有印装质量问题请与印刷厂质量科联系
联系电话:021-56401314

气候变化教育研究丛书
上海终身教育研究院　总主编
气候变化教育的中国行动（第一辑）

团队成员：

王丫珍	王巧玲	王 莹	王 萌	王 燕	平怡菁	付 蓉	龙超霞
巩淑青	朱 敏	刘美芳	孙 娟	孙童童	李书涵	李星月	李秋菊
李家成	苏 琳	张永红	张 杰	张 勇	张 婧	陈才英	陈亚兰
陈宝瑜	陈 爽	吴钰涵	林建芬	杨子仪	杨志平	杨懿凝	周晓娟
范漪猗	赵泽龙	赵 静	赵嫔婷	胡志文	姜 琳	姚 瑶	施一凡
顾伟伟	顾培培	顾维彬	顾惠芬	钱欢欣	徐倩华	郭友纯	倪 菡
黄 婕	盛晓燕	程育艳	曾玉芳	蒋丹妍	谢楠楠	蓝美琴	蔡 燕
蔡晓峰	潘 虹	藤依舒					

主持单位：

华东师范大学上海终身教育研究院

支持单位：

华东师范大学基础教育改革与发展研究所
教育部中学校长培训中心
联合国教科文组织中国可持续发展教育项目组
中国可持续发展教育研究共同体
北京教育科学研究院生态文明与可持续发展教育创新工作室
全国气候变化教育研究联盟
全国代际学习研究联盟

支持课题：

国家社科基金教育学重点课题"服务全民终身学习视域下社区教育体系研究"（AKA210019）

前　言

气候变化就在我们身边,而教育的力量并未缺席,中国教育工作者的贡献没有缺位。

在 2019 年上海加入联合国教科文组织全球学习型城市网络、担任可持续发展教育(ESD)全球集群的牵头城市之一的背景下,上海终身教育研究院聚焦全球气候变化问题,探索本土原创教育经验,助力可持续发展目标落地,参与全球对话与全球治理。特别是自 2023 年以来,上海终身教育研究院联合相关科研院所、大中小学等教育机构,进而联动企事业单位和社会组织,全力投入气候变化教育的实践探索与理论创新,秉持对全民终身学习的追求,形成了对从中小学生到老年大学学员的全年龄段的关注,形成了学校家庭社会协同育人的教育格局。

这一努力,在 2023 年 9 月 15 日举行的以"协同·赋能·创生"为主题的"学生暑假生活与学期初生活重建研究"全国现场会上,通过来自常州市新北区龙虎塘实验小学陈亚兰副校长、上海虹口区实验学校唐莺老师、深圳市宝安中学(集团)实验学校陈才英老师的三大实验性案例,有了突破,有了面向全国同行的第一次集中呈现。

2023 年 12 月 12 日,"第一届学校家庭社会协同开展气候变化教育研讨会"在上海市闵行区七宝明强小学举行。该次会议由上海终身教育研究院等多家单位联合举办,参会者来自全国各地,尤其是在与当地镇政府和区气象局的合作中,形成了多方协同的新格局。

2024 年 3 月 1 日,"第二届学校家庭社会协同开展气候变化教育研讨会"在江苏省常州市龙虎塘实验小学举行。该次会议集合了更多的政府部门、科研院所、学术机构的力量,尤其是得到生态环境部宣教中心专家的指导,且有了大量企业参与的经验积累。

2024 年 4 月 26 日,"2024 上海气候周气候变化教育论坛"在华东师范大学举办。更丰富的案例成果得以交流,多部门、多主体合作,面向从儿童到老人的全民终身学习,理论与实践密切互动的发展范式进一步清晰,并产生了积极的影响。尤其是在老年大学发展背景下开展气候变化教育,形成了有一定国内国际影响力的成果。

2024 年 5 月中旬,上海终身教育研究院正式提交了建立"华东师范大学气候变化教育实验室"的意向书。这一文科实验室建设的启动,推动了适合中国国情的气候变化教育模式的创新,将促进本领域的中国自主知识体系建设,并为国内研究人员和学生提供参与国际研究的机会,培养一批具有国际视野的气候变化教育人才。

本书中的案例来自中国各省市的教育工作者,致力于回应国际议题。中国教育实践者和研究者正在以开放的心态面对全球问题,将日常生活中的气候变化与参与全球气候治理联系起来,并更主动地交流分享中国的思想与实践。

这本书所呈现的教育实践,体现着中国担当。中国教育工作者特别重视从儿童到老人的全民终身学习,因此,将气候变化教育从面向青少年学生而扩展到了全民,而且,以学校家庭社会协同的方式进行,使得教育的力量延伸到了家庭和社会。这体现出中国教育工作者的大教育观。

全国各地的实践,传承着中国智慧。中华优秀传统文化滋养着中国人的生活,生态文明的思想直接促进着中国教育的变革。通过气候变化教育,更能体会到中国智慧的力量,也将在具体的代际学习和各类人群的共学互学中,体现蕴藏在民间的智慧。

这一阶段的努力,也为全球贡献了中国方案。在上海终身教育研究院的主持和推动下,上海开展气候变化教育的研究成果被作为典型案例,由联合国教科文组织向全球发布。上海终身教育研究院还积极主动地创设各类对话交流,提供相关课程资源,促进中国开展气候变化教育的思路、战略、策略、成效等不断被国外朋友了解和关注。

这本书的出版,就是上述努力的一次汇聚。在出版过程中,实践在继续一往无前,国际交流与合作不断深化,以教育的力量为全球气候治理所作的贡献也未停滞,发生在日常生产、生活和治理场景中的气候变化教育也在继续进行。

李家成、杨志平、顾惠芬

2024 年 6 月 1 日

目 录

第一篇　气候适应性导向下的校园建设 …………………………………………… 1
- 节能背景下教室智控灯光系统设计 ……………………………………………… 2
- 校园"鱼菜共生馆"建设及低碳教育实践 ………………………………………… 8
- 建一座不断成长的气候变化探索馆 ……………………………………………… 12
- "零碳校园"建设构想及教育实践 ………………………………………………… 20
- 🎓 对话专家 ………………………………………………………………………… 27

第二篇　基于学科教学的气候变化教育 …………………………………………… 29
- 小学语文教学与生态教育的融合实践 …………………………………………… 30
- 生态教育背景下小学语文单元教学活动设计 …………………………………… 34
- 小学数学项目化学习促进低碳生活的探索 ……………………………………… 38
- 英语课堂教学中的气候变化教育实践 …………………………………………… 44
- 基于科学实践活动的气候变化教育：以"降水量的测量"为例 ………………… 50
- 巧学·慧用·智创：信息科技教育与气候变化教育融合模式建构 …………… 56
- 生态文明建设背景下小学道德与法治学科教学活动设计 ……………………… 62
- 🎓 对话专家 ………………………………………………………………………… 70

第三篇　基于项目化学习的气候变化教育 ………………………………………… 73
- "失宠鞋"，你失宠了吗？——让循环使用思想融入小学生日常生活 ………… 74
- 关于牛奶盒回收再利用的研究 …………………………………………………… 79
- 基于 STEM 综合活动开展气候变化教育实践 …………………………………… 85
- 多学科融合的"咖啡渣的再生"项目设计与实施 ………………………………… 91
- 探究黄皮与气候，助力乡村振兴 ………………………………………………… 97
- 基于学生视角的气候变化与生物多样性关系探究 ……………………………… 103

- "水气相和 育万物共生"项目化学习设计与实施 ······ 109
- "气候变化与环境保护"科普项目化学习探析 ······ 115
- 对话专家 ······ 121

第四篇 基于综合活动的气候变化教育 ······ 123

- 寻长江口二号古船,扬长江口生命力 ······ 124
- 长江口重工业工厂的过去、现在与未来 ······ 131
- "绿色万里路",让减排被看见 ······ 138
- 中职学校实施气候变化教育的实践 ······ 144
- 访气象俗谚,探气候变迁 ······ 151
- 基于具身实践的"气候变化教育"学习新样态 ······ 157
- 对话专家 ······ 166

第五篇 学校家庭社区协同开展气候变化教育 ······ 169

- "银芽气候课堂"的学习路径设计与实施 ······ 170
- 绿色创新与气候教育:构建终身学习新典范 ······ 177
- "凤凰"展翅 "骑"乐无穷 ······ 185
- 传导"绿色生产力",打通气候教育"链" ······ 192
- 黄河口长江口学生协同开展绿色企业参观活动的个案研究 ······ 201
- 家庭绿色制冷原理探究及操作实践 ······ 209
- 新能源企业助力气候变化教育的经验 ······ 218
- 学在自然:"共同世界教育学"理念下的小学气候变化教育实践研究 ······ 224
- 长江探秘系列之船舶探索行 ······ 232
- 学校家庭社区协同,融创"双碳"教育新"气象" ······ 239
- 对话专家 ······ 247

附录 气候变化教育指导纲要(试行) ······ 248

第一篇

气候适应性导向下的校园建设

面对日益严峻的全球气候变化挑战,气候变化教育势在必行,学校发挥着不可替代的关键作用。学校是探索教育场所减排的主阵地,是提升气候适应性和韧性的实验室,是培育具有气候素养公民的学习场。

本篇所呈现的案例,围绕"节能减排设施改造""校园局部环境改善""气候主题场馆营建""零碳校园建设畅想"等主题展开,包含校园绿色建筑设计建造,照明设备调整升级,智慧节能教室建设,气候变化教育教学实验场所和主题场馆营建,校园"碳核查"、能源审计及规划探索等具体内容。特别值得一提的是,这些行动都是学校师生共同参与、亲身体验的,彰显出教育工作者应对气候变化自我革新、积极创新的勇气、智慧与担当。

节能背景下教室智控灯光系统设计

研究背景

在当前全球气候变化的背景下,节约能源成为一项日益重要的全球性任务。《巴黎协定》要求,"在本世纪下半叶,在公平的基础上、在可持续发展和努力消除贫困的背景下,实现温室气体人为源排放和汇清除之间的平衡"。①

目前,人类迫切需要创新性的解决方案来降低碳排放、加强能源利用效率,以减缓气候变化带来的影响,而"教育是应对气候变化的一个核心举措,通过影响价值观、思维方式、行为和生活选择,帮助学习者形成应对气候变化所需要的素养"。② 本案例围绕新兴科技在气候变化研究和解决方案中的创新应用,探讨如何利用算法来节约能源。

项目设计

在许多学校、家庭和办公场所,室内照明系统的能耗占据了相当大的比例。然而,传统照明系统没有智能控制功能,导致能源的浪费。本案例基于"我心目中的智控教室"一课,从学生在校园内发现的"节能痛点"出发,阐述如何指导学生设计智控灯光系统,并进一步优化迭代。

一、活动目的

(1)帮助学生结合生活经验,掌握逻辑运算在智控灯光系统中的作用。
(2)激发学生的思考兴趣,培养学生分析问题、提出假设、论证结论的能力。
(3)培养学生的实践操作与解决实际问题的能力。
(4)增强学生建设节能型校园的意识。

二、活动流程

(1)活动引入阶段:教师通过"节能痛点"引发学生思考,引出改造智控教室的教学内容。
(2)实验探究阶段:教师演示实验 A,介绍"非运算"的概念。
(3)交流讨论阶段:学生进行实验 B 和实验 C,讨论、探究影响智控灯光亮灭的条件。
(4)表达展示阶段:学生分享对改造智控教室其他设备的想法,进行自然语言描述和评价标准的讨论。

---

① 联合国.巴黎协定[EB/OL].(2015-12-12)[2024-05-21].https://www.un.org/zh/node/181391.
② 上海终身教育研究院.气候变化教育指导纲要(试行)[EB/OL].(2024-03-06)[2024-05-21].http://www.smile.ecnu.edu.cn/c8/53/c42519a575571/page.htm.

三、活动工具

(1)实验小程序:活动前设计了学生实验活动的程序和配套材料,包括实验程序、实验数据记录表,确保活动顺利进行。

(2)学习任务单:活动中提供学习任务单,帮助学生理清完成任务的思路与要求,引导学生有目的地参与活动。

(3)学习评价单:活动后请学生填写,用于学生自我评价在活动中的表现和学习情况。

项目实施

一、情境导入:学生离开教室忘记关灯

上课伊始,教师通过一个视频,呈现了学生离开教室去上体育课却忘记关灯的情景——即使教室内无人,但灯光依然亮着,以这个情景引发学生的思考:如何改造教室的灯光设备,实现智能控制,从而节约能源?由此引出核心问题——如何设计你心目中的智控教室?

二、实验模拟:理解"与""或""非"三种基本逻辑运算

在课前,教师根据"时间""光线强弱"和"室内是否有人"这三个因素编写了三个程序(见图1),作为本节课的实验教学素材。三个程序分别模拟了三种智控灯光的使用情境:实验A:满足"时间是晚上"则灯亮,反之,则灯灭;实验B:满足"教室内光线不充足且有人",则灯亮,反之,则灯灭;实验C:满足"教室内光线不充足或有人",则灯亮,反之,则灯灭。

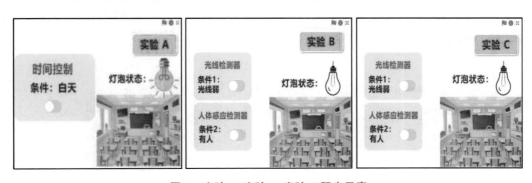

图1 实验A、实验B、实验C程序示意

首先,教师通过师生问答和学生自主讨论的形式,让学生明确影响智控灯光亮灭的可能因素,即室内是否有人、时间是否是白天、室内光线是否充足,以及检测这些因素需要用到的传感器(见图2)。

接着,学生打开提供的学习素材"智控灯光模拟实验程序",在教师指导下完成三个实验,并在表格中记录实验过程和结果。

实验A:教师引导完成实验A的操作过程。学生先自主控制看实验是否满足条件,再仔细观察在不同条件下灯光的亮灭情况,并在表格中记录实验过程和结果。实验A的情况是:不满足条件"白天",则灯亮;满足条件"白天",则灯灭。实验结果记录:定义符合条件记作

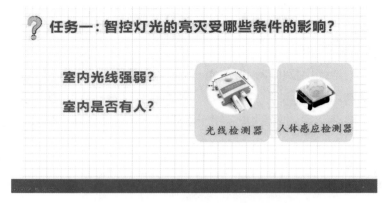

图 2　明确影响智控灯光亮灭的可能因素和所需传感器

"1",不符合条件记作"0";在亮灯情况一栏,灯灭记作"0",灯亮记作"1"(见表 1)。

表 1　实验 A 记录表格

条件:白天	亮灯情况
0	1
1	0

实验 B:学生与同桌配合完成实验 B。只有同时满足条件"光线弱"和"有人",灯才亮。

实验 C:学生与同桌配合完成实验 C。只要满足条件"光线弱"或"有人",灯就亮。实验 B 和实验 C 的结果记录如表 2 和表 3 所示。

表 2　实验 B 记录表格

条件 1:光线弱	条件 2:有人	亮灯情况
0	0	0
0	1	0
1	0	0
1	1	1

表 3　实验 C 记录表格

条件 1:光线弱	条件 2:有人	亮灯情况
0	0	0
0	1	1
1	0	1
1	1	1

通过对这三个实验的手动操作和沉浸式体会,学生深入理解了三种基本逻辑运算的规

律,明确了改造教室灯光系统的方法。在实验 A 中,他们认识到"非"运算,即条件和结果相反;在实验 B 中,他们理解了"与"运算,即只有两个条件同时满足时,结果为真;在实验 C 中,他们掌握了"或"运算,即只要满足其中一个条件,结果为真。

三、对比探究:从节约能源的角度考量

学生与同桌讨论,如何从节能的角度选择合适的方案来进行教室灯光系统改造。通过实验模拟及结合自身的生活经验,学生们一致认为实验 B 方案(即"与"运算的方案:满足"教室内光线不充足且有人",则灯亮,反之,则灯灭)更适合应用于校园环境下教室内的灯光系统改造,因为它能够根据教室内的光线情况和人员活动情况智能地控制灯光,从而更有效地节约能源。

经由合作讨论来选择合适的解决方案,并辅以数字化学习方式,学生们不仅增进了对节能概念的理解,还培养了团队合作的能力和解决问题的技能。他们在思考、选择最佳方案的过程中,不断提高着自己的逻辑思维能力与创新意识。

四、练习巩固:用不同的逻辑运算符控制不同的智控设备

首先,学生运用本节课所学的新知识——"与""或""非"三种基本逻辑运算,从节约能源的角度考虑,根据智控设备的特性和任务需求,自主思考并确定适用的逻辑运算符。然后,学生将这些设计想法记录在表格中,形成设计方案。最后,他们根据填写的表格,结合教师提供的句式,"智控设备是_____(智控设备名称),当_____(条件)的时候,_____(操作)",尝试用语言来描述他们设计的智控设备的工作原理和应用场景(见图 3)。

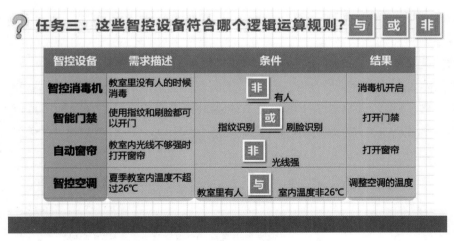

图 3 基本逻辑运算符的应用练习

五、拓展运用:探讨智控空调如何恒温

学生模仿上一环节给出的句式,用自然语言描述在自己心目中的智控教室里使用逻辑运算符设计智控设备的运行方案;从节约能源的角度考量,尝试找出给定的智控设备应该使用哪种基本逻辑运算,甚至哪几种逻辑运算组合进行操控。

在此过程中,学生产生了更深入的探讨。他们认为"调整空调温度"这个操作步骤不够具体,到底是往更高的温度还是更低的温度调整呢?教师适时引导:确实要分情况判断才能具体操作,可以考虑加入一个逻辑运算符"与"也可以考虑加入条件判断的语句。学生模仿句式用自然语言来表达,尝试给出了智控空调的调整方案,即当智控空调的传感器检测到教室内有人与室内温度高于26℃时,调低温度;当智控空调的传感器检测到教室内有人与室内温度低于26℃时,调高温度。这一想法非常好,但要用到条件判断语句。对于学生们来说,这是新的尝试与挑战,由此激发他们对后续设施设备改造课程内容学习的兴趣。

成效与反思

本案例基于学生在教室外活动时忘记关灯的问题,激发学生思考如何运用所学知识改造教室灯光控制系统。学生在实验模拟环节,通过合作实验和记录,深入理解了逻辑运算的规律;在对比探究环节,基于实验和实践,习得如何从节能减排角度选择合适的方案,强化节约用电和环保意识;在练习巩固环节,通过实践操作,系统整理所学知识,提升了问题分析表达与解决能力;在拓展运用环节,积极构思其他可行的智控设施改造方案,运用所学设计智控教室,展示出不凡的想象力与创新力。

教室智控灯光系统改造是学校智慧节能教育的一部分,其不仅增强了学生的学习体验,还大幅降低了学校的电力消耗。但若要全面实施智慧节能教育,仍需从多方面进行拓展。

一、学科新进展:智慧节能教育在小学中高年段的系统化探索

随着社会对可持续发展和环境保护的重视不断增强,智慧节能教育在小学中高年段的系统化探索显得尤为重要。在这个阶段,经过一学期信息科技学科学习的学生已经具备了一定的数字化认知能力和学习能力,可以接受更加系统化的智慧节能教育。

学校可以组织中高年段学生参与智慧节能实践活动,如制作节能科技作品、参与智慧节能游戏等,通过体验活动培养学生的智慧节能意识和智慧节能技术;同时,还可以在小学中高段开设专题课程或设立智慧节能教育课程模块(见图4),实现对节能教育的系统化教学和管理。

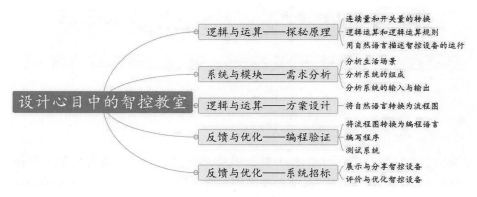

图4 智慧节能教育课程校本单元设计

二、校园数字化：智慧节能教育在参与校园管理时的拓展应用

校园是智慧节能教育的重要场所之一，建设气候智慧校园可以将智慧节能教育在校园管理中进行拓展应用。如利用数字化手段，建立校园智慧节能管理平台，实时监测校园能耗情况，提供智慧节能建议和管理建议，为数字时代的校园智慧节能管理提供科学依据。

上海市金山区前京小学已有这方面的规划，如针对校园内净水饮水机进行智能管理，使之按照程序定时自动开闭，既节约用电又节约用水；针对卫生间"小厨宝"进行智能管理，根据校园师生的作息时间，设定管理策略，使之在节约能耗的同时延长设备使用寿命，并降低常态使用引发的电气火灾风险；针对校园能源开展有效管理，降低能源消耗，减少碳排放。

三、家校社合作：智慧节能教育在家庭及社区范围的跨界应用

智慧节能教育不能仅限于校园内部，还需要与家庭、社区进行有效且常态化的合作。在课堂上，教师可以引导学生畅想、设计心目中的智控教室，也可以让学生回家与父母讨论、规划如何避免家中电器能源的浪费，还可以通过与学校周边社区的合作，将智慧节能教育融入社区建设和社区活动，如邀请专业人士进入学校课堂开展生态文明专题讲座等。

智慧节能教育课程的系统化研究、校园智慧化进程的持续性推进和家校社紧密合作是小学阶段智慧节能教育的三个重要探索方向。相信通过各方的合作和努力，我们一定可以促进学生智慧节能科学知识的学习与智慧节能意识的养成，为建设资源节约型校园，为世界可持续发展作出积极贡献。

（平怡菁，上海市金山区前京小学教师）

校园"鱼菜共生馆"建设及低碳教育实践

研究背景

在当前全球环境挑战日益严峻的背景下,学校作为培养未来社会公民的重要场所,有责任也有义务承担起气候变化教育引领之责。上海市金山区前京小学"校园'鱼菜共生馆'建设及低碳教育实践"便是在此背景下提出的。

环保与我们的日常生活紧密相连。培养学生的环保习惯,不仅能够减少对环境的负面影响,还能为构建可持续的生态环境贡献一份力量。校园"鱼菜共生馆"不仅为学生们提供了一个直观了解环保与生活联系的平台,还让学生们在实践中切身感受、体验环保的价值与意义,认识到环保是全球性议题,更与我们每个人息息相关。此外,教师们在此教育实践过程中,不断学习、探索新的环保教育方法,也让自身环保教育专业素养得到了提升。

项目设计

一、"鱼菜共生馆"整体设计

"鱼菜共生馆"位于学校"京京菜园"板块(见图1)。馆内建构了一套"鱼菜共生系统",主要由如下部分构成。

(1)基础设施建设:具体包含鱼塘、水耕栽培床、水泵和管道等。其中,鱼塘用于养殖鱼类,水耕栽培床用于种植植物,水泵和管道用以将鱼塘里的水抽到栽培床上,实现水资源的循环利用。

(2)鱼类养殖:一般选择适应性强、生长快的鱼类品种(如金鱼)进行养殖。通过合理的投饵和管理,确保鱼类的健康生长。

图1 上海市金山区前京小学"鱼菜共生馆"

(3)水耕栽培：利用鱼塘里富含营养的水进行植物栽培。一般选择生长周期短、产量高的植物品种，如生菜、青菜、芹菜等。

在此系统中，养殖鱼类的水中富含营养物质，经过处理后可用于植物灌溉。而植物则吸收这些营养物质，净化水体，为鱼类提供清洁的生存环境。这一系统不仅能实现资源的循环利用、减少化肥和农药的使用，还能降低环境污染、水资源的消耗，实现低碳、环保的农业生产。

二、学习目标设计

（1）知识习得方面：通过"鱼菜共生馆"的建设与运营，帮助学生深入理解气候变化的原因及其影响，特别是农业活动与气候变化之间的关系，了解"鱼菜共生系统"作为一种低碳农业模式，在应对气候变化、减少温室气体排放方面的重要作用。

（2）习惯养成方面：通过参与"鱼菜共生馆"的日常管理，引导学生形成节能、节水等环保习惯；培养学生对气候变化问题的长期关注与对环保及可持续发展领域的兴趣，不断探索新的环保理念、技术和方法，并将之应用于生活。

（3）思维培养方面：通过"鱼菜共生系统"的学习与实践，培养学生的系统思维，帮助学生更好地理解生态系统中各个组成部分之间的相互关系与影响，引导学生树立可持续发展的观念，认识到环境保护与经济发展的重要性。

项目实施

在"鱼菜共生馆"中，学校开展了各式各类低碳教育活动，列举如下。

一、"节能减排"主题教育活动

在"鱼菜共生馆"中，学生通过实验与观察，了解节能减排的科学原理与实际应用。学生们观察到，通过高效的系统设计和精细的管理，"鱼菜共生馆"能够在满足生产和生活需求的同时，显著减少能源消耗和碳排放。

学生们通过操作智能灌溉设备了解到，当智能灌溉系统设置为一小时灌一次水，每次灌水 10 分钟时，能提高水的利用效率，确保灌溉过程中的准确性和高效性，起到节能效果。这让学生们真实感受到节能减排的必要性，从而在日常生活中采取节能措施，减少能源消耗。此外，学生们还通过绘画的方式表达自己对于节能减排和环保的理解及感受。

二、"循环利用"主题教育活动

"鱼菜共生馆"中的水资源循环利用系统是一种重要的教育资源。在教师的指导下，学生们观察并参与到雨水收集、废水处理及循环再利用的每一个环节中（见图 2）。他们学习雨水收集器收集雨水的工作原理，了解收集到的雨水将储存在"鱼菜共生馆"地底下的蓄水池中并最终用于灌溉植物。通过观察，他们了解了废水经过处理、净化，如何重新成为可再利用水资源的全过程。学生还参与到设计与改进循环利用系统中，通过与家长交流或查阅资料的方式，寻找"鱼菜共生馆"中其他设计循环利用系统的环节。

图2 "鱼菜共生馆"中的雨水收集器

三、"生态环保"主题教育活动

"鱼菜共生馆"为学生们提供了一个观察和研究生态系统的平台。在这里,学生们了解生态系统的构成和运行机制,学习如何保护生态环境,参与一系列对于生态系统的观察和记录活动。

学生们分组合作,定期观察鱼类及植物的生长状态以及水质的变化等,并将这些观察结果记录下来。学校还于2024年3月15日邀请生态环保专家来到"鱼菜共生馆"开展讲座与专业指导(见图3)。专家生动的案例呈现和深入浅出的讲解,激发了学生了解人类活动对生态系统影响的兴趣。通过参与"鱼菜共生馆"的运营和管理,学生们更深入地了解了生态环保的重要性及其原理,并在日常生活中更加积极地践行环保。

图3 学校开展的生态文明讲座

成效与反思

经过一个学期的实践探索,校园"鱼菜共生馆"以其独特的魅力,模拟自然生态系统,将水产养殖与水培植物和谐地融为一体,不仅实现了资源的循环利用,更维系了校园局部生态的平衡,提升了校园环境整体质量。这一项目不仅为学生们提供了参与校园生态系统构建与管理的机会,还成为培养学生科学素养、实践能力和创新精神的宝贵载体。学生们在参与过程中,深入了解低碳理念,了解生态系统的运作机制,体会到了资源的珍贵与环保的重要性。

在项目实施过程中,学生的问题解决能力得以提升。在面对池水变脏这一现实挑战时,学生们积极应对,深入分析问题、查找原因、探求解决方案,最终通过养殖黑壳虾、用渔网捞绿藻等创新方法,成功破解了这一难题(见图4~图6)。

此外,在校园"鱼菜共生馆"建设中,学生的团队协作能力和沟通能力也得以增强。在团队中,每个学生都扮演着重要角色,他们相互协作、共同面对挑战,共同商讨制订解决方案,并深刻认识到自己作为地球公民所肩负的责任与使命。

图4 学生释放黑壳虾　　　　　　　　图5 学生用渔网捞绿藻

图6 学生铲除绿藻

在这次极富创新性与实用性的项目实施中,我们也发现了一些需要改进的地方。譬如,虽然学生们参与项目的积极性很高,但由于缺乏相关专业知识与经验的支撑,部分学生在实践过程中可能会感到困惑、无助而难以持续。因此,教师需要不断自我提升、深化学习,并加强对学生的系统化培训与个性化指导,以探索出更为高效的解决方案,如构建跨学科合作平台,促进多领域知识的交汇与创新思维的碰撞;实施循序渐进的分阶段培训策略,逐步巩固学生的专业基础并提升其专业技能水平。同时,设立项目导师制度,确保每位学生都能获得贴合其需求的个性化指导与及时的学术援助。通过这些综合措施,我们将有力推动"鱼菜共生馆"的建设与研究走向更深层次,取得更加丰硕的成果。

(胡志文,上海市金山区前京小学教师)

建一座不断成长的气候变化探索馆

研究背景

2023年，江苏省常州市作为"新能源之都"，迈入"GDP万亿之城"行列。喜讯发布时，各大媒体竞相报道。该话题也引起了学生们的关注：为什么要开发新能源产业？新能源产业的市场前景如何？基于此，学生们展开讨论、调查，了解了新能源产业与气候变化的紧密关联。

2023年10月，经常州市红梅实验小学的"红领巾议事厅"讨论表决，"探索气候变化，贡献少年力量"当选第十五届科技节的研究主题。学生们聚焦"气候变化"开展研究性学习，通过访谈、观测、实验等形式，了解气候变化给人类生活带来的影响。在科技节闭幕式上，学校组建了首个"气候变化教育导师团"，邀请市气象局的首席预报员、生态环境局的专家等担任导师。同时，还发布了寒假气候变化教育研究项目——"气候变化与人类生活"，并部署"亲子寻访记""宣传策划团""藏品征集令"等十大"寒假行动"。

2023年11月，学校加入华东师范大学上海终身教育研究院李家成教授领衔的气候变化教育研究联盟。2024年1月，学校响应《中共中央、国务院关于全面推进美丽中国建设的意见》，践行"鼓励社区、学校等基层单位开展绿色、清洁、零碳引领行动，充分利用博物馆、展览馆、科教馆等，宣传美丽中国建设生动实践"，持续深化气候变化教育研究。为推动气候变化教育研究由短期变为常态，将学生的研究过程材料进行收纳展陈，学校投入专项资金，启动校内"气候变化探索馆"建设。2024年1月，气候变化探索馆项目正式启动。同年4月，学校成为江苏省常州市天宁区中小学科学教育馆校课程项目领衔校。

项目设计

围绕气候变化这一研究主题，学校基于原有研究基础，引进多方资源，建设集科普宣传、实验探究、沙龙分享、趣味观影等功能于一体的气候变化探索馆，并将之与学校的课程建设、社区的科普教育相结合。项目设计追求如下目标：建设一所沉浸体验式的探索新场馆，打造一种馆校融合的课程新样态，成立一支科学探究的导师新队伍，构筑一方与时俱进的科普新生态。

气候变化探索馆为学校高质量实施气候变化教育、落实科学教育提供了教学资源与教学空间。它不是"器材展示角"，而是"探究实验坊"；不是"资料陈列站"，而是"思想聚集地"；不是"活动会议室"，而是"人才孵化所"；不是"班级授课处"，而是"混龄共长区"。学生在此探寻研究奥秘，共享研究成果，举行科研活动，获得科学启蒙；教师在这里设计跨学科活动，开发馆校课程；社区居民周末来这里参观，志愿者在这里开展生态文明宣传活动。可以说，这里是一个终身学习者共建美丽中国的实践基地。

项目实施

为了科学规划气候变化探索馆的建设，学校寻访新能源企业，走进气象局、生态环境局开展学术交流，走进科协等单位进行联动研讨；同时，以"参与式学习"理念为指导，积极引导学生参与到"气候变化探索馆"这一活动空间的管理、使用与设计中，使学习空间更丰富多样，也更具育人价值。

一、场馆设计：参与式共建

（一）教师参与，全员全学科融入

学校由校长领衔成立了馆校项目工作组，同时组建策划部、课程部、外宣部、活动部、后勤部等机构，形成相应的工作制度，规划工程推进，策划具体活动，保证设施设备，调配师资等，确保探索馆建设高效实施。

以课程部为例，各学科的教研组长、备课组长带头学习李家成教授等整理的气候变化教育文献，将各学科课标和教材中与之关联的内容进行梳理，列出了小学六个学年的学科融合气候变化教育框架图。同时，各年级确定每一学期的气候变化教育研究主题，策划年级活动，精备融合课例，设计跨学科作业，用好课堂主阵地，开发课外活动圈，将学科教学与活动实践紧密结合，确保研究的深度与广度。

（二）学生参与，全体全方位在场

学校从儿童立场出发，以儿童视角观察、思考环境建设，升级改造校园场馆，让校园充满生命成长的气息。我们将探索馆的建设计划告知学生，并提供场馆地图、数据比例等信息，统筹各学科教师，开展"我的场馆我设计"等活动。当学生成为学校教育场馆的设计者、建设者、研究者和使用者时，他们主动学习的参与感被激发，对学校的认同感得以生发。

在收集的学生建议中，王子谦同学建议选择"红梅"作为气候变化的观察对象。因为红梅不仅是学校的名称，也是中国文化意象的表达。校园里种植的多株红梅，非常便于学生们观察记录红梅的生长历程。在王同学的提议下，六年级组启动了"红梅的物候研究"，学生们通过研究红梅的生长荣枯，探寻每一年的气候、物候变化情况。

经师生反复商议，最终定稿的探索馆馆内面积近 $700m^2$，分气候变化科普区、气候变化探索区、气候变化实验区、气候变化教学区、应对气候变化畅想区等五大空间。馆外空间约 $300m^2$，有气候监测区、植物种植区、星空观测区等，满足了学生在气候变化中的研究与探索需求。场馆融科技、艺术、文化于一体，提供学生多维度的参观体验，各主题空间的内容展示也充分体现出科学性、思想性和人文性（见图1、图2）。

图 1　气候变化探索馆前方俯视图

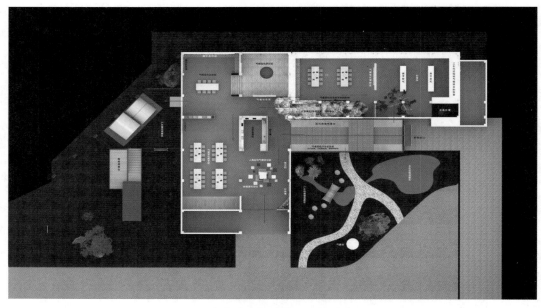

图 2　气候变化探索馆正上方俯视图

二、课程开发：跨学科共创

为开发利用好"气候变化探索馆"的馆校资源，学校以培养学生的科学素养和科学志趣为目标，以学生学习的发展节律为依据，完善课程元素，协同育人力量，在高质量实施国家必修课程的基础上，进一步延伸开发出气候变化探索的主题选修课程与特色体验课程，构建出"必修为主、选修为辅、专修体验"的三位一体"气候变化探索馆"课程图谱。

项目组系统梳理了小学阶段各学科的必修教材,初步架构了课程建设整体方案。课程内容涵盖师生在校生活的多个角度,充分体现跨学科、多学科的交织、连接与融通。具体有以下三个方面的举措。

(一)高质量实施"每日"必修国家课程群落

学校借助"气候变化探索馆"的组织与活动,整合国家课程中的各学科资源,形成气候变化教育主题国家课程群落,探索项目化学习、跨学科学习,提升学生解决问题的能力。

比如,在科学学科中,我们梳理出六年级上册的"二十四节气"知识点,与语文学科的节气歌、劳动课中的节气劳作等进行融合,设计了"二十四节气"主题活动任务群(见表1)。

表1 常州市红梅实验小学气候变化教育课程资源设计(科学学科)

年级	主题内容	目标	教学设计
一年级上册	天气知多少	・知道阴、晴、雨、雪、风等天气现象,知道天气变化对生活的影响	(1)结合生活实际,说一说身边所知道的天气情况。 (2)说一说常州一年四季有哪些显著的天气变化
一年级下册	观察天气	・在教师的指导下,通过口述、画图等方式交流对天气的观察结果	(1)观察一种天气现象,试着用语言描述、画图的方式向大家介绍。 (2)阅读一些与天气、气候有关的绘本
二年级上册	关心天气	(1)了解天气符号的含义,认识常见的天气符号。 (2)在阅读天气预报的基础上,有结构地描述气象要素。 (3)说出天气对人们生产生活和动植物的影响	(1)自己设计天气符号,认识常见的天气符号。 (2)观看天气预报,总结天气预报的内容要点,并尝试着自己做一位"小小播报员",播报天气。 (3)说说天气对人们的生产生活和动植物的影响
二年级下册	气候与动物	(1)知道不同的动物适应不同的气候条件。 (2)能观察并总结不同气候条件下的动物的身体特点和生活习性等。 (3)知道动物的生存与气候变化息息相关	(1)认识不同气候条件下的不同动物,讨论它们为什么能在这种气候下生活。 (2)观察并讨论不同动物适应气候变化的不同本领。 (3)挑选一种气候下的动物,制作科普卡片,介绍它们的生活习性等
三年级上册	地球上的风	(1)知道风的成因。 (2)知道季节性风的成因	(1)知道冷空气和热空气的特点,通过实验了解并解释风是如何形成的。 (2)小组合作解决问题:为什么夏天刮台风,而且都是从海洋向陆地?

(续表)

年级	主题内容	目标	教学设计
三年级下册	测量天气数据	(1)能熟练测量气温。 (2)会测量云量和雨量。 (3)知道什么是风向和风力。 (4)知道什么是天气和气候以及两者之间的区别	(1)学习使用气温计,测一测当天的气温。 (2)小组合作,测量云量。 (3)自制雨量筒,测量一场雨的雨量。 (4)测量风向和风力。 (5)说一说什么是天气、什么是气候,能够分清天气现象和气候现象,说一说它们之间有什么区别
四年级上册	气候与植物	(1)知道不同的植物会在不同的气候条件下生长。 (2)通过观察了解植物适应不同气候的结构特点和本领。 (3)通过栽种一株植物了解植物的生长与气候息息相关	(1)认识不同地区不同气候下生长的植物,讨论它们为什么在这种气候条件下生长。 (2)观察并讨论不同植物适应气候变化的不同本领。 (3)栽种一株郁金香,持续观察,了解气候对植物生长的重要性
四年级下册	气候与地表	(1)认识不同的地貌,并了解不同地貌的产生条件。 (2)通过模拟风对地表的侵蚀作用,意识到大自然的鬼斧神工。 (3)了解全球气候变化导致的全球地貌变化	(1)认识不同的地貌,了解不同地貌是由不同气候条件造成的。 (2)设计模拟实验,模拟风蚀地貌的产生。 (3)意识到气候变化对地貌的影响,了解全球气候变暖对地貌的改变
五年级上册	水滴的旅行	(1)能够描述地球上的水在陆地、海洋和大气之间不间断循环的过程。 (2)初步建立水循环的动态平衡意识,理解地球上的水处于一个动态平衡和不断更新的状态之中。 (3)能够用水循环的知识解释全球气候变暖带来的一系列后果	(1)通过交流"天上的水为什么总也降不完",初步建立水循环的概念。 (2)通过箭头和词汇构建水循环图。 (3)运用水循环的知识交流讨论,为什么全球气候变暖会导致冰川融合、海平面上升,并进一步推演接下来还会发生哪些自然灾害
五年级下册	四季的气候	(1)总结四季的气候特点,知道四季形成与地球围绕太阳公转有关。 (2)知道四季变化对地球上一些现象的影响。 (3)通过模拟实验了解四季的成因,以及地球上与之相关的有规律的自然现象的成因	(1)读图,讨论总结四季气候特点,知道一年中四季气温、降水、昼夜长短等都是有规律的。 (2)探究地球倾斜与直射、斜射的关系,探究直射、斜射对温度的影响。 (3)通过实验模拟地球自转与公转,加深理解

(续表)

年级	主题内容	目标	教学设计
六年级上册	气候了解史	(1)了解气象观测工具的发展历程,体验技术对气象观测的重要性。 (2)知道人类对气象的探索历程,简单了解气象站的工作原理。 (3)通过了解并背诵二十四节气,认识到古人的智慧	(1)收集资料,讨论总结人类对气象的观测历史和工具的迭代更新。 (2)认识气象站,了解气象站的工作原理。 (3)熟背二十四节气,了解其中的气象内涵,知道不同的节气有什么习俗,了解背后的内涵和意义
六年级下册	理想的家园	(1)了解全球气候变暖对人类生产生活和动植物带来的一系列影响。 (2)意识到全球气候变暖会带来不可挽回的后果。 (3)树立地球主人翁意识,为阻止全球气候变暖出一份力	(1)观看视频、图片,了解因为气候变暖而引发的一系列后果。 (2)通过实验模拟全球冰川融化导致海平面上升,意识到全球气候变暖的影响。 (3)讨论我们可以做些什么来应对全球气候变暖,撰写200字倡议书

(二)创意化开设"每周"选修主题课程群落

学校于每周的社团及课后服务时段,开设品类丰富、面向每一位学生的探索馆主题选修课程;同时,加强对科技社团和兴趣小组的指导,引导支持有兴趣的学生长期、深入、系统地开展有关气候的科学探究与实验。

为此,学校引进了优质的校外机构。随着社会专业师资的不断涌入和有序参与课后服务,越来越丰富的选修课程被开发出来,如"有趣的化学元素""气候与民俗""二十四节气之美食"等,并朝向"开放化、个性化、选择化"的高质量目标发展。

(三)序列化组织"每月"专修馆校课程群落

学校利用科学、综合实践、劳动等时段,每月至少开设一次馆校课程。教师基于教材中的拓展点和学生的兴趣点,采用任务驱动、教师指导、学生自主合作探究为主的科学主题探究和实践活动样式,开发馆校专修体验活动群落。

学校向内开发、向外拓展,建构馆校课程群落,开发完善"馆校课程圈",并合理利用气象站、科普馆等场地资源优势,联合共建单位开设学生教育活动基地,拓展学生活动空间,实现多元育人场域的互通。在此过程中,主要采用任务驱动、教师指导、学生探究的学习样式,推进基于自主实践的参与式学习,把社会当教材、把世界当教室,在丰富的情境浸润中,引导学生体验、思考、成长。

三、导师培养:多元式共长

学校通过"三大行动",着力打造一支科学探究能力强的导师新队伍。

其一,名师引路,指明前行方向。学校特别邀请国内气候研究领域的专家、省内知名研训员等担任"科研导师",定期组织教师团队培训,指导教师完善选题思路和研究方法,提出宝贵建议,给予专业引领,实现导师队伍的高质量发展。

其二,"民师"导行,指点具体实践。学校找寻了一批有专业、有特长、有情怀的行业能手,如花卉园艺师、建筑材料师等,充分挖掘这些"民师"的智慧,探索研究路径,优化研究过程,实现导师实践的可行性生长。

其三,"青师"自培,指南助力成长。学校组建了青年教师成长营,同伴协作、细化分工,用目标导向、任务驱动的模式,鼓励他们担任各子项目的负责人,在项目实践过程中,不断打磨,获得成长。

四、科普推广:数字化共享

学校打开校门做气候变化教育宣传,力争构筑一种与时俱进的科普推广新生态。

其一,打造一座"无围墙的气候变化探索馆"。学校打通校园与社区的边界,与居民共享气候变化教育的空间资源。周末及假期,社区居民可以来馆内参加科普活动。以此方式,推动社区居民终身教育与学校教育深度融合通联,打造无围墙学校,共建未来学习中心。

其二,打造一个"数字化的云传播科普平台"。学校尝试在气候变化教育的探索中,用数字赋能教学、赋能作业,打造教与学的广度、深度和精准度,使教学更加有效。同时,学校积极开发云端科普活动。如科学学科教师张璐丹开设"气候变化教育"视频专栏,围绕"身边的气候"开展专题研究,带领学生分析气候变化成因,并开展系列模拟气候变化实验,如模拟全球气候变暖、冰川融化导致海平面上升的过程等,让气候变化教育的理念和践行辐射更广范围、更多人群。

成效与反思

通过这一阶段的实践,已初步取得一些成果:

其一,整体规划了气候变化探索馆的课程系统,构建了一张课程图谱。学校秉持"全员参与、个性发展"的课程理念,完善课程元素,协同育人力量,在高质量实施国家必修课程的基础上,延伸开发出"必修为主、选修为辅、专修体验"的三位一体"气候变化探索馆"课程图谱。

其二,创造性开发了参与式馆校课程的学教样态,拓展了气候变化教育的新路径。学校基于"参与式学习"理念,在馆校课程活动中,关注每一个学生主动学习的参与状态,激发学生的好奇心、想象力和探求欲,引导学生广泛参与探究实践。

其三,打造了一个校—家—社多维协同的空间场域,实现了资源的富集与合力的凝聚。学校与属地街道、周边社区联系,让学生走出校园,走进社区,同时,不断进行场景式、体验式科学实践活动,将学校的探索馆变为居民普及生态文明意识的有效载体。

其四,设计了一套能极大激发个体参与、个性发挥的评价体系。学校基于学科核心素养指标,围绕学生对气候变化探索的学习与实践,对学生的学习表现进行表现性评价与终结性评价,并通过学校的"梅好储能卡""争章评比"等增值评价方式,激发学生的积极性与创造性。

其五,探索了一种全科全员全程的组织管理新机制,建立了一套保障体系。学校围绕探索馆主题学习与实践,建设"校级探索馆—年级探索区—班级探索营"三层组织结构;组织校内教师,引进校外专家学者、行业领军人物等组建"专家导师团",对探索馆的各类活动开展辅助和指导,并定期参与校级、年级、班级的主题学习和活动。

未来,我们将持续深化"气候变化探索馆"的场馆建设、课程开发、活动创设与文化营造,积极推动气候变化教育的落地落实。

(王燕,江苏省常州市红梅实验小学校长)

"零碳校园"建设构想及教育实践

研究背景

学生是未来社会的主人,培养他们的环保意识和低碳生活习惯,对推动全社会实现碳达峰、碳中和具有重要意义。相较于"绿色校园"建设,"零碳校园"更注重对碳排放源的计算,更强调师生行为和碳中和,更具量化感,也更重参与度。"零碳校园"建设是一个专业、复杂的系统性工程,需多方合力、持续探索。本项目以华东师范大学第五附属学校(简称华师五附)"零碳校园"建设为例,呈现设计构想、实施准备及相关教育教学实践,是一次阶段性的成果梳理。

项目设计

一、目标设计

华师五附"零碳校园"建设分近期与远期两类目标。近期"绿色行动"目标:通过课堂讲解、宣传画册以及组织环保主题的课内外活动等形式,向学生宣传低碳生活的重要性和实践方式,践行低碳生活,增强学校全体师生的环保意识。同时,以点带面,通过学校的示范效应,推动学校、家庭、社区共同助力节能减排,倡导"零碳生活"。远期"零碳校园"目标:通过节能减排、植树造林等形式,抵消学校师生生活产生的二氧化碳或温室气体排放量,实现正负抵消,达到相对"零排放",实现学校"零碳校园"目标。

二、方案设计

有研究者指出,部分"初中校园基本可以做到零碳运行,校园最具潜力实现零碳运行目标。充分利用校园太阳能资源,具有显著节能减排意义"[①]。"零碳校园"建设是指在校园范围内推广一些活动,如整合教育资源与优化课程设置、通过互动实践体验推广低碳生活方式、实行校园碳核查、构建零碳校园能源系统等,实现校园碳排放与吸收的自我平衡,培养学生的环保意识和低碳生活习惯,达到相对"零排放"的目标。

(一)整合教育资源、优化课程设置,加强"零碳校园"建设宣传

学校通过课堂讲解,发放宣传画册,组织环保主题的课内外活动,如七年级秦山核电站行走活动、全年级午会气候变化教育德育课程等,整合教育资源,向师生们宣传"零碳生活"的重要性和实践方式,让学校师生了解碳达峰、碳中和、零碳、净零排放、碳汇、碳足迹等概念,学习计算个人碳足迹、校园碳足迹、产品碳足迹等(见表1)。

---

① 张建忠,尹文龙,马凯,等.中小学校园零碳运行潜力分析[J].建设科技,2024(1):22-25.

表1　华师五附"零碳生活"嘉年华课程设计

课时进度	课程内容
第一课时	一"碳"究竟:碳中和的起源与发展
第二课时	"零碳校园"1:碳排放核算和碳减排设计
第三课时	专家说"碳":湿地和滨海湿地的碳汇作用
第四课时	"零碳校园"2:优化电能设施及电能管理
第五课时	一周"绿闻":零碳生活科普宣传周
第六课时	"零碳校园"3:垃圾分类与循环经济
第七课时	"袋"走垃圾:校园垃圾清理实践活动
第八课时	从"零"开始:共建"零碳校园""零碳家庭""零碳社区"

（二）设置低碳生活展示区、组织环保实践活动，推广低碳生活方式

设置低碳生活展示区，让学生亲身感受"零碳生活"的实践方式与成果。学校组织学生参与环保实践活动，如"节能从我做起"、垃圾分类、绿色出行、废物利用、减塑行动等，养成低碳生活习惯。组织学生参加环保知识竞赛、低碳手工制作废物利用、《寂静的春天》阅读推介等活动，增强环保意识和创新能力。

（三）实行校园"碳核查"

碳排放核算是实现"零碳校园"的第一步。"纵观国内校园'零碳'探索，降低校园能耗是一项重要的节能减排举措，通过节能项目改造、低碳系统调适、技术路径创新、校园规划策略、碳汇价值提升、绿色技术应用、可再生能源替代、绿色教育全员参与等举措实现校园减碳的目标。"[1]本案例主要借鉴天津一所新建医学高校校园碳中和实施路径的规划，通过制订校园碳排放源核算流程、碳减排和碳汇核算管理流程、消费端行为减排量化核算流程来实现零碳校园建设。

第一，制订校园碳排放源核算管理流程。将校园碳排放源分为实体、产品和活动三类。其中，实体类包括各种用途建筑物、服务点和停车位等物理实体；产品包括日常使用的水、电、燃料、纸张、塑料及其由此所产生的垃圾等；活动包括服务、快递、出行、会议、回收等。

第二，制订碳减排和碳汇核算管理流程。碳减排流程需要更多制度性校园社区文化建设和管理机制，以确保行动的有效性。依照减量、替代、循环再生等原则，总碳减排量由新能源替代、绿色交通、减少塑料和纸包装、减少废弃物等不同方式组成。

第三，制订消费端行为减排量化核算流程。个人绿色行为减排量核算主要依据循环经济的"3R"原则，即减量（Reduce）、再利用（Reuse）、再生展开（Recycle）。此外还有替代、共享、再制造和租赁等模式。绿色行为减排量是通过比较采取绿色行为减少的碳排放数据与

---

[1] 李菲,杨丝路,宋风暖,等."双碳"背景下的校园规划实践:以天津某新建医学高校为例[J].绿色建筑,2023,15(4):25-28.

对应的常规行为基线排放数据,核算由于减少能源、电力、额外包装、交通配送等行为,或采用可再生能源,或发展推动循环回收等,减少的碳排放量。

（四）成功构建"零碳校园"能源系统

"零碳校园"能源系统主要包括：通过太阳能、风能等可再生能源的利用,实现校园能源的自给自足；构建校园级能源碳监测网络,实现绿色低碳校园能源碳数据透明化,建立全校园碳排放图,创建能源和碳网络；建立碳/能源资产管理监测评价体系,对校园用电进行分类/逐项管控,发现通过碳/能源优化获得最佳解决方案；采用节能建筑设计,创办腰带农场,墙壁绿化设计等,降低校园能耗和碳排放。

三、项目进度安排

"零碳校园"建设项目包括培育阶段和实施阶段两个部分。

（一）培育阶段

培育阶段包含三方面工作："制订项目计划""宣传推广"和"培训指导"。"制订项目计划"是在项目开始前,制订详细的项目计划,包括项目的目标、实施方式、时间安排、人员分工等。"宣传推广"指通过校园广播、宣传栏、网络平台等渠道,宣传项目的意义和目标,吸引更多的学生参与。"培训指导"指组织相关的培训和指导,让学生了解低碳生活的重要性和实践方式,培养环保意识和创新能力。

（二）实施阶段

项目实施阶段分为五个方面：宣传教育、互动体验、实践活动、竞赛活动及总结评价。"宣传教育"主要通过课堂讲解、宣传画册、环保主题等课外活动,宣传低碳生活的重要性和实践方式。"互动体验"包含设置低碳生活展示区,让学生在这里亲身感受低碳生活的实践方式和成果、学习碳核查知识和技术、组织互动体验活动。"实践活动"指安排组织学生参与垃圾分类、绿色出行等低碳生活实践,培养学生的低碳生活习惯。"竞赛活动"指组织学生参加环保知识竞赛、低碳手工制作等活动,增强学生的环保意识和创新能力。"总结评价"是在项目结束后,对项目实施效果进行总结和评价,包括参与度评价、认知度评价、行动能力评价和社会影响力评价等,为后续活动开展提供参考和经验。

四、项目组组建

华师五附"零碳校园"建设项目由校长邱超、副校长刘珈宏,华东师范大学地理科学学院王东启教授领衔设计,由分别具有语文、数学、科学、道德与法治、心理等不同学科背景和特长的五位教师共同参与策划。在项目分工上,各位教师既充分发挥自身专长,又积极探索跨学科合作,共同推进气候变化教育课程开发、互动实践体验、校园碳核查、校园能源优化等方面的工作。

项目实施

华师五附作为一所创办于2023年9月的新学校,在上海市嘉定区政府和华东师范大学教育集团的规划和关怀下,学校所用建筑材料绿色环保,倒 V 型屋顶设计不仅具有独特的审美价值,更为未来清洁能源的使用提供了可能性。

学校"零碳校园"建设正处于起步阶段,目前已初步构建了低碳环保的校园框架,并在寒假气候变化教育、五一劳动节假期作业单、绿色出行等方面做了一些活动尝试与探索。

2024 年 1 月寒假期间,学生们共同踏上"风华少年游——寒假探索之旅",通过运动、阅读、艺术、语文、数学、英语、科学、社会、劳动等各版块的活动,全方位提升气候素养。

学校为六年级学生设计了"充满生机的第一周""别样世界的第二周""守岁传承的第三周""捕捉美好的第四周"寒假校本作业,引导学生在实践中学习、体验。下面以"充满生机的第一周""守岁传承的第三周"内容为例进行说明。

一、"充满生机的第一周"之主题任务版块

内容分必做任务和选做任务两类。其中,必做任务为"关注气候变化 · 红领巾在行动"。选做任务为观看纪录片《零碳之路》。必做任务具体如下:

(1)以小组形式前往生态基地参观,每人撰写参观笔记和感悟,600 字左右。每个小组完成一篇通讯稿,要求图文并茂,可以编辑为视频。教师择优在校级公众号推送。

(2)围绕"气候变化"大主题完成一份探究报告。每个中队分成 5 个小组,以小组为单位,校家社协同,从以下五个方向中择其一,走入社区、企业,参加社会实践活动,探究并撰写研究报告,制作成 PPT,开学作汇报评比。

①能源效益:如新能源汽车节能减排原理,建筑隔热,工艺改良,可再生能源,风能,太阳能,绿电等。

②碳达峰与碳中和:如政策背景,相关举措,林业碳汇,海洋碳汇,碳足迹,碳排放等。

③循环经济:如垃圾分类,废物利用,二手市场,减塑行动等。

④保护生物多样性:如了解濒危物种,保护濒危物种的原因与方法,气候变化与保护生物多样性的关系等。

⑤"零碳校园":如以构建零碳校园为目的,开展问卷调查、实践研究、访谈等,以多种形式对校园主要的碳排放进行调研,从新能源开发、低碳出行、新型材料的使用等方面入手,提出可行的减碳行动等。

前期选好主题后请与科学老师联系,听取指导意见,确保项目顺利开展。

· 横向评比:每个中队分成 5 个小组分工合作,进行中队内小组评选,获胜组获得"蓝玫瑰科学奖状"。

· 纵向评比:四个中队中同一方向的 5 个小组,组成一个大组合作探究,5 个大组进行全校大组评选,获胜组获得"蓝玫瑰科学勋章"。

各中队组队参观(见图 1~图 3),撰写通讯稿,填写活动记录表(见表 2),讨论气候变化主题探究报告。在老师的指导下,同学们形成了较完整的图文并茂的通讯稿件,并制作成课件在各中队讲演汇报。

图1　2024年1月,"华闵玫瑰中队"气候变化教育活动

图2　2024年1月,"华深玉兰中队"参观嘉定气象科普馆

图3　2024年1月,"华美向阳中队"在秦山核电站研学

表 2 活动记录表

中队名称	华约紫藤中队	小组名称	华彩雏鹰小队
小组长	冯同学	辅导员	马老师
时间	2024 年 1 月 28 日	地点	上海汽车博览馆
主题	参观上海汽车博览馆		
时长	□一天　　□上午　　☑下午　　□＿＿小时　　□其他＿＿		
参加人员	冯同学等		
缺席人员	孔同学		
社会实践	☑社会考察　□职业体验　□公益劳动　□安全实训		
活动内容	参观上海汽车博览馆,探索汽车中蕴含的人类智慧		
活动收获体会	今天我们华彩雏鹰小队一起来参观上海汽车博物馆,参观出来,我有特别大的收获,来分享一下。 　　上海汽车博物馆位于安亭博园路上,整个博物馆由历史馆、技术馆、品牌馆、古董车、临时展馆五部分组成。进馆前,我们带着华彩雏鹰小队的专属旗子在门口合了影。一进去就能看到汽车博物馆的全景,里面有各式各样的汽车。 　　首先,我们观看的是历史馆,它介绍了汽车的诞生和发展,里面有大众、凯迪拉克、林肯等品牌汽车。接着,我们来到二楼,这一层介绍了汽车在 20 世纪的发展历程。在那个时代,很多人买不起汽车,但人们喜欢坐在汽车里拍一张照作纪念,包括很多明星也喜欢这样做。而我很喜欢一辆银白色的保时捷,它产自 1958 年,小小的,十分好看,整体呈流线型,车身线条十分流畅。然后,我们来到了第三层。呀！看到一辆车分解拆开,挂在天花板上,原来这是在为大家展示一辆小轿车有多少零件,这让我学到了很多知识,知道了汽车是先装好底盘,再装车身,然后装座椅,最后安装车门,从而组成一辆完整的车。过后我们来到了四楼,这里有一个关于汽车的图书馆,我和同学们一起挑了汽车方面的书籍,津津有味地看了起来。到下午 4 点我才依依不舍地离开了汽车博物馆,跟同学道别。 　　中国汽车记忆:如今,中国的大地上驰骋着亿万辆形形色色的汽车,当我们回望四五十年前的岁月,买车曾是千万人的梦想。那时的汽车都带着深深的时间印记,有的已经成为一个时代的符号,长久地留存在我们的记忆里。当前,我们正享受着汽车文明带来的便捷,也承受着汽车社会高速发展带来的困扰,比如汽车尾气、路面拥堵、全球变暖等。 　　面对这些挑战,中国积极响应全球环保号召,大力发展新能源汽车。新能源汽车的普及不仅有助于减少环境污染、缓解交通压力,更是推动汽车产业转型升级、实现可持续发展的关键举措。如今,越来越多的新能源汽车在中国大地上行驶,它们代表着绿色、智能、未来的方向,为中国的汽车记忆增添了新的篇章。		

二、"充满生机的第一周"之阅读版块任务

阅读《寂静的春天》《鲁滨孙漂流记》《骑鹅旅行记》等作品,整体了解全书的故事梗概,重点品读书中某些场景或主要人物的某些言行等(见表3)。

表3　华师五附气候变化教育"风华少年游——寒假探索之旅"推荐阅读书目

年级	作品	作者	出版社
六年级	《寂静的春天》	（美）蕾切尔·卡森	人民教育出版社
	《鲁滨孙漂流记》	（英）丹尼尔·笛福	人民文学出版社
	《骑鹅旅行记》	（瑞典）塞尔玛·拉格洛夫	人民文学出版社
	《汤姆·索亚历险记》	（美）马克·吐温	人民文学出版社
	《这里是中国》	人民网、中国青藏高原研究会、星球研究所	中信出版社

三、"守岁传承的第三周"

设计"春节习俗，守岁传承"必做任务，如①"新年春联我来写"：送给家人、朋友，福气相伴；②"拜年问候我来发"：向每一位任教过的小学老师编辑真诚、有心意的祝福消息，向自己的亲朋好友发送拜年问候，不少于50字，可以中英文结合表达；③"家务劳动我来做"：除夕前主动给家里打扫卫生，并且每日把房间收拾整洁；④"拿手菜肴我来烧"：试着为新年团圆饭准备自己的拿手好菜；⑤"创意绽放我设计"：减少大气粉尘，思考不燃放烟花爆竹渲染春节的喜庆。龙年的精神与韵味，靠你的奇思妙想，绽放出无尽的光彩与热爱。

• 完成打卡任务"春节习俗，守岁传承"晒图，以上五个项目，每一个项目一张照片，开学进行"合群小龙人"评比。

成效与反思

华师五附的"零碳校园"建设及教育实践探索始终秉承着如下原则：

其一，多学科深度整合，即结合语文、数学、科学、道法、地理等学科，将环保教育与学科教学相结合，通过教学提高学生的环保意识和综合素质。

其二，多样化活动形式，即采用多种形式的活动，如宣传教育、互动体验、实践活动、竞赛活动、阅读推介等，吸引学生的参与兴趣，培养他们的环保意识和低碳生活习惯。

其三，多场景生活实践，即注重实践活动的组织和实施，通过引导学生参与垃圾分类、绿色出行等低碳生活实践，培养学生的行动能力和环保意识，推动全校乃至全社会的低碳生活发展。

其四，多维度目标评价，即采用多种评价方式，全面评估项目的目标达成度。通过参与度评价、认知度评价、行动能力评价和社会影响力评价等方式，对项目的实施效果进行全面评估，确保项目的实效性和可持续性发展。

在实施过程中，我们也面临诸多挑战。比如，每位项目组成员如何有效分工评价，如何在有限的课时中进行合理分配，如何在零碳校园建设项目中实现五育融合，如何更好地发挥学生的积极性与主动性，使其参与到活动的设计和策划中等。前行之路，并非坦途。我们将继续奏响"零碳校园"建设的华彩乐章，为地球的未来"碳"索新方向！

（黄婕，华东师范大学第五附属学校教师）

对话专家

气候变化是当前全球可持续发展战略事务中最为重要和紧迫的现实挑战之一。在2024年6月5日"世界环境日"上，联合国秘书长安东尼奥·古特雷斯发表《关于气候行动的特别讲话："直面事实的关头"》，动员全世界行动起来，努力兑现《巴黎协定》承诺，为应对全球气候变化采取切实行动，而解决的关键就在于我们自身。因此，通过气候变化教育，改变我们的认知、思维、价值观等来带动、维持和落实对周边环境的绿化改造是应对全球气候变化的重要举措。对于青少年来说，学校就是一个可及的天然实验室，更是一个学习气候变化素养的实践场。建设绿色低碳校园，正是联合国教科文组织在可持续发展教育领域所倡导的"全机构方法"的直接体现。

《节能背景下教室智控灯光系统设计》充分体现了基础教育最新一轮课程改革提倡的学科实践实施方式。该项目以校园真实问题为出发点，结合信息科技课程基于逻辑运算的初级编程学习任务，通过示范、小组实验、讨论、练习等方式，带领学生们在真实任务情境中完成了对教室灯光节能智控的设计，以科技力量解答了身边的节能问题。后续思路拓展至学校、家庭和社区的其他设备的智能化节能，具有很强的可操作性。

《校园"鱼菜共生馆"建设及低碳教育实践》定位于模仿与学习当前新兴的低碳农业新业态，将农业活动和气候变化之间的关系通过校园微生态建设予以直观显化。该项目通过组织学生实验观察、参与设计、查阅资料、专题讲座等，串联起节能减排、循环利用、生态环保教育，同时也有效改善了校园局部自然生态环境质量。

《建一座不断成长的气候变化探索馆》立志将气候变化教育由短期变为常态，通过对学校在各时段组织开展的气候变化教育各相关材料进行收纳展陈，为后续的再探索、再交流以及人才孵化和混龄学习提供实体基地，最终在参与式学习理念指导下形成了全师生、全学科、全社会共建的学校气候变化教育模式。

《"零碳校园"建设构想及教育实践》结合新时代国家"双碳"要求，聚焦"零碳"这个更为现实和专业的主题，以校园环境去碳为目标，通过开设专题课程、推广低碳生活方式、开展活动竞赛、实行学校"碳核查"管理、重建校园能源系统等软硬件双举措，构建出一个比较成功的现实"零碳校园"。

四则案例有力地展现了学校在气候变化教育领域的就地创意探索和良好发展样态，体现了气候变化教育背景下绿色低碳校园建设的高度、广度、深度和活跃度。我们发现，真实的问题缘起、师生的全员参与、与学科课程的巧妙整合、充分体验和实质参与、问题解决程度的现实评估、充分联动校外社会资源等都是普遍的有效经验。我们也认识到，气候变化教育乃至可持续发展教育本身所要求的高度专业性，需提醒学校、政府和社会：师资队伍的可持续发展意识和有保障的专业性尤为关键。这有助于引导和支持学生在可持续发展道路上长期稳步与坚定前行,并由此带动周边社区的良性链式发展,增强学校教育系统在全民可持续发展教育中的稳固力量。

(朱敏,华东师范大学教育学部副教授,上海终身教育研究院兼职研究员)

第二篇

基于学科教学的气候变化教育

气候变化教育的综合性、实践性与丰富性意味着所有学科都可以,也都必须发挥教育的力量。学生在气候变化方面习得科学知识,提升气候变化领导力、实践力、综合解决问题力、终身学习力,培育人与自然和谐共生、人类命运共同体、可持续生存与发展的理念与价值观,皆依赖于多学科的参与和跨学科的融合。

本篇案例呈现了语文、数学、英语、科学、信息科技、道德与法治等学科教学与气候变化教育的融合实践,带来气候变化教育在各学科教学中渗透、创生的可能之径与可行之策。这让我们看到,加强气候变化教育与学科教育的融合,能使不同学科充分发挥自身独特优势,为学生提供更为全面、系统、深刻的学习体验,实现协同育人的教育目标。

小学语文教学与生态教育的融合实践

研究背景

持续的气候变化给自然生态系统和人类社会带来负面影响。2023年7月17—18日，习近平总书记在全国生态保护大会上讲话指出，我国经济社会发展已进入加快绿色化、低碳化的高质量发展阶段，生态文明建设仍处于压力叠加、负重前行的关键期①。提升全民的生态文明意识，学校教育是最为基础也是最为重要的一环②。

小学语文是基础教育的重要学科之一，兼具工具性和人文性，是学习其他科目的基础。在小学语文课堂教学中融合生态教育，引导学生树立正确的生态价值观，提升学生生态文明素养，是学生成长之需，也是社会发展之要。小学语文教材中包含着极其丰富的生态文明教育内容。部编版小学语文教材选文中，体现生态文明教育思想和内容的篇目十分丰富。这些篇章涉及认识自然、敬畏自然、保护自然、人与自然和谐共生等方面，是语文教学中渗透生态文明教育，促进语文教学与生态文明教育深度融合的重要基础。在全球应对气候变化与生态环境保护的大形势、大趋势背景下，小学语文教师在课堂教学中也应结合生态教育的特点，充分依托教材，开展内容丰富、形式多元的生态文明教育活动，促进学生生态文明价值观形成。

项目设计

一、内容选择

本案例选取部编版小学语文教材二年级上册第一单元中《我是什么》这篇课文作为内容主题。该课文所在的单元主题是"大自然的秘密"。该单元从动物、天气变化、植物等不同角度生动介绍了大自然中一些事物的变化规律与科学现象，内容选择十分贴近学生的生活经验。

《我是什么》这篇课文采用第一人称叙述角度，以"我"为主人公，通过自述，引出水在大自然中的循环现象，生动有趣地介绍了"我"的千变万化（如汽、云、雨、冰雹、雪等），采用拟人化方法叙述"我"生活在哪些地方，"我"温和或暴躁的脾气等。文章用生动形象、充满童真童趣的语言，激发起学生的学习兴趣，帮助学生了解这些变化的规律及其科学道理，引导亲近、

---

① 新华社.习近平在全国生态环境保护大会上强调：全面推进美丽中国建设 加快推进人与自然和谐共生的现代化[EB/OL].(2023-07-18)[2024-05-22].https://www.gov.cn/yaowen/liebiao/202307/content_6892793.htm.
② 陈振林,王昌林.应对气候变化报告（2023）：积极稳妥推进碳达峰碳中和[M].北京：社会科学文献出版社,2023：212-225.

喜爱、观察大自然,探究大自然的奥秘。

二、目标设计

(1)引导学生了解水的形态变化,了解水在自然界的循环规律,促进学生更好地认识自然。

(2)引导学生了解水对人类正反两方面的意义,培育学生对自然的敬畏之情。

(3)引导学生用实际行动发挥水的价值,减轻水的灾害,提升保护自然的能力。

(4)促进学生形成人与自然和谐共生的理念。

三、方式选择

(1)直观演示法。教师借助插图、多媒体等载体,展示文中相关语句,带给学生更加直观生动的体验,帮助学生更好地理解自然知识。

(2)阅读感悟法。教师将"读"的训练贯穿于整个教学过程中,在朗读中,培养学生对自然之美的感知能力。除课内阅读外,教师还向学生推荐关于水的书籍,进行课外拓展阅读,引导学生更深入地了解自然生态问题。

(3)合作交流法。通过教师与学生、学生与学生面对面地听、说、读、问、评、议等,提高学生的思辨能力。

(4)画图梳理法。学生以"水的变化"为题,绘制连环画,画出水的不同形态,并用文字加以描述说明。

(5)实践体验法。通过开展多类实践活动,促进学生参与环保行动。

四、评价设计

本案例主要从学生的课堂练习、背诵表达与绘图展示三个方面展开评价。一是将课堂练习作为阅读教学的有机组成部分,在开展课堂教学任务后,即针对课堂练习情况给予及时评价[1]。二是创设星级挑战(见表1),引导学生对背诵表达情况作出精准的自我判断与评价。三是引导学生用绘图形式展现自己对于水的变化形态和循环规律的理解,给予评价。

表1 《我是什么》背诵星级挑战评价

评星等级	背诵要求
☆	我能正确地背诵自己喜欢的句子
☆☆	我能有感情地背诵自己喜欢的句子
☆☆☆	我能有感情地背诵自己喜欢的句子,而且能在背诵之前,用自己的话概括介绍背诵内容

---

[1] 方蓉飞.保底 分层 梯度——《我是什么》第一课时开放性练习设计[J].语文教学通讯,2013(11):31-33.

项目实施

一、深研文本，引导学生习得水的知识

《我是什么》一文重点介绍了"水"经太阳晒而变成"汽","汽"升到天空变成极小极小的点儿，连成一片变成"云","云"降温后变成"雨"或"冰雹"或"雪"，从天而降，"雨""冰雹""雪"变回"水"汇入池子、小溪、江河、海洋里的过程。通过研读文本，学生可以了解有关水的自然知识，比如：水有哪几种形态变化？水在自然界的循环规律是怎样的？

为了激发学生们的探究兴趣，教师创设情境，引发兴趣："今天有一位神秘的朋友来到我们的课堂，你们知道它是谁吗？"学生们带着疑问，进入文本研读。待充分阅读、熟悉文本后，教师提出问题——"小朋友，你们猜猜，'我'是什么？"学生反馈自己的阅读发现。

接着，学生带着"'我'变成了什么？"这个问题继续阅读，用"找一找，圈一圈"的方式，在文本中找到答案。在互动过程中，教师引导学生用自己的语言表达水的"汽、云、雨、雹、雪"这五种形态变化及其变化条件；引导学生展开想象，体会雨、雹、雪从天空落下来的不同特点，感受水的变化之奇，领略自然之美。

水在大自然中有其循环之道。为了让学生了解有关水循环方面的知识，教师引导其阅读课文第三自然段，让学生找找水都去了哪些地方，感知水在自然界不同环境中的独特之美。待学生充分了解水的循环知识后，可用背诵、绘制思维导图等方式梳理内化相关知识。

二、巧妙设疑，培育学生的自然敬畏之心

课文第四自然段运用对比描写的手法，突出了水的性格的两面性。教师适时提问，请学生结合自己的生活经验，讨论交流："你觉得水重要吗？如果没有水会怎样呢？"待充分交流后，教师用直观的图片展示出水对于人类的重要性。

尽管水对于人类的生存不可或缺，但我们又需要认识到它对人类造成灾害的一面。教师引导学生自由阅读第四自然段，关注"有时候我很温和，有时候我却很暴躁"这一句中的一对反义词，"温和—暴躁"，体会水的两面性。接着，教师通过拓展课外资料，向学生介绍相关自然灾害，如发生在 2004 年 12 月 26 日的印尼大海啸事件，①让学生深刻理解自然的运行之道及人类对自然需持守的敬畏之心。

三、促进探究，激发学生的自然保护意识

通过前一阶段的学习，学生了解了"水"既可以给人类带来好处，又会给人类造成危害。基于文中的"人们想出种种办法管住我，让我光做好事，不做坏事"这句话，教师可进一步向学生追问：有哪些做法可以让水给人类多带来好处，少带来害处？引导学生通过查询资料、求助专家等方式，展开课下探究。在此基础上，教师创设专题分享交流会，让学生联系生活谈谈自己的好办法以及如何在生活中从自己做起，做环保小卫士。同时，学校层面组织开展

---

① 天气网.回顾：2004 年印尼海啸死亡人数[EB/OL].(2015-09-17)[2024-05-25].https://www.tianqi.com/news/105958.html.

各类环保小活动,如"评选校园节水小卫士""评选校园环境美容师""评选校园小园丁"等,将环保行为落到实处。

四、创设活动,培育学生人与自然和谐共生的理念

自然之水形态多样,有自成系统的变化循环规律,对于人类世界的影响有利有弊。待学生了解上述知识后,教师提问"那人与水、与自然之间应建立起怎样的关系呢?"促进学生思辨。通过推荐阅读相关作品,如谢武彰的《水会变哦》、韩国作家申东卿的《水是从哪里来的》、英国作家马修斯的《水的故事》等,让学生进一步了解人类与赖以生存的水资源同属一个自然生态系统,人的生存发展与生态系统中的各部分紧密相连。

在该环节,教师创设实践活动"小瓶子里的大世界",指导学生制作一个"生态瓶"(见表2),让生态系统循环可视化。学生通过参与制作的过程和观察生态瓶的状态,体验生态环境对于自然生物生存的决定性作用,领悟人类的可持续发展有赖于人与自然和谐共生的道理。[①]

表2　生态瓶的制作记录

水中物品种类		物品名称、数量	设计图样
生物	动物		
	植物		
非生物			

成效与反思

本教学案例融知识性、科学性、趣味性于一体,在内容及方式选择上充分考虑学生的生活实际。从学生对自然的真切感知处着手,在提升学生阅读能力和语言表达水平的同时,通过情境想象、对比分析、归纳判断等,促进学生思维的敏捷性、独创性与批判性,锻炼学生整合运用不同学科知识、思维、方法来分析问题、解决问题的能力,很好地培养了学生对自然觉知和审美的能力,促进了人与自然和谐共生自然观的形成。

生态教育的最终目的是促使学生主动关注生态环境,关注气候变化问题,并有效地保护环境,成为有责任感的地球公民。在后续研究过程中,我们将在教育教学过程中,加强体验性活动的创设,打通课内课外,引导学生更多走进大自然,用实际行动保护大自然。

(王丫珍,天津经济技术开发区国际学校教师)

[①] 岳伟,徐洁.培育生态人格:生态文明建设的教育使命[J].教育研究与实验,2015(1):18-22.

生态教育背景下小学语文单元教学活动设计

研究背景

《义务教育语文课程标准（2022年版）》强调学生应积极观察，感知生活。其中提到"充分发挥课程资源的育人功能，优化教与学活动，课程资源的使用要以促进学生核心素养发展为目的，多角度挖掘育人价值"。[①] 小学语文教学对学生人生观、价值观的培养有重要作用，在气候变化与生态环境教育方面具有巨大优势。本案例基于部编版语文一年级下册第六单元开展大单元教学活动设计。

本单元主题为"夏天"，案例设计立足语文学科核心素养，基于语文学习任务群特点，整体规划活动内容，创设真实而富有趣味的情境，让学生在活动中体会夏天的美丽风景、美好意蕴，培养学生热爱自然的审美情趣，探索自然的科学兴趣，形成人与自然和谐共生的生态文明素养。

项目设计

一、内容选择

生态文明素养包含"生态情怀、生存技能、生态智慧、生态审美、生态实践"五个关键素养[②]。语文教材中的生态环境教育资源十分丰富，如在小学低年段的语文教材，就涉及天象变化和自然事物特点的相关话题。内容的呈现形式多样，有"四时诗"、小短文及书中插图等，都是培育学生生态文明素养的关键媒介。

本单元内容编排了《古诗二首》《荷叶圆圆》《要下雨了》这三篇课文和一个"语文园地"，从不同角度描绘出夏天的特点。案例活动设计以"夏日的旅行"为主题情境，整合"三个任务、六个活动"，囊括识字写字、课文阅读等教学内容，引导学生在读读、写写、演演中积累相关知识，感受语文学习的快乐，体察自然之美，陶冶审美情趣。

二、目标设计

(1)培养学生观察自然、亲近自然的意识，提升审美能力。

(2)引导学生了解气象常识，探索自然规律，感悟自然智慧。

(3)引领学生做生态环境的守护者，提升学生参与生态保护活动及绿色地球建设的

[①] 教育部.义务教育语文课程标准（2022年版）[M].北京：北京师范大学出版社，2022：6-54.
[②] 陈振林，王昌林.应对气候变化报告（2023）：积极稳妥推进碳达峰碳中和[M].北京：社会科学文献出版社，2023：212-225.

能力。

三、方式选择

案例活动方式包含"文本研读""环境感知体验""实践活动""展示分享"等。"文本研读"指学生通过阅读课文，获取相关生态知识，产生对大自然的喜爱之情。"环境感知体验"指学生走进自然、观察自然，亲身感受美好的自然生态环境对人类的重要作用。"实践活动"指通过组织创建"明朗教室"，构建"绿色校园"等活动，引导学生积极践行环保。"展示分享"即通过观点分享、作品展示、节目表演等形式，呈现学生观察自然、了解自然的感受与收获。

四、评价设计

教师采用多种评价方式以科学客观地评判、展现学生生态文明素养的提升情况，具体评价标准如表 1 所示。

表 1　主题活动评价

评价活动	评价标准	评价方式		
		自评	同学互评	师长评价
夏日摄影展	愿意与别人交流对夏天的观察成果	☆☆☆	☆☆☆	☆☆☆
	别人分享时能认真倾听，自己发言时能把句子说完整	☆☆☆	☆☆☆	☆☆☆
	对夏天以及自然产生兴趣，有自己的观察方法	☆☆☆	☆☆☆	☆☆☆
学习课文《要下雨了》	能正确、流利、有感情地朗读课文，了解气象特点，感受自然的美好	☆☆☆	☆☆☆	☆☆☆
	能用表演的方式，呈现出天气变化对小动物的影响，了解大自然的规律	☆☆☆	☆☆☆	☆☆☆
	别人发言时能认真倾听，自己发言时能把话说完整	☆☆☆	☆☆☆	☆☆☆
学习气象谚语	愿意通过与他人交流沟通，搜集气象谚语；乐于与他人分享气象谚语	☆☆☆	☆☆☆	☆☆☆
	对天气以及自然产生兴趣，能把学到的气象知识应用到生活中来	☆☆☆	☆☆☆	☆☆☆
采访长辈	愿意与他人交流气候与自然情况	☆☆☆	☆☆☆	☆☆☆
	能用自己的话说出气候变化的相关现象，表达自己对气候变化的感受	☆☆☆	☆☆☆	☆☆☆
	别人说话时能认真听，自己发言时能把句子说完整	☆☆☆	☆☆☆	☆☆☆

(续表)

评价活动	评价标准	评价方式		
		自评	同学互评	师长评价
环保活动"打造明朗教室"	积极参与活动,打造绿色教室	☆☆☆	☆☆☆	☆☆☆
	参加养护绿植活动,了解养护知识	☆☆☆	☆☆☆	☆☆☆
	养护的绿植生长茂盛	☆☆☆	☆☆☆	☆☆☆
	把环保行动应用到校园更大范围	☆☆☆	☆☆☆	☆☆☆

项目实施

一、开展"夏日摄影展""消夏晚会",培养学生观察、亲近自然的意识

本单元以"夏日"为主题,从不同角度描绘出夏天的季节特点。教师通过创设多元活动,让学生寻找夏天的特点,感受夏天的美好,生发对大自然的亲近喜爱之情。

其一,夏日摄影展。此活动旨在让学生走出课本、走出课堂,走向生活、走向大自然。首先,教师设置问题:"你在生活中,见到过课本中的夏日景色吗?请你拍摄与夏天有关的景、物,并担任一次'小荷导游师'。"学生们利用课后时光,用自己的眼睛、借助自己的相机发现、捕捉夏天的美好。然后将所拍摄的夏季特有食物、生活用品、活动、景色等(见表2),于课堂上展示分享,感受、回味夏日的快乐体验。

表2 "夏日摄影展"学生拍摄的作品内容

内容分类	内容表达
夏天吃的	葡萄、桃子、梨、冰沙、刨冰、奶昔、凉拌面、酸梅汤……
夏天穿的	短袖、短裤、薄裙、T恤、汗衫……
夏天玩的	游泳、打水仗、漂流、海边玩沙……
夏天的天气	打雷、暴雨、台风、炎热……
夏天的昆虫	蚊子、苍蝇、蜻蜓……
夏天的花	荷花、牵牛花、葵花、兰花、百合花、千日红、睡莲、紫薇……

其二,"消夏晚会"。教师以"联欢会"的形式,营造一个让学生能拓展、可展示的欢乐课堂。学生唱童谣《砍蚊子》,朗诵儿童诗《夏天是个娃娃》,分享《牛郎织女》的民间传说,叙说夏天里的快乐与烦恼。师生还一起仰望星空,一起倾听夏夜里蝉鸣、蛙声、蛐蛐声、风声组合成的"夏夜交响曲",领略夏日的独特魅力。

二、创设"雨前探秘"主题活动,激发学生阅读和探求自然奥秘的兴趣

《要下雨了》是本单元的一篇科学童话,它以浅显生动的语言文字揭示了下雨前的气象

特点与动物活动之间的关系。教师创设"雨前探秘"主题活动,激发学生的阅读兴趣,在阅读、表演中感受自然奥秘。

首先,教师设置层层递进的问题,引导学生发现动物在下雨前的异常行为变化,如:先"找一找故事里有哪些小动物?""下雨前,燕子、小鱼、蚂蚁在干什么?"待学生充分表达后,及时引导学生探究,寻得雨前的秘密:"下雨前,燕子低飞,鱼游水面,蚂蚁搬家。"紧接着,教师引导学生通过分角色朗读,理解下雨前天气变潮湿这一特点,及其与三个小动物异常表现之间的内在联系,感知生态系统中各个要素之间的紧密关联。最后,学生用分角色表演故事的方式演绎这种关联,加深对自然规律的理解。

基于本单元语文园地中日积月累的内容,教师引导学生通过"读一读,连一连,背一背"的方式,积累有关气象变化的趣味表达。基于"气象小寻访"课外实践拓展活动,教师引导学生咨询家人和朋友,积累相关谚语,再通过分享会分享。

三、设计代际交流活动,引导学生积极参与生态环境保护行动

为了增强学生的生态环境保护意识,教师设计了一场代际交流活动"小小记者团采访长辈"。学生围绕话题"现在夏天与过去一样吗?"与自己的长辈进行沟通,并思考人类活动是如何带来温室气体排放的增加,导致大气中温室气体浓度不断升高,使得地球表面温度逐年升高的。

保护生态环境,关键在行动。为引导学生从身边小事做起,成为绿色地球的建设者,教师创设了一个环保小活动"打造明朗教室",学生们用美丽的绿植装点教室,制作"绿植养护指南",定时定量地为绿植浇水,打造低碳环保的教室环境,营建绿色校园。

成效与反思

将生态教育和语文教学内容融合,是落实新课程改革的切实行动。在本项目中,依据大单元设计理念,教师高效整合单元内容,突出单元主题,创设更贴近学生生活的教学活动,让学生的生态文明素养在真实情境中得以提升。在实践过程中,教师不是用生涩的语言为学生讲授自然知识,而是引导学生走进自然,观察自然,感受自然的美好;不是将语言文字孤立地呈现给学生,而是在探访自然的活动中,让学生与美好的语言文化自然相遇,领略语言文字的魅力与深意。

此外,项目实施过程中我们还逐渐形成了一套校家社协同育人模式。营建的校—家—社多维协同空间场域很好地促进了学生更好地走出校园,走进社区,走入生活,在不同场景中开展体验式学习。

生态教育是一项综合性、长期性的系统工程。语文教师应有参与其中的育人自觉,不断提升生态教育的意识及于学科教学中渗透生态教育的专业素养与实践能力,使学生从小树立正确的自然生态观和环境保护意识,学会用科学的眼光认识世界,分析问题、解决问题,实现人与自然的和谐统一。

(赵静,天津经济技术开发区国际学校教师)

小学数学项目化学习促进低碳生活的探索

研究背景

数学来源于生活,数学学习与生活实践密不可分。节能低碳环保生活是未来家庭生活发展的总趋势,是积极应对气候变化的可行之策。身为一名小学数学教师,深知将气候变化教育融入数学教学的重要性。本案例基于上海市九年义务教育课本四年级第二学期数学练习部分第二单元"小数的认识与加减法"的实践作业"计算一周的家庭开支"展开。我们从学生提交的作业中发现,各项数据存在显著差异。为深入探讨这一现象,并协助学生为家庭提供解决方案,四年级组的数学老师们在充分考虑数学本体知识以及学生的年龄特点后,决定开展主题为"居家生活小管家"的项目化学习,帮助学生更好地理解和应对现实生活中的挑战。

项目设计

"让学生关注真实的世界,不仅仅是为了让学生深度理解和掌握概念,或者锻炼思维能力,同时也是为了引导学生敬畏自然与生命,理解何为社会责任"[1]。本案例从"如何合理购买生活必需品"问题切入,引导学生担任"生活小管家"角色,体会数学与日常生活的密切联系,感悟数学学习的价值与乐趣,发展运算能力、应用意识等学科核心素养[2]。同时,提升学生对绿色消费、低碳生活的认知与理解,形成应对气候变化的素养。具体目标如下:

(1)感受小数在日常生活中的广泛应用,并正确运用小数加减法解决生活实际问题,发展运算能力和应用意识。

(2)经历数据调查、观察比较和统筹规划等活动,提升团队合作、沟通交流和问题解决的能力。

(3)通过担当"生活小管家"角色,探索"如何合理购买生活必需品",感悟学习的价值与乐趣,培育绿色消费、低碳生活的观念与意识。

项目实施

一、基于真实生活情境,发现问题、提出问题

(一)引发猜想

教师首先向学生展示他们提交的小实践作业数据统计表(见图1)。学生们惊奇地发现,

---

[1] 夏雪梅.项目化学习的实施:学习素养视角下的中国建构[M].北京:教育科学出版社,2020.
[2] 教育部.义务教育数学课程标准(2022年版)[M].北京:北京师范大学出版社,2022.

有的家庭一周支出接近 2 000 元,而有的家庭仅约 200 元。这一显著差异引起了他们强烈的好奇心。学生们积极思考,提出猜想后进行验证,最终得到结论:在统计中,家庭开支较少的主要原因是家中已有较充足的存货或采取了节俭和预算管理的消费习惯;相反,支出较多的家庭是因为缺乏对日常消费的预算管理或具有冲动和过度的消费习惯。

 小实践。

你想知道家里一周要花多少钱吗?请你来当一回小管家。每天晚上问一下爸爸妈妈,并把每一笔开支记录下来。算一算每天花了多少钱,一周一共花了多少钱。

	星期一	星期二	星期三	星期四	星期五	星期六	星期日	合计
金额(元)	265.9	176	374.64	241	695	123.4	104	1979.94

	星期一	星期二	星期三	星期四	星期五	星期六	星期日	合计
金额(元)	147.8	208.9	69	189.8	46	396.8	100	1258.3

	星期一	星期二	星期三	星期四	星期五	星期六	星期日	合计
金额(元)	0.00	125.78	0.00	0.00	75.64	0.00	0.00	201.42

图 1　学生提交的"小数的认识与加减法"单元的小实践作业数据

(二)提供支架

教师让学生观察班级里一位学生记录的家庭 4 月份生活支出手账,引导他们展开讨论(见图 2)。

图 2　学生制作的手账

基于学生们的观察比较、猜测验证与积极讨论,教师确定本项目的本质问题:"如何调动学生生活经验,有条理地调查、统筹规划、制订计划,发展学生运算能力和应用意识,提高问题解决的能力?"驱动性问题:"你作为家里的生活小管家,如何合理购买生活必需品?"

二、头脑风暴,分析问题、解决问题

教师以在线的方式,组织头脑风暴讨论会,师生共同分析、讨论解决问题的步骤、关键要素,教师帮助学生有条理地制订购物计划。以下为讨论过程及在此过程中产生的四个子问题:

第一步:调查家里缺什么生活必需品。

子问题1:生活必需品有哪些?

学生们踊跃举手发言,并在互动消息板上积极留言,热烈参与讨论,最终将物品分为调味料类、乳制品类、肉类、蔬果类、主食类、生活用品类、学习用品类、药品类等。此外,学生们倡导选用环保包装和本地生产的商品,以减少运输过程中的碳排放。

第二步:如何制订一份完整的购物计划?

学生们从不同角度提出购物计划的重要组成部分,为制订购物计划提出合理方案,避免过度消费和浪费。例如一是列商品名称,确保所列商品符合实际需求的同时提倡优先考虑环保产品;二是查商品价格,建议比较不同渠道的商品价格。

子问题2:如何查询物品价格?

学生们共同探讨出查询价格的主要方法有线上App查询与线下商超查看。在此过程中,学生们讨论认为应在购买前确定商品数量,避免过度消费,减少浪费。

子问题3:买多少生活必需品?

学生提出,应针对购买物品的特点合理采购。保质期长或可以冷冻保存的商品,可以适当增加购买量以减少购物次数,降低碳排放。保质期短、不宜长期储存或易腐败的商品,如蔬菜与水果,要适量购买,并优先选择当季食品。还要考虑家里的储藏空间,如冰箱的空间是否充足,以避免过量购买导致食物浪费和资源过度消耗。此外,还要寻找合适的购买渠道,鼓励选择低碳的购物方式。

子问题4:在哪里买生活必需品?

对此,学生梳理出主要有线上、线下两种方式。线上购买,即通过手机App或小程序下单,且购买时尽可能将多个订单合并在一起配送,以减少单独配送带来的重复碳排放。线下购买,即到超市采购。线下购买时鼓励选择步行或骑行前往。如果通过团购可以享受批量优惠,还可以考虑与邻居凑单购买,既能共享优惠,又能减少每个家庭的碳足迹。在此过程中,就需估算总金额,确保购物计划不会超出家庭设定的预算。

第三步:评价改进、成果展示。

本项目给予学生充分的时间和空间,引导学生独立完成或是同小区内同学合作完成。学生们开展调查,结合当前家庭实际需求和市场状况,合理制订购物计划(见图3),确保既能满足日常需要又能合理控制预算。最后,根据购物计划进行采购,接受家长评价、同学互评,并记录过程中有意义的故事或"小管家"心得,在学校视频号中展示项目成果。

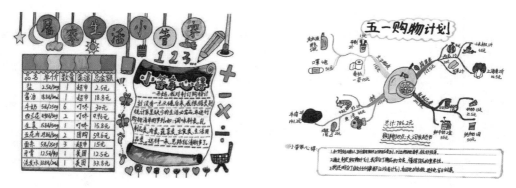

图3　学生制订的家庭购物计划

成效与反思

本次学习活动加深了学生对"小数的认识与加减法"的理解，发展了学生的运算能力。同时，通过倡导积极践行绿色消费，避免过度囤积、减少食物浪费，提升了学生及其家庭对低碳生活方式的认同感。

一、发展了学生的运算能力

在数学教学中，用情境表征算理算法是培养学生运算能力的重要策略之一。从图4可见，该生在购买物品时会主动看保质期和有效期，真正理解数据在现实世界中的意义和价值。在计算总金额时，能正确计算，并根据实际促销活动"满100减10"算出优惠金额。教师还了解到，该生为了计算精准，反复检查了好几遍，体现出精益求精、严谨求实的科学态度。

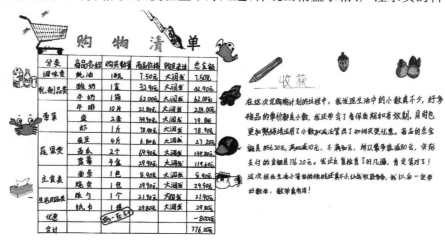

图4　学生制订的家庭购物计划及写下的项目学习心得

活动结束后，我们对该年级学生"计算掌握"版块的表现进行了持续追踪，分别统计了获得"优秀""良好""合格"和"须努力"评价的学生人数，并一直监测至他们小学毕业。如表1所示，毕业时，整个年级的平均优秀率超过了95%，特别是3班，优秀率达到了100%。整体而言，学生的计算能力呈现出明显的上升趋势。

表 1 学生"计算掌握"版块表现情况统计

班级	2021年12月 计算掌握								2022年6月 计算掌握								2023年6月（毕业）计算掌握							
	优秀	/%	良好	/%	合格	/%	须努力	/%	优秀	/%	良好	/%	合格	/%	须努力	/%	优秀	/%	良好	/%	合格	/%	须努力	/%
01班	34	82.93%	4	9.75%	0	0.00%	3	7.32%	36	87.80%	2	4.88%	2	4.88%	1	2.44%	39	95.12%	1	2.44%	1	2.44%	0	0.00%
02班	29	72.50%	6	15.00%	4	10.00%	1	2.50%	32	80.00%	7	17.50%	0	0.00%	1	2.50%	38	95.00%	1	2.50%	0	0.00%	1	2.50%
03班	34	82.93%	6	14.63%	1	2.44%	0	0.00%	36	87.80%	4	9.76%	1	2.44%	0	0.00%	41	100%	0	0.00%	0	0.00%	0	0.00%
04班	33	80.49%	5	12.20%	2	4.88%	1	2.43%	36	87.80%	4	9.76%	0	0.00%	1	2.44%	38	92.68%	2	4.88%	0	0.00%	1	2.44%
05班	27	64.29%	10	23.81%	4	9.52%	1	2.38%	34	80.95%	6	14.29%	1	2.38%	1	2.38%	40	95.24%	1	2.38%	0	0.00%	1	2.38%
06班	29	70.73%	7	17.07%	3	7.32%	2	4.88%	37	90.24%	2	4.88%	1	2.44%	1	2.44%	40	97.56%	0	0.00%	1	2.44%	0	0.00%
总计	186	75.61%	38	15.45%	14	5.69%	8	3.25%	211	85.77%	25	10.16%	5	2.03%	5	2.03%	236	95.93%	5	2.03%	2	0.81%	3	1.23%

二、增强了学生的低碳生活意识

从学生提交的"小管家"心得中可以发现,在推动可持续消费方面,他们采取了一系列措施,如避免购买非必需品、与邻居凑单、将多余物品分享给需要的人、实行理性消费、减少食物浪费等。这些行为强化了学生的绿色消费理念,增强了他们对低碳生活的认同感。而社区团购的持续实施也证明了人们正在逐渐接受、采纳一种更加绿色低碳的生活方式。

三、培养了学生的反思习惯

在学习过程中,学生们基于驱动性问题,充分调动自身生活经验,切身参与数据调查、统筹规划、做出预算,并最终为自己家庭解决了采购物资的问题。有了这些经历,相信学生们在今后的生活中,也能有条理地调查、统筹规划、制订计划、反思迭代,不断提升问题解决的能力。此外,这些学习经历还极大地促进了学生对日常生活的关注及对家庭生活的反思,增进了他们对低碳生活的认识与理解,并将这种影响传递至家庭和社区,产生更大的辐射效应。

当然,在本案例具体实施过程中,因主要采用的是在线形式开展讨论交流,师生、生生实时互动的次数相较于传统课堂有所减少。为弥补此不足,建议学生先自由组队,展开小组层面的讨论,再组织集体汇报交流,以此提高全体学生的参与度与互动性。此外,为了加深学生的理解,我们可以收集整理不同学生家庭的实际案例,展开深入分析和讨论,进一步引导学生做出避免过度消费和浪费的合理决策,让他们深刻地感受到低碳生活的意义。

(孙娟,上海市金山区前京小学教师)

英语课堂教学中的气候变化教育实践

研究背景

作为应对气候变化的重要力量,气候变化教育能够影响人的思维方式和行为选择。《义务教育英语课程标准(2022年版)》(以下简称"英语新课标")提出,英语课程围绕核心素养,使学生通过课程学习逐步形成社会所需要的正确价值观、必备品格和关键能力,增强学生的人类命运共同体意识和社会责任感。学生能够从"人与自然"的角度理解"热爱与敬畏自然,与自然和谐共生"的重要意义。此外,《义务教育课程方案和课程标准(2022年版)》(以下简称"新课标")还提出,跨学科主题学习应满足不少于10%的课时设计。教师应引导学生在英语学习的立场下,通过跨学科主题学习,了解学科之间的联系,结合实际生活,解决实际问题。[1]

鉴于此,气候变化教育与英语新课标中对于构建人与自然生命共同体意识的目标以及跨学科整合和项目化学习的途径相一致。本案例将气候变化教育融入初中英语课堂教学,创设跨学科主题学习活动,倡导联系生活实际,关注社会热点,合理应对气候变化,培养热爱自然的意识和保护环境的能力。

项目设计

本课例选自外研版英语教材七年级上册第六模块第二单元 *The tiger lives in Asia*[2]《老虎生活在亚洲》。语篇从外观、食物、栖息地和生活习性等方面介绍了大象、大熊猫、斑马、老虎和猴子等五种动物。该内容属于人与自然的主题范畴,主题群为自然生态,子主题为热爱与敬畏自然,与自然和谐共生。

结合语篇内容及特点、新课标要求及气候变化教育目标,教师鼓励学生以"制作保护东北虎的宣传手册"为主题任务,参考世界自然基金会等相关网站的信息,结合生活实际,分小组开展跨学科主题学习活动。

一、学习活动名称

制作保护东北虎的宣传手册。

二、学习任务

学生应邀参加世界野生动物基金会(WWF)保护野生虎的宣传活动,设计以"保护东北

---

[1] 教育部.义务教育课程方案和课程标准(2022年版)[M].北京:北京师范大学出版社,2022:11.
[2] 陈琳,(英)格里诺尔.义务教育教科书·英语(七年级上册)[M].北京:外语教学与研究出版社,2012:38-39.

虎"为主题的宣传手册,讲解东北虎相关的知识,介绍保护东北虎的方式和途径,宣传保护东北虎、维护生物多样性的重要意义。

三、涉及学科

英语学科:用英语撰写东北虎的介绍、保护东北虎的措施以及保护东北虎对人类生存和发展的重要意义。

美术学科:绘制东北虎元素的宣传图片。

地理学科:了解东北虎的地理分布、栖息环境等地理知识。

生物学科:了解东北虎的形态特征、生活习性等生物知识。

四、学习目标

(1)通过阅读语篇和相关资料,学生梳理整合关键信息,积累介绍东北虎的形态特征、生活习性、地理分布及栖息环境的英语表达和素材,了解生态系统及相关自然科学类知识以及生物多样性和气候变化的关系。

(2)通过查阅图书、互联网等相关资料,学生进行探究性学习,从多方面、多角度思考保护东北虎的举措,主动发现问题、分析问题和解决问题,培养与自然和谐相处的意识和思维方式,养成在日常生活中从我做起、保护动物、保护地球的行为习惯。

(3)通过小组合作设计保护东北虎的宣传手册,学生深入研究以东北虎为代表的野生动物现状,深刻了解动物与气候、人与自然的密切关系,培育人类命运共同体意识。

(4)通过展示保护东北虎的宣传手册,学生向身边人宣传保护野生动物的重要性和紧迫性,形成人人保护野生动物、维护生物多样性、共同应对气候变化的良好局面。

五、评价设计

"英语新课标"指出,教学评价应贯穿英语课程教与学的全过程。在本次学习活动中,教师采用教师评价、学生自评和学生互评相结合的方式,提高学习成效,落实"教—学—评"一体化。

(一)教师评价

根据课堂教学目标,观察学生回答问题的情况,了解学生的学习过程和学习中的困难,分析学生目前的学习水平和教学目标的差距。例如,在"根据思维导图复述课文"活动中,教师根据学生对语篇复述的完成程度判断学生是否完成教学目标,并提供针对性的指导和帮助,给予必要的鼓励和表扬(见图1)。

(二)学生自评

学生在回答问题后,结合教师所给出的评价量表开展自我评价和同伴互评(见图2)。在评价的过程中,学生通过自我评价,主动反思,促进自我监督学习。在同伴互评中,学生充分发挥主体作用,成为评价活动的参与者与合作者,并从中取长补短,促进自我成长。

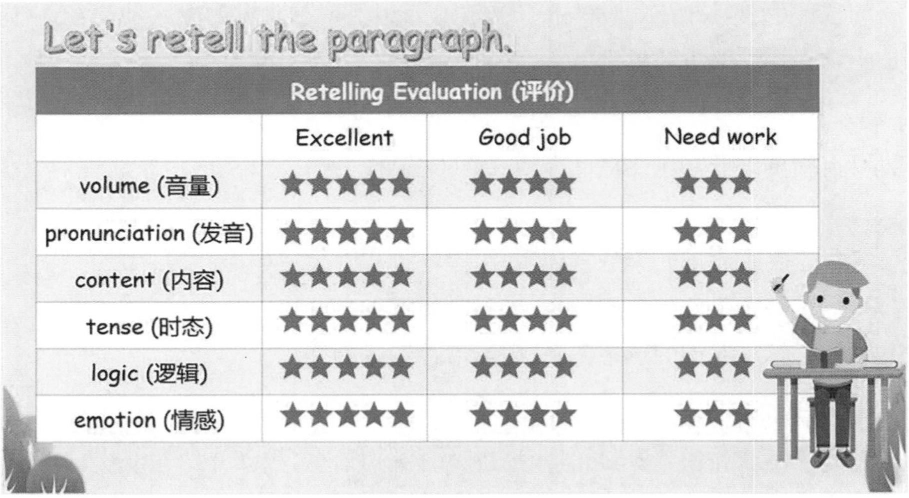

Let's retell the paragraph.

Retelling Evaluation (评价)	Excellent	Good job	Need work
volume (音量)	★★★★★	★★★★	★★★
pronunciation (发音)	★★★★★	★★★★	★★★
content (内容)	★★★★★	★★★★	★★★
tense (时态)	★★★★★	★★★★	★★★
logic (逻辑)	★★★★★	★★★★	★★★
emotion (情感)	★★★★★	★★★★	★★★

图 1　"根据思维导图复述课文"活动评价量表

Brochure Evaluation (评价)	Excellent	Good job	Need work
topic (主题清晰)	★★★★★	★★★★	★★★
contents (内容全面)	★★★★★	★★★★	★★★
suggestions (建议合理)	★★★★★	★★★★	★★★
design (布局合理)	★★★★★	★★★★	★★★
colour (色彩丰富)	★★★★★	★★★★	★★★

图 2　"制作保护东北虎的宣传手册"活动评价量表

项目实施

本次主题学习活动采用课堂教学和课后任务相结合的方式进行，具体由学生选题、规划设计、活动实施和提交作品四部分组成（见表1）。在活动实施过程中，教师将大任务拆解成学生可操作的小任务，让学生通过课堂学习、课后探究等方式来完成。这种方式既体现了教师的引导作用，又彰显出学生的主体作用。

表1 "制作保护东北虎的宣传手册"学习活动实施方案

学习主题任务	实践任务	课时
选题(创设情境,明确小组分工)	任务1:学生了解并分析主题任务,联系实际生活;根据学生特长和意愿,确定小组分工。	课时1
规划(确定设计思路,梳理并提取材料)	任务2:学生阅读课本语篇,提取并梳理相关信息,利用思维导图了解语篇结构和具体细节,从外观、食物、栖息地和生活习性等方面对动物进行全方位介绍,为后续保护野生动物提供知识基础。 任务3:通过课堂讨论完善设计思路。	
	任务4:学生阅读相关书籍、浏览世界野生动物基金会网站等互联网资源,搜索补充资料,从而认识到野生动物的存活和气候变化息息相关,并寻求保护措施。	课后任务
实施(小组合作)	任务5:小组分工合作,绘制宣传图,从东北虎的形态特征、生活习性、地理分布和栖息环境介绍东北虎,认识到气候变化对东北虎的影响,提出保护东北虎的措施,思考保护东北虎对人类发展的重要意义,并完成文稿撰写。 任务6:展示小组初稿,老师和同学分别对初稿进行评价,提出修改意见。	课时2
总结(提交作品)	任务7:结合老师和同学的修改意见对宣传图和文稿进行修改并装订成册。	课后任务

一、创设情境,明确小组分工

首先,教师向学生介绍本次学习活动的背景和任务。然后,学生以五人为一组,在教师的引导下对学习任务展开解读。在解读任务的过程中,学生结合生活中所见过的宣传手册,自主讨论,认为宣传手册的设计需分为两部分:一是在宣传文字的设计方面,内容要简练、结构要清晰,向读者传达最重要的信息;二是要图文并茂,能反映宣传主题,吸引读者。之后,学生根据自己的特长和意愿,确定基础分工:三人完成文稿撰写,两人完成手册中图画的绘制。

二、确定设计思路,梳理提取材料

教师通过精读、泛读策略让学生提取、梳理语篇关键信息,引导学生从外观、食物、栖息地和生活习性等方面进行信息整合,利用思维导图工具熟悉语篇结构、复述语篇细节,为后续介绍东北虎提供框架和内容。

起初,学生在复述语篇过程中,对大象、大熊猫、斑马、老虎和猴子等五种动物进行了初步了解。基于此,教师引导学生从形态特征、生活习性、地理分布、栖息环境等方面全方位了解这些动物。这为宣传图册中东北虎的文稿撰写提供了思路。

通过课堂讨论,学生进一步完善设计思路,并制定详细分工:一人完成对东北虎地理分布、栖息环境等地理知识的资料查找和文稿撰写;一人完成对东北虎形态特征、生活习性等生物知识的资料查找和文稿撰写;一人完成对东北虎的现状和保护东北虎的措施的资料查找和文稿撰写；一人完成宣传图册封面的绘制；一人完成宣传图册内页插图的绘制。在查找资料过程中，学生认识到：由于气候变化，东北虎的栖息地面积和生存质量发生了变化，猎物的数量也在慢慢减少；气候变化导致的森林病虫害和草原退化等问题也严重威胁到东北虎的生存和繁衍；重视气候变化，保护东北虎，维护生物多样性刻不容缓。

此外，学生还于课后完成相关补充资料的查阅，通过工具书、互联网等途径，了解东北虎的地理分布、栖息环境、形态特征、生活习性、现状、保护措施等信息，并及时展开交流探讨，从多方面、多角度思考保护东北虎的举措。在此过程中，学生们意识到气候变化的严峻局面，认识到保护野生动物、共同应对气候变化的紧迫性与重要性。

三、小组合作，推进实施

依据分工，学生完成文稿的资料搜集、撰写和图片的绘制。在这一过程中，教师给予必要的帮助和指导。例如，在资料搜集的过程中，地理老师和生物老师进行答疑解惑，英语老师指导学生进行文字翻译工作，美术老师指导学生进行图片的排版等。在撰写文稿的过程中，学生彼此协作，学习掌握生态系统及自然科学知识，培养保护动物、保护自然的意识。

此后，学生在课堂展示小组初稿，借助评价量表开展自我评价、同学互评和教师评价。师生以评促学，共同商讨、完善作品内容。

四、梳理总结，提交作品

学生结合同学和老师的修改意见利用课后时间完善宣传手册的内容，并及时提交作品（见图3、图4、图5）。

图3 "制作保护东北虎的宣传手册"作品展示（封面）

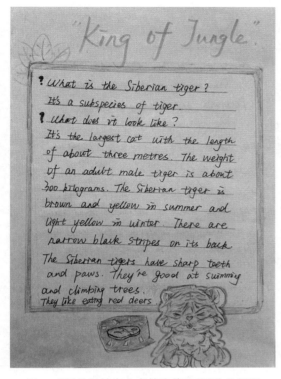 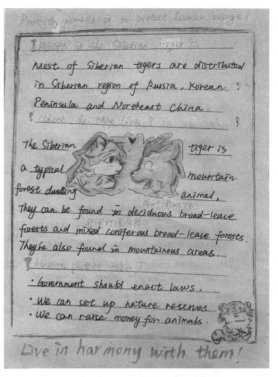

图 4　"制作保护东北虎的宣传手册"作品展示（内页 1）　　图 5　"制作保护东北虎的宣传手册"作品展示（内页 2）

成效与反思

气候变化教育与自然科学息息相关，但英语学习同样强调人与自然关系的重要性，且英语学科强调的跨学科主题学习以其综合性、丰富性和实践性与气候变化教育特征深度契合。本案例通过跨学科主题学习活动，为学生创设了真实的学习情境，将语言知识、学习能力、文化意识和思维品质的培养结合在了一起。学生在学习活动的过程中，感受到气候变化对动物和人类造成的巨大影响，领悟到保护野生动物、维护生物多样性的重要意义。

但气候变化教育不仅仅是传授知识，更是培养学生积极应对气候变化的素养和能力。因此，教师在设计本学科的学习活动时，要聚焦气候变化的现实问题，倡导学生亲身体验、实践探究、解决问题。例如，指导学生研究具体的气候变化案例，让学生从不同角度探讨气候变化问题，分析案例中的原因、影响和应对措施；鼓励学生参与社区举办的气候变化相关活动，让学生通过实际行动体验自身在应对气候变化中所发挥的作用，培养社会责任意识，促进自我成长。

（张杰，天津经济技术开发区国际学校教师）

基于科学实践活动的气候变化教育：
以"降水量的测量"为例

研究背景

准确及时的天气预报对生产生活、防灾避难有着重要意义。记录、共享气象数据，对研究大气现象的规律、探索大自然的奥秘、准确预测天气能发挥重要的作用[①]。自古以来，人们就有意识地记录气象信息。近200年来，随着气象仪器的发明和改进，气象数据愈加准确，且世界各地建立了越来越多的气象台站，越来越多的气象专业人员持续地观测、记录、共享气象数据。在此基础上，更加有效的气象理论和分析模型得以建立。

降水是天气的一个基本特征，也是"天气日历"中重要的记录数据。在预测天气时，很重要的一个方面是观测"水"的状况。在本项目开始之前，天津经济技术开发区国际学校的学生们已经在尝试经由感官来观察、判断降雨情况。本案例旨在引导学校3—6年级的学生通过"关注生命之水，珍惜水资源"科学实践活动，了解气象学家测量、记录、确定降水量的途径与方式，亲手制作雨量器，并尝试利用自己制作的雨量器持续观测、记录降水量。

项目设计

一、活动目的

在本次活动中，学生以一个月为调查周期，借助自行设计制作的雨量器，测量并统计所在地区的降雨量及雨水pH值。活动的设计有如下五个方面的目的：其一，帮助学生理解"降水量"科学概念；其二，教会学生制作简易雨量器，并习得如何用简易雨量器测量降水量；其三，提高学生整理数据、分析数据以及小组合作的能力；其四，培养学生探索大自然奥秘的兴趣，引导学生通过实践活动了解地球水资源的现状及变化趋势，提高学生的生态环境保护意识；其五，通过倡导并鼓励学生像气象学家那样观察和记录每天的天气信息，敦促学生参与一段较长时期的天气观察、记录和分析数据的活动，促进学生建立长期坚持不懈地进行科学活动的意识，培养认真细致的观察习惯与持之以恒的科学研究品格。

二、评价设计

（一）评价目的及原则

本活动的评价目的在于检验教学效果、诊断教学问题、提供反馈信息、引导教学方向以及调控科学实践活动进程，激励学生树立科学观念，形成科学思维，让他们乐于探究实践，善

[①] 教育部.义务教育科学课程标准（2022年版）[M].北京:北京师范大学出版社,2022:12-14.

于合作与分享。①

活动评价遵循客观性、科学性、发展性以及参与性原则。客观性原则体现为评价过程和结果皆基于事实和数据,尽可能减少主观判断,不受个人情感、偏见或立场影响,确保评价结果的准确性和可靠性。科学性原则体现为评价遵循科学方法,使用合理的评价标准和工具,确保评价过程的逻辑性和系统性。发展性原则体现为评价不仅关注当前的表现和成果,还着眼于活动的长远影响和改进的可能性,鼓励参与者共同学习进步,促进活动持续发展。参与性原则体现为鼓励、促进学生积极参与,提高评价的全面性和包容性,提高评价结果的可接受性和实施的可能性,使活动评价更加公正有效。

(二)评价内容

本活动的评价内容包含"学习态度""学习效果""学习能力"以及"团队合作能力"等方面,方式有学生自评、学生互评、家长评价和教师评价(见表1)。最终评选出的优秀者将获得学校颁发的"国际科技小达人"证书。

表1 科学实践活动"降水量的测量"学生学习综合评价表

	评价项目	学生自评	学生互评	家长评价	教师评价
综合评价	1. 对活动的兴趣、求知欲和探究精神				
	2. 对知识的掌握				
	3. 对活动过程、方法的体验				
	4. 与人合作、与环境相处的态度和表现				
	综述:				

项目实施

一、准备阶段:了解基本知识,组建活动小组

在准备阶段,教师用3个课时引导学生:学习、了解有关水的科学知识;掌握雨量器的使用方法,了解雨量器测量降水量的原理;通过动手实验,学习相关实验操作的基本知识与基本技能;制作简易雨量器,并学会用简易雨量器测量降水量。

学生根据活动方案,在6月30日至7月3日暑假之前以自由结合的方式,成立科学实践活动小组。每个学生活动小组准备好一套工具,包括最高温度计、最低温度计、雨量器和精密pH试纸。

---

① 喻伯军.义务教育课程标准(2022年版)课例式解读.科学[M].北京:教育科学出版社,2022:11-12.

二、实施阶段：制作雨量器，测量、记录降水量

（一）熟悉相关知识，引出核心问题

教师利用科学课向学生讲授有关水的科学概念和知识，如水的性质、地球水资源的储备与分布、地球水循环、酸雨的成因与危害、人类活动对于水循环和水资源的影响等。

首先，教师向学生介绍降水是天气的一个基本特征，也是地球水循环中一个非常重要的环节；降水的形式有多种，常见的有雨、雪、冰雹等。然后，与学生一起回忆最近的一次降雨，并提问"雨下得大还是小呢？""你们是怎样判断降水量的大小的？"鼓励学生回忆可以从哪里观察到雨的大小，如地面上水坑积水的深浅、放在外面的容器中雨水的多少。其次，教师提出核心问题："我们怎样才能准确地知道雨下得有多大？"并向学生介绍气象学家是如何利用雨量器测量降水量以及根据降水量的多少来区分降水等级的，同时出示气象学家测量降水量的资料、仪器等。

（二）自制雨量器，模拟测量实验

教师要求每个活动小组准备好制作雨量器的材料：一张制作方法说明书、一个厚底直筒玻璃杯或塑料杯、刻度尺、纸带、透明胶带、剪刀等。学生按照教师讲解的步骤制作简易雨量器。当各小组制作完成自己的雨量器后，向全班同学展示。教师适时提问并组织讨论，如"雨量器口径的大小是否对测量有影响？""是否能用不同大小的雨量器监测降水量？"等。

雨量器制作完成后，教师引导学生学会检查雨量器是否漏水，并开展一次实地模拟测量。学生到室外选择合适场地，放好雨量器，用喷壶模拟降雨。在此过程中，学生们经历一次收集、测量降水量的过程，并对照降雨等级标准判断测得的雨量等级。在读取雨量数据时，教师特别提醒学生相关注意事项并讨论、解释原因，如移动雨量器时需注意不要让雨量器内的"降水"溢出；读取数据时，雨量器需放置在水平桌面上；视线应与雨量器上的刻度数保持平行。

此后，小组各成员依次读取自己小组的雨量器数据，并汇报组内，经商量后达成一致的读数。各组学生代表在"我们的降水量填充图"上记录下降水量。教师组织讨论：每组的降水量为什么不同？如何判别大雨、中雨、小雨等不同雨量等级？接着出示气象学家根据降水量的多少区分雨量等级的数据表，学生据此对照，做出判断。

接下来，教师对学生课外降水量测量和记录活动予以指导，师生共同梳理注意事项，如确定雨量器摆放位置，尽量选择上面或附近没有遮挡物的较开阔的地点；测量前固定好雨量器，避免它被风吹倒；每24小时记录一次降水量，确定雨量等级；每次记录完毕后，须将雨量器内的水倾倒干净，避免影响下一次的测量与记录等。

（三）学生实地测量，教师及时指导

学生选择暑期内的一个月为调查周期，利用自己设计的和购买的雨量器分别测量并统计所在地区的降雨量及雨水 pH 值。对比自制雨量器和购买的雨量器的测量统计数据，分析产生误差的原因并据此不断改进。教师适时提供远程指导。

三、总结、展示阶段：分析数据，展示成果

经过为期一个月的科学实践活动，学生收获颇丰，也对活动过程及结果进行了多种形式的记录。各活动小组学生对记录的数据进行分析，总结天气变化的规律，并填写相关数据表格，上传至学校指定邮箱。对于过程性记录资料，各活动小组以PPT、美篇、文字等方式展示汇报。为鼓励学生积极投身科技创新实践活动，弘扬爱科学、学科学、用科学的精神，学校对所有参加此项科技实践活动的同学给予了表彰，并将获奖作品及时发布到了学校"科学飞书群"，展示在学校的微信公众号上。此举令学生备受鼓舞。

成效与反思

在此科学实践活动中，教师通过介绍气候变化的背景知识，引出自制雨量器的设计实践，进而引导学生亲手制作仪器并观测、记录降水数据（见图1、图2）。通过长期观测，学生了解到了降水量的季节性变化与年际变化。

图1 学生准备的实验仪器

图2 学生监测降雨量及天气信息原始记录

通过案例研究，学生了解到了气候变化对全球不同地区的影响。教师适时发出行动倡议，鼓励学生思考如何减少温室气体排放，并提出应对气候变化的策略。图3～图6是学生在活动中根据记录的数据制作的图。

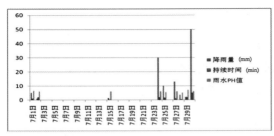

图 3　学生记录的降水情况柱形图

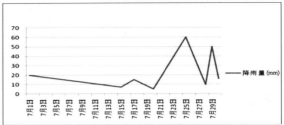

图 4　学生记录的降雨量折线图

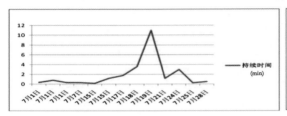

图 5　学生绘制的降雨持续时间折线图

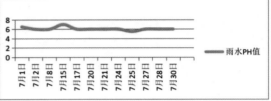

图 6　学生记录的雨水 pH 值折线图

有学生在活动心得中写道：

今年暑假，我和于同学投入"珍爱生命之水"科学实践活动中。我们的工作既繁忙，又有意义。通过一个月的测量和统计，7 月份天津市塘沽区杭州道共下了 12 场雨，每一场雨的酸碱性是不一样的。

酸雨对农作物有很大的危害，它破坏土壤成分，使农作物减产，使湖泊中鱼虾死亡，还腐蚀建筑物和工业设备，破坏露天的文物古迹。饮用酸化物污染的地下水，对人体有害。今年 7 月份我们这里就有一场酸雨，我们应该予以高度重视。另外，我们感到暑期气温越来越热，全球气候变暖已成不争的事实。全球变暖对水的循环和利用也将产生不利影响，进而危害全人类。

通过参与此次调查、测量和研究活动，我们不但了解到许多知识，锻炼了科学实践能力，还懂得了要珍惜水资源、保护环境。地球是最大的生物圈，是我们人类共同的家园。我爱我家，我们爱我们的家！

有学生发出感慨：

我们在实验的过程中，懂得了许多科学原理，还明白了一个重要的道理：要节约用水，珍惜每一滴水！

从学生提交的数据及研究成果来看，尽管学生在实践过程中还存在诸多不足，如有些观测、记录的数据准确性不高，有的学生未能坚持一个月的测量时间。但大部分学生在面对困难时，都能积极应对，并不断探寻问题解决之策。

在此过程中，他们学会了简单的气象观测和记录方法，了解到即使同在一个区，哪怕仅隔一条公路，降雨开始时间、持续时间、结束时间以及降雨量也可能不同，由此懂得气象数据多点采集的原因与意义所在。他们还增强了小组合作能力，提升了整理数据、分析数据的能力，切身感知到科学研究过程中保持严谨求实、实事求是、坚持不懈等精神品质的重要性。更为重要的是，他们养成了主动觉察天气变化的习惯，增强了对周遭环境信息的感知，理解了水源保护对于解决水资源危机的重要作用，提高了生态环境保护意识。

同时，也存在一些值得反思和改进的不足之处：其一是理论与实践结合得不够紧密。在活动过程中，学生虽然参与了实践操作，但对于相关理论知识的掌握不够扎实。对此，须采取一系列有效措施，加强理论与实践的结合。如：在活动开始前，安排专题讲座以及研讨会，确保学生掌握必要的理论知识；在活动进行中，设计更多基于实际情境的任务，让学生在实践中运用所学理论。其二是对于学生自主学习能力提升不足。为此，可更多设计一些具体的探究性问题，鼓励引导学生独立思考和解决问题，不断提高自主学习能力。未来，我们将持续改进，提升活动实效，促进学生素养的全面提升。

（张永红，天津经济技术开发区国际学校教师）

巧学·慧用·智创：
信息科技教育与气候变化教育融合模式建构

研究背景

中国作出力争在 2030 年前实现碳达峰、2060 年前实现碳中和的庄严承诺[①]，表明应对气候变化的决心与雄心。青少年儿童作为未来的主人，其气候变化意识和行动对于实现这些目标至关重要。信息科技学科凭借跨学科融合、数据驱动和技术创新的独特特点，为气候变化教育提供了丰富的资源和手段。借助信息科技，学生可以更加直观地了解气候变化的科学原理、影响及应对策略，同时也可以利用信息科技手段参与到气候变化的监测、分析和应对中。本案例旨在探索信息科技学科教学与气候变化教育的融合实践，从"巧学""慧用"和"智创"三个维度，实现气候变化教育的有效渗透和深入推进。

项目设计

一、目标设计

（1）培养学生对气候变化的科学认知，理解气候变化的成因、影响及应对策略。

（2）提高学生的数据分析能力和科学探究能力，使之能运用信息科技手段分析基础性的气候变化数据。

（3）增强学生的环保意识和社会责任感，使他们积极参与应对气候变化的行动。

二、内容设计

本案例的具体内容为：其一，气候变化相关基础知识的学习活动。相关知识包括气候变化的含义、成因、影响及全球应对气候变化措施等。其二，学生数据分析技能的提升活动。该活动主要引导学生使用信息科技工具（如数据处理软件、编程语言等）来分析气候变化数据（如温度、降雨量、海平面上升等）。其三，跨学科融合的学习活动。结合信息科技、自然、数学等学科资源，引领学生分析气候变化对自然环境、生态系统和社会生活的影响。其四，气候变化相关的科研项目、环保活动及社区的科普服务等。

三、活动方式

（1）跨学科主题探究。气候变化教育涉及多个学科领域，需打破学科界限，深度融合信息科技、数学、科学等学科，以教学内容的选择、教学方法的应用、教学资源的开发、教学评价的设计为关键环节，建立完整的跨学科主题项目，引导学生开展跨学科知识整合与问题

[①] 新华网.习近平在第七十五届联合国大会一般性辩论上的讲话（全文）[EB/OL].（2020-09-22）[2021-01-10]. http://www.xinhuanet.com/politics/leaders/2020-09-22/c_1126527652.htm.

解决。

(2) 实验教学。利用信息科技实验室或虚拟实验室,组织学生进行气候变化数据分析和模拟实验。

(3) 实践活动。组织学生参与实地考察、社会调查、科研项目等实践活动,让学生亲身体验气候变化的影响和应对策略。

(4) 线上学习。利用在线学习平台或社交媒体,为学生提供丰富的学习资源和交流机会,促进学生对气候变化问题的深入探讨。

四、评价设计

通过对学生"知识掌握""技能应用""态度与行动""跨学科融合"四个方面学习成效的评价,助推信息科技学科与气候变化教育的深度融合,提高学生的气候变化意识和行动能力(见表1)。

表1 信息科技学科与气候变化教育融合学习评价设计

评价项目	评价形式	评价内容	评价等第
知识掌握	考查、作业	对气候变化基础知识的掌握程度	☆☆☆☆☆
技能应用	项目报告、数据分析报告	运用信息科技手段分析气候变化数据的能力	☆☆☆☆☆
态度与行动	观察、访谈	态度:环保意识、社会责任感 行动:应对气候变化方面的行动表现	☆☆☆☆☆
跨学科融合	跨学科项目或综合实践活动	在不同学科间整合知识和思维方式的能力	☆☆☆☆☆

项目实施

气候变化教育独具综合性、实践性和丰富性,横跨多个学科领域。[①]《义务教育信息科技课程标准(2022年版)》围绕数据、算法、网络、信息处理、信息安全和人工智能这六条逻辑主线,构建了完整而系统的全学段内容模块。[②] 每个模块均精心设计了跨学科主题活动,这为气候变化教育在信息科技教学中的深度融合提供了强大的支撑和有利条件。

一、巧学:信息科技助力气候变化教育科普

气候变化与日常生活紧密相连。在学科教学中融入气候变化教育,旨在通过构建贴近生活的场景,引导学生在解决实际问题过程中,逐渐增强对气候变化的认知与行动意识。

[①] 上海终身教育研究院.气候变化教育指导纲要(试行)[EB/OL].(2024-03-06)[2024-05-21].http://www.smile.ecnu.edu.cn/c8/53/c42519a575571/page.htm.
[②] 教育部.义务教育信息科技课程标准(2022年版)[M].北京:北京师范大学出版社,2022:12.

二十四节气作为农耕文化瑰宝,是先人智慧的结晶。在"在线学习与生活"板块,教师设计了"二十四节气演说家"单元活动,引导学生在线收集整理与节气相关的信息,并通过团队协作深入探索其中蕴含的智慧,全面了解气候变化的影响,最后借助思维导图完成二十四节气的演说分享。这一教学方式的"巧"处在于激发学生兴趣,使知识贴近实际,引导他们认识气候变化的重要性和与日常生活的紧密关系。

学生在探索二十四节气过程中,提出"古人流传下来的气象俗语科学吗?"这一核心疑问,教师敏锐地捕捉到这个有价值的课堂生成资源,并设计了"数"说气象俗语综合活动。以"古人流传下来的气象俗语科学吗?"作为单元的核心问题,并分解出"用什么方法验证?""需要什么数据?""使用什么工具?""得出什么结论?"的问题链。对应问题链,建构活动串(见图1)。这种以问题为驱动的学习方式,能够培养学生的独立思考能力和解决问题的能力,是教学过程中的一种"巧"思。

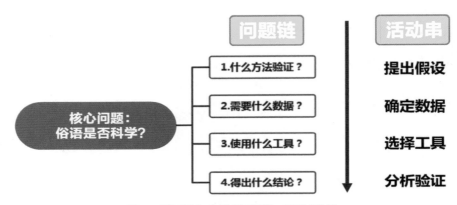

图1 "数"说气象俗语"问题—活动"设计

譬如,在判断"大暑小暑不是暑,立秋处暑正当暑"①这条俗语的准确性时,教学过程如下:首先,提出假设,"大暑小暑不是暑,立秋处暑正当暑"这条俗语是科学的,并选择数据对比验证。其次,确定数据,同时提问"对比一组数据还是多组数据",引发学生思考,并认识到多组数据的对比,更具规律性。再次,选择合适的工具,通过可视化图表呈现对比结果,使之更直观清晰。最后,分析验证,得出结论。如图2所示,除了2021年以外,上海地区近十年的小暑与立秋节气数据都符合俗语规律。学生们得出的判断结果是:这条俗语具有较强的科学性。

---

① 该俗语的意思是:大暑小暑还不是一年中最热的节气,立秋处暑才是一年中最热的节气。

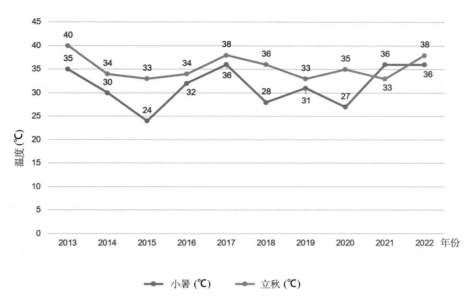

图 2　近十年上海地区小暑与立秋节气当天最高气温对比

而对于"大寒不寒,春分不暖"这条俗语①,学生通过数据验证,发现规律性并不明显,甚至不符合规律。如图 3 所示,学生提出困惑:从数据上来看,该俗语确实不符合规律,难道古人流传下来的经验不科学?

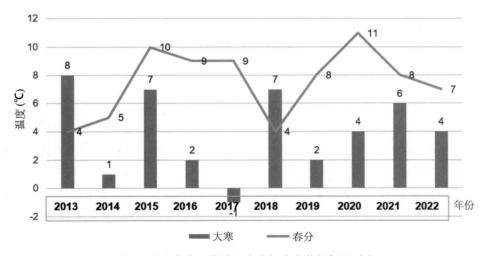

图 3　近十年中上海地区大寒与春分节气气温对比

对此,教师巧妙引导学生深入剖析这些俗语背后的原因,尤其是其中涉及的复杂全球气

---

① 该俗语的意思是如果大寒节气不寒冷,寒潮就会延后,到了下一个节气春分的时候,天气就不会很暖和。

候变化因素。气象俗语是古人多年积累的经验成果,地方性色彩浓厚。受全球气候变化影响,当下要综合俗语内容和科学的气象预测,融合各学科知识辩证地应用气象俗语指导生产生活。由此,学生更懂得应通过收集、整理和分析数据,用科学的方法来验证气象俗语的正确性。这种方法的"巧"在于能够培养学生的科学素养,让他们学会用科学的视角来看待问题,同时也能够让他们更加深入地理解气候变化的规律和影响。

二、慧用:气候变化探索成果智慧地应用于生活

气候变化教育强调通过亲身体验,以培养学习者的知识发现能力、创新能力和批判性思维。在新课标指导下,教师结合学生的生活实践,设计了信息科技与气候变化教育相结合的项目,引导学生通过实践体验,既收获知识成长,又将学习成果智慧地应用于日常生活,实现学习价值的最大化。

以"自制气象生活小贴士"为例,教师引导学生认识数据在日常生活中的重要性:通过校园气象科普数据平台收集数据,运用技术工具进行数据整理和可视化表达,制作实用的"气象生活小贴士",并将探究成果通过校园电视台、融媒体中心等多元平台,与全校师生同分享。这一过程促成了知识的传播与智慧分享,同时,学生在知识分享中接收到来自他人的反馈和建议,据此进一步完善了自己的作品。这种循环往复的过程是"慧用"的一种体现。

通过"气象生活小贴士"项目单元学习,学生提升了用数据说话、用数据解决问题的意识,同时增强了数据应用效能。有学生自主探究气候变化与衣食住行的关系,结合数据可视化制成科普展板,在学校组织的"气象嘉年华"活动中展示分享;有学生探索气候变化与穿衣的科技课题,撰写气候变化研究报告《考虑气象因素的上海市小学生校服穿着建议》,并将26℃穿衣法则在校园推广。如此,"慧用"还体现在将气候变化探索成果普及给更广泛的人群,提高其气候变化意识和参与度,让更多人了解气候变化的影响和应对策略,形成全社会合力。

三、智创:面向未来气候变化教育的智慧创新

在应对未来气候变化的教育之路上,仅停留于普及知识是远远不够的,应从小培养学生的问题解决意识,构建应对气候变化的必备素养。为此,学校信息科技社团依托独特的"太空舱火星农场"学习空间,创新性地开展了"智能种植"项目,为学生搭建"智创试验田"。

此项目旨在模拟非自然甚至极端天气条件下,通过编程技术智能控制种植系统,以优化种植流程,探索对抗未来极端气候的创新策略。项目融信息科技、自然科学和劳动教育等多个学科于一体,选取南瓜、樱桃萝卜、矮牵牛、矮番茄、草莓等多种植物作为研究对象,深入分析影响植物成长的关键因素。

在项目启动之初,我们提出了一个核心问题:"如何利用 Farmbot 机器人生态箱模拟极端环境,实现植物的智能种植?"并将之分解成多个子问题,如"哪些种子适合在太空舱生态箱内种植?""太空环境下影响植物生长的要素有哪些?""如何通过机器人及人工智能技术调控植物的生长?"等,激发学生的探索欲望。

随后，根据植物生长的六大要素——光、温、湿、水、气、肥，结合六维编程的核心任务（见图4），精心设计跨学科的教学单元。例如，在光照控制的学习中，学生们学习如何根据植物的高度调节光照板的高低；在温度控制的学习中，学生们学习如何结合生态箱内的温度合理启用加热扇或风扇；在肥料控制的学习中，学生们掌握如何根据植物的生长状态决定是否配制肥料，还习得使用专业土壤检测仪测量植物当前pH值的方法。

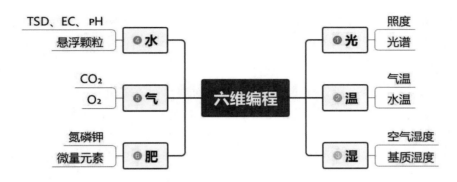

图4 "智能种植"项目六维编程

学校"太空舱火星农场"学习空间的微控生态箱配备了各种传感器，为学生提供了真实的智能种植探究环境，学生利用机器人技术和人工智能技术，对植物生长的每一个因素进行编程控制。这片"试错"的智能化试验田不仅让学生们在实践中学习、探索，更为他们未来积极应对气候变化问题提供了宝贵的思路和经验。

成效与反思

如今，信息科技在气候变化教育中的角色日益凸显。巧学、慧用、智创的教学模式，让学生们的学习力和问题解决能力得以显著提升，也让学生的环保意识得以增强。他们开始更加关注气候变化问题，并主动围绕日常生活，开展相关的科技课题研究。

然而气候变化教育并非学校一己之力就能完成的。校家社三方的紧密合作，是形成气候变化教育合力的关键。我们需要为学生打造一个立体化的气候变化教育空间，让学生与真实的世界建立更紧密的联系，全面参与、切身体验，更大程度地发挥他们的想象力与创造力。未来，我们将继续深化信息科技与气候变化教育的融合，培养更多有责任感和行动力的未来公民，为守护地球和绿色家园贡献智慧与力量。

（程育艳、孙童童、顾维彬，上海市闵行区七宝镇明强小学教师）

生态文明建设背景下小学道德与法治学科教学活动设计

研究背景

"道之以德,齐之以礼,有耻且格",生态文明建设需要以法规严守生态底线,更需以教化培育生态文明意识。与生态文明教育有机融合,促进生态文明社会实践,既是时代赋予道德与法治教育的新使命,也是生态文明建设的内在要求。

《义务教育道德与法治课程标准(2022 年版)》明确将发展学生核心素养确定为课程发展的根本方向,提出"帮助学生形成正确的价值观,涵养必备品格,增强规则意识,发展社会情感,提升关键能力"。[①] 本案例立足学生视角和现实生活,引导学生感受并探索大自然,进行人与自然的道德关系的重构,让"生态文明"从书本上、课堂中一个个看似深奥抽象的知识点,变为学生可亲近、可感知,深度融入学生真实生活的活动。

项目设计

一、活动目标

"人与自然"是小学道德与法治课程"人与自我""人与自然""人与社会"三条主线中的一条。经梳理,我们发现,道德与法治课程在 1—5 年级,是依据不同成长阶段学生身心发展及道德成长需要,以螺旋上升的方式引导学生逐渐形成人与自然和谐共存的自然观。《道德与法治》(五·四学制)教材中"生态文明"相关内容如表 1 所示。

表 1 《道德与法治》(五·四学制)中"生态文明"相关内容

学段·用书	单元	课文
一年级·下册	第二单元	第五课:风儿轻轻吹 第六课:花儿草儿真美丽 第七课:可爱的动物 第八课:大自然,谢谢你
二年级·上册	第二单元	第八课:装扮我们的教室
	第四单元	第十三课:我爱家乡山和水 第十四课:家乡物产养育我 第十六课:家乡新变化

[①] 教育部.义务教育道德与法治课程标准(2022 年版)[M].北京:北京师范大学出版社,2022:6-54.

(续表)

学段·用书	单元	课文
二年级·下册	第一单元	第四课:试种一粒籽
	第三单元	第九课:小水滴的诉说 第十课:清新空气是个宝 第十一课:我是一张纸 第十二课:我的环保小搭档
三年级·上册	第二单元	第六课:让我们的学校更美好
四年级·下册	第五单元	第十三课:我们所了解的环境污染 第十四课:变废为宝有妙招 第十五课:低碳生活每一天
五年级·上册	第五单元	第十五课:地球——我们的家园 第十六课:应对自然灾害
小学高年级·习近平新时代中国特色社会主义思想(学生读本)		第十讲:绿水青山就是金山银山

据此,我们进一步明确道德与法治课程生态文明主题实践活动的各学段目标:

第一学段(1—2年级):通过参与观察、种植或养殖等实践活动,学生能够识别常见动植物,了解它们的基本生活习性,亲身体验到与自然的亲近感,并学会以尊重与爱护的方式对待动植物。

第二学段(3—4年级):通过参与主题实践活动,学生了解自然与人类生活的紧密关系,认识到自然是我们生活的共同家园,懂得保护环境、爱护动物、节约资源的重要性。

第三学段(5年级):通过参与综合性的项目或研究,学生运用所学知识分析环境问题,提出可持续发展的解决方案;树立可持续发展的价值观,认识到实现人类、社会、环境和谐共存的重要性。

二、活动内容

本案例通过创设系列"生态文明"主题课后综合实践活动,引导学生在完成课堂学习后,依据任务单,以小组合作的方式展开探究。

一年级的"我与自然"实践活动旨在让学生亲近自然,感受自然的美好,使热爱成为"生态文明"主题教育的基调;二年级"我与家乡"的实践活动让学生进一步了解自己出生长大的地方,通过具体的事物建立起家乡与自己的联系,让学生对自然广博的爱有了更具象的落脚点,情感体悟更真切;三年级"我与校园"的实践活动意在激发学生"我也可以做些什么"的主人翁意识;四年级"我与可持续发展"的实践活动让"绿色环保"进入学生的真实生活,以"成就感"促学生的责任感与担当意识;五年级"我与灾害"的实践活动引导学生从另一个视角看待人与自然的共生,建立起正确的可持续发展生态观。

三、活动评价

本次活动评价从"活动兴趣""活动能力""活动成果"三个维度进行等第制评价。同时，评价主要以学生自评、同伴互评、教师评价相结合的方式进行(见表 2)。

表 2 综合实践活动评价

姓名_____

评价方式		自评	同伴评价	教师评价
活动兴趣				
活动能力	搜集资料			
	沟通表达			
	合作学习			
活动成果	任务完成度			
	交流汇报			

项目实施

一、创设"小海鲸与风儿交朋友"活动，引导学生感知自然

上海市金山区属于北亚热带，是东南亚季风盛行的地区。该区四季分明，气候温和湿润，但受冷暖空气交替的影响，台风暴雨时有发生。虑及一年级学生既有较强的好奇心和求知欲，又存在耐心及自我约束能力不足等特点，在"小海鲸与风儿交朋友"实践活动中，教师先结合《风儿轻轻吹》的课堂学习，让学生了解如何"捕捉"风的足迹，再结合综合实践活动任务单引导学生观察(见表 3)。为了增强学生对于"风"的感受力，教师首先引导学生思考"风"背后的气候变化，并让他们记录不同季节、不同地区"风"给人的感受；然后，让学生通过绘制"风"的样子及风给家乡带来的变化，将气候问题具象化；最后，引导学生化身家乡植物，创编一段与风的对话，感知家乡气候变化与植被间的关系。

表 3　"小海鲸与风儿交朋友"综合实践活动任务单

小海鲸与风儿交朋友		
寻风 导语:风,是我们最亲密的小伙伴,它行过四季,拂过万物。什么时候,你会去哪里找寻风呢?它带给你怎样的感受呢? ＿＿＿＿,我去＿＿＿＿找寻风的身影,它带给我的感受是＿＿＿＿。	解风 导语:"解落三秋叶,能开二月花。"李峤眼中的风,秋天带走落叶,春天吹开花朵,那你们找到的风会带给家乡什么样的变化呢?	谢风 导语:风儿不仅是我们的朋友,更是大自然的好朋友,请你选择家乡的一种植物,想象它会对风说些什么。 （　） （风）

二、创设"小海鲸漫游金山"活动,促使学生关注生态环境

金山区有万亩良田、水果公园及美丽的"花开海上"生态园,这些都得益于温和的气候环境。结合"家乡物产养育我"一课的学习,学校创设"行走的 LOVE"活动,通过"小海鲸漫游家乡",让学生在实地探访中与家乡生态建立起真实的情感联系。在活动中,学生根据自己设计的路线图探访家乡,寻找家乡特产,寻访家乡人的智慧,全面感知家乡生态环境(见表4)。

表 4　"小海鲸漫游金山"综合实践活动任务单

小海鲸漫游金山
导语:金山区位于上海市的西南部,物产丰富,生态宜人。选一个周末,和你的爸爸妈妈一起去感受古镇的悠然、果园的香甜、渔村的浪漫、田园的纯真……用你的双脚去丈量你的家乡,用你的双眼记录独属于你的家乡物产图,也请用真心的话语表达对孕育丰富物产的家乡的赞美之情吧。

(续表)

三、创设"小海鲸让校园添彩"活动,助推学生参与环保行动

金山区是上海几乎唯一汇集山、海、岛、城于一体的地方,适合发展都市农业。宜人的气候使金山拥有金山渔村、吕巷水果公园、廊下生态园等具有多功能、独特美丽的"田园综合体"群落。在学习完"让我们的学校更美好"一课后,学生通过"小海鲸让校园添彩"综合实践活动,借助教材中的《校园调查报告》思考如何进一步完善、打造更舒适的校园环境(见表5)。学生们还在学校"少代会"中提出了具体有效的方案。

表5 "小海鲸让校园添彩"综合实践活动任务单

小海鲸让校园添彩			
校园一角	存在的问题	改进措施	后续维护

(续表)

红领巾提案表	
提案人	所在集体
附议人	
提案类别	
提案名称	
提案内容	
有关建议	
提案答复	
备注	

四、创设"小海鲸之创意一刻"活动，形塑学生的审美意识与环保理念

在"变废为宝有妙招"一课的学习后，学生通过"小海鲸之创意一刻"综合实践活动，将"为菜园量身定做环保装饰品"活动纳入学校"京京菜园"的自主管理中，以行动响应环保理念，呵护田园之美（见表6）。

表6 "小海鲸之创意一刻"综合实践活动任务单

小海鲸之创意一刻

导语：京京菜园的种植季马上就要到来啦！请你们用勤劳的双手，利用生活中的材料为他们制作、绘制美丽的装饰品，让我们的菜园多彩起来吧！

- 准备材料

- 制作过程

- 用法范例

五、创设"小海鲸的避险手册"活动,增强学生的避险能力与安全意识

自然灾害教育是素质教育的重要构成,防灾减灾知识技能是学生必备的实践能力之一。教师结合"应对自然灾害"一课的学习,通过"小海鲸的避险手册"实践活动,让学生基于各项消防、地震演习中观察到的问题,进一步规划各班逃生路线,绘制"校园逃生图",树立危机意识,提升避险能力。并将这份"危机意识"迁移到家庭生活中,与家人一起绘制"家庭逃生路线图",全方位提升"安全意识"(见表7)。

表7 "小海鲸的避险手册"综合实践活动任务单

小海鲸的避险手册
导语:大自然给予了我们无尽的宝藏,但是,如果我们不爱护它,大自然也会用它的方式来告诫和惩罚我们,如果发生灾害,我们该怎么办呢?
• 校园逃生路线图
• 家庭逃生路线图

成效与反思

道德与法治课程是一门与学生生活紧密联系的综合性活动课程[①]。综合实践活动的设计是密切联系学生生活实际的关键。在开展此次"生态文明"主题综合实践活动的过程中,我们认识到三个重要性。

一是准确把握"生态文明"理念的重要性。教师要不断加强对生态文明教育的学习和研究,认真研读党和国家有关生态文明建设的政策,掌握思想内涵,并将这些思想融入教学实践。

---

① 孙彩平.怎样上好小学道德与法治课:基于全国优秀教学案例的分析[J].南京:南京师范大学出版社,2020(4):149.

二是让"生态文明"行走起来的重要性。对于小学生而言,"生态文明"与现实生活之间确实存在隔膜。需要教师整合生活资源,让学生在社会大课堂的"行走"中,在真实的探索中,构建起切实深刻的生态文明观。

三是全学科推行"生态文明"建设的重要性。2019年,中共中央、国务院印发《关于深化教育教学改革全面提高义务教育质量的意见》,明确将生态文明教育作为德育的重要内容之一。培养学生的生态文明素养,仅依靠小学道德与法治学科教学是远远不够的,需让更多的学科教师参与进来,共同营建多元多样的生态文明观念培育与行为引导课程基地。

(姚瑶,上海市金山区前京小学教师)

对话专家

教育部新颁布的《义务教育课程方案和课程标准（2022 年版）》中明确提出"热爱自然，保护环境，爱护动物，珍爱生命，树立公共卫生意识与生态文明观念。……关心时事，热爱和平，尊重和理解文化的多样性，初步具有国际视野与人类命运共同体意识"的育人目标要求。《普通高中课程方案和语文等学科课程标准（2017 年版 2020 年修订）》中，也充实和强化了"生态文明与海洋权益"等内容，强调了"尊重自然，保护环境，生态文明意识"的培养目标。可见，基于学科教学的气候变化教育成为基础教育课程教学变革的重要组成部分。

基于学科教学的气候变化教育是指以生态文明与可持续发展观提升学科教学内涵与品质，以气候变化素养与学科核心素养的迭代达成为目标，以可持续原则更新教与学方式，以学科视角分析气候变化议题的教学实践过程。

本篇展示了语文、数学、英语、科学、信息科技、道德与法治等学科教学与气候变化教育融合实践的优秀案例，呈现了基于学科教学的气候变化教育的三种实践样态，即学科教学中的气候变化教育、气候变化主题大单元教学、学科＋气候变化综合实践活动。其对学科教学与气候变化教育融合实践带来的启示在于：

第一，认真分析学科课程标准与气候变化教育的关联性，把握学科融合全局特征，是学科教学与气候变化教育融合的重要起点。如《义务教育语文课程标准（2022 年版）》不同学段"阅读与鉴赏"板块中均明确了"关心自然与生命，对感兴趣人和事物表达自己观点"的要求。英语课程标准中强调"家国情怀与人类命运共同体的文化意识、批判与创新的思维品质"的育人目标要求。物理课程标准中提出"亲近自然，崇尚科学，乐于思考与实践，具有探索自然的求知欲与好奇心。能关注科学技术对自然环境、人类生活和社会发展的影响，遵守科学伦理，有保护环境、节约资源的意识，能在力所能及范围内为社会可持续发展作出贡献，具有实现中华民族伟大复兴的责任感与使命感"育人要求。

第二，学科教学与气候变化教育融合需要把握学科回归逻辑，在坚持学科立场、素养目标迭代、内容融合优化与学科视角递进分析气候变化议题实施路径的基础上，实现学科素养与气候变化素养育人目标的螺旋回归。如本篇中的语文教学课例在坚持学科素养目标培育立场下，通过水循环的情境想象、对比分析、归纳判断等，很好地培养了学生对自然的觉知和审美能力，促进了人与自然和谐共生自然观的形成，提升了语文教学的品质与境界。

第三，气候变化主题学科大单元教学设计与实施关键在于以气候变化主题与教材单元的迭代重构为主线，帮助学生形成结构化思考习惯、关联性意义养成与价值内在性驱动的设计与实施过程。其中大概念提取是一个难点，通过气候变化主题与学科单元内容的优化，提炼大概念，形成学科视角分析气候变化议题的思维框架，在核心知识间生成意义关联，进而激发学生行动的内在驱动力。

第四，学科＋气候变化综合实践活动充分利用10%的实践空间，创设沉浸式体验与探究空间，提升学科视角解决气候变化议题的综合实践能力。如本篇语文＋气候变化教育、英语＋气候变化教育、数学＋气候变化教育、信息科技＋气候变化教育、道德与法治＋气候变化教育中，鲜活的真实性情境与沉浸式体验，都让学科素养达成更饱满。

联合国教科文组织在2024年6月5日世界环境日到来之际推出"教育绿色化"工具，呼吁增强对年轻人的环境教育，使他们能够更好地参与应对气候变化等问题。面向我国"人与自然和谐共生"的中国现代化建设与美丽中国建设，把气候变化与可持续发展教育融入育人全过程，强化绿色低碳创新人才培养，是广大教师助力强国铸魂的时代使命。

(王巧玲，联合国教科文组织中国可持续发展教育项目组秘书长、北京教育科学研究院终身学习与可持续发展教育研究所副所长)

第三篇

基于项目化学习的气候变化教育

我们在实践中发现,气候变化教育非常适合以项目化学习的方式开展,特别是如下三类项目:生活方式转型类、生产方式改善类、生态环境保护类。项目化学习也将前文中多学科的力量汇聚在了一起。通过学科教师的联合,以及项目负责教师的主动沟通,形成综合研究状态。

本篇所呈现的案例,就分布在上述三类项目中,呈现出中国教育工作者策划、组织本主题的项目化学习的过程,分享学生、家长和教师在此过程中提升气候素养的经验。

"失宠鞋",你失宠了吗?
——让循环使用思想融入小学生日常生活

研究背景

"看,我的新鞋漂亮吗?"

"颜色真好看,还没有鞋带!"另一位孩子露出了羡慕的眼神。

一下课,我就发现孩子们正在看一双新潮出品的运动鞋。那鞋的独特设计是没有了传统的鞋带,没有了粘贴扣,也没有磁吸抽绳扣,而是一个一拉就可以放松鞋带、一按就可以旋转按钮收紧鞋带的潮流按钮,加上红白相间的鞋身与国潮图案,一双喜庆、时尚又方便的国潮鞋引来了大家的目光。旧鞋换成了新鞋,脚上的旧鞋又该怎么处理呢?

随着时尚潮流的迅速更迭,鞋柜中的旧鞋子数量也在不知不觉中增加。这些曾经伴随我们走过春夏秋冬的"小伙伴",一旦失去了原有的光泽和功能,就会成为"失宠鞋",最终成为环境的负担。据统计,全球每年销毁鞋子数量高达 240 亿双,其中 90% 的鞋子要么被填埋,要么被焚烧。[①] 这些废旧的鞋子主要由橡胶、塑料、纺织品等不易降解的材料构成,在自然环境中分解需要数百年时间,且在分解过程中会释放有毒物质,对土壤和水源造成污染。若焚烧,所产生的有害气体,也会增加废气的排放量,对大气造成污染。而这些,都会影响全球气候变化。如果我们能够对这些"失宠鞋"进行合理利用和处理,循环使用,不仅可减少学生对新鞋的需求量,还能降低温室气体排放,减缓全球气候压力,更为可持续发展的循环经济作出贡献。

2023 年 9 月中旬,我们就此对本校五年级学生展开调查,发现,48.51%~49.79% 的学生选择把鞋直接扔掉或者放在橱柜角落里,只有 30.64% 的学生选择循环使用。可见,学生的循环使用理念还是比较薄弱的。那么,小学生该如何形成"失宠鞋"循环使用的理念呢?

项目设计

本项目以循环使用、应对气候变化为着眼点,立足"失宠鞋"循环使用现状,以五年级学生为主体,通过问卷、动手设计与绘制等方法,提升项目的趣味性与实用性。其总目标在于树立学生的循环使用意识,激发学生的竞争与合作意识,促进物品循环使用理念深植日常生活。同时,激发学生潜能,探寻生活中循环使用的多种新途径,让循环使用成为一种生活习惯。

[①] 搜狐网. 每年销毁鞋子 200 亿双!一场制鞋业的战争[EB/OL].(2023-07-27)[2024-06-18]. https://www.sohu.com/a/706780822_121123884.

项目实施

一、趣味实践,让"失宠鞋"变"受宠鞋"

老师们调查发现,很多孩子喜欢用手绘涂鸦、DIY 手工装饰等方法将"失宠鞋"打造成自己喜欢的款式。基于此,学生们集思广益,将"自己设计一双别出心裁的失宠鞋"这一想法变为现实。

(一)设计构思,树立循环使用意识

2023 年 10 月,学生们开启思维碰撞,根据平时各科所学,借助课外画刊,为构图设计寻找方法、拓展思路。同时,讨论并绘制出大家喜欢的富有主题图案的设计构思图(见图1),在元素、主色调、风格等方面更是极尽巧思。小小的自主创新设计,让"失宠鞋"变"受宠鞋"。而这些动手实操、切身参与的体验,也让学生们在心里种下了一颗循环使用、呵护环境的种子。

图 1 "失宠鞋"设计构思

(二)巧手绘制,践行循环使用理念

2023 年 10 月下旬,活动进展如火如荼,孩子们对如何做到既经济环保又便于循环使用的主意也是层出不穷。

"老师,我突然想到,要绘制一双环保的鞋子,我们可以选用水溶性强的颜料。"

"对,还要那种无毒无铅的颜料或者水彩笔,这样就算鞋子扔了,颜料也不会污染环境了。"

"除了颜料上的选择,我们还可以自行搭配,比如有些学生鞋子不能穿了,但是鞋带还是很好的,那我们可以搭配起来,既循环利用了又美观。"

"这个创意不错,还有鞋垫也可以再利用,搭配到不同的鞋子里……"

不断激发的创意思维与循环使用的意识,让学生们创意迭出。经过三个星期左右的设计与实践,一双双集美观、环保理念于一体的"受宠鞋"终于诞生了(见图2)。它们在孩子们

的劳动创造中换新颜、展新生。

图 2 循环使用的手绘鞋

（三）模特秀展示，深化循环使用理念

学生们通过自己的劳动创造，不仅绘制出个性化的专属受宠鞋，还举办了一场"受宠鞋"模特秀，让"失宠鞋"再次受宠（见图 3）。此外，学生们还打造了集环保、创新、审美于一体的个性化"失宠鞋"展柜。随着活动的不断深入，学生们循环使用的理念得以不断增强，对新鞋子的期望也降低了。

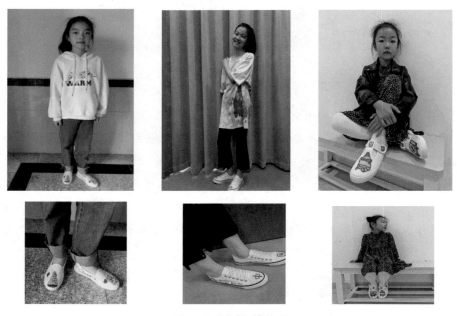

图 3 "受宠鞋"模特秀

二、循环使用，开辟衍生效应新路径

循环既是终点，亦是起点。学习之余，学生不仅探讨如何把冷落的物品进行再创造，延续物品的使用寿命，还将循环使用的意识潜移默化地变成了一种生活习惯。班级里、校园里探讨物品循环使用的氛围逐渐浓郁了起来。

"生活中，除了鞋子能循环使用外，还有什么能循环利用，从而减少温室气体排放，减缓气候变化？"在"失宠鞋"活动末期的一次拓展课上，我问道。

一位同学答道："我经常看到同学们把没用过的草稿纸撕成小纸条或者揉成团，扔来扔去，很是浪费。我觉得，我们可以把爱惜纸张作为一种习惯。纸头不要乱撕，就算写完了，也不要扔，回收起来，与家里的快递盒、报纸等一起作为废纸卖，也是一种循环使用。"

当其他学生听到纸张能循环使用后，紧接着回答："网课期间，家里打印练习时，妈妈经常给我双面打印，这也是减少纸张浪费，循环使用了吧？"举起的小手越来越多，思维也越来越开阔。有的说，家离学校比较近，上学就选择步行或者乘公交车的方式，可以减少汽车尾气的排放；有的说，要改掉不关灯的习惯，节约用电；更有同学说可以把家里不穿的衣服循环利用，变"失宠衣"为"受宠衣"，制作出其他新的物品，这不仅能培养创新力，更能保护生态环境。

随着师生们循环使用思维的不断拓展，我们的实践行动也在不断衍生。譬如，通过"雏鹰假日小队"宣传循环使用的意义，了解保护环境、治理气候变化的重要性及措施；组织学生聆听关于当地和社会环境变化的讲座，懂得保护环境的重要性；举办阅读关于环境、气候类书籍的读书分享会；绘制气候变化宣传图，向社区居民宣传循环使用的重要性等。最可喜的是，越来越多的家庭参与进来，越来越多的家长以身作则，践行循环利用。

成效与反思

一、经由动手实践，树立了循环使用的意识

循环使用是一种理念、一种态度、一种生活方式，也体现和影响着每一个孩子的气候变化应对能力。小学阶段的孩子对新鲜事物充满好奇，对环境问题有着天然的敏感性和同情心。但因缺乏及时引导，很多孩子环保意识显得很薄弱：班级餐巾纸到处可见，厕所洗手液玩出泡沫吹一吹，卫生纸一沓沓地扔，衣物玩具喜新厌旧……通过此次"失宠鞋"大变身活动，学生们了解了环境污染、气候变化等问题的严重性及其对人类生活的影响，认识到每个人的环保行为都与气候变化息息相关，很好地树立了循环使用的意识。

在此过程中，我们认识到，趣味性的活动能够大大吸引学生的注意力，激发他们的兴趣和好奇心，从而提高他们的参与度和积极性。此次"失宠鞋"大变身活动，就是集趣味与实践于一体，营造了积极向上的活动氛围，让学生们感受到愉悦的同时，意识到保护气候变化是一种有乐趣、有意义的行动。当学生在手绘主题图案时，自己就是"小小设计师"的快感油然而生；当变成"小小模特"时，那种被关注和被肯定的感觉让他们更自信；当践行循环使用理

念,在社区平台不断拓展新渠道时,那种挑战性和刺激性,让他们对复杂问题、复杂环境的应变能力和心理承受能力都得以增强。

二、经由竞争与合作,增强了团队的凝聚力与协作能力

在此次活动中,竞争与合作发挥着非常重要的育人功能。在构思设计时,学生间的竞争意识不断激发着他们的创新思维,最终成就一幅幅独特又富有创意的设计稿。这种竞争意识让他们在发现自身优点与不足的同时,更好地调整自己的学习方法和策略。而每一双"受宠鞋"从设计到实现,又离不开团队成员的合理分工与紧密合作。由此,团队凝聚力、协作力得以提升。

三、经由多元主体参与,让循环使用理念得以更大范围传播

学生在实践中体会到了循环使用的意义,而参与项目的学科教师、家长也强化了循环使用的意识。

教师在日常教学中随时渗透环保理念,如通过绘画、音乐、自然等课程,呼吁保护环境,以让学生在感受美的同时,增强保护气候环境的意识和决心。又如在语文、数学、英语、道德与法治等课程中分享关于自然和环保的故事,启发孩子们思考人与自然和谐共处的可行性。

家庭是深化循环使用意义的重要空间。家长的循环使用意识深刻影响着孩子的言行,而家长的支持也是对孩子行动的支持。很多家长在陪伴孩子参与活动时,逐渐意识到要通过日常生活中的点滴,为孩子做好榜样。如以步行或骑行代替驾车,以循环使用的帆布袋代替一次性塑料袋购物,将一张餐巾纸对折后多次使用等。这些小小的举动都潜移默化地影响着孩子。

本项目的开展,既是学生、家长、教师对循环使用的一次深入了解,也是学生在生活中养成循环使用习惯的一次生动实践,但这一项目也存在一些亟待提升之处。如许多学生都热衷于设计图案后自己作画,所以"失宠鞋"大多选择的是白色,相较多彩的鞋子,虽然发挥了学生作画的想象力,但是不利于多元素实物搭配的循环使用。在项目实践中,可以更针对性地让一部分学生进行剪剪贴贴缝缝等不同元素的搭配大改造。其次,项目完成后,还可以继续跟进学生生活中循环使用理念的持续性与实践的习惯性。

循环使用是一种生活方式,是一种社会责任,更是一种可持续发展的理念。树立循环使用的意识,养成循环使用的习惯,有助于学生认识、理解气候变化,提升应对气候变化的能力与素养。

(盛晓燕,上海市金山区前京小学教师)

关于牛奶盒回收再利用的研究

研究背景

在学校,一般学生们会自带牛奶。据我观察,每天、每个班都能收到整整一垃圾袋的牛奶盒。若一个班每天平均产生 42 个牛奶盒,学校现有 36 个班,就会产生 1 512 个牛奶盒。而这些牛奶盒最终都会被丢进垃圾桶。

我将观察到的这一现象,在班级里与学生分享,学生们也表示很关注。通过查阅资料,我们进一步了解到,当前人们的生活与消费水平提高了,喝牛奶的人也越来越多。仅 2022 年,我国奶类产量就达到了 4 026.5 万吨,需要消耗百亿个包装盒,而这些包装物大多是与生活垃圾一起被填埋或者焚烧[①]。但牛奶盒含有大量的不可降解材料,如果被当成垃圾烧掉或埋掉,不仅会造成资源和能源的浪费,还将污染环境,危害人类健康。

大家都觉得牛奶盒虽小,但废弃的牛奶盒垃圾却不容忽视,应该让更多的人了解乱扔牛奶盒的危害性,号召大家一起加入环保的行列。2023 年 10 月 10 日,同学们商议一起开展一个"关于牛奶盒回收再利用"的研究,让牛奶盒垃圾能够发挥它真正的价值。

项目设计

本项目让学生:通过了解牛奶包装盒的种类和制作材料,明白牛奶盒虽然小,却也会造成不小的环境问题,要学会观察思考;通过问卷、采访、实地考察等手段,增强与人交往的能力,学会合作式学习;通过对牛奶包装盒不当回收处理方式的认识,提出合理优化建议,树立环境保护与节约意识;通过对研究成果的宣传,促使更多的人关注牛奶盒的回收再利用,呼吁更多的人成为环保行动的一分子。

项目实施

一、初步认识牛奶盒的种类和制作材料

2023 年 10 月 12—16 日,学生们首先通过某宝旗舰店,对"中国奶业三巨头"——光明、蒙牛和伊利的不同包装牛奶销量进行了调研,发现不论是哪一种品牌,盒装牛奶的销量都远远超过了罐装的,而且这三家旗舰店里都没有售卖袋装牛奶。看来,盒装牛奶的需求量真的很大。

那么,一只小小的牛奶盒,为什么回收率这么低?它里面又有哪些不可降解的材料呢?

---

① 中国奶业协会,农业农村部奶及奶制品质量监督检验测试中心(北京).中国奶业质量报告(2023)[R].北京,2023.

通过上网查询，学生了解到，现在市面上使用较多的牛奶包装盒主要有"利乐包"和"屋顶盒"两大类。其中，"利乐包"又分为"利乐砖"和"利乐钻"。于是，大家对它们的结构、保质期等展开了对比分析。

学生发现这几种包装的材质都使用了塑料和铝。在包装行业，像牛奶盒这类包装有一个学名，叫"复合纸包装"，是一种由纸浆、聚乙烯塑料、铝以及印刷油墨和涂料合成的6层复合结构。什么是聚乙烯塑料呢？通过请教科学老师，大家了解到，其主要成分是聚乙烯，需要从石油中获取，属于石油化工产品。而铝属于重金属物质，如果被掩埋，在土壤中不能分解，对土地、牲畜、水源都将产生极大的污染破坏。

随着研究的深入推进，大家又发现，当前我国已经有不少技术专门针对牛奶盒这类复合纸包装进行回收再利用。其中，比较成熟的回收再利用技术主要有水力再生浆技术、塑木技术、彩乐板技术和铝塑分离技术。简单地说，就是经过专业的再生利用加工处理方式，把废牛奶盒等还原为纸浆、塑料粒子和铝粉，使之成为其他行业的原材料，从而达到100%再生利用。但是，像这样的复合结构，很难作为单一材料被处理。对我们而言，缺乏专业技术的支持，根本无法达到再利用的要求。

二、学生调查了解周边人对牛奶盒的日常处理方式

2023年10月20日，为了全面地了解人们日常对牛奶盒的处理方式，同学们展开讨论，和老师一起商量制定了调查问卷，并通过班级微信群等平台发放了问卷。

通过本次调查，同学们了解到：

（1）从每户家庭每天要喝的牛奶数量来看，相应产生的牛奶盒数量比较庞大，牛奶盒垃圾不容忽视，值得我们关注和研究。

（2）虽然大多数人知道随意丢弃牛奶盒会对环境产生影响，但认识不深刻，需要加大宣传力度，让大家真正地意识到废弃牛奶盒的危害性，从而实现牛奶盒再利用。

（3）当前对于牛奶盒再利用的相关宣传途径五花八门，没有形成系统的路径。针对不同人群该采取怎样的宣传手段，我们可以借助哪些平台进行宣传，都是我们在研究中需要积极探索的。

为了更好地了解周边的人对牛奶盒再利用的看法，大家还在老师的帮助下一起设计了采访单。于2023年10月25—27日，走进各社区，对居民进行了更详细的采访（见图1），全面了解牛奶盒再利用的情况，获得如下信息：

人们对于牛奶的需求量比较大，而且一般会选择盒装牛奶，平均每个家庭产生的牛奶盒在2~3个。此外，喝完牛奶以后，大部分人会直接扔掉，对于牛奶盒的回收再利用考虑得比较少，或者几乎没有考虑过。

图 1　学生在小区进行采访

三、集思广益，制定不同的牛奶盒回收再利用小妙招

针对调查反映的问题，2023 年 11 月 6—15 日，同学们对当前国内外牛奶盒的处理方式进行了对比研究，了解到：

(1)国外。借助网络资源，大家发现如下几个国家对牛奶盒的处理方式值得学习。比如在西班牙巴塞罗那，斯道拉恩索集团利用最先进的热解技术依次分解像牛奶盒这样的旧饮料盒中的纤维、塑料和铝。① 在日本，一般采用方形纸杯包装牛奶，这种纸杯所用的纸张属于特殊的优质纸，所以有较高的回收率。同时，牛奶盒的外包装上会有明确的标识，教你怎么分解包装盒。②

(2)国内。从 2002 年开始，在我国浙江、广东、山东、北京、福建等地，陆续出现了一些应用废牛奶盒等纸塑铝复合包装物的造纸企业。③ 中国台湾目前采取"四合一"制度，即结合住宅小区居民、回收商、地方管理部门及回收基金等四方面力量，实施资源回收、垃圾减量工作。④ 2010 年上海世博会上，人们把回收的利乐包装制作成了 1 000 多张造型时尚、简洁实用的广场座椅，放置在活动场馆，供游客休息。2019 年，随着垃圾分类在上海的正式推行，光明乳业开始在全国推广牛奶纸盒回收公益行动。2021 年，第十届花博会上，这些收来的牛奶盒被制成了观花长椅。⑤

那么这些回收处理方案，有没有值得学习和借鉴的呢？大家进行了优缺点对比梳理，达

---

① 伍安国.斯道拉恩索利用热解技术处理饮料、牛奶包装盒[J].纸和造纸,2015(8):78-78.
② 肖慧琳.浅析日本牛奶盒包装设计的现状[J].艺术品鉴,2016(10):82-82.
③ 陈维西.废牛奶盒的"重生"之路[J].百科知识,2020(5):32-35.
④ 吴亚明.给电子废弃物找到归宿:台湾的环保之路之二[J].台声,2013(12):58-59.
⑤ 严远,轩召强.502.7 万只废弃牛奶盒能做什么？花博园里两排观花长椅火了[N].新民晚报,2021-04-29.

成了如下三点共识：

其一，增加牛奶盒集中回收点。目前我们对牛奶盒的处理大多是随手丢弃或者以家庭为单位将之与其他垃圾混在一起扔掉。如果能像日本那样，有集中的回收站点，处理起来就会方便快捷许多。

在研究中大家发现，光明乳业已经开始开展"牛奶纸盒绿色回收大行动"，并在全国设立了3 000多个站点。但这些站点大多设立在上海、南京、济南这样的一、二线城市，在三、四线城市比较少见，且站点间距离比较远，城乡间也存在着很大差距。以常州为例，最近的站点在宜兴。因此，同学们认为，可以在相近的几个小区设置一个站点，也可以在牛奶需求量较大的学校附近设置牛奶盒回收点，便于废弃牛奶盒的集中处理。

其二，优化牛奶包装盒的设计。大家调研了几款常喝的牛奶包装盒，发现某产品的包装上印有"请勿乱扔空包 保持环境清洁"字样，另一款生牛乳产品包装盒上注明了"可回收物""饮用后请清洁压扁后 投入可回收物容器"。两相对比，后一款产品就更为重视包装盒的回收再利用。与其追求华而不实，不如在牛奶包装盒的外观设计上融入一些知识点，比如回收牛奶盒的好处、牛奶盒的回收方法等，提高人们的回收意识。

其三，加强对牛奶盒包装回收利用的宣传。当前，人们对于这一话题普遍了解不够。唯有让大家充分认识到随意乱扔牛奶盒的危害，懂得回收牛奶盒的重要性，大家才能有意识地采取回收手段。这就需要我们加大宣传力度，扩大宣传范围了。同学们觉得可以从班级到年级，再到学校，最后辐射到社区。

四、变废为宝，让牛奶盒焕发新生机

2023年11月24日，经讨论，同学们一致认为"变废为宝"是当前关于牛奶盒回收利用我们所能采取的最佳方式。基于此，开展了"牛奶盒变变变"系列活动。

一是利用班队会，详细了解掌握清理废弃牛奶盒的步骤。简要归纳为以下四步：剪开喝完的牛奶盒—清洗—放置晾干—统一丢进回收站。

在实施过程中，还遇到了这样一个问题：由于有些同学喝不干净，在剪牛奶盒的时候，牛奶残汁会流到地上。大家便通过网上学习和实践，发现了这样一个小妙招：喝完利乐砖包装牛奶后，把包装盒的四个角展开，由下往上卷推盒子，这样就能排出剩余的残汁了。残汁掺水稀释后还能浇花呢！

二是动手制作牛奶盒回收箱，并安置于教室的卫生角。

三是制订废弃牛奶盒认领办法。通过讨论，大家最终确定，根据同学们平时的小岗位表现，老师、组长给予小红花奖励，获得5朵小红花，就能到站长高同学处兑换一个牛奶盒。

四是发动"牛奶盒变变变"环创作品征集令。同学们可以根据废弃牛奶盒认领办法，领取相应的牛奶盒，进行再设计（见图2）。

图 2　学生把废旧的牛奶盒"变废为宝"

五、宣传动员,小手拉大手,共建环保世界

2023 年 11 月 27 日至 2023 年 12 月 22 日,"牛奶盒变变变"活动得到了大家的积极响应,这极大地增强了同学们要让更多人加入牛奶盒守护队伍的信心。

2024 年 1 月 12 日,在老师的帮助下,同学们联系到学校"红领巾电视台"负责人,利用"红领巾电视台"向全校学生科普了乱扔牛奶纸盒的危害,分享了正确回收牛奶盒的方法,并呼吁更多学生参与,并带动家人一起加入"回收牛奶盒,一起做环保"行动。

为了让更多的人加入回收牛奶盒、保护环境的队伍,同学们还向所在小区物业发出了增设牛奶盒回收点的倡议,建议可以通过兑换机制,比如集满 50 个牛奶盒兑换 1 盒鲜奶等,鼓励居民积极参与。

成效与反思

本次废旧牛奶盒项目,在校园、家庭、社区掀起了一股热潮。

一是学生的综合素养在提升。在本次研究中,每一个同学都全身心投入,为活动的顺利推进出谋划策。同学们的交流表达能力和合作能力都有了一定的进步。同时,通过学习,同学们对废弃牛奶盒的危害性有了更深的了解,掌握了正确清理牛奶盒的方法,更懂得气候变化是关系到人类生存和发展的根本问题。我们要有一双善于发现的眼睛,细心观察、用心思考,用自己的力量助力环境新发展。

二是废旧牛奶盒的回收利用率在提高。在同学们的号召下,学校很多班级都开展了"牛奶盒大变身"活动,让原本堆积如山的牛奶盒,改头换面。这不但增加了回收利用率,还有效减轻了保洁阿姨的工作量。废牛奶盒的"重生"之路,成为校园里一道靓丽的风景。

三是家校社携手,共绘垃圾分类同心圆。我们通过"小手拉大手"的方式,号召更多人加入"废旧牛奶盒回收再利用"的行列,让更多人了解到牛奶盒回收再利用的重要价值,增强了全民环保意识。

当然，本项目还有许多亟待完善深化之处。譬如，在研究后期，同学们走进社区，让身边人普及牛奶盒的相关知识，号召大家掌握正确回收的方法。但这项工作，还只是停留在宣传阶段，未能进一步创设相关活动，让"牛奶盒"重生之旅落地生根。又如，此项目主要集中在调查研究、班级实践阶段，对于周边资源的利用率并不高。后续，我们将组织学生走进专业的牛奶盒回收处理厂进行实地参观，学习了解专业的处理废旧牛奶盒的技术，通过校企合作方式，开展更多的活动，为气候变化教育添砖加瓦。

(苏琳，常州经开区小学学生部主任)

基于 STEM 综合活动开展气候变化教育实践

研究背景

党的十八大以来,我国将生态文明建设纳入"五位一体"总体布局,放在全局工作的突出位置。中国低碳试点城市——广东深圳也在水土保持、污水治理和净水工作等方面取得很大成就,对国内水污染和减缓气候变化的协同治理具有重要借鉴意义,有利于迈向未来的"碳中和"社会。

为积极推进学校家庭社会协同开展气候变化教育、共商应对气候变化挑战的教育策略、共谋人与自然和谐共生的教育之道,广东省深圳市宝安中学(集团)实验学校(以下简称"宝中实验学校")成为"全国气候变化教育研究联盟"发起单位之一。在当前教育改革背景下,如何做好科学教育加法,成为一线科技教师聚焦的工作热点。而如何基于 STEM 综合活动开展气候变化教育实践,成了宝中实验学校气候教育变化跨学科团队的重要任务。

为了树立青少年应对气候变化的意识,培养青少年正确对待气候变化问题的科学知识、科学方法、科学态度、科学精神和科学品质,鼓励更多青少年参与气候行动,宝中实验学校基于校本科创特色[①]和"1+3+5 框架"的气候变化教育方案,即:①一个目标——建设有科创特色的气候变化教育实践学校;②三个品牌——气候变化校本读本开发、每年 10 月气候变化学习月、每年 4 月气候变化教育教学成果展示;③五个维度:社会——引入气候变化的社会教育资源;学校——建设气候变化教育科普基地;家长——建立气候变化的家校教育联盟体;教师——开发气候变化教育的校本课程及读本;学生——人人争做气候变化科普宣讲达人。面向小学生,宝中实验学校积极探索基于生物园农耕田和温室大棚的 STEM 种植和节能减排项目研究;面向初中生,开展新能源、节能减排的科技发明与项目实践活动,引导学生关注水土保持、水资源保护主题,进行气象气候和双碳主题项目实践,加强对气候变化和生态文明的认知,参与低碳环保行动。

宝中实验学校的科技教育特色课程分为三大系列:科学普及课程、科学行动课程和科学超能课程。气候变化教育跨学科团队在此三个维度都设置了相关项目活动,并利用现有创客空间不断改造、优化。其中,气候变化教育跨学科团队通过设置不同项目,聚焦三月植树节,四月地球日,五月劳动实践月,六月世界环境日,九月、十月节能减排等科学调查活动。信息技术课则引入了包含节能环保、编程、3D 建模、人工智能等内容的"未来生态创客教育"。在中小学社团、素养课等学科拓展课程中,还设置了气候变化教育相关的活动。如在

---

① 林建芬,王瑜,曾玉芳,等.青少年科技创新课程体系的构建与实践:以深圳市宝安中学(集团)实验学校为例[J].教育与装备研究,2020,36(5):35-38.

小学的"小小发明家""小小实验家"等社团活动中增加了气象气候科普解说项目、巧改废弃物等实践活动;在初中的"科学盒子"社团课中引入了水科技、生命科学、工程与技术科学、资源与环境科学等学科领域的课程。在竞培课程活动,如人工智能探究、AR/VR科幻体验、气候STEM项目、双碳项目、生物多样性项目、白名单赛事培训中,同样增加了多种多样的气候变化教育项目及活动。

项目设计

STEM教育提倡基于生活实验任务的活动,要求学生在这些活动中制作一系列实体模型,在制作过程中探索、应用、修正科学、工程、环境等领域的知识。STEM实验活动中使用的材料简单,但能让学生在活动中体验科学家、工程技术师的思考及工作方式。基于STEM教育理念,学校开展"水质差异探究及净水器研发"项目(见表1),提倡学生自己动手设计并完成实验,充分发挥创新精神,把所学的知识、技能与生活实际紧密联系起来,让科技活动在校园、家庭和社会开花结果。

表1 "水质差异探究及净水器研发"项目设计

3—9年级	总课时	6×60分钟
项目设计框架:		
课时	重点、难点	解决方法
第1课时	水的概念、水污染的原因危害以及防治措施、水质检测的指标、软硬水的概念及鉴别、硬水如何软化	1. 教师生动讲解 2. 教师及时答疑解惑
第2课时	1. 利用钙镁检测试剂进行软硬水鉴别 2. 利用软水树脂、活性炭、分子筛三种材料,对自来水进行软化,探究和比较三种材料对自来水的软化效果	1. 教师示范讲解 2. 教师及时答疑解惑 3. 教师PPT示范
第3课时	利用水质检测盒对市面上常见不同品牌饮用水进行检测,主要包括余氯含量、固体溶解物含量的测定以及检测是否为矿物质水	1. 教师示范讲解 2. 教师及时答疑解惑
第4课时	了解水土保持、水质净化原理,对市面净水器进行产品调研,对自然界中水的成分、除杂所需的必备材料和原理有所了解,掌握过滤的操作	1. 学习净水原理 2. 比较常见的净水器
第5课时	在工程设计过程指导下,STEM实验小组开展自制简易净水器的项目实践和产品优化	1. 工程项目实践 2. 产品测试和优化
第6课时	1. 答辩PPT的学习和制作 2. 答辩PPT的汇报与修改	1. 学生汇报 2. 教师指导
项目阶段目标:		

(续表)

(一)第1课时
　　了解水的概念,水污染的原因、危害以及防治措施,水质检测的指标,软硬水的概念及鉴别,硬水如何软化等知识。
(二)第2课时
　　(1) 通过探究不同水样是否为软水实验利用钙镁检测试剂进行软硬水鉴别探究。
　　(2) 通过探究自来水软化方法及效果实验,利用软水树脂、活性炭、分子筛三种材料,对自来水进行软化,探究和比较三种材料对自来水的软化效果。
　　(3) 学生掌握汇报PPT的基本要点,动手做部分汇报PPT。
(三)第3课时
　　(1) 通过"探究不同品牌饮用水的水质差异性"实验,学生学会利用水质检测盒对市面上常见不同品牌饮用水进行检测,主要包括余氯含量、固体溶解物含量的测定以及检测是否为矿物质水。
　　(2) 读取实验数据,记录实验现象,分析总结实验结果。
　　(3) 学生掌握汇报PPT的基本要点,动手做部分汇报PPT。
(四)第4课时
　　班级学生以3~4人形式组建STEM实验小组,用一两张纸的文章篇幅概括介绍小组在实验主题相关概念或原理上的学习成果及查阅、引用的文献资料,其中需附2张所研究内容的图片。教师要为学生提供资料查阅的路径和方法,组织学生召开资料查阅总结报告会,使学生学会收集信息,学会提出问题,学会撰写资料查阅报告。
(五)第5课时
　　通过查阅文献知道水土保持、人为净水的重要性,了解净水原理,对市面净水器进行产品调研。知道生活中常用的净水器主要有前置过滤器类产品、PP棉过滤类产、超滤膜过滤类产品、反渗透膜、纳滤膜过滤类产品,并对他们的作用、优点、缺点进行比较。开展自制简易净水器的STEM研发项目活动。
(六)第6课时
　　(1) 学生提前制作答辩PPT、进行项目汇报和净水器产品展示。
　　(2) 通过教师指导,学生能够改进自己的不足之处,使自己的汇报PPT更加完善和精美。

项目实施

一、学生了解掌握关于水的基本概念

在第一课时,教师讲解第一个科学知识点——水的概述,增进学生对水的认识。首先,教师引出水污染的问题,引导学生认识水污染的危害和防治工作对应对气候变化的重要性,了解"以防为主、防治结合"方针及综合治理理念。

接下来,师生共同研讨水质检测的物理指标、化学指标和微生物指标。一方面,学生基于生活经验进行相关水质优劣的判断。具体为:一看:水杯接水,对着光线看有无悬浮在水

中的细微物质；静置三小时后观察杯底是否有沉淀物。二闻：水龙头接水，是否有漂白粉（氯气）的味道；如能闻到漂白粉（氯气）的味道，说明自来水中余氯超标。三尝：热喝白开水，有无漂白粉（氯气）的味道；如能闻到漂白粉（氯气）的味道，说明自来水中余氯超标。四观：自来水泡茶，隔夜后观察茶水是否变黑；如茶水变黑，说明自来水中含铁、锰严重超标。五品：品尝白开水，口感有无涩涩的感觉。如有，说明水的硬度过高。六查：检查家里的热水器、开水壶等，内壁有无结一层黄垢。如有，也说明水的硬度过高。另一方面，在教师指导下学生使用专门的设备和检测试剂，检测不同类型和品牌的水质，了解检测水质的各项指标，如余氯、水样酸碱性、水样中是否含锌元素、固体溶解物。

最后，学生利用肥皂水法、加热法、蒸发法等进行"软水"和"硬水"的鉴别，并通过实验室煮沸和蒸馏的方法将"硬水"转化为"软水"。

二、开展实验探究，了解水质差异

学生利用一个课时的时间，在教师指导下完成几个实验探究。一是探究不同水样是否为软水；二是探究自来水软化方法及效果。

基于此，教师设置生活情境，引发学生思考，并提出需要研究的问题——"探究不同品牌饮用水的水质差异性"。师生共同分析该实验设计的物理量——自变量：不同的饮用水；因变量：余氯、导电性（矿物质）、固体溶解物。学生项目小组设计实验方案，探究不同品牌饮用水的水质差异性。

三、设计"自制净水器"实验方案

该实施环节主要由以下三个方面内容构成。

其一，教师阐述实验目的：水资源短缺成为当前世界最主要的问题之一。我们此前已学过过滤、吸附、水的净化、保护水资源等概念或原理，对自然界中水的成分、除杂所需的必备材料和原理有所了解，对过滤的操作也已掌握。那么，本实验我们的主要目的就是自己动脑、动手制作一个简易的净水装置。

其二，学生掌握文献研究法：以3~4人为一组，组建STEM实验小组。用一篇短文简要介绍本小组对于相关概念或原理的学习成果，查阅、引用的文献资料，并附两张所研究内容的图片。在此过程中，教师为学生提供资料查阅的路径和方法，组织学生召开资料查阅总结报告会，引导学生学会收集信息，学会提出问题，学会撰写资料查阅报告。

其三，学生设计实验探究方案：以小组为单位，合作研究净水器的原理、构成及功能，并提出净水器制作方案，展开头脑风暴，探讨可行性方案。基于此，着手制作原型，并通过实验探究对原型加以测评，不断改进。最后交流展示。

四、实施"自制净水器"实验及展示实验成果

按照工程设计过程，STEM小组开展自制简易净水器的研发项目活动。流程为：①研究过滤、水的净化等相关知识；②针对如何制作净水器开展头脑风暴；③绘制一份净水器的设计图；④着手制作环保型净水器；⑤查阅相关科学书籍或其他资料，检测模型的准确性和有

效性，记录检测对比的结果；⑥对模型进行评价，分析净水器各部分的成分与功能；⑦确定装置设计的改进方案；⑧对改进后的设计重新进行检验和评价；⑨展示小组学习成果。

最后，每个小组需要在教师规定的时间内，提交下述成果，并以小组为单位进行答辩汇报。汇报内容包括：①一篇1~2页的论文，总结对净水器原理的研究结果，标注引用文献，论文中要包含两张图片；②一张环保型净水器制作的设计图；③一份净水器性能检测后的记录、分析和说明；④一份对实验任务和实践过程的总结，应包含任务的目标、检验过程的简要说明以及对检验结果的解释；⑤一份"自我评价量表"和"反思记录"，要求STEM项目小组将自制净水器带回学校，在实验室准备好净水器，下端与接水的烧杯相连；把污水沿玻璃棒倒入净水器中进行观察，并汇报净水效果、改进策略以及经验教训（见图1）。

图1　环保小卫士自制净水器

成效与反思

"水质差异探究及净水器研发"项目是一个立足于工程技术、自然科学、社会科学领域，由教师引导学生就如何改善水环境质量、加强水资源节约和水生态保护、提高废水处理能力等内容开展的发明创新研究，能够有效落实国务院发布的《水污染防治行动计划》中"加强宣传教育，把水资源、水环境保护和水情知识纳入国民教育体系"的相关要求，对提高青少年生态环境保护意识及科技创新能力，引导青少年积极参与水生态环境保护和节约用水行动有积极的推动作用。下一阶段，我们将通过中国青少年科学调查体验活动，深入地引导学生开展水资源节约保护、水生态修复、水环境现状和公众水环境意识等主题的科技实践调查活动。

过去六年，宝中实验学校在推进青少年应对气候变化行动的过程中，鼓励学生采取撰写相关文章或个人演讲等方式，在学校、家庭、社区进行低碳行为宣讲，加深青少年对气候变化的认识和理解；鼓励学生通过环保主题的科学发明、编程、算法模型等，提出针对气候变化、低碳的解决方案、测算模型或实物发明。青少年作为未来世界的主人，也是应对气候变化的重要推动力量。青少年的气候变化教育是可持续发展教育的重中之重,其不仅能够帮助青少年了解和应对气候危机的影响,还能让他们具备作为变革推动者所需的知识、技能、价值观和态度。

[陈才英、龙超霞、曾玉芳、姜琳、蔡晓峰,广东省深圳市宝安中学(集团)实验学校教师;林建芬,广东省深圳市南山区蛇口学校教师]

多学科融合的"咖啡渣的再生"项目设计与实施

研究背景

上海是我国咖啡消费最为领先的城市。美团数据显示,截至 2023 年 3 月,上海拥有 8 530 家咖啡馆,每万人拥有咖啡馆数达到 3.45 家,位列全国第一。据测算,目前国内咖啡年人均消费杯量 10~12 杯,上海年人均咖啡消费杯量超 20 杯,是国内平均水平的近 1 倍。据此测算,上海平均每日产生咖啡渣高达 15 吨。如何利用这些咖啡渣,如何减轻垃圾处理压力,既是一个需直面的现实问题,也是一个值得探讨的研究课题。①

自 2018 年起,上海市第一中学启动中学生"全球胜任力"培育探索,学校各学科教师共同关注联合国 17 个可持续总体目标下的"全球性议题",如可再生资源、负责任的消费、气候行动等,尝试探索主题式多学科融合发展项目,在自主拓展课开设了"9+1"研修工作坊。其中,环保工作坊由化学、生物、地理三门学科的教师组建,旨在关注学生环保意识的建立、现实生活经验的增长、课题研究技能的提升以及问题解决能力的增强。

项目设计

一、理论依据及整体思路

"咖啡渣的再生"项目设计考虑科学(science)、技术(technology)、社会(society)、环境(environment)四方面的教育,整体遵循"发现问题→提出问题→分析问题→解决问题→升华结论"的思路进行(见图 1)。

图 1 "咖啡渣的再生"项目设计思路

---

① 易永坚.消费行业新消费研究之咖啡系列报告:中国现磨咖啡市场有多大&瑞幸的天花板在哪?[EB/OL].(2023-08-08)[2024-06-18].https://www.doc88.com/p-35329440676097.html?id=2&s=rel.

二、主题、课时及内容安排

"咖啡渣的再生"项目包含多个子课程。以"咖啡的故事"子课程为例,其主题框架、课时计划和内容安排如表 1 所示:

表 1 "咖啡的故事"子课程安排

主题	课时	内容
1. 常见市售咖啡的分类及数据测评	3	分组采购常见市售咖啡,进行分类和多种数据测评,形成研究性报告并展示汇报
2. 咖啡包装材料的分类与回收处理	3	对咖啡的包装材料进行分类,不同的材料属于哪种垃圾,研究不同材料回收处理的技术并实际测试
3. 咖啡制作工艺的演变和实际制作	2	邀请星巴克的咖啡师来校,指导不同咖啡的制作方法,学生实际体验多种花式咖啡的制作
4. 咖啡渣培养土的测定与种植测试	4	咖啡渣植物营养土的实验研究,包含对咖啡渣的测定、细菌培养、配比种植实验等
5. 咖啡渣气味改良剂的制作和测试	2	利用咖啡渣香气,设计制作香囊,改善冰箱、鞋柜和教室等处的异味
6. 咖啡的起源和咖啡的历史小故事	2	咖啡的历史和各种有趣小故事

需要特别说明的是,上述主题的先后次序并非固定不变,而是每个学期依据不同学生的自主意愿等实际情况进行调整。课时长短也是一个概数,因为有些章节的教学工作是多学科教师共同参与,学生分组同时进行的,无法精确统计。

项目实施

下面,以主题"4. 咖啡渣培养土的测定与种植测试"为例来说明本项目是如何实现多学科融合的。

在主题 4 的实施中,由化学、生物、地理三个学科的教师在不同的实验室进行指导。班级学生被分成三组,每组学生在三个实验项目之间循环流动。如图 2 所示,第一组学生在化学老师的指导下进行化学实验 A-DIS 测定,第二组学生在生物老师指导下进行生物实验 B-微生物培养,第三组学生在地理老师的指导下进行地理实验 C-咖啡渣营养土的肥力对比。每组学生完成试验和数据之后,换到另一个实验室。

这样的安排既保障了教学需求,又提高了教学实效。从实验操作的维度看,细菌培养和菜苗种植的实验往往需要持续数天甚至数周,学生需要经常到实验室进行测量和记录。从学生参与的维度看,每节课上,每组学生都可以完成多个学科的实验操作与记录,提高了效率。从教师施教的维度看,每节课可专注于不同小组的同一个项目,既可以提升指导效果,

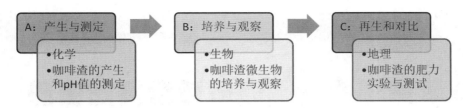

图 2　化学、生物、地理三个学科的任务流程

又可以减轻负担。

以下为本主题中的三份学生活动记录单示例。

(1) 学生活动 A：垃圾分类游戏，咖啡和咖啡渣的 pH 值测定（见图 3），由化学老师指导。

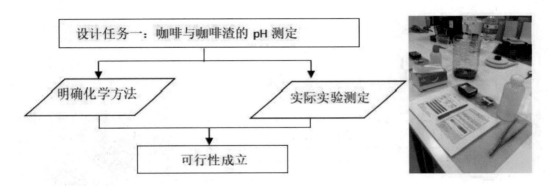

图 3　测定咖啡渣的 pH 值

学生活动记录单-A				
【活动 A-1】咖啡和咖啡有关的物品是什么垃圾？				
易拉罐	咖啡袋	玻璃瓶	塑料瓶	塑料搅拌棒	木质搅拌棒
咖啡	奶精、糖	咖啡豆	手提袋	咖啡纸杯	过期咖啡粉
【活动 A-2】溶液酸碱性的测定方法：＿＿＿＿＿＿＿＿＿＿＿				
【活动 A-3】咖啡渣是酸性还是碱性？				
可乐	苏打水	牛奶	糖	现磨咖啡	咖啡渣
参考方案 1-pH 试纸法：				

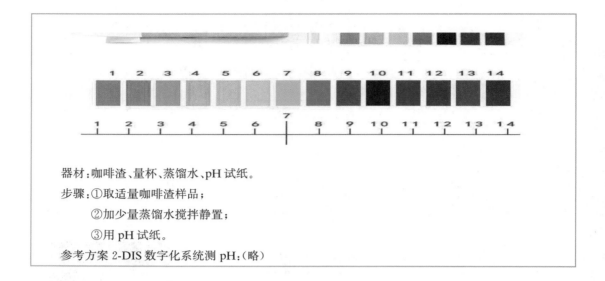

器材:咖啡渣、量杯、蒸馏水、pH 试纸。

步骤:①取适量咖啡渣样品;

②加少量蒸馏水搅拌静置;

③用 pH 试纸。

参考方案 2-DIS 数字化系统测 pH:(略)

(2)学生活动 B:咖啡渣微生物的培养和观察(见图 4),由生物老师指导。

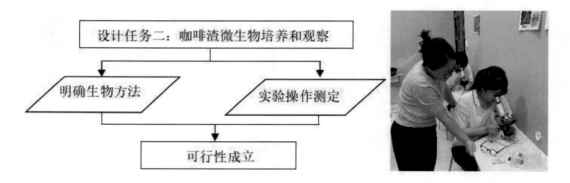

图 4 观察咖啡渣上的菌丝

学生活动记录单-B

【活动 B-1】琼脂平板制作(前期准备)

两茶匙半白糖、两汤匙半琼脂粉、两杯低盐牛肉汤。两杯水。放入锅中加热沸腾。倒入培养皿中冷却至 60℃。放置冰箱保存备用。

【活动 B-2】接种

将咖啡渣少量用棉签接种在琼脂平板上,放置在温暖潮湿(20~30℃)的环境中(培养箱中)培养。

【活动 B-3】制作临时装片

取一片载玻片,在中央滴一点清水,用解剖针或镊子挑少许咖啡渣产生的菌丝放在水滴中,盖上盖玻片,制成临时装片。即可在显微镜下观察。

【活动 B-4】数码显微镜观察
①放置手机在显微镜目镜上,调光,直至出现白色明亮的圆形视野。
②放置临时装片在载物台上,调整距离,直至在视野中出现菌丝物象。
③仔细观察在青霉菌丝的顶端,可以看到扫帚状结构,在分支上还生有成串的孢子,成熟的孢子呈青绿色。

【活动 B-5】拍照、绘图
用铅笔在活动纸上绘出你用显微镜看到的霉菌的菌丝和孢子。

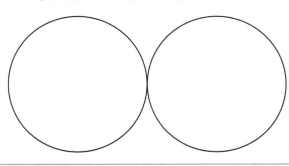

(3)学生活动 C:咖啡渣营养土配制和菜苗种植实验(见图5),由地理老师指导。

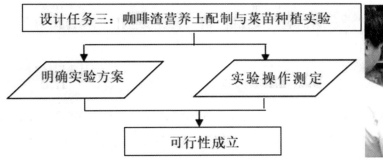

图 5　指导学生移植菜苗

学生活动记录单-C					
【活动 C】咖啡渣肥力实验观察记录					
	盆1	盆2	盆3	盆4	盆5
	100%咖啡渣 0 营养土	75%咖啡渣 25%营养土	50%咖啡渣 50%营养土	25%咖啡渣 75%营养土	0 咖啡渣 100%营养土
种植数					
出苗数					
出苗率					

(续表)

盆1 100%咖啡渣 0营养土	盆2 75%咖啡渣 25%营养土	盆3 50%咖啡渣 50%营养土	盆4 25%咖啡渣 75%营养土	盆5 0咖啡渣 100%营养土
长势__				
长势__				
长势__				
长势__				
长势__				
长势__				
长势__				
……				

器材:咖啡渣、育苗盆、营养土、青菜种子、喷壶、保鲜膜。

步骤:活动前提前将青菜种子在水中浸泡6~8小时。

①在不同颜色育苗盆中,分别装入不同配比的咖啡渣与营养土。

②浇透水后,在育苗盆中放入5~6粒青菜种子,再覆盖一薄层细土。

③使用喷壶保持湿润,并罩上保鲜膜。结合持续养护2~4周。

④观察不同育苗盆种子发芽及生长状况,及时拍照,完成试验报告。

成效与反思

本项目从学生现实生活的真实问题出发,注重学生的主动参与,不仅具有实际意义,且很有趣味性,学生的参与兴趣比较浓厚。而这样的多学科融合课程活动的实践,也有利于教师跳出单一学科的局限性,增强与其他学科教师间的合作。

但在课题实施过程中,仍然存在一些局限性。比如这一项目每周只安排了一个课时,但植物及微生物的状态每天都在发生变化,这就需要老师和学生寻找合适的时间去测量和记录。而人工记录的频次与数据精准度往往很难达到高标准。希望以后能用到更多的技术设备,弥补人工记录的不足。

咖啡渣真的是垃圾吗?其实不然。咖啡渣还可变身为新型生物燃料、吸臭除湿剂,甚至可以被制成好看的杯子……而这些创意的实现,需要教育者的不懈努力,培养出更多关注全球议题、秉承绿色发展理念、具备气候变化素养的未来建设者。

(顾伟伟,上海市第一中学教师)

探究黄皮与气候，助力乡村振兴

研究背景

紫南小学处于广东省佛山市禅城区的一座村庄内，拥有着丰富的乡土资源，尤其是紫洞黄皮闻名遐迩。紫洞黄皮以果大无毛、汁多核小、皮脆味甘、口感清甜而闻名。自古以来，紫洞村家家户户都会在房前屋后种植黄皮。这是开展气候教育的一个绝佳点。2024年2月，新学期开始，我们结合乡土特色，盘活资源，在班级中开展综合实践活动，探究黄皮与气候之间的关系，助力乡村振兴，并将活动与"你好·周末"结合起来，利用周末时间进行一部分的探究活动，实现"育人天地宽，成长不放假"。在综合实践活动课的启动环节，孩子们围绕"紫洞黄皮"展开讨论：紫洞黄皮这么好吃，是由怎样特殊的气候条件造成的呢？全球气候变化对黄皮的生长有影响吗？如何打造好黄皮品牌，助力乡村振兴……孩子们热情高涨，师生共同创设活动方案，一系列富有乡土特色的气候变化教育活动徐徐展开。

项目设计

"探究黄皮与气候，助力乡村振兴"项目的总目标为：通过深入了解和研究黄皮与气候之间的关系，直观感受气候变化对植物生长的影响，学会关爱地球、尊重自然、珍惜资源，形成可持续发展的观念，挖掘和利用黄皮的产业潜力，助力乡村振兴。具体可分为如下子目标：

（1）认识黄皮与气候的关系，增长科学知识。

（2）通过黄皮产业宣传当地文化，增进对自己家乡的了解，提升参与社区基层治理的归属感与认同感。

（3）加强自身及公众的环保意识，提高对黄皮产业与环保关系的认识。

（4）经由同伴间的共同协作，完成参观、采访、观察、记录等任务，培养团队精神、沟通能力和问题解决能力。

该项目的内容设计、实施方式和评价方式如表1所示。

表1 "探究黄皮与气候，助力乡村振兴"项目内容、方式及评价

项目名称	内容设计	实施方式	评价方式
美味黄皮，气候所赐	1. 观察黄皮生长 2. 记录黄皮的生长环境和气候条件 3. 收集当地的气候数据，找出与黄皮生长相关的规律	1. 前往黄皮种植园实地考察 2. 采访园区工作人员 3. 在图书馆搜索资料并分析	针对"参与度、积极性、合作精神"进行自评、小组互评，送上"大拇指"贴纸

(续表)

项目名称	内容设计	实施方式	评价方式
气候变化，黄皮危机	1. 了解黄皮种植的经验和气候影响的具体例子 2. 验证气候对黄皮生长的具体影响	1. 前往紫洞村采访果农 2. 在科学老师的指导下做对照组实验	针对"参与度、积极性、合作精神"进行自评、小组互评，送上"大拇指"贴纸
保卫黄皮，守护家园	思考应对气候变化和保卫黄皮的措施	1. 头脑风暴思考对策 2. 开展绘制宣传海报等活动	评选"环保宣传大使"
打造品牌，助力振兴	思考打造黄皮品牌，助力乡村振兴的对策	1. 实地研学 2. 采访群众 3. 总结乡村振兴金点子并演讲	评选"优秀小村长"

项目实施

一、阶段一：美味黄皮，气候所赐

2024年初，学生们组成研究小组，前往紫洞黄皮种植园进行实地考察，观察黄皮的生长情况，记录黄皮的生长环境和气候条件（见图1）。在园内，学生们认真研读了黄皮生长的资料，采访了园区里的工作人员。回到学校后，他们利用图书馆的信息网络收集当地的气候数据，包括温度、湿度、光照、降雨量等，并进行分析，找出与黄皮生长相关的规律。

图1 学生在黄皮种植园观察

同学们研究发现：紫洞村处于我国广东南部，属亚热带，气候温暖湿润，年平均气温在20℃以上，冬季平均温度达12℃以上。紫洞村所处位置还拥有充足的光照，有助于黄皮进行光合作用,增强树体养分的积累,扩大挂果面积,提高果实的品质和产量。

同时,同学们通过采访还有一个意外的收获:原来种黄皮树是紫洞的一个传统,假如有村民盖了新房子,很多都会在门前或者院子里种黄皮树。因为村里的人认为,黄皮树有辟邪的作用,且不惹虫,够干净,易打理。

同学们通过本次活动直观地感受到了气候对黄皮生长的重要影响,这有助于他们建立起对自然科学的基本认知,并激发他们对科学探索的兴趣。同时,在活动中,同伴之间共同协作,顺利完成了参观、采访、观察、记录等任务,这不仅能够培养他们的团队精神,还能够提高他们的沟通能力和解决问题的能力。

二、阶段二:气候变化,黄皮危机

气候变化会对黄皮生长产生影响吗?这是同学们急迫想了解的。项目组的同学走进紫洞村,采访了当地的农业专家和种植经验丰富的农民,了解黄皮的种植经验和气候对其的影响(见图2)。种植经验丰富的农民大多数是同学们的爷爷奶奶。俊朗同学的爷爷告诉他们:"之前南庄镇的陶瓷厂污染严重,气候变差,黄皮树产果量一度减少。后来随着镇内和村里的一些陶瓷厂陆续搬迁,环境就慢慢开始变好了,黄皮树也开始开花结果了,现在,村里黄皮的收成可谓一年比一年好。"

图 2 学生请教专业果农

除了采访调查,同学们还在图书馆查阅气候变化与植物相关的书籍资料,并在科学老师的帮助下设计实验,模拟不同气候条件下黄皮的生长情况,如设计对照组实验,观察温度变化、降雨量变化、光照变化等对黄皮生长的影响。

同学们通过一系列研究后发现:气候变化会对紫洞黄皮产生多方面的影响。这些影响直接关系到紫洞黄皮的生长周期以及最终的产量和品质。

首先,气候变暖会导致紫洞黄皮的生长季延长,开花提前。高温也会抑制紫洞黄皮的光合作用,降低其生长效率。温度的波动容易导致黄皮生长乏力甚至死亡。其次,洪水灾害会造成紫洞黄皮的死亡。雨水过少则导致紫洞黄皮的叶片和根系出现干旱、水分亏缺以及营养失衡等问题。最后,工业化导致的二氧化碳浓度升高也会对黄皮生长产生严重影响。

气候变暖不仅影响植物本身的生长和适应性,还会影响生态系统的平衡。通过了解气候变化与黄皮生长的关系,同学们更加深入地认识到保护环境、减少污染的重要性。

三、阶段三:保卫黄皮,守护家园

保卫美味黄皮,我们能做些什么?面对气候变化,我们又能做些什么呢?同学们观察生活,围绕议题进行思考与讨论,并将讨论出来的措施写下来,请教科学老师是否可行。他们把可行的金点子通过自己的巧手化作宣传标语或宣传海报,在学校、社区等场所张贴宣传,号召身边人齐心协力应对气候变化,保卫家乡美味黄皮。

同学们建议:

(1)从日常小事做起,比如节约用水、用电,记得关掉不用的电灯和电器,用节能灯代替普通灯泡,这样既能节约资源,又能减少碳排放。

(2)学习垃圾分类和回收知识,将可回收的物品进行分类投放,降低对自然资源的过度开采和环境的污染。

(3)多参加植树造林活动,绿化校园和周边环境,树木能够吸收二氧化碳并释放氧气,有助于缓解全球变暖。

(4)向家人、朋友和同学宣传全球变暖的危害性和应对方法,鼓励大家一起节约能源、减少废物产生、多采用骑自行车或步行等低碳出行方式。

保卫黄皮活动帮助同学们树立了正确的环保观。他们从中学会了关爱地球、尊重自然、珍惜资源,形成了可持续发展的观念。

四、阶段四:打造品牌,助力振兴

紫洞黄皮这么好吃,如何打造成品牌,助力乡村振兴呢?教师通过开展"假如我是小村长"活动,引领学生以"小村长"的身份深入紫洞村进行实地研学(见图3),采访群众,收集意见,思考如何打造黄皮品牌,助力乡村振兴,以"小村长"的身份提出自己的想法和建议。

图3 学生在调查群众意见

对此，学生们提出了许多具有创意和参考价值的建议，如：

（1）举办黄皮采摘节等活动，让游客们参观黄皮种植园，体验采摘乐趣，品尝新鲜的黄皮果实。举办黄皮文化节，组织紫洞黄皮摄影大赛、黄皮美食品鉴会，多举措提高紫洞黄皮的知名度和美誉度。

（2）积极与电商平台合作，拓宽销售渠道，让黄皮产品走出乡村，走向更广阔的市场。

（3）发展深加工产业，将黄皮加工成果汁、果酱等产品，创作黄皮咖啡、黄皮千层糕、黄皮饼等新式美食，延伸产业链条。

（4）开发黄皮文创产品，制订有效的营销策略和推广计划，利用社交媒体、线上线下活动、合作推广等方式进行宣传和推广。同时，可以与相关文化机构、旅游景点等合作，将黄皮文创产品作为特色商品进行销售和推广。

在这个活动中，我们通过儿童视角，推进儿童友好理念融入乡村规划建设，让孩子们了解乡村，提升青少年儿童参与社区基层治理的归属感、认同感，带动青少年儿童群体发挥自身力量，多参与、多宣传，为家乡、社区建设贡献力量，实现孩子与社会共同发展。

成效与反思

本项目自 2024 年 2 月开展至今，虽时间不长，但形成的成果比较丰富。在教师成长方面，项目组汇编了一本《探究黄皮与气候，助力乡村振兴》的研学手册。项目主导者之一陈宝瑜老师于 2024 年 4 月带着气候教育项目经验在"上海气候周"教育论坛进行了分享交流。

学生成长方面，该项目让学生对黄皮这一本地特色农作物有了更深入的了解，包括其生长习性、与气候的关系等。这种直观的学习方式比传统的课堂教学更能激发学生的学习兴趣，也使知识更加生动具体，学生学习更深入。学生们的各类实践技能也得以提升，习得了如何观察、采访、记录和分析数据，如何做对照组实验等，实践能力和科学素养得以增强。活动还加深了学生们对乡村文化的了解和认同，认识到乡村不仅是他们的根，也是他们学习和成长的重要场所。这种文化认同有助于推动学生们为乡村振兴作出贡献。此外，学生们在实践中更加关注自然环境和生态保护，了解了气候变化对农作物生长的影响，环保意识也得以提升。

活动还促进了学校、家庭和社区之间的合作与交流。家长们积极参与孩子的实践活动，提供了强而有力的支持和指导。社区也为学生们提供了实践场所和资源，形成了良好的教育合力。

2024 年 4 月，项目组老师在学生中进行问卷调查，94%的同学喜欢并认同这次项目式活动。教师也进行了相关采访，家长、老人和村委会工作人员反馈良好。

张同学妈妈说："这个活动太好了，从学校小课堂到社会大课堂，孩子们的体验更丰富了，也更懂得环保，现在上学、上班都建议我们绿色出行呢。"罗同学爷爷说："没想到我这么老了还能发挥作用，把我多年来的经验分享给同学们，我很支持这个活动，很有意义。"

紫洞村村委会工作人员罗女士说："同学们提出的打造黄皮品牌的措施很有参考意义，

让人感到惊喜,我们内部会好好考虑研究的,感谢同学们积极参与本次活动,让这股新生力量加入我们乡村振兴的行列中。"

当然,本项目仍存待改善之处。如尚未建立起更加科学、全面的评价机制来评估活动的成效和学生的学习成果。后续,我们将更注重对学生自我评价和反思能力的培养,持续跟进学生的成长和发展情况,通过定期回访、组织相关活动等方式来保持与学生的联系和互动,鼓励他们将所学知识和技能应用到实际生活中去。同时,我们也将持续深入关注乡村振兴的发展情况,为学生提供更多的实践机会和资源支持。

(陈宝瑜,佛山市禅城区南庄镇紫南小学教师)

基于学生视角的气候变化与生物多样性关系探究

研究背景

昆明位于云贵高原,海拔较高,冬季受北方冷空气影响较小,夏季受到印度洋西南季风的影响,带来丰富的降水,使得这座城市四季温暖如春,素有"春城"之美誉。

可近两年,孩子们发现昆明的天气开始变得不那么宜人。有孩子说:"夏天特别炎热,在教室上课都要用扇子;冬天又很寒冷,父母都购置了新的羽绒服。"还有孩子注意到:持续不降雨,让学校原本滋养着很多植被的土壤变得干裂,许多植被都已枯萎。这一系列的现象,引发了孩子们的联想:持续高温干燥的气候对生物种类有什么影响?气候变化会不会影响生物多样性?如何应对气候变化?作为小学生,我们可以做什么?带着以上问题,笔者设计了以探究气候变化与生物种类关系为主题的项目式学习。

项目设计

一、目标设计

本项目的总目标为:了解气候变化与生物多样之间的关系;了解应对气候变化的日常生活行为。

据此总目标,我们又分别针对四、六年级学生能力现状进行了子目标的确定。其中,四年级活动目标为:①通过查阅资料、询问家中长辈、实地考察、整理资料等方式,学生了解并且说清楚云南果群与气候变化的关系。②通过此次项目研究,学生知道日常生活中自己可以做什么来改善极端天气带来的生物种类减少,进一步提高保护生物资源的意识,增强个人保护生态环境的社会责任感。

六年级活动目标为:①通过了解气候变化的知识,学生科学地认识气候变化与人类活动之间的关系,从而让活动更具科学性。②通过持续开展活动,学生养成主动的环保行为,并带动更多的人将环保落实到日常行动中去。

二、内容设计

依据活动目标,笔者分别设计了四年级与六年级的两类主题学习内容。其中,四年级的主题为"探究气候变化与水果种类的关系",六年级的主题为"探究气候变化与人类的关系"(见图1)。

图 1　项目内容设计

该活动的设计原则为：

其一，以学生问题为导向，致力于让学生在项目化的探究中，综合运用多种方法尝试对问题进行解决。

其二，以关联探究为核心。在活动期间，指导学生对自己所处的区域特色资源，关联式地进行感受、探究。在关注气候变化或生物多样性的同时，依据两者本身就密不可分这一特点去进行关联探究。

其三，以日常行动为根本。在了解了气候变化与生物多样性的关系后，本内容实施会对应对气候变化与保护生物多样性有明确的日常环保行动指导，携两个年级学生、教师及家长之力打造生态圈。

项目实施

一、项目启动：确定活动主题与聚焦的问题

为了更好地推进项目落地，笔者将其纳入学校层面的活动中。学校层面活动的大主题为"协力打造生态圈"。2023年12月8日，笔者组织校区全体班主任召开了项目启动会，并于会上学习了气候变化及气候变化教育的相关信息。老师们了解到，《联合国气候变化框架公约》缔约方会议第28届会议（简称"COP28"）于2023年11月30日至12月12日在阿联酋迪拜世博城正式举行。11月30日，即COP28开幕当日，在COP28中国角举办了以"生物多样性保护与应对气候变化协同机制"为主题的边会（以下简称"边会"）。"边会"在中华人民共和国生态环境部生态司指导下，由中国绿色碳汇基金会、中华环保联合会联合主办，世界自然保护联盟（IUCN）、自然资源保护协会（NRDC）、大自然保护协会（TNC）、野生救援（WILDAID）等机构协办。有了这些背景知识的了解，班主任们就学生们提出的问题展开了讨论，并积极研究应对之策。会议最终达成共识，即确定以四、六年级为载体，开展关于气候变化和生物种类的关联式探究。其中，四年级聚焦"果群与气候"，六年级聚焦"人类与气候"。

二、活动准备：策划活动内容，设计活动方案

启动会后，四、六年级的班主任们带领学生们围绕主题策划了具体的活动内容（见图2）。

在此过程中,班主任与班级、年级家委及时介入指导,助益活动落地实施。图 3 为六年级的一份项目方案。

图 2　教师带领六年级三个平行班学生策划活动

图 3　六年级学生策划的项目方案

在活动方案设计过程中,我们特别留心不同年级段学生能力培养的差异性。四年级着重培养学生的资料收集与整理能力,六年级组着重培养学生的环保意识与能力。

三、宣传推广：促进更多主体参与其中

通过前期的准备，学生对于目标设计、活动内容已经比较熟悉了。2023 年 12 月 28 日至 2024 年 1 月 12 日，我们利用家长会、学校公众号等向家长介绍项目相关的内容与要求，充分利用家校平台保障本次活动的有效推进。

四、实施保障：利用假期，实地考察

2024 年 1 月 12 日，孩子们的寒假生活如期而至。以四年级为例，部分学生回到自己的老家，如大理、临沧等地，了解当地近几年的气候变化情况并做好记录。教师要求记录时间不少于一周，记录要素包含观察地点、观察日期及时段、观察方法等。观察结束后，学生还可画出气候变化折线统计图。

随后，孩子们由家长带领深入大自然，观察应季的水果生长；走进水果市场、果园调查当季上市的非大棚水果；前往植物研究院询问工作人员云南水果的生存条件。在探访过程中，学生们及时将观察所得、研究过程等用照片、文字的方式记录下来，这一活动持续了近一个月时间。

五、呈现展示：方式多元，载体多样

2024 年 2 月 10—14 日，学生分别以美篇、立体书、宣传手册、立体模型还原及项目研究报告等形式呈现了个人研究成果（见图 4）。

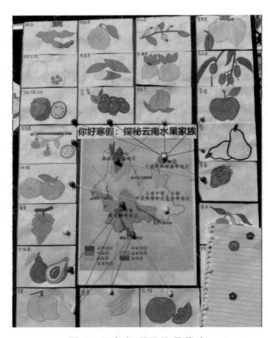

图 4　四年级项目作品代表

六、评价设计：素养导向

当学生们历经前五个阶段的活动后，笔者组织了系列评价活动。其中，关于学生素养的

评价主要从"学会学习""责任担当""实践创新"三个维度展开。具体评价指标还做了低年段、中年段和高年段的区分(见表1)。关于学生作品的评价,则通过校级与年级的作品展评会来进行(见表2、表3)。

表1 项目评价设计

学生素养	低年段	中年段	高年段
学会学习	能在同伴、教师、家人的引导下,基本知道当下全国或全球关于"生物多样性与气候变化"的前沿话题	能在同伴、教师、家人的引导下通过网络、报刊等媒介了解当下全国或全球关于"生物多样性与气候变化"的前沿话题	能自主通过网络、报刊等媒介了解当下全国或全球关于"生物多样性与气候变化"的前沿话题
责任担当	1. 会在教师、家人、同伴的引导下实施环保相关的行为(垃圾分类、恢复一个区域、不食用野生动物、不购买无环保认定的产品等) 2. 守秩序,爱公物,能在家人的帮助下有效节约资源	1. 会自主实施环保相关的行为(垃圾分类、恢复一个区域、不食用野生动物、不购买无环保认定的产品等) 2. 能遵守公共秩序,在活动中形成初步社会公德	1. 会自主实施环保相关的行为(垃圾分类、恢复一个区域、不食用野生动物、不购买无环保认定的产品等) 2. 能积极参与活动并带动更多群体参加
实践创新	1. 能在教师、家人的帮助下,初步进行简单的口头策划,并明确自己的职责 2. 能在教师、家人的引导下,按活动要求进行活动,遇到困难会寻求家人、教师的帮助	1. 能自主进行较为完整的策划,并形成个人方案,要素需包含:时间、地点、人员基本分工 2. 能自主按活动方案进行活动,在过程中会根据实际情况有针对性地寻求帮助	1. 能自主进行完整的策划,并形成组内或班级总方案 2. 能自主按活动方案进行活动,在实施过程中,会根据实际情况灵活调整

表2 四年级项目作品评价标准

评价标准	同学评价等级
作品围绕主题	☆☆☆
图文并茂且内容完整	☆☆☆
作品富有创意	☆☆☆

表3　六年级项目作品评价标准

评价标准	同学评价等级
作品紧扣主题	☆☆☆
装订成册	☆☆☆
图文并茂，活动过程持续性强并可在生活中持续实践	☆☆☆
作品中的收获真实、具体	☆☆☆

评价过程中，四年级的一位老师感慨说："本次项目实践活动带给同学们无尽成长，听到同学们的讲解，我们仿佛也跟着他们进入一个个五彩斑斓的水果乐园。令人惊喜的图册，仿佛能嗅到水果的香气。通过这次活动，四年级同学带领着家长去尝试打造生态圈，共同探究气候变化与生物多样的关系，争做生态保护的代言人。"

学校也及时给展评投票数高的孩子们颁发了"优秀作品奖"和"活动实践达人"荣誉称号，鼓励孩子们继续积极主动地参与活动，并带动更多人加入生态环境保护的队伍中来。

成效与反思

在进行以上活动一段时间后，我们分别对学生、教师及家长进行了活动效果调查，下发调查问卷874份，回收有效问卷654份。调查结果显示：近95%的家庭对于前期的研究活动高度认可并表示会持续参与后续的活动，98%的学生认为活动内容比较多元化，85%的家长表示："通过活动，孩子与自己的关系更亲密了。"94%的教师认为孩子通过长期开展的保护生物多样性的活动，在搜集、整合资料，宣传方面都有了明显的转变，同时，学生的活动作品也越来越丰富了。

基于现阶段研究成果，笔者学习有关气候变化的政策后发现：一所小学，如果要达到长程性地引导学生关注气候变化与生物种类的关系并持续践行环保行动，仅靠学校和家庭的力量是无法实现的。因而，在接下来的研究中，我们将寻找相关政府部门及本地企业给予活动场地或教育支持。同时，我们也将继续从生物研究角度展开探索，如与中国科学院昆明植物研究所、中国科学院昆明动物研究所、中国科学院昆明分院、昆明气象局等相关科研单位取得联系，充分利用地理优势开展活动。

气候变化会直接导致生物种类的减少，而生物种类的减少势必会带来生物链的断裂，那随之而来的就是人类生存受到威胁。本项目研究及后续拓展，都是为了更科学、更实际，也更长程地引导学生探索"生物种类与气候变化"的关系。关注气候变化，关注生物多样性，是全人类的共同话题，也是全人类的共同责任。

（李星月，昆明市西山萃智御府学校学生发展部部长）

"水气相和 育万物共生"项目化学习设计与实施

研究背景

水是气候变化最直接、最重要的影响领域之一,气候变化的影响可以通过水来反映。"世界气象日"又称"国际气象日",定于每年的 3 月 23 日。从 1961 年到 2022 年,我们已经历了 62 个"世界气象日",其中有 9 次的主题直接与水相关。2023 年,恰逢世界气象组织成立 150 周年,以"天气气候水 代代向未来"作为主题,呼吁全社会增进对全球变暖背景下气候系统的了解。我国一向高度重视水资源的合理开发与利用,将水资源管理列为国家战略的重要组成部分。

在此背景下,我们开展以"水气相和 育万物共生"为题的项目化学习活动,培养学生的环保意识,提升社会责任感,积极采取行动以有效应对气候变化,自觉践行生态保护的使命。

项目设计

一、整体流程设计

项目化学习是以项目为载体的学习方式,不论项目大小,都要遵循项目周期的常规流程。依据北京师范大学项目式研究团队提供的"四大阶段、八大模板"流程,笔者设计了本次项目式学习流程图,具体包含立项组队、计划筹备、项目实施与复盘结项四个阶段(见图 1)。四阶段循环往复,形成一个不断迭代,不断升级的项目式学习系统。

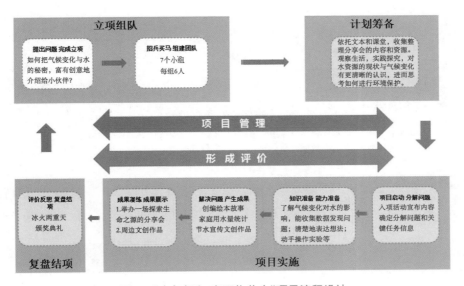

图 1 "水气相和 育万物共生"项目流程设计

二、目标设计

首先应进行目标设计(见表1)。

表1 "水气相和 育万物共生"项目学生培养目标设计

	维度	能力表现
PBL核心素养目标	具备社会责任感,形成正确的价值观与合理的思维方式	能够在家庭环境中进行简单的污水过滤实验,了解水的净化过程;能够了解水资源的现状;能够设计并参与节约用水活动,提高社会责任感;发展应对气候变化的相关自然科学类知识
	有效沟通与合作,发展学习者的领导力、实践力、综合解决问题力	善于倾听和表达,能做到认真听取他人观点;能够通过有效的交流(通知、说服、询问、激励等)达成自己的目的;能够与他人合作,共同实现一个目标
	主动探究和实践,养成可持续发展的生产、生活方式与习惯	具备好奇心和主动性,能够了解水的循环原理,能够运用科学知识解释实验结果,理解水的性质;保持探究热情,掌握探究方法;能够参与水的科学实验,观察并记录实验结果,在实践中应用知识、验证想法、探索世界
	审美创造	能够通过语言、艺术和设计等方式来表现美和创造美,能够通过美术作品表达对气候变化或者水的情感和认知
	技术使用和信息素养	能够通过技术手段为自己的学习和探索提供支架;能够应用信息,并通过信息进行交流和解决问题

三、内容设计

接下来则为内容设计(见表2)。

表2 "水气相和 育万物共生"项目内容设计

核心驱动问题	总任务	最终成果
如何把气候与水的秘密,富有创意地介绍给小伙伴?	举办一场探索生命之源分享会	分享会及周边文创作品
分解驱动问题	**主任务**	**主产品**
面对如此熟悉的水,你对它有多少了解?要完成分享会,要准备哪些宣讲内容和资源?	走近气候与水:依托文本,品读语言,查阅相关资料,知道水的千变万化,并用一两句话进行介绍	关于小水滴变形记的PPT,手抄报

(续表)

分解驱动问题	主任务	主产品
如何展示"气候与水"的秘密更有趣？	气候与水的探秘之旅：观察生活，实践探究，通过问卷了解人们对水资源的认识和使用习惯，通过记录家庭用水量，了解水在生命活动中的重要用途，对水体污染有更清晰的认识，进而思考如何净化水	节水标语、调查报告、统计表、水的科学实验、节水三句半
如何通过分享会宣传水资源保护，提升人们的生态环境保护意识？如何让我们的分享会具有更好的推广价值？	整理资料，准备分享会，展现自己的发现	完善本组作品和分享稿

四、评价设计

项目评价围绕学生"气候变化与水循环作品""家庭节水调查统计与节水妙招"以及"探索生命之源分享会"三方面内容展开。评价目标、评价证据、评价方式及评价类型具体如表3所示。

表3 "水气相和 育万物共生"项目评价设计

评价内容	评价目标	评价证据	评价方式	评价类型
气候变化与水循环作品	主动阅读材料，乐于思考，初步了解"气候和水"的关系，并能用不同的方式向别人介绍；在多样的情境中利用自己学到知识，解决实际问题	设计作品，并能向大家清楚地介绍气候变化与水的关系	评价量规自评互评	形成性评价终结性评价
家庭节水调查统计与节水妙招	设计问卷调查表，通过问卷了解人们对水资源的认识和使用习惯，通过实验的方法，对改善水资源的利用提出意见和建议。寻找到更多节水护水的办法，通过绘制手抄报、撰写文章等宣传将废水再利用从而实现珍惜水资源	动手实践，了解科学实验的原理。生活中宣传并做到节约用水	评价量规自评互评	形成性评价终结性评价
探索生命之源分享会	增强对气候与水的了解，增进对天气和气候的关注，提高防灾减灾意识，以实际行动维护好我们美好的家园	分享会展示成果交流	评价量规自评互评	形成性评价终结性评价

项目实施

本项目结合学校"种子成长"特色课程,探索以"学科+"为主要特征的项目式学习,聚焦学生真实生活,展开跨学科学习,提高孩子们分析问题、解决问题的能力。

面对驱动问题"如何把气候与水的秘密,富有创意地介绍给小伙伴?",我们确定核心任务为"举办一场探索生命之源分享会",并基于此进行了问题与任务的分解。

(一)子任务一:走进气候与水

该任务的成果为"气候与水的手抄报及演示文稿"等。学生通过对《我是什么》《雾在哪里》《雪孩子》三篇课文的学习,感受到了水的神奇变化。此外,小组成员还一起共读了绘本故事《水的旅行》《思考世界的孩子》,观看了《气候与水》《关注气象 保护水资源》等科学纪录片,对水有了较为全面的认知。基于此,教师引导学生根据课文内容,发挥想象力,并以故事的形式讲述小水滴的旅行故事,最后创作自己的作品,画一画水的三态变化以及雾、雪、冰雹等形成过程,并用一两句话予以介绍(见图2)。

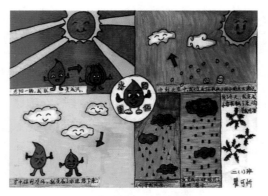

图2 学生作品《水的变化》

(二)子任务二:气候与水的探秘之旅

该任务的成果为"家庭节水调查 节水妙招"。在了解到水资源于我们的作用后,孩子们走进生活,探秘气候与水的奇妙关系。令人印象深刻的是实验探究环节,孩子们在科学老师和家长的指导下展开了各项有趣的水实验,并写下了自己的所见所想:水宝宝是个"魔术师"——让画在纸上的图案反转;水宝宝是个"体操运动员"——136滴水能够"站"在一枚小小的硬币上……一个个小实验让孩子们见识到了水的特性:水会使光线发生折射、水具有张力、"毛细现象"、水油不相溶……

面对节水意识淡薄的问题,师生们共同设计了一份调查问卷,了解人们对气候对水的认识。调查结果显示:大家对于气候与水普遍都了解,懂得节约用水的方法也有很多,但是很多人没有意识到要去二次利用水资源。此外,学生通过对家庭用水数据的收集、整理与分析,能更具针对性地选择节水措施,树立良好的节约用水习惯与绿色环保理念。

如何在实践中更好地保护水资源,提高水的利用率呢?孩子们通过观看视频,了解了自制净水器的制作与原理,便纷纷投入实验(见图3)。

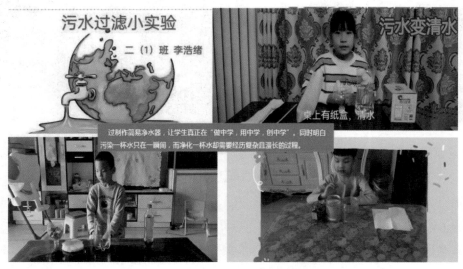

图3 学生作品"自制净水装置"

(三)子任务三:提升推广分享会的质效

在探索生命之源的分享会上,精彩纷呈的作品(见图4)展现了同学们的收获,也丰富和升华了学生们对气候与水的认知。项目最后,依据评价表,开展了总结和反思活动,还评选了多种不同类型的奖项。

图 4　学生成果汇报展

成效与反思

本项目对于学生分析思维、创新思维与实践思维的提升都起到了很好的作用。譬如,在分析思维方面,经此项目历练,很多学生在面对生活中的许多相关问题时,都会产生一种自觉的思考意识,有积极的思考动力。慢慢地,问题分析能力也在逐步提升。在创新思维方面,纵观整个项目式学习过程,学生们都是用自己喜欢的方式去发现、去实践并记录自己的过程与感想。老师和家长也给学生们提供了许多可以发挥自由想象、施展自我创作的空间以及科学适用的学习支架,帮助学生逐步构建起对主题及核心问题的知识逻辑结构,推动学生思维素养水平的提升。在实践思维方面,本项目在"气候变化与水的关系""如何进行家庭污水过滤实验""如何获得水资源现状""如何了解大家对气候变化和水的认识"等问题的分析与探寻中,都给予了学生充分的讨论空间和丰富的实践体验,学生们的动手实操能力得以增强。

同时,我们也进行了复盘与反思,本次跨学科项目式学习的融合性还有待改进,如学科间教材的分析,需要更加深入,找准契合点,提炼相通知识点,进行项目式学习主题设计,再根据内容设置,设计以学生学情为出发点的问题驱动,进而在课程开发中实现学科整合优化,避免跨学科教学有"拼盘"或串烧的感觉。

一泓清水,是生命之源,也是生态之源。每一个人、每一寸土地、每一个国家,其繁荣发展皆离不开气候与水的恩泽。"水气相和　育万物共生"项目仍在持续、改进。绿水青山、万物生长,需你我携手呵护。

(王萌,北京市房山区良乡镇良乡中心小学教师)

"气候变化与环境保护"科普项目化学习探析

研究背景

面对气候变化这一全球性挑战,各国政府和国际组织正积极采取行动来减缓它的影响。我国建设性地提出了可持续发展战略、新发展理念,为解决以能源危机、环境变化、自然灾害为主要表现的气候变化问题贡献了中国智慧。

这一议题延宕至教育教学领域,便成为教育工作者的责任。作为中学教师,我们通过"气候变化与环境保护"科普项目化学习活动,引导学生探究气候变化的成因、影响及应对措施,提高初中生对气候变化的认识,培养其环保意识和创新能力。

项目设计

本项目的设计遵循渐进性、统整性、理论与实践相结合的原则,分为认知、发现、分析、合作探究、创造等多个维度,具体如下:

其一,研究中国历代气温变化规律,让学生了解气温变化的原因和影响,明确气温变化对环境带来的影响,树立保护生态环境的社会责任和科学意识。

其二,分析气温变化对环境的影响,尤其聚焦全球变暖,极端天气事件增多,探讨环境保护的重要性和措施,提高学生对环境保护的认识和意识,培养他们的社会责任感和行动力。

其三,结合历史、地理、科学等多学科知识,提高学生跨学科学习能力和团队合作精神。

项目实施

一、建构学习小组,分配学习任务

为了更好地保障学生的学习效果及项目推进质量,我们将学生分为四个小组,对任务进行了分配,每个小组领取一个研究任务(见表1)。

表1 "气候变化与环境保护"科普项目化学习小组任务

组别及负责内容	具体任务
第一组:历代气温变化	任务1:查阅中国知网文献,了解历代气候的变化。
	任务2:查阅中国古代文献、诗歌、散文、笔记等关于气候的记载。
	任务3:以自己的家乡为例,测量今年的气温,并在当地气象局查阅20世纪以来当地的气温变化情况,绘制气温变化柱状图。
	任务4(项目成果):绘制历代的气候变化图。

(续表)

组别及负责内容	具体任务
第二组:气候变化给生态环境带来的影响	任务1:查阅互联网、中国知网,了解气候变化对自然环境的影响。
	任务2:查阅互联网、中国知网,了解气候变化对生物多样性的影响。
	任务3:查阅互联网、中国知网,了解气候变化对农业生产的影响。
	任务4(项目成果):举办图片展览。
第三组:各国如何应对气候变化带来的环境问题	任务1:查阅联合国、欧盟等国际组织如何应对气候变化带来的环境危机。
	任务2:查阅欧洲发达国家如何应对气候变化带来的环境危机。
	任务3:查阅以中国为代表的发展中国家如何应对气候变化带来的环境危机。
	任务4(项目成果):撰写小论文。
第四组:中国智慧与中国方案	任务1:查阅人类命运共同体等相关理念。
	任务2:查阅可持续发展理念。
	任务3:项目成果——制作宣传品。

第一小组的任务是研究中国历代气温变化。组员们运用地理、历史、数学统计、英语和语文等多学科知识,对中国历史上的气温数据进行整理与分析,并通过对不同历史时期的气温数据的对比,探寻气温变化的规律和趋势,以及这些变化对社会和自然环境产生的影响。

第二组的任务是研究气候变化对生态环境带来的影响。组员们从生物、地理、英语等学科角度出发,探讨气候变化如何影响生物多样性、生态系统平衡以及人类居住环境。此外,还将通过实地考察、实验研究和文献综述等方式,深入了解气候变化对生态环境的具体影响机制,并提出应对策略。

第三组的任务是研究各国如何应对气候变化带来的环境问题。组员们结合英语、道德与法治、生物、地理、历史等学科知识,了解各国在应对气候变化方面的政策和实践。对比分析不同国家的策略和效果,总结经验和教训,为中国应对气候变化提供借鉴和参考。

第四组的任务是给出问题解决的中国方案并制作宣传品。组员们结合美术、道德与法治、语文、生物、地理等学科知识,提出针对中国气候和环境问题的解决方案。同时,他们还利用创意和艺术手段制作宣传品,通过视觉和语言的双重冲击,呼吁社会关注环境变化问题,增强公众的环保意识和参与度。

在整个项目的开展与学习过程中,学生接触到包括历史、地理、气候、环境科学等多个学科的知识,不断提升多学科跨学科的研究能力,加深对环境保护问题的理解。最终,他们撰写研究报告,展示研究成果,并进行小组讨论和交流,进一步深化对气温变化与环境保护关系的认识(见表1)。

二、依循学习路径,推进项目实施

每一个任务的完成过程大致可分为七个阶段。第一阶段主要是教师设计、确定研究主题。自第二个阶段开始,皆由师生共同完成。以"中国历代气温变化与环境保护"子任务为例,主要历经"制订研究计划"—"搜集资料"—"分析与讨论"—"撰写报告"—"展示与交流"—"反思与改进"等过程(见图1)。

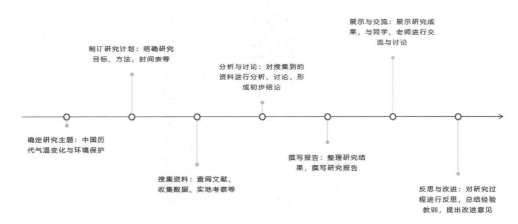

图1 "气候变化与环境保护"科普项目化学习过程路径

首先,教师给出研究框架,如当代地理科学研究路径图、自然科学研究常用方法及相关论文等,学生依照研究的目标、方法,制订出详尽、可操作的时间表等。其次,学生在教师的指导下通过查阅资料、调研考察等方式对"中国历代气候变化"问题展开探究,获得气候变化的相关数据,探讨可能的原因。基于此,学生开始分成小组对所获得的数据展开分析、讨论,利用合理的证据形成自己的初步结论,并整理研究结果、撰写研究报告。在小组形成之后,各个小组此时基本完成了探究工作,获得了较为自洽的研究结果和结论。再次,是展示与交流。在此环节,各小组需采取合适的方式进行交流展示,老师作为顾问、指导角色参与其中,以获得全面而科学的研究结论。最后,学生还需要对一些结论的二律背反现象进行反思与改进,对研究过程、研究方法和实验结论得出的路径进行复盘,以促进后续研究的不断优化。

成效与反思

相较于传统的学习形态,项目化学习是一种更为开放的学习模式,它直指学生的创新思维和实践能力。本项目中每一个小组的学生都以绘画、音频、视频、文字等不同的形式呈现了各自的研究成果,还策划组织了场馆考察、垃圾分类等实践活动。

封晴菡同学设计了一幅画(见图2),画面主体为无数破碎的动物图案组成的一个人,也可解读为一个人分裂成了无数个动物,其寓意是人与自然和谐共生。"万物有灵,不仅仅人类是万物之灵,其实并没有万物灵长的概念,有的是人类中心主义的泛滥。"封晴菡同学如是说。之所以采用画作的形式来表现,是因为她认为,图案相比文字对于我们来说显得更为震撼。

图 2　学生封晴菡的作品

2024年1月23日,学生们来到位于南山的垃圾分类科普体验馆,开启了一场关于环保和垃圾分类的深度探索之旅(见图3)。在体验馆中,学生们遇到了一位热情的讲解员——一位满头白发的老人。他以丰富的知识和生动的语言,向同学们详细介绍了垃圾的生产过程、对环境的危害性、垃圾分类的重要性,以及垃圾经过科学处理后如何转化为有价值的肥料、能源等,深深地感染和打动了在场的每一位同学。

图 3　南山能源生态园活动留影

本次项目化学习收获了一系列成果(见图4—图7),呈现出联动性、广延性与开放性。我们将此次项目化学习安排在寒假,学生们在假期中深入调研、分析数据、整理资料,为开学后的集中汇报展示做好了充分的准备。这种联动课上与课下的方式,不仅使学生们能够更好地理解和掌握知识,更能够培养他们的实践能力、创新精神和团队合作精神。

图 4　学生制作的"环保"艺术品

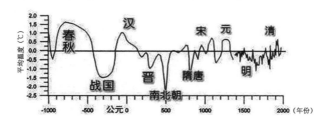

图 5　第二小组绘制的中国中原地区冬季温度变化

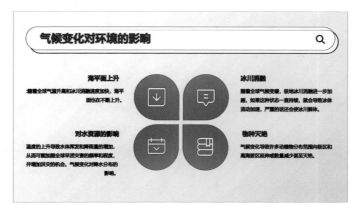

图 6　第三小组制作的 PPT：气候变化对环境的影响

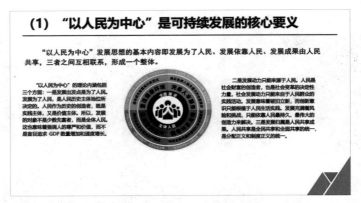

图 7　第四小组梳理总结的气候变化与环境保护议题的中国智慧与中国方案

总的来说,这次项目化学习是一次富有成效和意义的探索。通过深入研究、团队协作和成果展示,学生们不仅增长了知识、锻炼了能力,更培养了责任感和使命感。这种以学生为中心、以实践为导向的学习方式,提高了学生们的学习兴趣和动力,为他们未来的学习和生活奠定了坚实基础。

(刘美芳,深圳市宝安中学外国语学校教师;
赵泽龙,华东师范大学第三附属中学教师)

对话专家

习近平总书记指出:"建设生态文明,关系人民福祉,关乎民族未来。"党的十八大以来,党中央将生态文明建设纳入中国特色社会主义事业总体布局,生态文明建设的战略地位更加凸显。习近平总书记在党的二十大报告中强调了大自然是人类赖以生存发展的基本条件。尊重自然、顺应自然、保护自然,是全面建设社会主义现代化国家的内在要求。因此,全社会需要以习近平生态文明思想为根本遵循,牢固树立和践行绿水青山就是金山银山的理念,大力推进美丽中国建设,建设人与自然和谐共生的现代化。《"美丽中国,我是行动者"提升公民生态文明意识行动计划(2021—2025年)》中明确了大力推进学校生态文明教育并将其纳入国民教育体系,旨在深入学习宣传贯彻习近平生态文明思想,引导全社会牢固树立生态文明价值观念和行为准则,进而助力实现建设美丽中国、实现中华民族伟大复兴的中国梦。

学校是开展生态文明教育的关键阵地,教育是实现生态文明建设目标的重要路径之一,在生态文明建设中起着关键基础作用。2021年联合国教科文组织正式颁布了《2030年可持续发展教育路线图》成果文件,重点强调了教育对实现可持续发展目标的贡献,提出了将可持续发展教育和17项可持续发展目标(SDGs)全面整合进国家政策、学习环境、教育工作者能力建设、赋权青年、地方(社区)行动等优先行动领域中,其中包含了可持续发展目标13(SDG13)——气候行动。因而,其强调全民终身学习、以学习者为中心,号召各国在国家政策的支持下运用全机构模式推进可持续发展教育。

项目式学习的本质在于利用真实、富有挑战性的问题培养学生创造性思维、批判性思维以及团队合作等终身学习能力。教师需要引导学生持续地探索,尝试创造性地解决问题,最终形成相关成果,在尝试解决问题的过程中培养学生的可持续学习能力,培育学生的生态文明素养。本章呈现的八个有关气候变化教育的优秀案例,体现了以气候变化教育为主线,涵盖了多个可持续发展目标。生态文明与可持续发展教育项目式学习案例的特点主要体现在以下几个方面:

一是凸显跨学科融合:学生在学习过程中需要运用多个学科的知识和技能,如自然科学、社会科学、人文科学等,以全面理解和应对生态文明与可持续发展的问题。

二是凸显了问题导向:以真实、具体的问题或挑战为起点,这些问题与生态文明和可持续发展密切相关。学生需要围绕这些问题进行探究,通过收集信息、分析问题、提出解决方案等过程,深化对生态文明和可持续发展理念的理解。

三是凸显实践性:气候变化项目式学习案例注重实践性和操作性,鼓励学生亲身参与实践活动,如实地考察、社会调查、实验设计等。通过这些实践活动,学生能够更直观地感受生态环境与可持续发展的现状,增强对问题的认识和解决能力。

四是凸显团队协作:案例中采用了小组合作的方式,学生在团队中分工合作,共同完成任务。这种学习方式不仅有助于培养学生的团队协作能力,还能促进不同背景、不同观点的学生之间的交流与碰撞,从而拓宽视野,激发创新思维。

五是凸显评价与反思:在气候变化项目式学习案例中,显示出了评价和反思环节的不可或缺性。教师、学生需要对学习过程、成果和合作情况进行自我评价和反思,同时接受来自同伴的评价与反馈,从而更深入地开展气候变化教育项目式学习与研究。

面向未来,全社会开展气候变化教育应把生态文明与可持续发展教育融入育人全过程,为生态文明建设提供全方位的人才、智力和精神文化支撑。以生态文明与可持续发展教育助力实现美丽中国与2030可持续发展目标,增进全球人类福祉,为全球生态治理和教育治理提供中国智慧和中国方案。

(张婧,北京教育科学研究院终身学习与可持续发展教育研究所
生态文明教育研究室主任)

第四篇

基于综合活动的气候变化教育

在学校教育场景下,综合活动不同于学科教学,但扎根于中国班级建设、学生工作中。很多班主任、年级组长等积累了丰富的经验,将气候变化教育作为突出的内容融入考察探究类综合活动、设计制作类综合活动、文化传承类综合活动、主题研学类综合活动。

本篇所呈现的内容,反映出本领域的相关探索,反映出中国班主任、年级主任及家长等的实践智慧。

寻长江口二号古船，扬长江口生命力

研究背景

气候变化对长江口有着深远的影响。据报道，2020年，长江口海域夏季出现大面积低氧区，高海平面抬升风暴增水的基础水位，加重了致灾程度。① 上海市浦东新区潼港小学位于高桥镇，紧邻长江口，孩子们和这片热土朝夕相伴，气候变化对于长江口的影响，也会间接影响孩子们的生活。

上海长江口二号古船打捞和安家的新闻，牵动着潼港小学每位学生的心弦。2022年11月21日，上海长江口二号古船被成功整体打捞出水，超600件水下文物被清理入库。11月25日，长江口二号古船在"奋力轮"和四艘拖轮的护卫下，来到杨浦上海船厂旧址1号船坞"安家"，开启文物保护与考古发掘新阶段。长江口二号古船被发现之处正处长江"黄金水道"的入海口和中国南北海岸线的中心点。自古以来，在这繁忙的航线上和复杂的水域里，埋藏有不计其数的水下遗珍和未解之谜。

青年参与全球气候变化治理是新时代的全球治理重要议题，也是全球青年事务的重要发展方向。② 基于此，关注长江口二号古船的前世今生，探索长江口可持续发展刻不容缓，"寻长江口二号古船，扬长江口生命力"项目应运而生。

项目设计

本项目以潼港小学三年级学生为主体，从长江口二号古船探索切入，通过校、家、社联合，探索长江口可持续发展。

该项目意在激发学生的创造力和思辨力，培养学生尊重古迹、顺应自然、保护自然的生态文明理念。同时，建立以学校为主体、以家庭为支柱、以社区为依托的多功能育人网络，促进学校、家庭和社区三方的联系和交流。该项目还将努力带动社区、学校发展，普及长江口可持续发展知识，践行可持续发展行动。

项目实施

一、初寻长江口二号古船

2024年1月20日至2月25日，学生们积极响应"寻长江口二号古船，扬长江口生命

---

① 国家海洋信息中心. 中国气候变化海洋蓝皮书(2021)[M]. 北京:科学出版社,2022:前言.
② 王思丹. 青年参与全球气候变化治理的路径和特点[J]. 中国青年研究,2023(6):24-32.

力"项目的号召,于寒假,通过查阅相关资料、绘制古船图画以及撰写关于古船的文章等,深入了解长江口二号古船的历史背景和文化价值,感受长江口独特的生命力。

首先,查阅资料,了解信息。学生们充分利用图书馆、互联网等资源,搜集关于长江口二号古船的历史背景、建造工艺、历史地位等方面的信息。通过阅读和研究,学生对长江口二号古船有了更深入的了解,对其在历史上的重要地位和影响有了更清晰的认识。

其次,画古船,知雄伟。学生们发挥想象力和绘画技能,通过画笔描绘了长江口二号古船的形象。这些画作不仅展现了古船的雄伟壮观,也体现了学生对古船文化的热爱和传承。在绘画过程中,学生们进一步加深了对古船结构和细节的了解,增强了对传统文化的感知和认同。

再次,写古船,扬情感。学生们撰写了关于长江口二号古船的文章,从不同角度探讨了古船的历史价值、文化意义以及对现代社会的启示和影响。通过写作,他们锻炼了自己的文字表达能力,深化了对古船文化的理解和思考。

最后,开班会,总结交流。2024年3月,班级组织了一次班会,对寒假期间的项目活动成果进行了总结和交流。学生们积极分享了自己查阅资料的心得、绘画作品和文章创作体验。通过这次班会,学生们不仅展示了自己的成果,也感受到了其他同学的知识和见解。

二、成立"潼绘未来"气候变化小分队

在学生们对长江口二号古船的各方面有了初步认识后,"潼绘未来"气候变化小分队正式成立。气候变化小分队名称"潼绘未来"寓意队员们将心往一处使,致力于探索气候变化对长江口二号古船的影响,创造长江口可持续的未来。

如何寻求气候变化对于长江口二号古船的影响呢?经过学生讨论、教师总结,"潼绘未来"气候变化小分队决定从"沉船奥秘我探寻""文物打捞我守护"和"古船介绍我能行"三方面进行调查研究,并以表1所示过程性评价为考量标准。

表1 "潼绘未来"气候变化小分队过程性评价

评价维度	评价内容	自评	组评	师评
个人	参与小组合作,表现出积极参与,团队协作完成任务的态度			
	在小组内讨论交流,分析问题及解决问题			
	思考与倾听,表达清晰、精简有逻辑			
	从全球气候变化的角度,发表自己的看法			
小组	设计出完整的汇报、倡议书			
	发挥团队力量,科学、严谨地完成气候变化系列活动			
	有团队意识			
自我建议:				
组内建议:				
教师建议:				
注:自评、组评、师评均以 A(优秀)、B(良好)、C(一般)、D(较差)评定				

(一)沉船奥秘我探寻

"潼绘未来"气候变化小分队分为两组,共同寻找长江口二号古船沉船的奥秘。由于目前考古界对于长江口二号古船沉船原因尚未形成统一说法,对于小分队来说,就需要通过查阅古籍、上网寻找资料、观看纪录片等方式,来合理推测沉船的奥秘。

在已知的官方报道中,长江口二号古船是一艘清朝同治年间的贸易商船,也是中国水下考古发现的体量最大、保存最为完整、船载文物数量巨大的木质帆船。

小分队第一组猜想:沉船可能是受台风天气影响。他们在华东师范大学官方网站找到了一篇《长江口二号古船因何沉没?华东师大河口海岸地质团队解密》的报道,其中有上海地方志记录同治时期风暴事件,从表2中可以发现同治三年至同治七年间,上海有"飓风""海啸"和"大风雨"的极端天气。虽然目前没有直接证据指向是因极端天气沉船,但上海地方志的存在,让这一推测变得有可能。①

表2 上海地方志记录

1864年	同治三年	六月初十	嘉定、川沙、徐汇	飓风大作,拔树倒屋,瓦飞如燕,徐家汇公学被吹倒,川沙南北门吊桥飞入城壕,泊舟有覆溺
1866年	同治五年	八月初八	川沙	海啸,二三时许始息
1867年	同治六年	秋		飓风,咸雨
1868年	同治七年	五月二十九		大风雨,拔木倒屋,田庐积水数尺

小分队第二组猜想:沉船可能和木质沙船的特点有关。木质沙船虽然在近海航行方面性能优越、载重量大,但是受水面积大、破船能力差,遇到突发情况,可能来不及应变。长江口二号古船的沉船原因,官方没有给出定论。对于学生而言,答案不是重点,重在思考过程。猜想不是一味地凭空想象,而是摆事实讲道理,通过查阅相关资料,做出合理的推测。

新泽西州三年级跨学科教学案例提到学生合作设计解决极端天气事件(如洪水、热浪、森林火灾、干旱和/或飓风)下现实世界问题的方案。学生先选择与当地环境相关的天气事件,从多个角度研究问题,确定个人和技术在降低天气相关风险方面所起的作用。然后,设计解决方案来应对气候变化带来的当地天气威胁,并向相关人员提出建议,以促进现实世界的变化。受此启发,教师引导孩子们利用课余时间把古时和现今应对极端天气的做法进行对比,以小组形式进行汇报。

---

① 河口海岸学国家重点实验室.长江口二号古船因何沉没?华东师大河口海岸地质团队解密[EB/OL].2021[2024-05-25].http://english.sklec.ecnu.edu.cn/node/7246.

学生汇报小结

- 在古代,沙船航行在广阔的大海上,面对狂风暴雨、巨浪滔天等极端天气,船长和船员们往往只能依靠经验和智慧来应对。他们通过观察天象、海象,以及感知风向、水流等自然规律,来预测和规避可能发生的危险。在船体设计上,古人们也充分考虑了抵抗极端天气的性能。沙船通常采用坚固耐用的木材建造,船身厚重,底部扁平,以适应在大风浪中稳定航行。此外,船上还会配备一些应急设备,如救生艇、救生筏等,以备不时之需。
- 然而,尽管古人们在应对极端天气方面展现出了不俗的智慧和勇气,但与现代科技相比,其手段和效果仍显得相对简陋和有限。
- 在现代社会中,人类拥有先进的天气预报系统、卫星导航技术、船舶自动化控制系统等一系列高科技手段,使得应对极端天气的能力和精度得到了极大提升。
- 现代船舶在设计和建造过程中,不仅注重船体的稳定性和耐用性,还充分考虑了环保、节能等因素。同时,现代船舶还配备了各种先进的导航和通信设备,使得航行更加安全、高效。此外,在应对极端天气方面,现代船舶还配备了各种先进的应急设备和技术,如自动救生艇、自动灭火系统等,大大提高了船舶在遇到危险时的自救能力和生存率。

学生们的汇报精彩纷呈,制作的幻灯片内容丰富、生动有趣(见图1)。教师在学生展示的基础上总结道:古时与现在应对极端天气对比,展示了人类科技与社会的巨大进步,也提醒大家要珍惜现代科技带来的便利和安全保障;同时,也应该认识到,在面对自然灾害等极端情况时,人类的智慧和勇气仍然是最宝贵的财富。

图1 小组代表对古时和当今应对极端天气做法进行汇报

(二)文物打捞我守护

随着全球气候变暖,长江口地区的海平面逐渐上升,海水温度、盐度等环境因素也在发生变化,这些变化给古船文物的打捞和保存工作带来了挑战。学生们开始思考气候变化对文物打捞的影响。

首先,海平面的上升直接威胁到长江口二号古船的安全。古船遗址位于水下深处,海平面的上升可能导致遗址周围的海底地形发生变化,进而对古船的结构稳定性产生影响。此外,海水的侵蚀作用也会因海平面的上升而加剧,对古船的木质结构造成更严重的损害。

其次,气候变化引起的海水温度和盐度变化,对古船文物的保存状况产生重要影响。水温的升高会加速古船遗骸的腐烂过程,而盐度的变化则可能导致船体材料的腐蚀加剧。这些因素都会使古船文物的保存状况进一步恶化,增加打捞工作的难度和风险。

为了应对这些挑战,老师和同学们思考了一系列措施来保护长江口二号古船文物。一是加强对古船遗址的监测和研究,及时掌握遗址的保存状况和变化趋势。二是制订合理的打捞方案,采用先进的打捞技术和设备,确保在保护文物的前提下进行打捞工作。同时,加强对气候变化的研究和应对,采取措施减缓海平面上升和海水环境变化对古船文物的影响。

总之,气候变化对长江口二号古船文物打捞的影响不容忽视。学生们意识到,在保护文物的同时,需要积极应对气候变化带来的挑战,为保护长江口地区的历史文化遗产贡献力量。

成效与反思

本项目中,学生们在老师的指导下,齐心协力完成了"潼绘未来"气候变化倡议书和"寻长江口二号古船,扬长江口生命力"宣传册。

(一)"潼绘未来"气候变化倡议书

气候变化教育是个长久、持续的过程,"潼绘未来"气候变化小分队撰写"潼绘未来"气候变化倡议书,呼吁学生、家长乃至社会共同关注长江口二号古船和长江口的环境保护,为保护历史文化遗产和维护生态平衡贡献一份力量,共同采取以下行动:

(1)加强长江口水域生态环境的监测和评估,掌握气候变化对长江口生态环境的影响,为古船的保护和考古发掘提供科学依据。

(2)推动长江口生态环境保护政策的制定和实施,加大生态环境保护力度,减少人类活动对长江口生态环境的破坏。

(3)加强长江口二号古船的保护和考古发掘工作,采用科学的方法和技术手段,保护好古船及其周边文物,为后人留下宝贵的历史文化遗产。

(4)加强公众宣传和教育,提高公众对气候变化和文物保护的认识和意识,形成全社会共同关注长江口二号古船和长江口环境保护的良好氛围。

这一行动不仅体现了学生们对环境保护的深刻认识,也展示了他们以实际行动参与社会、服务社会的责任感。

(二)"寻长江口二号古船,扬长江口生命力"宣传册

"寻长江口二号古船,扬长江口生命力"宣传册由学生提供素材,教师进行汇编整理。该宣传册力求多维度、多角度宣传长江口二号古船,特设"气候变化"专栏,包含"古时和现今应对极端天气对比"和"思考气候变化对文物打捞的影响",让学生、家长以及社会人员能看到"潼绘未来"气候变化小分队做出的努力。

学生们通过查阅资料、绘画、撰写相关文章等多种方式,全面了解了长江口二号古船的历史背景、文化价值及其在长江口地区的独特地位,交流汇总后设计了宣传册,不仅锻炼了资料搜集、绘画和文字表达能力,更在亲身实践中感受到了中华文化的博大精深,增强了团队合作和文化自信(见图2)。以"潼绘未来"气候变化小分队过程性评价为评价标准,使孩子们不盲目地为完成任务才进行各项活动,而是在活动中反思,在活动中成长。

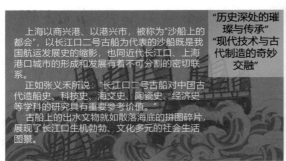 

图 2 "寻长江口二号古船,扬长江口生命力"宣传册

长江口二号古船项目不仅是一次跨学科育人实践,更是一次对学生全面发展和社会责任感的培养。学校和家庭的共同支持,为项目的顺利进行提供了有力保障。虽然家校之间并没有直接的交流,但从孩子们的热情反馈中,我们感受到家长对老师的肯定和对项目的重视。这种默契和配合,为孩子们的成长创造了更加和谐的环境。

当然,这一项目也面临了诸多挑战与亟待提升之处。其一,气候变化教育专业能力需提升。在给学生渗透气候变化教育的过程中,教师发现自身专业能力有待提升,有必要多积累、多补充相关方面的前沿知识。其二,宣传辐射范围要推广。可以利用学校微信公众号、短视频等传播手段,创新宣传形式,让更多的人知晓气候变化教育,了解长江口二号古船的历史价值和文化内涵。其三,拓展公共资源要联动。联动整合校家社的力量,共同深化走实。后续,我们将与地处黄河口的一些学校联动开展系列活动,深入推进长江口黄河口协同可持续发展。

(郭友纯,上海市浦东新区潼港小学教师)

长江口重工业工厂的过去、现在与未来

研究背景

科学研究认为,人类活动,特别是工业革命以来人类活动是造成目前以全球变暖为主要特征的气候变化的主要原因。[①]在当前全球气候变化日益严峻的背景下,工业发展对环境的影响成了不可回避的问题。上海市浦东新区高行镇东沟小学位于上海市浦东北路附近,周边重工业工厂林立,这些工厂在20世纪80年代发展迅速,为区域经济发展作出了巨大贡献。与此同时,对当地生态、气候变化产生了一定影响。

为深入了解工业发展与气候变化之间的关系,学校启动了相关项目化学习,让学生深刻认识工业发展对气候变化的影响。

项目设计

本项目立足学校所在区域的重工业产业状况,以三年级学生为参与主体,通过寻访、调查研究、收集资料、实地参观工厂等实践活动,深入了解学校周边工厂的发展历程、环境污染问题以及绿色转型的现状,探析工业发展对气候变化产生的影响。

项目实施

一、入项阶段:分解驱动问题

2023年12月25日,教师课前向学生们提问:"你们知道学校周围有哪些重工业工厂吗?"围绕学生的回答,继续抛出问题引发学生思考,如:这些工厂过去对人们产生了哪些影响?这些工厂的现状是怎样的?未来我们该怎么改进,减少工业发展对气候变化的影响?

在互动中,学生尝试分析环境污染的成因和对气候变化产生的影响。借此机会,教师提出项目的驱动性问题——"我们可以设计出什么样的'恒温'绿色工厂,减少工业发展对气候变化带来的影响?"以此激发学生思考。

二、探究阶段:形成实施计划

本阶段,学生围绕关键问题开展研究。由于学生尚处三年级,需要补充许多未掌握的知识。为更好地支持小组开展学习,教师以"参与式"活动形式引导学生亲身体验解决气候变

---

① 2008年2月,来自伦敦地质学会地层委员会的扬·扎拉谢维奇等21位成员联名在《今日美国地质学会》上发表论文认为,人类活动——特别是工业革命以来的人类经济活动对气候和环境造成了全球尺度的影响。Zalasiewicz, J., Williams, M., Smith, A., et al. Are we now living in the Anthropocene? [J] GSA Today, 2008, 18(2): 4–8.

化实际问题。

首先,学生根据各自的研究兴趣完成项目分组。然后,教师组织学生确定调查范围,选择学校周围有代表性的工厂为调查对象。2024年1月寒假前,根据学校大队部布置的"寒假集福活动"子活动——"探究福"的要求,学生实地走访,与工厂老员工进行深入交流,收集大量的第一手资料,了解学校周边工厂的历史沿革、生产状况以及环境影响。2024年3月1日,学生在课堂上以小组为单位分享过去一段时间观察收集到的资料。

有小组调查发现,学校周围曾经有工厂,总是有一股难闻的气味,让经过的人难以呼吸。他们猜测可能跟工厂原料有关。有小组通过询问家中长辈得知,工厂周边的河道曾经又黑又脏,并且烟囱里不间断地冒烟,经过工厂时,感觉温度比较高。

通过分享,学生们直观地感受到了工厂过去的辉煌与变迁。为了使之后的研究更加聚焦,教师引导学生们思考气候变化视角下重工业发展带来了什么问题、现阶段需要解决什么问题、未来可能遇到什么问题。各组交流讨论下一步要研究的关键问题。

三、行动阶段:完善知识建构

本阶段通过"体验式"活动的方式,完善学生对城市发展、工业发展、气候变化的知识建构。

首先,直观了解浦东的开发开放历程。2024年3月22日,教师组织学生参观浦东展览馆,沉浸式感受浦东开发开放30年来波澜壮阔的历程。学生们从中了解到学校周围分布的重工业工厂曾经为区域经济发展作出了巨大贡献,深刻意识到作为新时代的接班人,不仅要继承前人的智慧和经验,更要勇于探索、敢于创新,为未来的发展贡献自己的力量。

其次,学校积极与周边工厂企业等合作伙伴进行沟通和交流,通过"播下创新种子,点亮心中奇梦"的活动,共同推动项目。以中国石化上海石油化工研究院为例,2024年4月2日学生通过实地走访(见图1),深入了解了现在石化工业发展的实际情况和改进与创新,深入了解了研究院的发展历程,从最初为解决人民穿衣问题,到如今聚焦国家重大需求和产业需要,研究成果在多个领域绽放光彩。深入浅出的讲解激发了学生对石油化工的浓厚兴趣,也为学生设计未来城市"恒温"绿色工厂提供了贴近实际的建议。

植树造林是减缓气候变化的重要举措之一,为积极响应"绿水青山就是金山银山"的号召,2024年3月12日,学生和老师共同参加了高行镇浦江学堂组织的"情系生态环境,共建美丽高行"林长制活动(见图2),为爱绿护绿贡献自己的力量。学生了解到植树造林,增加森林覆盖率,可以有效地减少温室气体排放、调节气温等,有助于减缓全球气候变暖的趋势,为人类的生存和发展创造更加宜居的环境。

在本阶段,各小组在实践中不断建构自己对于关键问题的理解,认识到工业发展是气候变化的重要因素之一,但可以通过一些行动来保护环境,减缓气候变化的影响。

图 1　学生参观上海石油化工研究院

图 2　学生参与林长制活动

四、设计阶段:形成项目作品

为了在下一阶段更好地展示项目作品,2024 年 4 月 15 日,各小组根据前期的学习所得,制订了未来的城市"恒温"绿色工厂设计方案。小组成员以绘画、作文等方式呈现设计作品(见图 3)。通过小组讨论,学生们得出了未来绿色工厂的系列设计要求,如注重技术创新和绿色生产方式的应用;通过采用清洁能源等措施,实现工业生产与环境保护的协调发展等。有的学生还考虑了如何使工厂与周边环境融合的问题,提出了建设生态友好型工厂的构想。

图 3　学生绘画

五、评价总结阶段：展示研究成果

本项目的最终成果由个人成果和团队成果组成。团队成果是学生以小组团队形式制订未来的城市"恒温"绿色工厂设计方案。个人成果是学生画出未来工厂的设计图、撰写想象未来工厂的作文。

本项目以过程性评价与终结性评价相结合的方式展开（见表1、表2），以过程性评价为主，以终结性评价为辅，关注学生的组间评价和组内评价。每一个阶段性任务，都有相应的成功标准。

表 1　项目活动评价指标

评价指标	评价描述
团队合作	理解自己和合作伙伴的角色，基于成员擅长的内容领域，制定合适的参与规则，妥善处理交流沟通中的问题
沟通分享	通过合适的方式与同伴或者其他人交流想法，表达形式丰富、观点正确、内容新颖、个性鲜明
创意表达	对已有观点资料从不同的视角进行分析和论证，形成不同以往的策略、技术方法或问题解决方案

(续表)

评价指标	评价描述
信息收集	合理地对信息进行分析、加工和处理,从中提炼出自己的观点,为说明观点提供证据支撑
问题解决	理解所要解决问题的约束条件,确立目标并采取合适的行动完成任务,持续监控、优化或迭代成果

表2 项目活动团队合作表现评价

评价量规			组外成员评价	组内成员评价	自我评价
优秀	一般	需努力			
主动参与团队活动,经常发言提供建议	能在他人鼓励下参与团队活动,较少发言	基本不参与团队活动,不发言			
积极参与项目实践活动,收集信息并提出个人建议	参与部分项目实践活动,收集信息但不完整	未参与各类信息收集或项目实践活动			
参与制作项目成果,完成自己的任务并协助同伴	参与制作项目成果,能完成自己的任务,但较少帮助他人	不主动或未参与制作项目成果,没有完成自己的任务			

成效与反思

本项目通过深入探究学校所在区域重工业产业与气候变化之间的紧密关联,促进了学校教育的深化,使学生、家长、社会更加关注并理解气候变化的紧迫性。

一、学生树立了气候变化教育理念,增强了环保意识

通过项目化学习和实践活动,学生们深入了解了工业发展和气候变化之间的联系、影响及应对策略,形成了扎实的气候变化知识基础。学生对环保问题的认识从表面走向深入,不仅知其然,更知其所以然。

各种环保活动,如植树造林、垃圾分类、节能减排、环保宣传等,让学生在实践中体验到保护环境的重要性,从而自觉增强了环保意识。学生们开始关注自身行为对气候的影响,并主动采取环保措施,减少碳排放,实现低碳生活。

学生们意识到自己是地球村的一员,积极参与社区爱绿护绿活动,向更多的人宣传绿色理念,扩大了教育的影响力。随着项目的深入,学生们也逐渐养成了良好的环保行为习惯,如节约用水、用电,减少一次性用品的使用,增加循环利用等。这些习惯的养成不仅有利于个人减少对气候环境的影响,也对周围的人产生了积极影响。

以下是学生们对参与本次项目活动的感言:

小李：当我看到工厂烟囱中冒出的滚滚黑烟，心里总是感到一丝沉重。这些排放的废气正是导致全球气候变暖的元凶之一。如果没有改善，人类未来的生存环境将难以想象。

小王：我的老家是一个工业化程度较高的城市，那里的天空常常笼罩在雾霾之中。我深知工业污染对气候的影响，它不仅仅改变了我们的空气质量，更对整个地球的生态系统造成了严重破坏。

小刘：工业发展对气候变化的影响是全方位的。它不仅改变了大气成分，导致温室效应加剧；还改变了地表形态，加剧了城市热岛效应。这些变化都对我们的生活产生了深远的影响。因此，我们必须正视这个问题，采取有效措施来应对气候变化。

工业发展对气候变化的影响深远且广泛，本项目通过持续的教育和引导，希望学生在未来的学习和生活中都能保持对气候变化问题的关注和参与。

二、校家社联动，提升了育人实效

本项目获得了家长的深度参与、学校的积极引导及社区的广泛支持。

家长通过直接或者间接参与项目活动，对工业环境对气候变化的影响有了更深刻的认识，有些家长就在学校周围的工厂上班，他们的讲述和指导，对孩子的学习和探究形成强大的家庭支持。

学校不仅在课堂上开展工业发展与气候变化教育，还通过组织实地考察、讲座、参观等形式，激发学生对环保的兴趣和热情。

社区为项目活动提供场地、资源等支持，同时也通过项目活动，向居民普及环保知识，提高整个社区的环保意识。

以下是一些家长关于工业发展对气候变化影响的活动感言：

王爸爸：工业发展是时代进步的必然，但我们不能忽视它对气候变化的负面影响。作为家长，我时刻关注着孩子的健康和未来。我希望政府能够加强监管，推动绿色工业的发展，让我们的孩子呼吸到清新的空气，享受到蓝天白云。

张奶奶：我记得小时候的天空是那么蓝，水是那么清。作为家长，我希望教育我的孙子孙女们珍惜环境，保护我们的地球。同时，我也希望社会更加重视环保教育，让更多的人意识到工业发展对气候变化的影响。

刘爸爸：工业发展是现代社会不可或缺的一部分，但我们不能让它成为破坏环境的元凶。作为家长，我会引导孩子关注气候变化的问题，让他们了解工业发展对环境的影响。同时，我也会以身作则，减少浪费和污染，为环保尽一份力。

通过学校家庭社会联动、资源共享和协同育人，项目活动形成了良好的氛围，为应对气候变化、保护地球家园贡献了力量。

当然，在此过程中，也遇到了诸多阻碍与挑战。譬如，随着项目的深入推进，学生对工业发展的了解不够深入，由于一些工厂的特殊性，实地考察的难度较大，需要在后续的活动和实践中进一步加强。学生的项目成果创意度不够，设计比较单一。关于项目结合气候变化的教育过程也确实面临不少困难。由于气候问题本身的复杂性和抽象性，对于非专业的小学教师来说可能会感到吃力,这要求教育者需要具备相关的专业知识,并能用简单易懂的方式传达给学生。

针对这些挑战,未来将采取以下措施:一是加强与相关部门的合作,争取更多的支持和资源;二是加强与企业的合作与交流,完善并推动绿色工厂的设计;三是采取多种措施来推广气候变化教育。例如,加强教师的培训和专业发展,提高教学能力和综合素质;开发跨学科的教学资源和课程,让学生从不同角度了解气候变化问题。

从气候教育的视角来看,本项目不仅是对工业发展与气候变化关系的一次深入探究,更是学生进行气候教育的生动实践。让我们反思如何在教育中更好地融入气候教育的内容。可以结合当地的工业发展情况,设计更加贴近实际的气候教育活动;也可以利用现代化的教学手段和技术,如虚拟现实等,让学生更加直观地了解气候变化的影响。

综上所述,本项目不仅是对工业发展与环境保护关系的一次有益探索,更是学生进行气候教育的重要实践。项目的实施和反思,更加坚定了我们推动工业绿色发展和加强气候教育的决心和信心。我们将继续深化这一领域的研究和实践工作,为推动气候教育做出更大的贡献。

(蔡燕,上海市浦东新区高行镇东沟小学教师)

"绿色万里路"，让减排被看见

研究背景

全球气候变化已成为当今社会人们普遍关注的话题，气候变化导致极端天气普遍发生，对人类社会带来了诸多不利影响，如何应对气候变化更是人类科学研究的课题之一。联合国教科文组织将气候变化教育界定为"面对风险、不确定性和快速变化的学习"，目的是建立对气候变化以及全球变暖对生物多样性的影响的理解和应对能力。[①]

长期以来，中国高度重视气候变化问题，坚定走好生态优先、绿色低碳的高质量发展道路，将碳达峰、碳中和纳入经济社会发展全局，以降碳为生态文明建设的重点战略方向，努力建设人与自然和谐共生的现代化。[②]当前，面向青少年开展"天气变化教育"具有重要意义。作为未来社会的主人，青少年应当努力成为应对气候变化和环境保护的积极参与者，树立应对气候变化和环境保护的观念，培养环保意识，提升社会责任感，积极采取行动以有效应对气候变化。[③]

项目设计

笔者作为小学四年级的班主任，在常州市特级班主任的引领下，依托华东师范大学上海终身教育研究院的专业支持，借助研究团队的力量，指导班级多名学生利用寒假组建"绿色万里路"玩伴团，进行有组织、有计划的项目化研究实践。

学生参加或了解过一些节能减排的活动，但往往停留在宣传、行动的层面，并没有形成减碳数据，参与者并不知道自己的行为到底能对节能减排产生多大的影响。因此，学生通过讨论研究，形成了班本化的四（4）班"气候变化"项目化学习活动计划（见表1）。学生自发组建主题为"绿色万里路"的项目化学习研究玩伴团。

表1 四（4）班"气候变化"项目化学习活动计划

活动时间	活动方式	学习内容	成果展示
2024年1月6—7日	信息检索	气候、气候变化	主题小报、思维导图

[①] 唐科莉，周红霞，李震英，等．气候变化教育的全球发展趋势[N]．中国教师报，2023-02-15．

[②] 李卉，李怡芸，王华，等．"双碳"背景下中国青少年应对气候变化教育模式探索：基于江苏省气候变化教育示范项目[J]．环境教育，2022(10)：48-51．

[③] 唐莺．家校社协同开展项目化学习的有效路径：以"气候变化教育"项目为例[J]．现代教学，2023(22)：70-73．

(续表)

活动时间	活动方式	学习内容	成果展示
2024年1月13—21日	信息检索	了解全球气候变化的现状及影响	主题小报、思维导图、班级交流
2024年1月27—29日	信息检索	全球气候变暖的原因及温室气体的成分	主题小报、思维导图、班级交流
2024年1月30—2月18日	绿色出行	学习在路上	出行路线截图、标注、上传班级群相册
2024年2月19—2月25日	数据处理	出行方式的分类及里程换算	数据统计表、统计图
2024年2月29日	期初展销、课堂展示	成果学习	成果汇报、心得分享

项目实施

一、资料查找真学习

学生通过查阅网络和气候变化类科普书籍，逐步了解气候及气候变化的含义，知道全球气候变暖主要是由人类活动造成的。全球变暖会使全球降水量重新分配、冰川和冻土消融、海平面上升等，既危害自然生态系统的平衡，更威胁人类的食物供应和居住环境。

根据这些信息，同学们绘制气候变化小报，研究气象与气候的关系以及古今气象对比，感受气候变暖对物候的影响。随后，学生与家长、老师一起参观了位于学校附近的光伏产业园——天合光能科普馆，了解光伏发电板用于建造房屋、充储电站及制造电动汽车等应用场景。

学生通过研学发现，燃油汽车尾气含有大量二氧化碳，每燃烧1升汽油大约产生2.4千克二氧化碳，而汽车在城市道路行驶的油耗量比在高速或乡村要高得多，那么城市中大量燃油车的使用必然导致温室效应的加剧。因此同学们自发组建了"绿色万里路"玩伴团，打算从自己做起、从减少燃油私家车的使用做起，为减少二氧化碳排放做实事。

二、绿色出行真需求

在以往"绿色出行"活动中，活动人员多采用额外拟定的非需求性行动路线，无任何实际的"减排"作用，反而是增加了"碳排"。考虑到这一点，大家决定以真实的出行需求为研究抓手，实实在在地用自己和家人的行动来"减排"。

大家每一次绿色出行都以截图的方式将路线上传班级群相册，并标注好出行方式和里程。寒假结束后，玩伴团成员(见图1)共同统计汇总了所有原始数据，得到全团绿色出行统计表。学生通过分类统计，发现共发生绿色出行里程4 000余公里，"0碳排"的步行与自行车出行占2%、私家电动车(包括电动自行车和电动汽车)出行占29%、公共交通(包括公交车、地铁、高铁等)出行占69%。

图 1　绿色出行玩伴团

学生通过查找资料，按照常规私家车油耗量 7 升/100 千米计算，减少汽油使用量 295 升。按照每升汽油燃烧大约产生 2.4 千克二氧化碳计算，本次"绿色万里路"玩伴团共减少排放汽油燃烧产生的二氧化碳 709 千克。这可真是个振奋人心的结果。

三、多级展示"涨新粉"

2024 年 2 月，依托"你好，寒假！"期初展销会，班级将本次研究成果向全校进行分享。当其他同学看到 709 千克的二氧化碳减排成果时，都很惊奇，纷纷感叹"从没想到绿色出行真的可以减排如此大量的二氧化碳"，并纷纷表示要加入进来，为应对气候变化出一份力。

常州电视台也对这次活动进行了班级采访，将"班级经验"向全市观众进行推广，将研究成果向公众进行宣传，让更多伙伴想要加入这支"绿色小队"。

3 月，孩子们将自己的发现分享给全班同学的家长，并展开激烈讨论，完成了本项目校家社合作的闭环（见图 2、图 3）。

图 2　同学向全班分享自己的心得　　图 3　家长与孩子们展开激烈讨论

4月,这支"绿色小队"还受邀参与"上海气候周",在"青少年气候行动科创作品宣讲会"上分享他们的行动经验、展示研究成果(见图4)。笔者作为班主任也参与了"气候韧性社区论坛"的圆桌会议,分享了校家社在气候变化教育中的合力效应。

图4　青少年气候行动科创作品宣讲会

四、真体验才有真感受

活动结束后,孩子们纷纷发表他们的真实感受。他们表示这样的项目活动很有意义,在活动中不仅提升了综合能力、增强了社会责任感,还为应对气候变化贡献了自己的力量。

王同学说:"这个寒假,我们班组织了特别有意义的活动——绿色出行。我是从骑自行车、步行和坐公交车做起的。这次活动让我对绿色出行的意义有了更深刻的理解,同时我也意识到绿色出行需要大家的共同参与与努力,只有大家齐心协力,才能让我们的城市更加美丽宜居。"

孙同学表示,随着假期的开始,她参与了"绿色出行,从我做起"的班级活动。首先,绿色出行是一种相对环保的出行方式,既能节约能源、提高能效、减少污染,又有益健康、兼顾效率。同时,绿色出行也是一种具有象征意义的理念、一种可持续的环保观念,所以,选择绿色出行既是选择了一种健康的出行方式,也是选择了一种积极的生活态度。

张同学认为:"步行是一种低碳出行方式,既保护了环境,又锻炼了身体,还可以欣赏沿途的风景。未来要从我做起,从身边做起,低碳出行,为保护环境尽自己的一份力。"

郭同学也认为绿色出行,既不堵别人,也不被别人堵。"当堵车的人们望着望不到头的车时,我骑着心爱的小自行车,优哉游哉穿行而过,那滋味别提多美了!既锻炼了自己也为减少污染贡献了绵薄之力。"

黄同学也分享了他在寒假期间参与的绿色出行活动,他是从坐地铁、步行和坐公交车做起的。在不降低出行效率的情况下,既能节约能源、提高能效、减少污染,又有益于健康。他呼吁大家从我做起，从现在做起，将绿色出行作为出行的首要选择，为环保做出一点贡献。

刘同学表示，她在寒假期间参与了绿色出行活动，深刻体会到了环保的重要性。她选择公共交通，低碳出行，减少碳排放，为地球减负。沿途的风景让她更加珍惜自然之美。这次活动，让她明白了每个人的小小努力，都能为环境带来积极变化。

还有许多同学表示很高兴能为保护环境出一份力，将继续自觉遵守"135"出行方案，即1公里内步行，3公里内骑自行车，5公里内乘公交的出行方式，尽可能采取步行、骑自行车、乘公交车的出行方式，为低碳出行做表率。

家长们也认为，寒假绿色出行让他们体验到了环保的重要性（见图5）。每一次选择绿色出行，都是对地球的一份贡献。它不仅仅是一个口号，更是一种生活方式，一种需要我们每个人去实践、去坚持的生活方式。骑行、步行、乘坐公共交通出行，不仅能减少环境污染，还有助于身体健康。他们希望更多人选择绿色出行，用实际行动来捍卫地球的环境！

图5　家长分享活动感受和对活动的支持

成效与反思

解决问题的最佳方式是开始行动。不仅要知道为什么行动，如何行动，还要真实、科学地评估行动的效果和意义。用数据将真相展示在大家面前，提高行动效果的说服力。

绿色出行，是一种出行方式，也是一件小事。但大家一起做，就未必是一件小事了。它避免了全球变暖，冰山融化，避免了许多无辜的生命的灭绝。

本次班级微项目的推进，以点带面，从一个小队到一个班的卷入，最后向全校、社区、全市、全国进行经验分享和成果展示。从影响力来说无疑是一次较为成功的项目化活动。但依然存在以下不足：

一是学生的绿色出行是否属于"必要出行"，是否存在出行数据是为了活动而产生的情况。

二是学生的出行方式中有很多是电动汽车或电动自行车,那么燃烧煤炭发电的过程中会产生大量温室气体,未能体现在换算中,这可能是下一步的研究方向。

三是本次项目活动仅限于减少碳排放,而没有关注增加碳汇,后续的活动中可能会向增加碳汇延伸。

(付蓉,江苏省常州市新北区龙虎塘实验小学教师)

中职学校实施气候变化教育的实践

研究背景

中等职业教育的重要使命是培养应用型人才。怎样让中职生意识到气候变化对人类经济社会发展、生命健康以及自然生态系统产生的严重影响，如何结合学生所学专业，在职业教育中开展气候变化教育综合活动，成为摆在所有中等职业学校面前的课题。

习近平总书记在东北考察调研期间明确指出"保护生态环境的意义是战略性的"，强调"生态就是资源，生态就是生产力"。[1] 2023年冬，哈尔滨市政府关注特殊环境对人们的吸引力，大力发展冰雪旅游产业。寒冷的气候特点是东北发展旅游经济的重要条件，暖冬的出现无疑会影响东北地区冬季旅游业的发展。提高青少年对气候的关注，树立环境保护观念，对于东北地区的可持续发展至关重要。如《柏林可持续发展教育宣言》所强调的，可持续发展教育能培养学习者的认知和非认知技能，如批判性思维，协作能力，解决问题的能力，应对复杂情况和风险的能力，坚韧性，系统性和创造性思维[2]。作为护理专业的学生，需要特别关注气候变化对人类健康的影响，从而将所学专业应用到实际生活中，树立崇高职业理想，服务人群，关爱生命。

2023年12月，哈尔滨市卫生学校以2023级护理10班为试点，开展气候变化教育综合活动。在两个月的时间里，学生们开展了调查、研究、家庭访谈、社区宣讲等一系列活动。问卷调查显示，该项目不仅使学生在自然科学知识上有所收获，在合作和沟通能力等方面也有显著提高，满意度达90%。

项目设计

教师根据中职生的性格、兴趣及年龄特点，将气候变化教育设计为以学生为主体的探究综合活动（见图1）。

该活动的总目标是：通过气候变化教育综合活动，促使护理专业的学生掌握一定的环保知识，树立正确的价值观和态度，通过学习气候变化对人类健康的影响，用自身专业知识和技能服务人群，树立职业理想。

具体目标为：学生在活动中了解东北地区气候变化现状，提升环保意识；学生在活动中

---

[1] 公方彬.习总书记东北调研三大意义 为地区发展"找路子 谋方法"[EB/OL].(2016-06-02)[2024-02-20]. http://cpc.people.com.cn/xuexi/n1/2016/0602/c385477-28405715.html.

[2] Berlin Declaration on Education for Sustainable Development[EB/OL].2022[2024-02-30.] https://unesdoc.unesco.org/ark:/48223/pf0000381228.

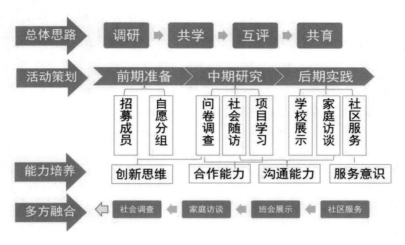

图 1　综合活动设计框架

掌握应对极端天气的方法,提高生存技能;学生在活动中探究气候特点与疾病的关系,提升专业知识水平。

项目实施

一、开展气候变化教育调研活动

2023 年 11 月 7—12 日,以 2023 级护理 10 班为试点,我们进行了活动招募和活动分组。基于气候变化教育要遵循贴近实际、贴近专业、贴近生活的原则,教师将气候变化教育综合活动的主题确定为"与气候相伴,用专业服务"。

2023 年 11 月 5 日,东北地区受大范围强雨雪天气影响,多地停课、停航。教师以学生刚刚经历的极端天气为气候变化教育的契机,向班级全体同学发出招募书。

气候变化教育综合活动招募书

亲爱的同学们:

　　我们刚刚经历了东北地区的大范围强雨雪天气,过去的两天,我们不得不停课在家,学习和生活受到极端天气的严重影响。连续两日的强降雨降雪导致路面结冰、积雪,车辆打滑,停运、停航,市民出行受阻。这样的气候变化是怎样形成的呢?气候变化又会对人类的生活产生哪些影响?作为一名在东北地区成长和学习的中职生,我们对自己家乡的气候了解吗?近些年家乡的气候发生了哪些变化?让我们一起加入"气候变化教育"的学习活动中,与上海、江苏、山东等地的伙伴们,共商"大事",为地球的未来,作出贡献!

　　"与气候相伴,用专业服务",同学们,快来加入气候变化教育综合活动吧!

哈尔滨市卫生学校
2023 级护理 10 班
2023 年 11 月 7 日

气候变化教育综合活动以尊重学生意愿,学生主动报名的方式为原则,只要学生有意向,都可以参加。综合实践活动的设计立足于促进学生的发展,重视学生主体参与意识和能力的培养,教师设计问卷调查、社会随访和项目研究三项活动,尽可能地让学生可以找到发挥自己特长的活动。学生根据自身的兴趣、性格和能力,自主选择综合活动组别,协商选出各组负责人。因项目研究参加人数多,故将其分为4个小组(见表1)。

表1 项目招募及分组情况

组名	问卷调查组	社会随访组	项目研究1组	项目研究2组	项目研究3组	项目研究4组
人数	7	5	5	5	5	5

二、开展气候变化教育探究活动

2023年11月13日—12月3日为本次气候变化教育探究活动的实施期。

第一步,我们开展了问卷调查和社会随访活动。首先,教师布置问卷调查小组活动内容,学生查阅资料找出关于气候变化的20个热点问题,然后大家集体讨论,确定10个与日常生活息息相关的问题。教师鼓励学生运用网络小程序制作调查问卷、发放调查问卷,并指导学生对调查结果进行分析,撰写调查报告。学生利用自身资源,积极转发问卷,以取得更多的可参考的调查数据。

第二步,教师布置社会随访小组活动内容。学生集体讨论确定5个气候变化热点问题(见表2)。采访开展前,他们找父母或者同伴演练,并确定访问对象男女比例、受访者的年龄段。访问时使用录音和录像设备,记录访问过程。访问结束后整理访谈记录。

表2 社会随访提纲

1.您感觉近些年东北气候有哪些变化?
2.您感觉气候变化给您的生活带来了哪些影响?
3.您认为影响气候变化的人为因素有哪些?
4.面对气候变化的一系列问题我们应该采取什么措施?
5.请您描述经历过的一个极端天气的场景,您是如何应对那次极端天气的?

第三步,教师与学生根据问卷调查结果和访谈实录,集体讨论,确定项目学习研究主题,分组讨论,设计项目研究框架(见表3)。教师重视探究学习方式的运用,教会学生检索文献、查阅资料的方法,培养学生的合作意识和团队精神。教师采用反馈互动的指导思路,在研究过程中设置两次阶段性组间学习汇报,开展组间互学,互评。每次在小组阶段学习汇报后,教师及时给予点评,提出参考建议,各小组不断丰富,深入学习研究内容。

表 3　项目研究框架

主题	研究框架	成果呈现方式
东北地区的气候	东北地区地理环境概述 东北地区的气候特点 东北地区气候变化规律 东北地区气候的未来发展与影响预测	运用文字、图片、视频等资料，制作演示文稿
东北地区极端天气	沙尘暴形成原因、危害及应对方法 暴雪形成原因、危害及应对方法 洪涝、干旱形成原因、危害及应对方法 高寒形成原因、危害及应对方法	
东北地区的气候变化趋势	东北地区近 50 年气温变化趋势 影响气候变化的因素 气候变化对生存环境的影响 保护气候环境的应对措施	
南北方的气候与健康	气候变化对人类健康直接和间接影响 适应气候变化，保护人类健康策略 南北方居民气候饮食差异与疾病 保持健康个人行动与生活方式调整	

三、开展气候变化教育宣传活动

2023 年 12 月 4 日至 12 月 17 日，综合活动成果在班级、学校展示。

第一，项目学习小组在气候变化教育主题班会课上与全班同学分享研究成果，开展共学、共进、评比活动。全部展示后，由班级同学投票选举其中 1 组代表班级向全校同学做观摩展示。班会结束后，鼓励学生们将所获得的气候变化自然知识、应对极端天气或保护气候环境的措施，制作成健康与环保海报。第二，延伸社会随访活动，号召全班同学参加家长访谈活动，在访谈中与家长互学气候知识。第三，在 2024 年 5 月 10 日，护士节即将到来之际，班级 10 名学生走进兆麟社区，用护理专业技能，为居民提供测量血压、血糖的志愿服务，在此过程中，宣讲气候环境保护与健康卫生知识，建议社区居民调整饮食结构和生活方式，保持身体健康。

气候变化家长访谈记录单

受访家长：＿＿＿＿＿＿＿＿
1. 您认为气候变化教育重要吗？为什么？

2. 您认为我们现在面临的主要气候问题是什么？我们该采取什么措施应对？

3. 请描述您年轻时遭遇的一次极端天气场景,并讲述当时是怎样应对的？

今天我是气候知识传递者：_____
应对气候变化,保护环境我们应该这样做：
1.
2.
3.
家长访谈时间：
家长访谈地点：

成效与反思

区域性气候变化教育综合活动的开展,有助于提升中职生对气候变化的关注,培养其协作能力、创造性思维,并积极采取负责任的行动,成为德才兼备、技能卓越的卫生专业人才。

本次综合活动自 2023 年 11 月开展到 2024 年 5 月,虽然时间不长,但研究成果丰富。2023 级护理 10 班共 45 人,参加综合活动 32 人,约占班级总人数的 70%,在项目展示评比过程中,未参加实践活动的 13 名学生与各小组组长成立评选委员会,投票选出杰出表现小组。

本活动达成全员主动参加,全员学习受教的目标。问卷调查组网络收回问卷 193 份,完成调查数据分析报告;社会随访组随机访问 10 人,完成录像、拍照,访谈记录的整理。以上两组数据,为项目学习组提供了选题依据。

项目研究 1 组学生通过学习地理知识,掌握东北地区所处的经纬度,地形地貌,气候特点及规律。

项目研究 2 组学生查询资料,对东北地区极端气候变化进行研究。历史资料表明,1949—2007 年间东北地区发生特大洪涝和特大干旱各 3 次。[1] 学生运用视频资料,教会同龄人应对洪涝、干旱、高寒、暴雪、沙尘暴的方法。

项目研究 3 组通过查阅文献对东北地区气候变化进行研究。赵宗慈等根据 23 个全球气候模式预估 21 世纪后期东北地区气温将明显变暖 3℃ 以上,降水可能增加。[2] 曾小凡等利用 ECHAM5/MPI-OM 模式得出 1980 年后松花江流域平均气温持续升高。[3]

项目研究 4 组学生通过研究气候变化对身体健康的影响得出高温天气影响睡眠质量,

---

[1] 汪金英.东北粮食主产区的旱涝灾害及对策分析[J].学术交流,2009(6):115-118.
[2] 赵宗慈,罗勇.21 世纪中国东北地区气候变化预估[J].气候与环境学报,2009,23(3):1-4.
[3] 曾小凡,李巧萍,苏布达,等.松花江流域气候变化及 ECHAM5 模式预估[J].气候变化研究进展,2009,5(4):215-219.

可能导致疲劳、焦虑和抑郁等问题。气候变暖将会加重过敏症状，推高皮肤癌等疾病的患病风险。南北方由于气候和饮食不同，导致心脑血管、高血压、糖尿病患病率不同。学生根据护理专业知识给出适应气候变化，保护人类健康的策略，提高公众对气候变化所造成的健康影响的认知度。此外，该小组回收气候变化家长访谈书 25 份，另有 10 名学生走进社区志愿服务，发放《气候变化与健康》宣传单 30 份。

在活动结束后，学生们表达了他们对气候教育综合活动的喜爱与肯定。

石同学说："这段时间，我们组用微信视频一起研究东北气候问题，研讨结束后意犹未尽，我们还一同复习课业内容，这次周测我还得了 100 分。"

李同学说："我终于成功地用小程序制作了调查问卷，做事情真的需要严谨，学好语文基础知识太重要了，感谢老师不断帮我校对问卷。"

曹同学说："我记得采访那天特别冷，刚开始对陌生人采访时，我遭到了好多次拒绝，但是我没有放弃，我相信坚持下来就一定会有完成任务。"

任同学说："第一次参加志愿服务，给社区居民测量血压、宣讲气候与健康知识时，我心里可骄傲了，今后我可以用我的专业知识和技能帮助更多的人。"

在实施气候变化探究综合活动后，我们也有一些反思：

一是综合活动的设计要符合学习者的学情，持续关注学习者的积极性，激发动力。在气候教育综合活动中，我们设计了不同类型的活动使学生发挥其强项，胜任并且出色地完成了学习任务。在项目学习中以分组的形式营造宽松、和谐、自由的学习环境，尽可能使每个学生的学习自主性都得到发挥，个性得到张扬。

二是确立学习者的主体地位，使其学会自主探究，得渔且用。教师建议学生前往图书馆、书店，通过百度文库、知网等，线上线下相结合查阅资料。学生在周末自觉分工，查阅资料，各组之间互相推荐可查阅资料的网站。学生围绕任务展开探究，形成知识的意义建构，并转化为可迁移的学习能力，通过项目学习，学生对校内课程的学习动力和学习效率得以大幅度提升，在期末考试中，参加项目学习的学生，合格率显著高于班级其他同学。

三是突破学科边界，多学科课程综合设计，将探求认知学习和社会性成长相结合。我们将解决问题贯穿于项目学习全过程，要求学生按照一系列任务框架持续学习，像学科专家一样思考与探索。东北地区气候变化未来趋势如何？气候变化会受哪些因素的影响？需要采取什么措施保护环境？结合护理专业知识，怎样帮助东北居民解决寒冷气候下的健康问题？研究性学习、实践性学习、社会化学习，促进学生将学到的多学科零碎知识转变为探究世界的综合力量，形成学习和生活的一致性和整体性理解。

四是在综合实践活动中提出具体要求，提升学生的综合能力。在规定期限内上交高质量的作业需要学生在小组内合理分工，这一要求极大地增强了学生的团队合作能力及沟通能力。为了更好地呈现研究结果，学生熟练掌握 PPT 制作方法、问卷调查小程序的应用、微视频的录制与剪辑，信息技术能力稳步提升，为今后更好地适应信息化社会奠定了基础。

五是对于东北地区气候变化的问卷反馈大部分来自黑龙江地区，教师和学生应适当地与东北地区其他中职学校沟通合作,形成气候变化教育共同体,增强调查问卷的信度和效度,使更多的中职生认识到保护环境的重要意义。在项目学习过程中,个别小组任务分工不明确,出现组长任务较重的现象,为避免此类情况,教师需要分组听取小组任务分工汇报,确保组内全员参与。此外,气候变化社会随访问题的深度和广度有待提高,下次可采取先广泛征集问题,再筛选确定实施的方式,提高学生的深度思考能力。

(陈爽,哈尔滨市卫生学校讲师)

访气象俗谚，探气候变迁

研究背景

《中国青年报》指出：许多老师在教学实践中发现，当与学生们谈及全球气候变化时，他们通常觉得这离自己非常遥远，而在学习气候变化时，学生也很难有深切的体会，这往往让老师有种无力感。所以他们认为气候变化教育要从娃娃抓起。[①] 浙江省武义县武阳中学深以为是。他们通过开展气象俗谚考证工作，推动"气候变化教育进校园"，提升青少年的气候变化意识、科学素养和人文情怀。实践证明，在初中生中开展气象俗谚调查，能够拉近学生与气候变化教育的距离。

气象谚语源于我国古代的农耕文化，是劳动人民在长期生产劳动实践中通过观察、总结和积累形成的。这些谚语世代相传，至今仍然闪耀着智慧的光芒，成为我们应对气候变化的宝贵财富。

武义县隶属浙江省金华市，位于浙江省中部，属亚热带季风气候，四季分明，是典型的山地城市。其地理呈"八山半水分半田"的基本格局，山水林田湖草资源要素齐全。武义早在新石器时代，就有人类居住，属万年"上山文化"圈的典型代表区域之一。武义方言有武义话、宣平话、永康话、畲族话、福建话、南金话等，武义地方方言的复杂性带来了民间俗语的丰富性与多样性。

"气候变化教育进校园"是时代赋予当代中小学基础教育的新课题与新任务。为了切实有效地提高学生气候变化认知水平与应对建设人类命运共同体的实践能力，我们组织初中生从他们身边大量生动鲜活的地方性气象俗谚入手开展调查与研究，以此为抓手提升同学们气候变化意识。

项目设计

气候是指一个地区大气多年的平均状况，主要反映的是该地区冷、暖、干、湿等特征。它是个长期稳定的天气现象，让学生直接感知气候变化比较困难。截至2024年，武义县气象站建站只有63年，有关天气信息的数据记录也只有63年，直接查阅这些数据，可以参考的时间跨度也较小，气候变化不明显。而气象谚语源于我国古代的农耕文化，是劳动人民在长期生产劳动实践中通过观察、总结和积累形成的，且世代相传，时间跨度大。将气象谚语与现代气象记录数据相对比，能发现气候变化的足迹。据此，我们设计让学生通过调查和考证

---

① 李瑞璇,邱晨辉.应对气候变化 教育要从娃娃抓起[N].中国青年报,2023-12-25(008).

气象谚语,认识气候变化的发生,从而拉近气候变化和他们的距离。

项目实施

"气候俗谚调查研究"是一项以学校为中心,家庭为单位,联动气象部门、科协共同完成的项目式学习过程,分为活动准备和活动实践开展两个部分。这一项目由武阳中学发起,在七年级同学中开展气象谚语调查考证项目化学习工作。武阳中学联合气象局、科协,让参与项目实践活动的学生在调查挖掘乡土气象俗谚文化资源的项目式探究学习活动过程中领会深刻的气候变化的相关气象知识,学习生存和生活技能(见表1)。

表1 "访气象俗谚,探气候变迁"项目实施进程及产出内容

第一部分:调查学生的元认知,培训学生调查技术		
一、调查学生对气候变化的元认知	调查学生及其家长对于当今世界气候危机、气象俗谚、武义本土气象俗谚的了解程度	67%的学生了解当今世界正在面临气候危机;51%的学生知道两条以上气象俗谚;94%的学生想了解武义本土气象俗谚;98%的学生想为全球变暖做有益的事,而家长对本活动的开展支持率达到100%
二、科普气象俗谚,增加他们的认知	科普气候变化特点和气象俗语,让学生知晓更多信息	老师在课堂中向同学们科普气象俗语
三、记录与标注气象俗语采集的过程信息	根据主体、时间、空间、事件、背景、心理、功能要素,标注气象俗语的应用场景	制作出一份气象俗语记录表格
四、发出倡议,做好宣传和策划	印制倡议书、编写活动策划书	在班级内发布并留有相关文稿
第二部分:活动正式实施过程		
一、依托"寒假·你好"活动,开展正式调查	2024年寒假,学校下发气象俗语调查表,鼓励运用文字记录、视频记录等方式,向长辈了解气象俗语,并发至微信群中	收集到本地气象俗谚99句
二、小组头脑风暴,确定要考证的气象俗谚	2024年2月22日,以武阳中学七(4)班的同学为主要参与成员,分成不同的小组,选出组长。组内商讨出需要考证的谚语,并前往气象局考证	"地上暖,天上孵雪卵" "霜降未降,廿天稳当" "立春雨,一春雨" "八月半,乱打扮"等
三、赴气象局查阅数据,考证谚语,撰写研究报告	2024年2月23日,各小组向气象部门工作人员考证俗语的可靠性,带队老师启发他们思考气象数据背后展现的规律	十个小组,十份调查报告(美篇呈现)
四、及时做好交流与反思	同学们反思调研过程中的收获与不足	老师、同学、家长的感想

（续表）

	第三部分：做好活动总结，分享活动感悟 从学生、家长和老师的角度挖掘三方应对气候变化意识、技能的提升
学生认为	活动增强了自己对于谚语文化、传统智慧的了解； 活动锻炼了自己的团队合作、人际交往能力； 活动锻炼了自己的策划、写作和探究能力
家长认为	能够和孩子们一起参与到活动过程中，感到非常幸运
老师认为	学生是气候变化教育的受教者，也是气候变化教育氛围的创造者

一、气象俗语调查工作的准备阶段

本阶段主要包括调查学生对于气象俗谚的元认知，培训学生掌握调研技术，发出倡议，做好宣传和策划工作等四个部分。项目组老师首先调查学生气象俗谚的知识储备，了解他们对于气象俗谚的元认知。而后，解说气候变化带来的挑战和气象俗谚知识，并借此机会扩充同学们的气象俗语知识储备。同时，为了更好地推进调查，项目组老师向同学们介绍记录与标注俗谚的方法。在准备工作完成后，项目组印制宣传册和活动倡议书，分发给班级同学，鼓励其参与。

二、气象俗谚调查工作正式展开

本阶段主要包括在2024年寒假期间下发气象俗谚调查表，由孩子们向长辈调查气象俗语，并做文字和视频记录。在寒假结束返校之后，班内同学分组，每组确定一或两句感兴趣的气象俗谚进行考证（见图1）。经学校与气象局对接，孩子们在气象局下班后前往值班工作人员处考证气象俗谚（见图2），同学们认真调查并撰写相关报告。在活动结束后，同学们反思调研中的不足之处。

图1　2024年2月22日，学生在校园内开展气象俗谚考证

图 2 2024 年 2 月 23 日,前往气象局进行俗谚考证

三、气象俗谚调查工作的总结阶段

本阶段主要是在调查工作结束后,学校向学生、班主任以及家长征求活动反馈,了解活动过程中的收获与改进意见。从本班参与的学生、家长、老师的感悟可以看出,三方应对气候变化的意识、技能均有提升。

赵同学:参加了这次活动,我深刻体会到气象与我们的生活密切相关。学习和掌握一些基本的气象谚语可以帮助我们更好地应对气候变化,减少自然灾害的影响。气象谚语的学习和实践让我更加深入地理解了气象对于国家和个人生活的重要意义。

邵同学:在这次气象民间俗谚调查实践研究中,我们以气象谚语为研究对象,搜集了身边大量鲜活的气象谚语,查阅了一些文献资料,结合前人优秀的研究成果以及自己的并不丰富的知识储备,对其进行整理。我觉得这不仅增强了我对谚语文化的理解,也让我了解了我国古代劳动人们的丰富知识以及对气象的认知。

张同学:在这次气象俗谚的活动中,我是一名组长,在确定组员时,好怕她们不参加……终于还是找到了合适的组员。这次活动让我收获许多,认识到良好的人际关系也是很重要的,所以在平常的学习之中,我们同样不能忘记朋友,要互相帮助,互相学习。

陶同学:在本次活动中,我们没有在调查之前进行规划,导致我们需要多次查阅数据。在气象专家和老师的指导下,我们知道了只有一组数据易出现偶然性,不具有普遍规律。在这次调查中,我们小组不仅明白了做事之前要有规划,也体会到了只有像科学家一样持之以恒,才能得到科学的结论。

张同学妈妈:这次活动,我不再是一个全程参与的"旁观者",而是活动的参与者。我看到女儿从接收任务到组织活动的圆满完成,孩子从一个小白慢慢成长起来,真的很开心。还有我和女儿一起学习气象知识,感知气候的变化,提升了气候变化意识;我和女儿一起学习、成长很开心。感谢有爱的班级,让我和女儿一起在活动中成事、成长、成人。

马同学爸爸:孩子回家和我讲气象俗谚,是我人生第一次知道有这种谚语。在活动中,我帮助同学们解决数据收集、分析问题，同学们带我学习气象俗谚知识，体验气候变化对生活、生产、经济发展等的影响。这次活动给我补了一课气候变化教育，对气候变化的重要性认识有很大的提高。

班主任：我是科学老师，是从小在农村长大的武义人。虽然科学教学中有气象知识，从小也听过不少气象俗谚，但把气象俗谚和气候变化教育联系起来还是第一次。同学们对研究气象俗谚的渴望，对小组组建的热情，对数据获取的执着，向专家请教的虚心，得到验证结果的谨慎，令我印象深刻。可见，学生不仅仅是气候变化教育的受教者，更是一场浓浓的气候变化教育氛围的创造者，参与其中的学生、老师、家长、气象局的工作者，都受到了一次具体的气候变化教育。

成效与反思

经过项目的落地实施，本项目在提升学生气候变化技能、协同校家社开展气候变化教育和打造气候变化教育样本方面取得了预期成效。

一是提升了学生的气候变化技能和科学素养。本项目旨在挖掘乡土气象俗谚文化资源并应用于气候变化教育中。项目在教师与专家的协同帮助之下，由学生自主选定研究目标，以真实的乡土气象俗谚文化描述的气象特征与环境物候特点为学习资源，挖掘关键学习素材，并通过团队成员的通力合作，解决开放式的气候变化问题。在这一项目式学习过程之中，学生主动应对挑战和不确定性的能力得到培养。例如如何获取气象与气候变化的知识；如何计划基于某一气象俗谚引发的气象与气候变化的调查探究项目；如何控制项目的实施；如何加强小组沟通和合作等。通过小组组建、构建假设、设计探究方案、收集证据、交流反思、再重建等过程得出结论，完成探究。这一过程不仅培养了学生的科学素养，还让学生真实地看到气温、降水等数据变化，了解到俗谚中应对气候变化的技能和方法，实现了学生应对气候变化技能和科学素养提升的真实发生。

二是协同校家社开展气候变化教育，凸显学习的全面性和终身性。基于气候变化的复杂性，需要发展多面向气候变化教育。武阳中学的气候俗谚调查研究，形成了以学校为中心，家庭为单位，借助气象部门等社会力量进行气候变化教育的新途径。在俗谚调查中，同学们向家人普及气候变化教育知识，老年人向孩子讲述当地的气候历史、自然条件和文化传统等。几代人因俗谚一线牵，共同学习气候变化教育知识，使气候变化教育深入各个家庭，凸显全面性、终身学习性。气象部门作为公共事业单位，在气候变化教育中发挥专家指导作用，实现了多元主体参与气候变化教育。

三是打造了气候变化教育的新样本。气候变化教育的目的在于减缓和适应气候变化。因此，追溯气象谚语并与当今的气象表征进行比对，能够帮助研究者发现气候变化的蛛丝马迹，从而加强他们对于气候变化的认知和关注程度。武阳中学通过挖掘地方气象俗语，总结出一套开展气象谚语调查研究的过程，形成了很好的示范作用，为后来者开展相关工作提供了参考。基于本项目，提炼出气候变化教育项目式学习应当具备的三个做法：发展在地化议题、打造规范化流程、发挥学生主体性。发展在地化议题指的是根据当地特色挖掘有价值的气候变化教育项目。基于项目设计规范化的实施流程，包括活动筹备、活动实施和总结。其间要注重发挥学生主体性，始终围绕着启发、引导和鼓励学生完成相关调查的原则，促进他们的自主学习。

着眼未来发展，本项目还有进一步改进的空间。经过气象俗谚调查和研究项目的开展，学生、教师、家长储备了一定的气候变化知识，对气候变化教育的重要性有了不同的认识，学生对气象俗谚也产生了浓厚的兴趣。但此活动学生局限于听长辈讲、进气象局查数据，而没有亲身感知俗谚中的气候变化，因此，在接下去的探究中，可以从以下三个方面开展：第一，积极带领学生走进田间地头，感受物候变化与气候变化之间的关系；第二，带领学生开展实地观察和调研，通过一段时间连续的气象要素的记录来验证俗谚的指向性；第三，对原来考证过的俗语进行深入研究。由于气象俗谚具有区域地方性、较强的凝固性和文学性等特点，之前的俗语放在当今和未来社会是否适用，仍然需要持续地追踪观察。

（周晓娟，浙江省武义县武阳中学教师）

基于具身实践的"气候变化教育"学习新样态

研究背景

人与自然和谐共生,教师应肩负起使命,加强气候变化教育,鼓励学生深入了解气候变化。培养拥有使命感和责任感的学生,才能推动气候变化教育迈向新的高度,共同守护地球家园的绿水青山[①]。纵观传统教育模式,学校教育更倾向于让学生端坐在书桌前,面对电脑或教科书,花费大量时间去理解老师的经验传授和抽象的学科术语,通过机械重复记忆,应对考试所需的知识和技能。在这种"久坐不动的教育"[②]中,老师成为知识的"搬运工",学生成为知识的"接收器",不仅不利于学生的身心健康,也不利于知识的吸收与转化。

本项目基于"具身认知"理论,主张把身体学习融入"气候变化教育"中,倡导身心活跃的教学,将难以被学生所理解的专业内容转化为学生的具身体验,促进抽象知识的充分理解和掌握。

项目设计

本项目整体包含项目启动、项目计划、项目实施、项目评价等阶段,分为"群学放学、信息检索、问卷调研""校企协同、方案设计""企业参观、专家讲座、航行模拟""成果形成、评价反思"等步骤。整体思路如图1所示。

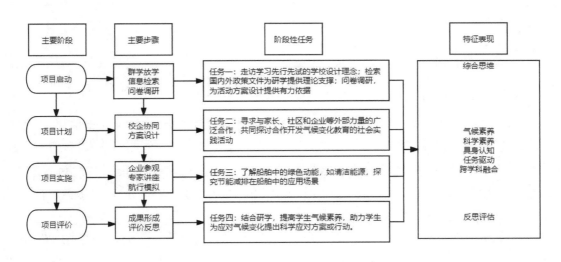

图1 项目整体思路框架

---

① 岳伟,余乐.应对气候变化:教育的时代使命与行动路径[J].湖南师范大学教育科学学报,2024(5):26-30.
② 叶浩生.身心二元论的困境与具身认知研究的兴起[J].心理科学,2011,34(4):999-1005.

一、基于问卷调查,了解班情学情

在本次活动前,教师采用问卷调查法了解班情学情。具体包括:在参观学习前,调查初中生是否了解气候变化?是否了解父母的工作环境与气候变化相关联?是否知道企业在气候变化领域做出的贡献?在上海船舶研究设计院参观学习后,有哪些感想或者体会?是否愿意继续加入类似的活动?

以此了解学生的真实想法和教育需求(见图2),同时为活动反思提供信息资源,便于活动方案改进和深入开展研究①。

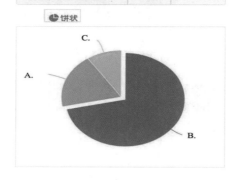

图2 数据分析示例

二、增强校企联动,共绘教育蓝图

秉承"校家社协同育人"的理念,教师尝试通过"校企合作"推动"气候变化教育"。上海船舶研究设计院对新能源有着深入研究,其大力发展绿色节能和智能船舶技术,所研发的高效适应扭曲舵 SATR、扇形导管等节能装备,均体现了该院对绿色、智能科技的执着追求。2024年1月29日,上海市浦东新区新场实验中学的13名学生及3位导师,以绿色出行的方式,一起走进了上海船舶研究设计院。

首先,参观研究院展厅与实验室,了解研究院发展历程、科研成果展示,参观船舶模型、节能零件等。其次,邀请研究院专家基于船舶设计、燃料选择、路线方案等方面开展节能减排的专题讲座,与学生互动交流。最后,分组实践,在导师指导下模拟航船。

---

① 唐莺.家校社协同开展项目化学习的有效路径:以"气候变化教育"项目为例[J].现代教学,2023(22):70-73.

三、侧重具身认知,体验气候变化

本次活动的参加对象是七年级学生,他们的逻辑思维能力正处于启蒙阶段,受限于传统教育"课堂"的"禁锢",缺乏主动学习意识。

通过上海船舶研究设计院的研学实践活动,让学生深度参与气候变化相关内容的学习,激发他们的学习热情和自我内驱力。由此构建"具身认知"的学习框架(见图3);通过"研学实践→合作探究→思维提升"的三维结构,以"提出问题、分析问题、解决问题"为主要步骤,通过"主题式、跨学科、项目化"的学习方式,提高气候素养[①]。

具体目标:在活动中,了解企业文化,提高科学素养;聆听专家讲座,学习船舶企业应对气候变化所采取的措施,拓宽视野,激发创新思维;通过与研究院的专家交流对话,更清晰地了解船舶设计行业的绿色发展,为应对气候变化打下科学基础。

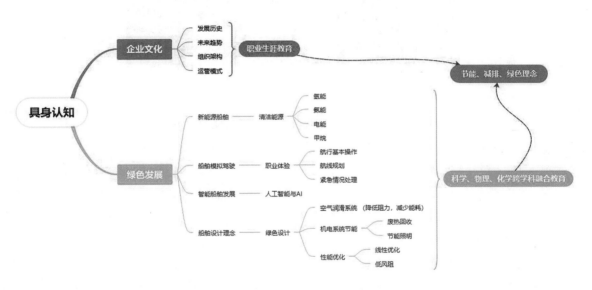

图 3 基于"具身认知"的学习框架

项目实施

一、初遇挑战,探索船舶文化

在学生对气候变化有了一些了解后,教师带领他们走进上海船舶设计研究院,深入地探索船舶与气候之间的关系。为了确保此次研学活动的针对性和有效性,教师在活动开始前就明确了核心问题:"寻找企业中的绿色能量",并基于此设计一份研学实践活动方案。

在设计院内,学生们在导师的引领下,分组进行具身学习和实践探究(见图4)。他们首

---

① 郝鹏翔,施美彬,刘媛,等.开展跨学科深度研学,实现"双碳"目标教育:能源与全球气候变化项目式主题学习[J].地理教学,2023(3):36-40.

先根据研究院的展览布局,有序地参观了各个展区。在参观过程中,学生们认真观看了各种船舶模型和实物展示,详细记录了不同船舶设计的特点和理念,特别关注了绿色节能产品在船舶上的具体应用。

图4　探索船舶文化

学生面对"能源转型促低碳,设计创新降能耗"这一船舶设计理念时,显得有些迷茫。他们纷纷表示,虽然大致理解这一理念的"节能减排"目标,但对于如何具体实现以及其中的技术细节并不十分清楚。为了帮助学生们更好地理解这一船舶设计理念,三位导师对此进行了简单的解释和指导。

二、困境破冰,认知技能提升

在上海船舶研究设计院的二楼,船舶设计专家系统地阐述了船舶行业的脱碳背景、本质节能、绿色动力和智能发展(见图5)。这些内容广泛涉及多个学科领域的知识,实施气候变化教育需要依托多学科融合方式加以开展。讲座结束后,学生们积极参与了互动交流,提出了一系列问题,如"清洁能源的定义是什么?""船舶烟囱排放的白烟包含哪些气体成分?""船上的生活废水和垃圾是如何处理的?是否直接排放入海?"等。这些问题充分反映了学生们在具身认知过程中的求知欲和探索精神。

针对这些问题,船舶设计专家结合七年级学生的学科基础,逐一进行了解答。在解释"清洁能源"时,涉及九年级的化学知识;在讨论废气废水处理时,引入了相关的物理知识。这种跨学科的解答方式,不仅有助于解决学生的认知困惑,还促进了他们对跨学科知识的融合与理解。对于未能及时解答的问题,我们鼓励学生记录问题、求助导师、询问家长、查阅资料,以寻求解决之道,并引导他们更全面、深入地理解气候变化和环境保护等全球性议题。

此外,学生们还在船舶航行体验馆,具身体验了船舶的航行过程,亲自操作船舶,进行航行决策,从而更直观地了解了航海工作的工作内容、工作流程、工作环境以及所需的知识与技能(见图6)。这种实践体验不仅激发了学生对船舶职业的兴趣,还帮助他们在操作中提升

图 5　专家讲座《船舶脱碳路径探索与智能船舶发展趋势》

了自我效能感。研学过程中,我们始终关注跨学科学习的严谨性和系统性,确保所传达的信息准确无误。通过此次活动,学生不仅拓宽了认知的视野,还为未来学习和职业发展打下了坚实基础。

图 6　船舶航行体验

三、研学共进,实现多维成长

回到学校后,我们根据学习兴趣和疑难问题进行分组,通过查阅资料、学做实验、求助导师、咨询家长等方式,先后解决了"清洁能源中的化学公式""船体设计与物理"等相关问题。

在评价展示环节(见表1),有的小组通过制作"未来社区发电系统"来展示清洁能源的可持续性。有的小组把新能源船只通过互动交流的方式,介绍给班上其他同学。大部分学生的评价反馈表上,都提到"切身感受到气候变化教育的重要性""学到了企业在绿色发展中作出的努力""我也要加入宣传气候变化的行列"。

表1 小组合作学习评价表

小组名称						
组长			组员			
活动中收获了什么:						
活动中遇到什么困难:						
活动心得,对于今后的启发:						
自我评价表:						
评价内容	分工合作(20分)	参与程度(20分)	互助互学(20分)	解决问题(20分)	自主探究(20分)	总分(100分)
学生A						
学生B						
学生C						
学生D						
导师评价:						

学生的自我评价和小组评价充分体现了"具身认知"的动力性[①],具体表现为:其一,把"气候变化教育"从情感上的认知,提升到了身体上的认知;其二,学生不局限于"船舶文化"的学习,还上升到跨学科乃至跨学段学习;其三,学生尝试用不同的形式和方法,让更多人加入气候变化教育。

四、多元评价,提升科学素养

第一,思维拓展,产生智慧火花。常规的思维方式和固有的教学模式往往会限制学生的想象力,使之难以接触更广阔的天地和作出更深入的思考。气候变化教育要求不满足于浅显的理解和解释,而是深入探究问题的本质和事物之间的联系。如本次活动中"清洁能源"问题,教师不但希望学生了解"清洁能源",更希望学生能整合、开发、利用它来解决气候变化问题。

---

① 叶浩生."具身"涵义的理论辨析[J].心理学报,2014,46(7):1032-1042.

彭同学拓展思维边界，进行了科学探秘。他对清洁能源产生浓厚兴趣，设想了一套高效、环保的未来发电系统（见图7）。这套发电系统充分利用"可再生能源"，模拟能量转换和储存技术，减少对传统化石燃料的依赖，缓解能源短缺问题。他还将目光投向了新能源汽车领域，他深知新能源汽车对于减少环境污染、推动能源转型的重要性，完成《上海市新能源汽车充电设施调查》，并提出有益的思路与参考，迸发出智慧的火花。

图7　未来发电系统模型设计

第二，提升科学素养，进行自我超越。应对气候变化不是纸上谈兵，需要学生在实践中不断探索和尝试，独立思考、深入分析，敢于提出自己的独立见解和解决问题的方案。

王同学积极参与各类科技竞赛，成功研发了"室内外空气质量监测比对系统"。这一系统，能够实时监测室内外的空气质量，为人们提供准确、及时的数据支持，有助于改善居住环境和提高生活质量，在"上海市第十五届头脑奥林匹克创新学习活动创新擂台赛"中荣获"金擂奖"。他撰写的《改善室内空气，共筑健康生活》，对提高环保意识和健康生活有着重要意义（见图8）。

第三，未来导航，坚守绿色信念。在交流展示活动中，有的同学介绍新能源船舶，阐述其在减少污染、保护水域生态方面的重要作用，畅想科技与自然和谐共生的美好愿景；有的同学巧妙地利用可回收材料，制作模型，不仅体现了环保理念，更展示了创意与智慧的完美结合（见图9）。科技畅想行动，传递出同学们对绿色生活的热爱与追求。

为了让学生对自己的表现有一个清晰的认知，教师采用过程性评价和结果性评价相结合的方式进行组内的自评和互评，帮助学生回顾自己在活动中的表现，展望未来的行动。激励他们面对日益严重的气候与环境问题，一定要坚守绿色信念，推动可持续发展，这已经成为时代的必然选择。教师作为未来世界的导航者，应该肩负起这一历史使命，推动美好中国的建设与发展。

图 8　王同学成果介绍

图 9　利用旧报纸制作模型

成效与反思

本项目通过校企合作协同创新,实现了个体、群体的共同成长。活动中,学生参观企业,学习企业的"绿色文化",了解"绿色科技"的发展和应用。学生的创新精神、实践能力及科学素养均得以提升。

此外,在项目实施过程中,学校还围绕"生态文明"主题,开展了系列校园建设。如综合融通气候变化的相关知识,设计了丰富多样的学习场域,初步形成了绿意盎然的生态校园格局;开展社团课程如植物鉴赏、乡村小导游等将生态文明融入课程教学中,让学生在学习知识的同时,感受自然,培养环保意识;通过校园广播、宣传栏、校园网,推广生态文明理念和环保知识;建立监督机制,设立节能环保监督小组,定期对校园的用电用水、垃圾分类等情况进行检查和评估等。

通过研究性学习、社会实践及学校气候变化教育浸润,学生们意识到自己对地球、对未来负有责任。他们愿意为保护环境付出更多的努力,在日常生活中,自觉减少浪费,节约能源和水资源,积极参与"垃圾分类"等环保行动,并且主动向家人和朋友宣传环保理念,带动更多人参与"环保行动",让大家逐渐理解"人与自然、社会、自我"的关系,认识到维护生态平衡的紧迫性和重要性。

然而,在实施过程中仍需注意学生的参与度、活动的组织管理和评价反馈等细节。

其一,明确认知目标。具身认知鼓励学生在特定的学习环境中,根据自己的理解和感受去解决问题。重要的是这些问题并没有通用的、一成不变的答案,如"如何去设计船体,让它做到低碳环保"等。相反,它们是为了激发学生的创造性思维和解决问题的能力,鼓励学生从多个角度思考问题,并产生多样化、有创意的解决方案。

在后期的评价反馈中,教师发现有部分学生非常佩服船舶设计员的精湛技术和航海人员在海上环境中展现的应对能力。这种兴趣有助于拓展学生的视野并可能激发他们对相关领域的探索,但当学生的注意力被船舶设计和航海等具体职业所吸引时,需要更加明确地将这些内容与气候变化的大背景联系起来。例如,可以强调船舶设计和航海技术在减少碳排放、提高能源效率方面的重要性,以及这些技术在应对气候变化挑战中的潜在作用。教师需要不断调整和优化教学方法和内容,确保学生能够全面、深入地理解气候变化问题,认识到不同职业在这个问题上的责任和角色。

其二,整合学习资源。为了完成活动任务单,学生可以利用多种资源。这些资源可能包括书本文献、网络资源、实验设备、教育软件、教师指导、同学讨论等。认知不仅仅是对新信息的接收和处理,更是在现有知识和经验的基础上进行建构和重构。在本次研学实践活动中,学生尚未对"清洁能源"进行前移学习,活动重心更多是"什么是清洁能源",若把这一内容作为研学前的前移学习,可以在活动中更深入地探讨和理解这一话题,如"清洁能源如何应用于日常生活""清洁能源的发展前景"等。因此,学生在开始学习之前,进行前期的资料收集和必要的知识储备是至关重要的。这些准备工作为学生提供了必要的认知基础,使他们在解决问题时能够更快速、更有效地找到相关信息和解决方法,从而为学习过程提供便利。

(施一凡,上海市浦东新区新场实验中学教师)

对话专家

在学校教育中,综合活动是培养学生综合素质的重要途径,它们能够跨越学科界限,让学生在实践中学习和成长。本篇的综合活动案例覆盖中小学和中等职业教育等不同阶段,以丰富的活动形式将气候变化教育融入其中。例如上海地区的考察探究类综合活动,通常涉及实地考察和深入研究,让学生亲身体验环境问题,从而深刻理解气候变化的影响;设计制作类综合活动能鼓励学生运用所学知识,设计并制作具有环保理念的产品或方案;气候谚语俗语等文化传承类综合活动能将中华优秀传统文化与环保理念相结合,让学生在传承文化的同时,认识到保护环境的重要性;主题研学类综合活动则围绕特定主题进行深入研究,让学生在项目式探索中学习气候变化的相关知识,了解气候变化的科学原理和影响。

总体来看,不同国家的气候变化教育各有侧重点,但都强调学生的参与和体验,以及跨学科知识的综合应用。例如,英国的气候变化教育更注重户外实践和亲身体验,学校通过组织户外教学活动,如参观风力发电场、参与植树活动等,让学生亲身体验可再生能源的使用和气候变化的影响。美国的气候变化教育更侧重于科学研究和数据分析,在其基础教育阶段的气候变化综合实践活动会通过与科研机构合作,让学生参与气候变化的研究项目,如分析当地气候数据、参与气候模型的构建等。荷兰的气候变化教育注重培养学生的批判性思维和解决问题的能力,学校经常通过模拟游戏和活动让学生理解气候变化的复杂性和紧迫性,如模拟联合国气候变化大会等。

在设计以气候为主题的综合活动时,教师需要考虑多个维度以确保活动的有效性、趣味性和教育性。与国际上其他国家的气候变化教育实践相比,中国的综合性活动在结合本土文化、历史与现实方面具有独特优势。例如,通过考察长江口二号古船、气候谚语等项目,学生能够将气候变化与本地的历史文化联系起来,这种结合有助于加深学生对气候变化影响的文化理解和传统价值观念的传承。本篇的活动案例提供了很好的参考,这些经验体现在:

第一,了解学校所在地理区域的特点:预先分析学校所在地区的气候特征,如温度、降水、季节变化等;考虑当地的气候相关问题,如极端天气事件、海平面上升、生态系统变化等。

第二,以学生为中心展开综合活动设计:根据学生的年龄、性格和兴趣定制活动,确保活动能够吸引学生的注意力并激发他们的好奇心。设计活动时考虑到学生的认知水平和实践能力,确保活动既不过于容易也不过于困难。

第三,考虑活动设计的跨学科整合:通过跨学科项目,让学生从不同角度理解气候变化的多维度影响;将气候变化教育融入不同学科的教学中,如地理、科学、数学、艺术课程之中。

第四,强化教学方法的实践性与探索性:优先设计实践活动,如天气观测、碳足迹计算、社区绿化等,让学生在实践中学习;鼓励学生通过观察、实验、调查等方式主动探究气候变化问题。

　　第五，充分发挥校家社活动合作机制优势：邀请家长、社区成员和气候专家参与活动，增加学生的社会互动和学习机会；促进学生之间的合作，通过小组讨论、项目报告等方式提高学生的沟通技巧。

　　第六，建立系统的评价与反馈机制：地方教育管理部门需要设计有效的评价体系，评估学生在活动中的学习成果和参与度，才能更好地确保活动设计具有一定的灵活性，能够根据学生的反馈和气候变化的最新研究进行调整，将气候变化教育作为一个持续的过程，而不是一次性活动。

　　第七，注意综合活动过程中的情感带动与气候行动行为激发：实现可持续发展目标，需要从孩子抓起，在学校教育中，需要教师鼓励学生将所学知识转化为实际行动，如参与环保项目、推广气候友好行为等。

　　这些独具创意特色的气候教育综合活动，可以将气候变化教育融入日常教学中，让学生在亲身体验中认识到环境保护的重要性，培养他们的环保意识和责任感。然而，与以教师为主导，特色鲜明且种类丰富的气候教育综合活动相比，其背后的气候教学资源支持略显匮乏，可能导致综合活动的可持续开展和系统性发展面临挑战。这些挑战可能存在于综合活动覆盖面的局限，尽管综合性活动在气候变化教育方面取得了一定成效，但由于资源、时间等限制，这些活动的覆盖面往往有限，不能确保所有学生都能参与；很多学校在开展综合性活动后，缺乏对学生环保行为和实践效果的长期跟踪评估，间接影响了气候类综合活动的持续改进和深化；由于缺乏与社区、企业等外部资源的有效合作，综合性活动的内容可能与实际环保需求脱节，导致教育效果大打折扣；最后，许多教师在气候变化教育方面的专业知识和实践经验相对不足，限制了综合性活动的质量和效果。

　　为了更有效地通过综合性活动促进气候变化教育，建议加强教师培训、优化资源配置、提高学生参与度，并建立长效的评估机制。同时，借鉴国际经验，推动跨学科合作和国际交流，以全面提升气候变化教育的质量和影响力。

（藤依舒，北京师范大学人文和社会科学高等研究院副教授）

第五篇

学校家庭社区协同开展气候变化教育

气候变化教育的综合性,要求教育的时空与人们的生产和生活时空相一致,要求教育的对象面向全民,也要求教育的开展充满协同性。

在本篇中,包括了代际协同开展气候变化教育、企业和社会组织参与气候变化教育、学校家庭社区协同开展气候变化教育等多个案例,也呈现出本领域的多主体、多机构合作的特征,更彰显着气候变化教育的独特性。本篇中的案例也来自不同地域,而且实践还在发展,探索与创新的空间也越来越大。

"银芽气候课堂"的学习路径设计与实施

研究背景

应对气候变化是联合国的整体战略。①《联合国教科文组织：以"全学校"路径推动气候变化教育》一文指出：气候危机是全球所面临的现实，没有教育就没有解决办法。② 随着气候变化及其影响的加速，青少年参与了解气候变化显得尤为重要，气候变化教育是推动年轻一代了解气候变化并采取积极行动的关键策略。课堂作为教育的主阵地，拥有独特的优势来推动这一进程。祖辈作为社会的重要力量，他们的经验和智慧可以与孙辈的创新精神和学习能力相结合，创生出新的教育模式——"银芽气候课堂"。"银芽气候课堂"是"银芽联盟"的重要课程之一。③

项目设计

一、项目目标

总体目标：通过"银芽气候课堂"学习路径的设计与实施，提升祖孙应对气候变化的意识与能力，提升气候素养，促进环境教育普及。

具体目标：

(1)以"银芽气候课堂"为阵地开展气象知识共学，增强学习者环保意识。

(2)通过"班会＋家庭学习"等多维度互学融合，提升"银""芽"防灾减灾和应对气候变化的能力。

(3)挖掘校内外资源，开展主题实践，促进理论学习的实践转化，提升"银""芽"科学精神与气候素养。

二、学习路径设计

"银芽气候课堂"的路径设计是一个具有综合性和创新性的过程，它包括祖孙共学互学课程体系、代际互动教学、实践共同体项目等。具体内容包括祖辈指导孙辈观看教育视频、共同进行户外观察和实验、孙辈教导祖辈游戏化学习、长幼共同阅读相关书籍等，实践项目由祖孙共同参与实施，具体内容包括组织实地考察、外出宣传、模拟演练、制订家庭气候行动

---

① 学习型社会读书小组.气候变化教育指导纲要[EB/OL].(2024-03-01)[2024-05-25].https://mp.weixin.qq.com/s/LlIFamqFO49-XO7eX2qD2w.

② 陶广璐.联合国教科文组织：以"全学校"路径推动气候变化教育[J].上海教育，2023(24)：57.

③ "银芽联盟"创建于2021年，是由武义县熟溪小学与武义老年大学合作开发的品牌，它致力于通过引进优质老年人资源，实现代际互学，是一个以互学互助为核心，旨在促进跨年龄交流与知识传承的合作共同体。

计划等,同时鼓励学生、教师、家长以及社区成员共同参与气候课堂的活动,形成多元互动的学习共同体。

三、项目发展阶段

第一阶段是 2023 年 12 月,项目教师参加武义县教育局组织的"气候变化教育进校园"专题培训,以此为起点,建立研究团队并进行项目设计。

第二阶段是 2024 年 1 月至 2024 年寒假结束,重点是开展祖孙共学互学活动,围绕基础类知识、应对策略方面的内容开展课堂共学活动,以祖辈教孙辈和孙辈教祖辈的形式展开互动教学。

第三阶段是 2024 年春学期开学至今。以挖掘校内外资源,促进理论学习的实践转化,提升参与者的科学精神与气候素养为目标开展丰富的实践活动。

项目实施

《柏林可持续发展教育宣言》强调,可持续发展教育应该成为各级教育系统的基本要素,将环境和气候行动作为核心课程内容,并强调以整体观对待可持续发展教育。[①]"银芽气候课堂"是一个富有创意和深远意义的教育项目,它的实施需要细致规划和精心准备,以下是具体的实施过程。

一、教师参加培训,组建研究团队并制订实施方案

2023 年 12 月 22 日,"癸卯·寒假,你好!"气候变化教育进校园暨武义县中小学气象民间俗谚调查实践研究项目研讨活动在武义县熟溪小学举行,笔者以"银芽联盟"骨干教师的身份参加了本次活动。该项目得到了华东师范大学终身教育研究院副院长李家成教授的支持与指导。华东师范大学终身教育研究院特聘研究员、武义县教科所原所长雷国强详细解读了《踏雪寻谚迎春到——"癸卯·寒假,你好!"气候变化教育气象民间俗谚调查实践研究进校园行动方案》。

以本次培训为新起点,学校召开"银芽联盟"双方负责人会议,经过研讨确定了学习目标、学习内容及实践项目,研制了如图 1 所示的项目思路。

二、共学互学活动的开展

应对气候变化迫在眉睫,对于教育而言,其功能主要体现在能够通过传授气候相关知识、树立正确的价值观、提升应对技能等,引导学生主动适应气候变化和积极参与减缓气候变化的行动。[②]

为积极推进气候变化教育进校园,提升青少年的气候素养,笔者以"银芽气候学堂"为平台开展了形式丰富的祖孙共学互学活动。基础类的共学内容包括:气候与天气的区别、温室

---

① 陶广璐.联合国教科文组织:以"全学校"路径推动气候变化教育[J].上海教育,2023(24):58.
② 岳伟,余乐.应对气候变化:教育的时代使命与行动路径[EB/OL].(2024-01-17)[2024-05-25].https://mp.weixin.qq.com/s/mrZ7EfK3h4Ewyqv-2JDnRA.

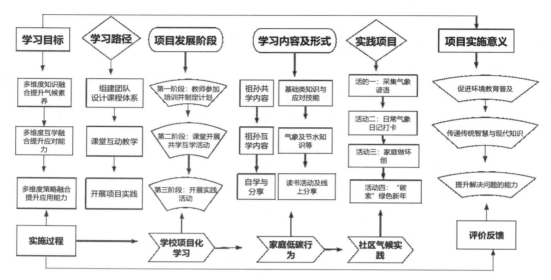

图 1 "银芽气候课堂"学习路径设计与实施

效应、碳足迹、自然界的碳循环、可再生能源与不可再生能源、气候变化的影响、应对气候变化的国际协议等；应对气候变化策略的共学内容包括：节能减排、交通出行、废物处理、可持续生活方式饮食选择、绿色购物、水资源保护、气候变化科学与影响、探讨气候变化的本地影响、社区参与行动、社区绿化、环保活动等。

在"银芽气候课堂"中，老人们以丰富的人生阅历化身气象知识讲解员、乡土文化传播者参与课堂，分享他们的生活经验和对环境变化的观察，而年轻一代则带来最新的气候变化科学知识和行动策略。老人们向孙辈讲述了自己多年来的农耕经验，如何通过观察天象、动物行为等来预测气候变化，以及如何利用传统智慧来适应这些变化，分享了自己如何在干旱年份选择耐旱作物、在雨季来临时做好防洪准备等经验。孙辈们则利用自己的现代科学知识，向爷爷奶奶们介绍了全球气候变暖的原因、影响以及应对措施。他们通过图表和数据展示了温室气体排放对气候的影响，并解释了如何通过减少碳排放、植树造林等方式来减缓气候变暖，还分享了一些现代科技在应对气候变化方面的应用，如智能农业、气象预报等（见图2）。

图 2 祖孙共学、互学气候知识现场

三、实践活动的开展

实践学习活动的开展目的在于提升多代人的气候变化意识与紧迫感,提高学生应对气候变化的能力与解决问题的能力。围绕着这样的目标,"银芽气候课堂"开展了丰富的实践活动。

气候课堂进民间——气候谚语采集。气象谚语具有鲜明的地方特色,承载着丰富的民间智慧和生活经验,武义地方方言的复杂性带来了民间俗语蕴藏的丰富性与多样性,为我们开展"气候变化教育进校园"提供了大量生动鲜活的气象俗谚。笔者充分发挥武义气象谚语多、分布广的优势,开展"民间气象俗谚"收集活动。孩子们利用拜年的契机向长辈众筹气象智慧,呈现结果的方式很丰富:有的用图画的形式进行记录,有的录了视频,还有的亲子合作以美篇的形式呈现。全班共收到102条谚语。

气候课堂进日常——气候行动日记打卡和家庭一周情况记录。学习知识最重要的意义在于转化与应用,学生们将所学的气候变化知识与能力体现在日常生活中,以日记打卡的形式把自己日常践行气候行动的点滴记录下来。全班45位同学,每人坚持绘写气象日记,共写气象日记900余篇。他们的举动影响了一个个家庭,也发生了很多感人的故事。在进行日记打卡的基础上,学生们从"衣、食、住、行、观察"五个方面对家庭一周的情况进行记录,如表1所示,通过监测家庭成员低碳行为的变化趋势,及时调整和改进家庭低碳策略。

表1 气候变化与低碳行为家庭观察

	(第 周): 年 月 日至 月 日	
	学校: 班级: 姓名:	
类别	家庭成员行为或物品使用记录	数量
衣	1. 新买的衣服/裤子/鞋子总数(单位:样)	
	2. 用洗衣机洗衣服的次数(单位:次)	
食	1. 全家叫外卖的次数(单位:次)	
	2. 全家买各种肉的数量(单位:千克)	
住	1. 垃圾扔了几袋(以垃圾袋基本装满为标准)(单位:袋)	
	2. 外出购物拿回的袋子数量(单位:只)	
	3. 家里产生的瓶子、纸杯数量(单位:只)	
	4. 电视开机的小时数(单位:小时)	
	5. 自己玩手机/平板的小时数(单位:小时)	
	6. 电磁炉/燃气灶/烧的小时数(单位:小时)	

(续表)

行	1. 打的/开自家车的次数（单位：次）	
	2. 坐公共交通（地铁、公交车）的次数（单位：次）	
	3. 走路的次数（自己走路超过10分钟，计为1次）（单位：次）	
	4. 本人坐电动车、摩托车次数（单位：次）	
观察	1. 晴天天数（单位：天）	
	2. 雨雪天数（单位：天）	
	3. 极端天气（单位：次）	

"气候课堂"进家庭——预支红包做环创。同学们在课堂上学习了气候变化知识之后，回家跟自己的祖辈们分享低碳迎新春的"金点子"。在彼此的交流过程中，小朋友们发挥自己的灵动创造力，有了大胆的设想。说干就干，他们在家长的支持下，每人成功预支到100元钱（见图3），这样一笔巨款让他们既兴奋又小心翼翼，大家积极地参与到活动中来，因为他们需要对自己的资金负责。在老师的协助下孩子们在网上订购了帆布袋，印制了宣传单，与老师一起在广告公司做了宣传牌，每位参与者都在为保护环境而努力行动。

图3　学生向家长预支红包

"银芽气候课堂"进社区——"碳索"绿色新年活动。2024年2月5日下午，大家聚集到超市门口，拿着帆布袋，举着宣传牌，一边喊一边分发袋子和宣传单："爷爷奶奶、叔叔阿姨，大家快来领啊，这是我们用自己的零花钱定制的帆布袋，进超市购物用我们的袋子更环保。""垃圾分类做'捡'法，快乐聚餐请光盘，出门拜年'绿'出行，绿色新年要节俭。"老班长们也耐心地为市民们讲解倡议书的内容。他们的行动与诚意感动了很多人，很多市民表示愿意在春节期间尝试开启低碳生活方式，以此表达对环境保护的支持（见图4）。

随后，小朋友带着袋子，用剩余的70元钱跟着家长去超市购物，他们在践行绿色购物的同时，还向市民分发低碳行动倡议书，他们的行动影响了很多在超市购物的人们。

图 4　祖孙与家长参加绿色新年活动现场

成效与反思

代际共学气候变化知识不仅对家庭与个人成长具有重要意义,也对整个社会的可持续发展产生了积极影响。

一是促进了环保观念与环境教育的普及。"银芽气候课堂"通过讲解气候变化的影响和应对措施等知识,增强人们的环保意识。尤其对儿童来说,他们正处于价值观形成的关键时期,通过参与这样的课堂活动,他们更加深入地了解了气候变化问题的严重性和紧迫性,从而树立了正确的环保观念。李同学跟外婆生活在一起,外婆告诉我们:自从上了"银芽气候课堂",一家人都进入了低碳生活,且相互监督、相互学习,在节能照明、绿色出行、节约用水、减少食物浪费等方面增长了很多知识。几代人通过参与这样的课堂活动,更加深入地了解了这些科学知识,并提高了自己的科学素养。这对于推动环境教育的普及和提高公众对环境问题的认识具有重要意义。

二是双向传递了传统智慧与现代知识。"银芽气候课堂"作为一种独特的教育形式,能够跨越世代,将传统智慧与现代知识相结合,一方面,老一辈通过传统智慧的传递,可以帮助年轻一代建立与自然的和谐关系,理解人类与自然之间相互依存和相互影响的关系。另一方面,年轻一代能够将现代科学知识传递给年长一代。现代科学知识是解决气候变化问题的关键,年轻一代可以把学习到的科学原理、社会知识以及应对措施等现代知识传递给长辈,帮助他们更好地理解和应对气候变化带来的挑战,为未来的可持续发展作出贡献。通过这样的课堂学习,祖孙两代人不仅增进了彼此的了解和信任,还共同为应对气候变化贡献了智慧。

三是激发了创新思维,提升了解决问题的能力。面对气候变化这一全球性挑战,代际共学可以激发家庭成员解决问题的创新思维。"银芽气候课堂"涉及多个学科领域,通过融合不同学科的知识和方法,这种跨学科的思维方式有助于打破思维定式。"银芽气候课堂"不仅关注知识的传授，更注重问题解决能力的培养。通过参与课堂实践活动、讨论和案例分析等，参与者能够学会如何运用所学知识解决实际问题。这种以问题为导向的学习方式有助于培养参与者的问题解决能力。

"银芽气候课堂"通过传授气候相关知识、培养学生的跨学科思维以及适应能力来提升他们的气候素养，在提高两代人适应与应对气候变化能力、促进跨代沟通与合作方面进行了积极的探索。但由于一部分知识的学习相对比较枯燥，未来在教学方式、方法上有待改进：需要采用足够的视觉辅助材料、实例和故事，设计角色扮演和模拟大会等形式来帮助他们更好地理解与应用知识；需要持续关注学习者的积极性、持续加强课程的系统性。未来，我们将继续完善学习路径设计，拓展实施范围，推动更多家庭和社会成员参与到气候行动中来，共同为应对全球气候变化持续贡献力量。

（蓝美琴，浙江省武义县熟溪小学教师）

绿色创新与气候教育：构建终身学习新典范

研究背景

首先，政策鼓励。在全球气候变化的严峻挑战下，绿色创新和气候教育成为应对环境危机的关键途径。临港新片区作为中国（上海）自由贸易试验区的重要组成部分，积极响应国家战略，致力于打造绿色低碳的发展模式，并通过教育的力量培育公民的环保意识和行动能力。《中国（上海）自由贸易试验区临港新片区生态领域绿色低碳发展实施方案》指出，要引导全民实践，鼓励动员全民广泛参与生态领域的绿色行动，增强节约意识、生态意识、绿色生活意识，把绿色低碳理念转化为全民自觉行动，提高全民参与绿色低碳发展的积极性，加强低碳基础理论研究和科普宣传，提升民众对节能低碳内涵的把握，提高人员业务能力，通过线上线下对群众进行科普，有效促进公众对节能低碳理念的关注和理解[1]。

其次，社区可持续教育项目支持。南汇新城老年大学与上海交通大学中英国际低碳学院携手，在临港新片区打造创新型社区可持续发展教育项目。南汇新城老年大学，作为面向老年群体的公益性教育机构，以绿色科创与低碳科普为发展特色，致力于打造开放融合的终身学习平台。上海交通大学中英国际低碳学院自2017年成立以来，专注于低碳科技研究和人才培养，推动低碳科学成果的普及和应用。同时，整合高校、社区资源，通过多样化的教育和科普活动，提升居民对气候变化的认知，鼓励社区成员积极参与绿色、循环、低碳的可持续实践，为居民提供了宝贵的学习资源和参与机会，推动了临港新片区的绿色发展。

最后，是代际共学应对气候变化的需要。不同年龄层次对气候变化的认知和反应存在显著差异。老年人尤其脆弱，帮助老年人理解和适应气候变化带来的健康风险至关重要。《柳叶刀人群健康与气候变化倒计时2022年中国报告》指出[2]，65岁以上老年人更易因气候变化面临死亡威胁；中年人作为家庭的中坚力量，既是学习者也是教育者，需要提升自身环境意识，并掌握向子女传授知识的技能。教育内容应包括气候变化基础知识、家庭节能减排方法、家庭环境教育策略，以及激发子女环保参与的技巧。通过中年人的示范和教育，可以在家庭内部营造绿色生活方式，引发社会整体关注；青少年是未来的决策者和行动者，他们的气候变化教育尤为重要。教育应注重培养环境意识、科学素养和创新能力，使其理解气候变化的紧迫性，并积极参与气候行动。通过参与式学习和实践活动，青少年能够提高解决气

---

[1] 中国（上海）自由贸易试验区临港新片区管委会办公室.关于印发《中国（上海）自由贸易试验区临港新片区生态领域绿色低碳发展实施方案》的通知[EB/OL].(2023-11-28)[2024-5-15].https://www.lingang.gov.cn/html/website/lg/index/government/file/1729926955738894337.html.

[2] Cai, Wenjia, et al. The 2022 China report of the Lancet Countdown on health and climate change: leveraging climate actions for healthy ageing[J]. The Lancet. Public health, 2022, 12(7): e1073-e1090.

候变化问题的能力,形成可持续生活方式和消费习惯。

综上所述,本文拟构建一个从家庭到社会的全方位教育体系,促进多代人分层次、多样化的共学互学,实现终身学习,从而有效应对气候变化挑战,推动社会的可持续发展。

项目设计

总目标:期待通过家社校全人群共同开展气候变化教育活动,构建具有影响力和号召力的绿色创新与气候教育特色品牌,为个体提供全方位、持续性的教育学习服务,促使多代人共同学习,推进绿色与可持续发展社区建设,为建设美丽中国和应对全球气候变化作出贡献。

具体目标:

(1)提升公众的环保意识和气候变化认知。通过开展多样化、实践性的绿色创新与气候教育活动,促使个体深入了解气候变化对环境和社会的影响,增强其对绿色生活方式和可持续发展的认识。

(2)培养科技创新能力和创新精神。通过引入前沿科技和绿色发展理念,激发个体对科技创新的兴趣和热情,提升其科技素养和创新能力,使其能够在不断变革的社会中应对挑战。

(3)激发参与绿色行动的积极性和责任心。通过教育活动和社区共建,引导个体认识到自身行为对环境和气候变化的影响,激发自觉践行绿色生活方式和参与环保行动的愿望和动力。

项目实施

该项目的课程体系设计旨在全面应对气候变化挑战,促进社会可持续发展。多元化的课程包括科普、研学和实践课堂,以满足不同年龄和背景的学习者需求。讲座课程由上海交通大学中英国际低碳学院专业教师授课,内容涵盖气候科学、低碳技术、环保技能和可持续发展理念等。项目于2023年启动,面向主要人群为青少年、中年人和老年人,强化资源整合和跨界学习合作,通过一系列独具特色的教育活动,促进社区环保意识和绿色生活方式在不同群体中的普及。

一、面向青少年及家庭开展应对气候变化科普进社区系列活动

2023年,我们面向社区青少年及家庭,开展了"科学'碳'索,'气'向未来:应对气候变化与低碳发展科普进社区"系列活动(见表1),围绕科技创新和环保意识展开,旨在培养青少年及其家庭的科技创新精神和环保意识,同时提升他们的科学素养和动手实践能力。

表 1　2023 年项目安排计划

项目名称:科学"碳"索,"气"向未来:应对气候变化与低碳发展科普进社区项目(见图 1)	
参与对象:公开招募 30 组"一大一小"家庭("小"限定二年级及以上小学生)	
项目主题	具体安排
主题 1:低碳能源	应对气候变化科普课堂:应对气候变化背景下的低碳能源发展趋势 应对气候变化研学课堂:走进临港弘博新能源,"碳"究屋顶光伏对应对气候变化的贡献 应对气候变化科学"碳"索实践课堂:自制小小气象站模型
主题 2:弃物资源化	应对气候变化科普课堂:电子废弃物资源化技术的进步 应对气候变化研学课堂:走进上海生活垃圾科普展示馆(见图 2),带你了解垃圾的"前世今生" 应对气候变化科学"碳"索实践课堂(见图 3):利用回收塑料瓶自制桌面吸尘器

图 1　科学"碳"索,"气"向未来:应对气候变化与低碳发展科普进社区活动合影

图 2　研学课堂:青少年参观上海生活垃圾科普展示馆

图 3　科学"碳"索动手实践课堂

本阶段的策略与思考是：

（1）动手实践与创意结合：通过实际操作，让孩子们将废弃的塑料瓶变成有用的小工具，提升他们的动手能力和创造力。课程设计侧重展示如何将废弃物转化为有价值的物品，培养学生环保意识和资源再利用观念。

（2）家庭共同学习环保知识：适合家庭参与，增强亲子沟通和合作，共同增强全家人环保意识。通过废弃资源的再利用实践，让孩子们认识到每个人在日常生活中都可以通过小行动为环境保护作出贡献，认识到个人在应对气候变化中的角色和责任。

（3）普及科学知识：在动手制作过程中，让孩子们学习科学知识，提升科学素养，理解科技创新在应对气候变化中的重要性；通过探索废弃物的潜在价值，激发孩子们的创新精神和解决问题的能力；通过实践活动，让孩子们学会用创新的思维解决环境问题，成为未来应对气候变化的积极参与者和推动者。

二、面向老年群体开展绿色科创教育活动

2024年3月，项目以老年人为目标群体，特别关注提升其环境适应能力和生活质量，强化环保意识和社会参与度，激发思考与创新，期待他们为国家双碳发展献计献策（见表2）。

表 2　2024 年项目安排计划

项目名称：智慧老年——绿色科创与气候变化教育特别项目	
参与对象：每场活动面向社区公开招募 50 名学习者	
项目主题	具体安排
主题1：低碳智慧交通	特色课程项目启动——序章 科普讲座：上海交大中英国际低碳学院董雪副教授讲授《低碳智慧交通领域低速无人驾驶的关键技术》（见图 4） 走进高校实验室：走进低碳学院无人驾驶实验室，"碳"究科技助力应对气候变化发展（见图 5）

（续表）

	特色课程项目启动——研学篇章
主题2：智慧生活	企业研学：走进上海交大智能制造研究院，体验智慧生活。参与者走进上海交大智能制造研究院参观智能制造新技术、新突破，领略工业美学，通过体验协作机器人、心狗健康检测等项目，了解临港新片区新发展及科技创新技术，体验智慧生活带给人们的便利。 实践课堂：参与者将活动剪辑成一段研学视频，以智慧生活和气候变化教育研学活动为素材，主讲教师通过理论讲解、实用操作技巧介绍、现场提问等方式，介绍短视频的拍摄和制作；课程旨在提升老年人的数字素养和数字化学习能力，助力老年人跨越"数字鸿沟"，共享数字化发展成果，享受智慧生活
主题3：低碳能源	科普讲座：上海交大中英国际低碳学院张振东副教授讲授《应对气候变化背景下的低碳能源发展趋势》。讲座着重介绍探讨我国能源技术、能源结构与能源管理等发展路径
主题4：低碳监测	科普讲座：上海交大中英国际低碳学院任涛副教授讲授《辐射换热、温室效应与碳监测》。通过介绍太阳辐射、地球辐射与大气层的吸收之间的关系，进而谈到温室效应的形成，倡导大家积极探索低碳科技，践行低碳生活。 走进高校实验室：遥感二氧化碳监测实验室平台参观

图4　上海交通大学董雪副教授科普授课

图 5　参观上海交通大学无人驾驶实验室

本阶段的策略与思考是：

（1）专业性、实用性的课程内容设计：通过开展"绿色科创与气候变化教育"系列活动，如"科普乐学堂""绿色低碳探索团""智慧生活科技课堂""临港科创游学""绿色生活实践课"和"老年智学团建设"，积极推进构建终身学习的新典范。课程涵盖智慧生活、绿色低碳、智能科技等知识领域，旨在提升老年人的科学素养、实践能力和创新意识。

（2）互动性、实践性的教育方法：根据主题来设计系列易于理解的讲座、实地考察、动手实践等多样化活动，收集参与者反馈及时调整教学策略，增强老年人的适应能力和参与感，同时确保教育内容的趣味性和实用性，有效促进老年学员的积极参与和知识吸收。

三、面向家庭构建全方位的分层次气候教育体系

2024 年 4 月底，项目面向家庭一老一大一小，构建从家庭到社会的全方位教育体系，促进多代人共学互学，实现终身学习的目标，从而有效应对气候变化的挑战，推动社会的可持续发展（见表 3）。

表 3　代际学习项目安排

项目名称：绿色科创与气候变化教育特别项目	
参与对象：面向社区公开招募 30 组家庭"一老一大一小"共同参与	
项目主题	具体安排
循环经济	代际学习新篇章——小手牵大手 科普讲座：上海交大中英国际低碳学院李佳副教授讲授《电子废弃物资源化技术的进化》。通过介绍电子废弃物资源化回收的基本过程，以及电子废弃物回收在加速物质和能源循环、创造经济价值和环境价值过程中的重要作用。 走进高校实验室：《上海交大"低碳天团"硬核成果让"垃圾"变成"宝"》，手速快、目力好、节成本，超视觉垃圾分拣机器人"解放"人工助力垃圾分类

(续表)

科技"星场秀"	以轻松幽默的方式结合生动、有趣、直观的科学实验内容,融入舞台表演元素,传播科学知识和科学思想 ①再创"液"(全年龄段):科普密度差、伯努利原理以及液体表面张力等相关原理。 ②"振"能量(全年龄段):科普与传播正能量,科普震动的声波,可以是悦耳的歌曲,也可以是锋利的手术刀,取决于如何使用。 ③"导"亦有道(全年龄段):一起走进空间站,天宫地面体验舱,展示空间站热控系统、减振降噪以及天地通讯原理,通过科普能明白"导"中之道。 ④再"唱"东方红:在声学专家介入后,通过展示声音灭火器的威力,解释声音在火星中的特性,并用激光将声音可视化,让大家了解东方红可以在火星上奏响。 ⑤"天问"环游记:借助实验演示,科普"中国天眼""玉兔二号""天宫号"空间站霍尔推进器等大国重器背后的科学原理

本阶段的策略与思考是:

(1)多样化活动:通过讲座、互动讨论、动手实验操作、竞赛抢答等多种形式,激发参与者的学习兴趣和积极性。

(2)全社区覆盖:通过社区教育活动,广泛宣传气候变化知识,动员社区成员共同参与,形成全社区参与的良好氛围,推动绿色生活方式的普及。致力于在不同年龄层次中推广气候变化教育,培养环保意识和行动力,构建绿色创新与气候教育的终身学习新典范。

成效与反思

项目自2023年10月启动,至2024年4月,气候变化教育特色课程通过12场系列活动,成功吸引了2 000余人次的参与,对提升公众的环保意识、科技创新能力和绿色行动产生了积极影响。项目由政府、社区、企业、高校多方联合开展,增加了活动的丰富度,创新了社区可持续发展教育模式,也展示了社会各界对气候变化教育和绿色创新的广泛关注和支持。

(1)搭建平台:通过联合政府、高校、企业等多方资源,精心设计多元化的气候变化教育课程,并成功搭建了信息共享平台,推动合作服务。这一平台的建立打破了传统教育单一来源的模式,促进了各方资源的共享,还为社区居民提供了更多参与社区教育的机会。通过多方联动,项目在区域内取得了实质性的成果。学校家庭社区联动的合作学习不仅带动了绿色社区协同发展,也为未来教育项目提供了有力的合作基础。

(2)认知提升:通过一系列绿色创新和应对气候变化教育的社会参与活动,培养居民的环境责任感,确保课程内容贴近居民的日常生活和实际需求,激发社区居民生态情怀,提升个体生态素养,从而促使理论知识在实际行动中得以贯彻。项目显著提高了参与者对社会发展趋势与国家前沿低碳科创技术助力气候变化发展的认知水平。

(3)行为改变:通过创新实践探索气候变化教育与社区教育的融合,参与者亲身感受到低碳科技的影响,逐渐形成了采取低碳、环保生活方式的习惯。这种研学方式不仅提高了参与者对气候变化的认知水平,还在行为层面产生了积极的改变,为社区的可持续发展注入了新动力。此外,参与者通过分享自己的经验和见解,为社会创新和发展贡献了自己的力量,为社会的多元发展提供了宝贵的智慧。

气候变化教育特色课程项目在提升公众气候变化意识、培养科技创新能力和激发绿色行动积极性方面取得了显著成效。通过学校家庭社区联合育人,促进政府、高校和企业的跨界合作,项目实现了资源共享,推动了社区教育的共建和多元化发展。项目的成功实施不仅在教育层面取得了成效,还在社区层面推动了环保意识的普及和绿色生活方式的践行,为构建临港新片区终身学习新典范提供了有力的实践案例和宝贵经验。

然而,项目实施过程中也存在一些值得反思的地方。首先,覆盖面有待扩大:尽管项目吸引了很多社区居民参与,但仍有一部分人群由于各种原因未能加入;后续项目组会着重考虑如何更好地吸引和包容更广泛的群体,尤其是针对那些对气候变化教育不太了解或不太感兴趣的人群。其次,项目实践仍需不断创新:如何更有效地利用现有资源并不断挖掘新资源,或者寻找更多的绿色伙伴和资金支持,以实现更大的影响力和覆盖范围,也是我们需要进一步思考和努力的方面。

(李秋菊,上海市浦东新区南汇新城老年大学副校长;
王莹,上海交通大学中英国际低碳学院合作交流与高端培训部项目主管)

"凤凰"展翅 "骑"乐无穷

研究背景

党的十八大以来,气候变化教育呈现出风起云涌的可喜局面。习近平总书记指出,气候变化带给人类的挑战是现实的、严峻的、长远的。[①] 放眼国际,气候变化教育的最新进展表现为聚焦绿色课程、绿色教师、绿色能力与绿色社区四个方面;瞩目国内,气候变化教育则在理念先进、政策跟进、学校推进、研训并进等方面形成了中国方案和实践特色。推动形成绿色发展方式和生活方式已摆在突出的位置[②],如何培养和践行健康生活方式,已成为我国教育界乃至全社会需要高度重视的问题。因此,前京小学结合区域特点力求探索新形式,开展气候变化教育。

一、学校发展需求——寻求新生的校园特色项目

前京小学作为一所致力于满足未来教育需求的学校,始终秉持创新精神,不断探索与时俱进的教育模式。为进一步拓宽教育领域并深化学生对环境保护的认知,学校正计划推出一系列独特的校园特色项目,由社会各界人士提供宝贵建议与合作机会。

目前,学校正在积极评估多个项目构想,包括对"京京菜园"进行扩展升级,使其成为涵盖可持续农业教育的综合教学模块;对"小海鲸新闻传媒中心"进行升级,增强其在气候变化与环境保护宣传方面的专业性。同时,我们也在考虑引进一套互动式气候变化模拟系统,让学生在亲身体验中学习应对各种气候情境。

为实现这些新项目的落地,前京小学正积极寻求与地方企业的合作,例如与凤凰牌自行车公司探讨如何在校园内推广绿色出行方式;又如,与上海艾录包装股份有限公司、金山区漕泾化工厂等多家企业建立共建关系。我们期望通过此类合作,既能支持环保事业,又能为学生提供实践操作的平台。

社区成员、企业及教育工作者的积极参与和贡献,不仅让学校开发出了许多新的教育项目,也让学生通过这些特色项目学习到知识,亲身参与到环保的实践中。我们坚信,这将有助于培养学生对气候变化和环境保护的深刻理解与长期承诺。

二、企业发展需要——寻求转型的老品牌

凤凰牌自行车是金山区的知名品牌,拥有超过百年的历史。随着全球对气候变化的关

[①] 习近平.推动形成绿色发展方式和生活方式是发展观的一场深刻革命[J].党建,2019(5):15-17.
[②] 习近平主持中共十八届中央政治局第四十一次集体学习[EB/OL].(2017-05-26)[2023-10-20]. http://www.xinhuanet.com/politics/2017-05/26/c_1121028286.htm.

注日益增加,以及自行车文化的普及和健康生活理念的兴盛,凤凰集团认为有必要在教育合作中加入对气候变化的教育。因此,该企业与学校合作,在校内和校外成立了"凤凰驿站",开展了一系列以气候变化教育为主题的活动,引导学生将环保融入日常生活,如骑行代步,减少碳排放,从而对抗全球变暖。此举不仅能提升学生们的身体素质,也能增强他们的创新能力和对环保的责任感,为更加绿色的未来贡献力量。

项目设计

"凤凰驿站"项目致力于提升学生环境保护意识,通过学习自行车的发展历史和骑行实践活动,强化学生对低碳交通的了解。项目通过多样的教学模式,包括项目化学习、讲座、实践体验及多媒体互动,培养学生的科学探索、团队协作精神及社区责任感,并通过家长和企业的反馈确保教育的持续性发展。

一、目标设计

(1)通过了解自行车的发展历史及其对可持续交通方式的贡献,提高学生气候变化及环境保护意识。

(2)培养学生的实践能力和团队合作精神,提升科学探索能力和问题解决能力,增强公共参与意识和社区责任感。

二、内容设计

(1)理论学习内容:气候变化的科学基础及其对环境的影响;自行车发展历史及其在都市交通中的角色和未来发展。

(2)实践活动内容:头盔的美化、自行车的组装和维修工作坊;设计和参与环保骑行活动,包括路线规划与安全指导。

(3)文化教育内容:自行车在不同文化中的地位与影响;可持续发展案例研究,如世界各地的自行车友好城市。

三、方式选择

教学方式:采用"讲授+实践"的模式,通过讲座与互动讨论传播理论知识,通过实地操作和体验加深理解。利用多媒体和虚拟现实工具,增加教学的互动性和趣味性。

学生参与方式:分组进行不同的项目任务,如自行车维护、环保骑行策划等,让学生在活动过程中担任不同角色,如项目负责人、记录员等。

企业参与方式:邀请"凤凰驿站"专业教练参与课程讲座和户外骑行活动。组织公开课,推广气候变化意识。

四、评价设计

过程评价:学生方面,设立检查点,确保学生在理论学习和实践操作中都能达标,通过问卷和小组讨论收集学生对活动的反馈,及时调整教学计划;家长方面,通过与企业合作,收集家长的反馈,评估项目对学生的影响,同时设计家长问卷,了解教育项目的长远影响,确保持续性发展。

结果评价：结构性评估，通过考试和实际操作测试学生的知识掌握和技能应用情况，评价学生是否能将所学知识与实际问题结合，如策划并执行一个小型气候行动项目。

"凤凰驿站"项目通过这样全面、系统的规划，旨在提升学生对气候变化的认识和应对措施，同时增强学生实践能力，创造持久的环境保护动力。

项目实施

一、设施建设与资源整合：共同出"资"

延承"前京里"体验课程的一贯思路，开展以自行车骑行为载体的、能亲身感悟的实体场馆建设。经过双方几轮讨论，设计了"京·凤凰驿站"、户外骑行装置和凤凰主题教室三个体验场馆。在寻求专家建议后，学校添置了自行车体验场地，陈列各种类型的自行车，供学生参观和试骑。在此基础上，凤凰主题教室外围布置了自行车历史文化展示区。这些让学生不仅能动、能玩、更能学。

考虑到学校的实际情况，凤凰集团解决了必需的资金、设施、设备问题，学校主要是提供场地，以地为"资"。此次校企合作项目还得到了区政府、教育局的大力支持，"凤凰驿站"项目被列为当年金山区的重大实事项目，成为金山区落实融合教育的典型案例，为校企双方以自行车文化教育为契机共同开展气候变化教育打下了坚实的基础（见图1）。

图1　校外"凤凰驿站"

二、活动开展与教育实践:共同出"力"

一是激发学生学习的内驱力。为激发学生对气候变化教育的兴趣,学校多次携手凤凰集团开展校园趣味骑行活动,"京—凤凰"自行车模型拼搭活动(见图2)、"老品牌、新精彩"上海老品牌寻访活动、冬日童趣游园迎新活动等凤凰进校园系列活动。凤凰集团紧密配合,积极参与每一次活动,为骑行活动赞助自行车;派专家指导模型拼搭;把学生请进企业,派人引导开展老品牌寻访,从而使学校的各类活动都取得良好的效果。

图2 "京—凤凰"自行车模型拼搭活动

二是强化协同育人新模式。在实践过程中,我们探索各学科学习与凤凰骑行活动的有机整合,让低碳环保融入教育教学过程中。体育学科开展的骑自行车比赛得到了凤凰集团的大力支持,不仅提供车辆,还派专业技师来讲解"如何骑自行车更快"的技术问题,从而使活动圆满举行,有效提高了学生的自行车体育运动水准(见图3)。自然学科中设计了发电花鼓的实验活动,通过自行车的车轮驱动来发电,所产生的电能提供给手机、平板电脑等电子设备,起到了节能减排的作用。

三、课程开发与学术探索——共同出"智"

课程是教育的蓝本,驿站建成之后,我们便马不停蹄地开展课程建设。在设计阶段,校企双方共同出"智",各自抽调人员成立了编写组。通过多次研讨,从名称到结构、从内容到步骤、从形式到教学,反复推敲,终结硕果。

课程被命名为"凤凰骑行",成为学校和凤凰集团合作开发的特色课程,利用合作资源,我们开展兼具体验性和参与性的特色活动,打造交互性的学习平台,深受学生及家长的欢迎。"凤凰骑行"特色课程分启蒙、进阶、定制三个阶段实施。

(1)启蒙课程:学生了解"凤凰"这一老品牌的历史沿革、自行车种类等,从最基本的理论

图 3　校园骑行

展开教学,将凤凰集团文化融入校园文化,共同聚焦课堂,落实课堂主阵地作用。

(2)进阶课程:主要是技能性学习,引导学生探索技能方面的理论知识,以自行车模型的零部件组装、自行车头盔的美化等为基本内容。

(3)定制课程:本阶段主要通过项目化学科融合的方式让凤凰集团的文化落实到课程的方方面面,同时也充分发挥融合教育的优势,形成各科、各家齐动的可喜局面。如学校项目组孙老师和龚老师开展了"'骑'乐无穷——融合育人视野下的项目化学习探索""低碳环保亲子小调查活动"等。

五年级学生围绕驱动性问题"未到可以上路骑行年龄的你,学会骑自行车了吗?去哪里骑?请你设计一份安全、健康的骑行方案"展开。在探索、解决真实情境问题的过程中,不断提出问题、分析问题、解决问题。学生在计算骑行路程、时间、速度以及消耗的热量或 BMI 指数后,规划健康、合理的骑行方案。最终,我们将学生提交的骑行方案进行整理后将其分为运动类骑行方案、观光类骑行路线推荐、上下学最佳路线推荐三种。每一种方案都各有特点,不同的出行目的,设计方案也不相同。在此次项目化学习中,学生不仅锻炼了身体,体会了骑行带来的乐趣,也进一步巩固了数学学科的知识与技能,体会到了数学与日常生活的密切联系。

四、教师培训与双师模型——共同出"人"

首先,是"双师"制的设置。鉴于学校教师无法解答一些关于自行车的技术性问题,提出了"双师"制的方式,即每一次活动,由凤凰集团派出一位专家,学校派一位教师来具体实施。实践中,凤凰集团与前京小学紧密合作,共同推进双师教学。在这种模式下,两位教师共同参与教学过程,校内教师负责课堂教学和学生学习管理,校外导师负责答疑解惑和实践操作,共同提高教育质量,保障教学活动顺利开展(见图4)。

图 4　校外导师答疑指导

其次,是充分挖掘"双师"制的优势。在校企双师指导下,以凤凰牌自行车为活动载体,开展了"探寻'凤凰'百年史""我家的'凤凰'故事""我为'凤凰'支新招""百年'凤凰',一路相伴"等课堂探究活动和校外亲子活动、小队活动、中队活动等具身实践。凤凰集团的专业人员讲解凤凰自行车的历史、发展以及其他专业问题,取得了良好的活动效果。

自然学科的胡老师深入浅出地传授中国自行车的历史发展和车型分类等理论知识,而校外导师则以其丰富的实践经验,教授学生安全骑行的技巧,实现了知识与技能的和谐统一。此外,校企合作精心策划了"2024 趣味游园迎新活动""自行车头盔创意设计大赛"以及"国潮进校园,凤凰助成长"等一系列别开生面的活动。这些活动不仅极大地提升了学生的实践能力、创新思维和团队协作能力,更为校园文化生活注入了新的活力与色彩,让每一位小学生都能在丰富多彩的体验中快乐成长。

校企联合的"双师"制能有力促进活动的有效性,深刻展现教育的协作性。特别是双方产生文化共鸣后,无论是校园文化还是企业文化,都可以更好地传承与发展。

成效与反思

与凤凰集团的携手,使学校以自行车为载体的气候变化教育得以有效展开,成为学校一张亮丽的名片。在项目开展的过程中,学生们探寻自身需求、采集真实数据,大大激发了参与热情,也深刻地意识到了绿色出行的必要性。同时我们也感受到了校企合作开展气候变化教育的极大魅力。对于后续的展望,我们也有几点思考。

一、推动与凤凰的合作项目向纵深发展,做强"凤凰驿站"品牌

进一步融入科技:我们应紧跟时代科技的发展,让互联网科技加持项目的开展,使我们

的项目"骑"上互联网,设立校企共建"凤凰互联网网站",鼓励更多的学生、家长、工人以及社会人员一起为倡导"绿色出行"贡献力量。

进一步落实主题:加强"自行车上看气象"的活动主题聚焦,提高学生的学习兴趣,尝试以学校、家庭、社会的协同学习,实现从学校到社会的推广,不断扩大气象科普的影响力。

二、推广与凤凰合作项目的成功经验,做亮"气候教育"品牌

进一步挖掘原有场馆:学校特色项目"前京里"中,还有许多有待我们挖掘和发展的资源,可以用于气候教育的开展。比如"京京菜园"场地,就与气候变化有着天然的联系,金山又是农业高度发达的地区,有很多的企业从事着蔬菜的种植、经营产业。为此,本着校企合作,协同推进气候变化教育的思路,我们将走出校园,寻找合适的蔬菜企业合作开发学校的"京京菜园",使之成为我校第二个校企合作的典范。

进一步开发新的体验教育场馆:我们将以强化气候变化教育为宗旨,不断开发新的体验场馆,在针对性上着力,设想建设"前京小学气象站",供学生随时参与、体验和探究。

气候变化教育需要我们以昂扬的姿态不断耕耘、不停探索。希望我校能够依托"凤凰"展翅,吸引更多企业加入气候变化教育的队伍,为努力实现习近平总书记倡导的"重视生态文明教育"而协同奋进。

(钱欢欣,上海市金山区前京小学校长;
蒋丹妍,上海市金山区前京小学教师)

传导"绿色生产力",打通气候教育"链"

研究背景

气候变化正在发生,人类活动是其主要原因。早在1992年联合国大会通过的《联合国气候变化框架公约》就已明确提及人类活动这一影响因素,"'气候变化'指除在类似时期内所观测的气候的自然变异之外,由于直接或间接的人类活动改变了地球大气的组成而造成的气候变化"[1]。因而,气候变化教育的重心就在于教授和改变人的活动方式,生产、生活作为人的两类主要活动,都应成为学习的重要议题。

应对气候变化需"从娃娃抓起",小学具有不可替代的扎根力量。而当下的小学气候变化教育,往往只关注学生在校在家的生活方面,对于生产活动因为觉得与学生很"遥远"而涉猎不多。事实上,没有生产方式与产品的改变,人类就很难迎来适应气候变化的生活方式的改变。小学教育要培养出有意识、有能力参与气候行动的年轻一代,也必须强化当下育人与未来产业用人之间的关联思维,增强气候变化教育的系统性。

常州市新北区龙虎塘实验小学(简称"龙小")从街道层面着力,在学校家庭社区合作方面形成了特色,自2023年5月启动气候变化研究以来,也逐步形成了学校家庭社区协同开展的独特样态。龙虎塘街道位于常州高新区,基于镇域特点把做大做强做优产业作为头号工程,致力于锻造强大的"产业链",形成了特色鲜明的"一特一新一现代"产业集群。在这样的情境下,校企合作开展气候变化教育进入了学校的研究视域,并聚焦气候变化与生产活动及推动其发展的生产力之间的关系,清晰了两大研究问题与空间。

一、绿色生产力:气候变化教育"链"有待打通的薄弱"环"

习近平总书记指出:"绿色发展是高质量发展的底色,新质生产力本身就是绿色生产力。必须加快发展方式绿色转型,助力碳达峰碳中和。"绿色生产力是一种对"环境友好"的可持续生产力,也是气候变化教育所要研究、倡导、促进的关键因素之一。绿色生产与绿色消费、绿色生活组成了人们应对气候变化的行动链条,又因为其是后面两者的"输入端"而起着先决性作用,理应成为气候变化教育的重要内容。但在当下的小学气候变化教育中,绿色消费、绿色生活已经得到了普遍关注,"绿色生产"却往往被忽略,其研究现状的薄弱也意味着具有更大的研究潜力与探索空间。

二、校企合作圈:小学气候变化教育有待开发的关键"场"

在当下的研究与实践中,学校家庭社区协同育人往往被简化为学校家庭协同,尤其是小

---

[1] United Nations. United Nations framwork convention on climate change[EB/OL].1992[2023-12-04].https://unfccc.int/resource/docs/convkp/conveng.pdf.

学。龙小虽然在学校家庭社区合作方面已经取得了实质性进展,但也只是把社区、政府部门作为主要合作对象,因而,自开展气候变化教育研究以来,校外合作的主阵地也是放在家庭和社区,涉及的内容主要是绿色理念、绿色消费、绿色生活方式的宣传与践行,而对企业及其生产的教育价值开发并未重视。

此外,在中国知网(CNKI)检索包含"气候变化教育""合作"等主题词的期刊文献,5年来的情况如表1所示。从中,亦可以看出,研究者对气候变化教育的关注与投入呈现上升趋势,但在小学,从合作的视角,尤其是校企合作的维度,是有待开创性补白的重要场域。

表1 与"气候变化教育""合作"等主题相关的文献数量趋势(2019—2023年)

年度	气候变化教育	气候变化教育＋小学	气候变化教育＋合作	气候变化教育＋校企合作
2019	24	0	2	0
2020	19	0	0	0
2021	28	0	0	0
2022	38	2	0	0
2023	57	1	1	0

项目设计

从上述研究情境和研究问题出发,本项目聚焦校企合作,遵循促进气候变化产业和学界跨学科合作的主体思路,以共建共享、共学互学为基本原则,致力于完善生产生活一体化的气候变化教育内容体系,为气候变化教育的"全机构"实施探索路径,并在这一过程中进一步影响学生、企业员工的家庭及所在社区、相关部门等,以唤醒更多学习者的气候责任,不断传播可持续发展文化,从而形成了如下的研究目标与思路。

一、研究目标

(1)基于气候变化的人为原因,厘清绿色生产力与气候变化教育的关系,建构生产、生活一体化的气候变化教育全要素链条。

(2)基于校企合作的深度探索,相互促成气候变化教育的"全学校""全企业"式落地,创新学校家庭社区协同育人的路径与样态。

(3)基于全民终身学习视野,发展学生、教师与家长、企业员工、社区社会人士等的绿色教育伙伴关系,在合作开展气候变化教育的过程中互学共创,促成多主体的素养一体化培育与可持续发展理念的传播。

二、研究思路(见图 1)

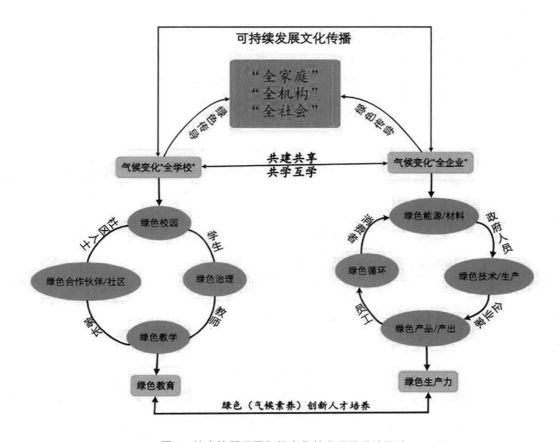

图 1　校企协同开展气候变化教育项目设计思路

项目实施

一、调研梳理：聚焦绿色生产核心要素，构建校企试验共同体

首先，项目组通过网上数据查阅、咨询街道经济和科技发展局等途径，了解龙虎塘企业的整体布局，尤其是高新企业的特点。由此获悉，至 2023 年，注册企业约 7 000 家，有效期内的高新企业 103 家。

随后，我们依据适应气候变化的绿色生产关键要素——能源或材料、技术、产品、循环经济，锚定了与之相关的企业，一类是此前与学校有过合作的，一类是在某领域特别具有典型性的，通过家长、政府领导、公益组织等资源进行衔接与沟通，建立合作意向。

接着，项目组成员通过去企业走访、与企业管理人员交流等，相对深入地了解企业的文化、队伍、生产或产品特点(见图 2)。

图 2　项目组成员走访天合光能有限公司、温康纳机械制造有限公司

在此基础上,学校结对了第一批气候变化教育企业共同体,并对合作研究的方向、内容形成整体思考(见表 2)。

表 2　气候变化教育校企合作共同体及研究方向

研究(绿色生产)要素	合作企业	气候功能
绿色能源/材料	天合光能股份有限公司/天合富家能源股份有限公司 聚合新材料股份有限公司	清洁能源,低碳化无碳化
绿色技术	温康纳机械制造有限公司	减排、增"汇"
绿色产品	森萨塔科技有限公司	低碳节能
绿色循环	瑞赛环保科技有限公司 光大环保能源有限公司	节约资源,低碳循环发展

二、打造样板:聚焦碳中和首要目标,探索能源革命教育化

百年来,为了获得电、热等能源,人类早已习惯了燃烧煤炭、石油为主的化石能源。但其

产生的二氧化碳、甲烷等温室气体成为气候变化、全球变暖的主因。因此，碳中和的目标首先指向了能源革命。由此，学校借助家长资源，与街道最具代表性的新能源企业"天合光能股份有限公司"率先开展深度合作，以期探索出可行性的基本模式再加以推广。天合公司已经形成了光伏产业集群，考虑到小学生的认知特点，我们进一步聚焦到其旗下主要开发家用光伏产品的"天合富家能源股份有限公司"（下文简称"天合富家"），目前合作已进行至第四阶段。

第一阶段：构建合作愿景。2024年年初，学校项目组一方面走进"天合富家"进行参观、学习；一方面邀请公司管理团队来校座谈、研讨，并促成了街道、生态环境局等各部门人员，家委会、高校专家等的共同参与（见图3）。校企双方将各自的文化追求加以融合，形成了"传导数智能源零碳生产，共育零碳生活绿色未来"的气候变化教育合作愿景，并基于各自的实践资源和经验，初步达成了近期合做一个活动、中期合建一个基地、长期合成一个样板的三个"一"行动计划。

图3　多方共同参与的校企互访、座谈、共研

第二阶段：打造经典活动。2024年寒假，学校与"天合富家"联合组织了"走近清洁能源，创构绿色未来——龙矗矗富家杯全国科创画大赛"，鼓励学生以家庭代际共学互学（尤其关注与企业员工家庭间的联动）、玩伴团活动研学等方式，线上线下探究清洁能源与气候变化的关系、了解太阳能产品，并能基于家庭过年生活的需求与体验设计新能源产品，最终用科创画的形式进行成果展示。3月1日，在学校承办的"'学校家庭社会协同开展气候变化教育'第二届全国研讨会"上，我们举行了活动的颁奖仪式，公司颁发给获奖同学的奖品亦是系列光伏产品。随后，活动成果也通过学校、企业的平台，用小视频、宣讲等方式向更多人传播（见图4、图5）。

图 4　龙龘龘富家杯全国科创画大赛中的玩伴团研学、颁奖仪式

图 5　龙龘龘富家杯全国科创画大赛的部分作品

第三阶段:策展系列合作。全国科创画大赛的开展与效应,为学校与"天合富家"有限公司合作开展气候变化教育积累了经验、坚定了信心。"天合富家"把学校举荐给了天合光能股份有限公司。2024年4月9日,天合公益基金会负责人、天合富家品牌主管来到学校,与项目组核心成员就长期的、可持续合作展开研讨,并具体商讨了本年度的合作计划(见图6)。后经双方几次细化、论证,形成了《光伏＋双碳教育——2024气候变化教育合作行动(天合公司＆龙虎塘实验小学)》方案,主要从基地建设、课程打造、活动开展、成果传播四方面开发系列合作活动(见图7)。"天合富家"的技术人员也随即来校进行勘测,并设计好了气候变化教育基地建设中的光伏项目方案。

图6　校企双方就年度合作计划进行研讨

主题版块	具体内容	实施方式	时间
基地建设	生态秘境建设	气候、气象观测设备的引进	4~8月
	校园光伏建设	图书馆及校园各类屋顶的光伏发电建设。打造空中绿色庭院	
课程打造	依托学校建模俱乐部的特色社团,合作开发"智能家居设计与制作"普及清洁能源知识、推广天合理念和产品、创新产品。	1. 校本课程开发 2. 比赛模型设计	6~8月
活动开展	组织绿色生态玩伴团,家校社企多方卷入,常态合作开展气候变化教育主题实践活动	1. 参观柯林双碳劳动基地 2. 新能源职业初体验	4~6月
成果传播	多种渠道传播合作成果	1. 合作参与或主办气候变化教育研讨会(国内/国际)	4月 上海气候周气候变化教育论坛
			9月 你好暑假
		2. 学术成果发表(论文、案例、著作)	9~12月

图7　《光伏＋双碳教育——2024气候变化教育合作行动》方案书(部分)及校园光伏项目建议书(部分)

第四阶段:复盘合作经验。依托4月25—27日举行的"2024上海气候周"活动,学校推动师生、"天合富家"管理人员共同复盘、总结了合作经验。在气候周的气候变化教育论坛上,校长顾惠芬、天合富家能源股份有限公司公共关系与品牌部负责人蒋宁分别从学校、企业的角度分享了合作成果;付蓉老师和学生小队也分别在"气候韧性社区论坛""青少年气候行动科创作品宣讲会"上介绍了相关合作研究与实践(见图8)。"新华网"作了专题报道——《常州:校企共赴"气候周",点亮"新经验"》。

图8　学校师生、"天合富家"管理人员在"2024上海气候周"传播校企合作经验

三、多"环"实践：聚焦生产生活的一体化融通，做强气候变化教育的全民性

与天合富家能源股份有限公司的合作，初步实现了"能源革命"的"气候变化教育化"。学校以此经验为观照，整体梳理了与其他企业的前期合作，与温康纳机械制造有限公司、森萨塔科技有限公司、光大环保能源有限公司等也开始了实践探索，分别聚焦"绿色技术减排增'汇'""绿色产品低碳节能""绿色循环分类再生"等绿色生产关键环节、核心价值拓展教育主题。

在实践过程中，项目组致力于打通绿色生产与绿色生活（含绿色消费）之间的壁垒，构建一体化的气候变化教育"链"，主要通过三大行动策略实现：

一是以学校、企业的双向传导为主要路径。师生、家长以玩伴团研学等方式走进企业，企业人员以公益服务等形式走进学校；学生、家长向企业人员倡导"绿色生活方式"，企业人员向师生、家长传递"绿色生产"的理念、技术与产品。以与森萨塔科技有限公司合作开展的"六一"活动为例，5月31日，公司团队进入校园，与科学教研组合作，和孩子们一起"摆摊设点"，传播"传感技术，低碳产品"；6月1日，学生组队走进公司，联合举办"STEM公益课堂"，企业讲师教授新能源新技术，工程师们与学生合力制作遥控电动车模型并进行小组赛，学生向企业员工宣讲关于气候变化的研究。

二是基于学生立场促进"场景化"融通。学校、家庭、社区是学生生活的主要场域，从学生身处其中的实际需求出发，增强校企合作开展气候变化教育的多主题整合与多场景融通。如在读书节活动中，学校与森萨塔公司合作开展了"以'书'易'蔬'"公益活动，传播循环经济理念。又如，与光大环保公司等合作的"垃圾分类"活动，也会同时与周边的社区合作来开展。

三是在终身教育背景下推动更多主体的共学互学。项目组将学校"多力驱动，多环交融，多学赋能"的学校家庭社区合作模式沿用至气候变化教育的校企合作中，着力推动生态环境部门、气象机构、终身教育系统等的融入，家庭内的三代间、学生家庭与企业员工家庭之间的联动与互学，让更多群体卷入应对气候变化的学习与行动。

成效与反思

一、完善了内容体系，但还需进一步增强气候变化教育的结构内生力

校企合作，让小学气候变化教育跳出了常规视野，从局限于气候变化的理解与认知、绿

色种植、绿色生活方式等内容,进入了生产领域,让教师从人类活动这一影响气候变化的主要因素出发,对教育内容加以完善,也让学生对新时代所召唤的"绿色生产力"亦即"新质生产力"有了感知,这有利于他们从小埋下绿色创新的种子,将来成为绿色创新人才。

但要通过结构调整释放内生动力,真正激发气候变化教育的创新活力,还需要进一步对生产领域内部、生产与生活之间、生产生活与气候变化等维度进行系统梳理,以校本课程、公益课堂等形式增强教育内容的整体建构;更需要从育人价值开发的角度,寻找到学习对象的最近发展区,增强他们的发展内驱,尤其是要在学校的绿色素养培育与未来绿色产业用人之间建立长程视野,为创新人才培养机制作出贡献。

二、拓展了合作系统,但还需进一步激活气候变化教育的生态赋能力

通过校企合作开展气候变化教育,学校家庭社区协同育人的格局得到了延展。在以往以学生家庭、事业部门、社区为主的基础上,学校与各级生态环境局乃至生态环境部、终身教育研究机构、服务于企业的经科局等政府部门、企业及其内部的公益基金会或社会服务组织都建立起了合作关系,也有效增强了气候变化教育的影响力。

但是,这样一个庞大的合作系统,要真正成就一方绿色生态、减缓气候变化,就不能止步于一项活动、一次课堂、一个基地的创研学,还需要把散落各方的气候变化教育资源联接成一张"地图",把有效的合作经验转化为一套常态行动机制,把偏重于一方的教育效益扩大为多方共享……

三、创新了互学网络,但还需进一步提升气候变化教育的全民行动力

学校在十多年的学校家庭社区合作研究中,已经初步建立起了横向学校家庭社区三方主体之间、纵向家庭三代之间的共学互学网络。而在气候变化教育的校企合作中,又加强了与生态环境、经济发展相关的政府、机构人员以及企业员工的共学互学,尤其重视学生家庭与企业员工家庭的联动,并通过企业公众号、全国气候变化教育联盟群等线上方式与更多"民众"共学互学……从学生的学习心得与成果评价,企业、政府人员等的活动反馈来看,低碳生活、绿色发展的意识均得到了强化。

但要让应对气候变化的意识真正化为学习者日常的点滴行动,还有很长的路要走,需要学校、企业在合作中进一步探索"全学校""全企业"的实践模式,提炼出"全机构"式的实施路径与具体策略,从而去带动更多单位、单元的在地化实施,让更多工作、生活在其中的人担当起气候责任,实施气候行动。

(顾惠芬,江苏省常州市新北区龙虎塘实验小学校长;
陈亚兰,江苏省常州市新北区龙虎塘实验小学副校长)

黄河口长江口学生协同开展绿色企业参观活动的个案研究

研究背景

2023年5月,在华东师范大学上海终身教育研究院的推动下,"黄河口长江口联动开展可持续发展教育"项目成功确立,参与研究的两地教育工作者、研究者在气候变化背景下,通过学校、家庭、社会协同,黄河口、长江口联动,创新开展可持续发展教育[1]。此举让黄河长江以另一种方式交融相通,也为山东省东营市胜利孤岛第一小学与上海市第六师范学校第二附属小学两所入海口城市的小学搭起了友谊之桥。

山东省东营市作为胜利油田的主产区与所在地,因"油"而建,因"油"而兴。山东省东营市胜利孤岛第一小学(简称"胜利孤岛一小")位于东营市河口区孤岛镇,曾是胜利油田孤岛采油厂的一所子弟学校,现在仍有近半数的学生家长从事与石油有关的职业。

减缓和应对气候变化是国际共识,是全人类共同的责任。而煤炭、石油、天然气等化石燃料燃烧产生大量温室气体,成为造成气候恶劣的主要原因。根据《巴黎协定》第二条所述,21世纪要把全球平均气温升幅控制在2℃之内,并努力限制在1.5℃之内。这就给化石能源企业提出了绿色低碳转型的要求。2021年10月21日,习近平总书记来到位于黄河入海口的胜利油田考察,勉励大家要端牢能源饭碗,集中资源攻克关键核心技术,加快清洁能源高效开发利用,提升能源供给质量、利用效率和减碳水平[2]。

2023年12月,胜利孤岛一小"芦芽儿"低碳环保社团的同学们了解到,在阿联酋迪拜举行的《联合国气候变化框架公约》第28届缔约方会议(COP28)讨论的重点包括制定减排方案、减少石油及煤炭等化石燃料生产、增加可再生能源使用等议题时,立刻想到父母所在的企业——胜利油田孤岛采油厂。作为化石能源企业,孤岛采油厂在"双碳"背景下,是如何向绿色低碳高质量发展转型的?带着这个疑问,胜利孤岛一小与孤岛采油厂对接,开展了东营、上海两地联动,线上线下同步的研学实践活动。

项目设计

一、项目目标

东营、上海两地学生通过线上线下同时参观胜利油田孤岛采油厂采油管理七区的生产

---

[1] 陈振林,王昌林.应对气候变化报告(2023)积极稳妥推进碳达峰碳中和[M].北京:社会科学文献出版社,2023:212-225.
[2] 大众日报.情满黄河心系海岱——习近平总书记在山东考察回访记[EB/OL].(2021-10-23)[2024-06-19]. http://www.shandong.gov.cn/art/2021/10/23/art_97902_508890.html.

运行情况,了解化石能源企业为应对气候变化推进绿色低碳转型发展的做法。以此为契机,为两个入海口城市的少年搭建共学互学平台。具体来说,要做到:

学生方面:了解石油企业在光伏发电、能源数字化等方面的低碳节能生产情况,关注可再生能源,感受石油人创新发展的精神,并在项目化学习中共学互学,学会思考、学会探究、学会学习、学会合作。

企业方面:面向学生、教师、家长开放,提供实践机会,介绍化石能源企业低碳转型做法,普及新能源知识,增强学习者对气候变化的实际理解,承担社会责任。

学校方面:探索学校与企业的合作路径,健全学校、家庭、社会协同推行气候变化教育机制。

二、项目流程(见图1)

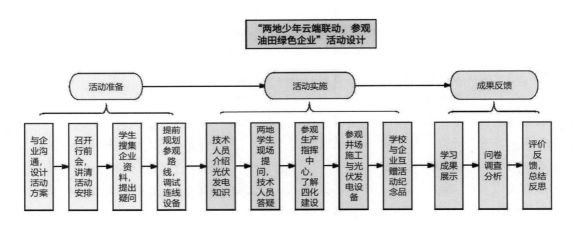

图1 "两地少年云端联动,参观油田绿色企业"活动流程设计

三、评价设计

根据活动中学生积极主动、学习思考、团结合作、成果展示、总结反思等方面的表现,对学生进行星级评价,最终授予"低碳环保小达人""低碳环保小卫士"综合奖,以及针对"学习思考"和"成果展示"环节评选出"应对气候变化探究奖""应对气候变化行动奖"等单项奖。

项目实施

一、活动准备阶段

为制订详实的活动方案,2024年元旦之后,孤岛采油厂团委与胜利孤岛一小多次进行沟通,商讨活动主题、时间、地点、内容等细节,制订了《大手拉小手,参观绿色企业》活动方案》。后在华东师范大学上海终身教育研究院执行副院长李家成教授的鼓励与指导下,实现了与六师二附小开展同步云参观的活动设想,方案升级为《两地少年云端联动,参观油田绿色企业》。

绝大部分学生是第一次到油区现场参观,对油田生产充满好奇,但对具体事项缺乏认知,所以学校编写了《参观油田绿色企业"明白纸"》,告知活动时间、活动内容、活动流程、活动时长、组织形式等,于1月12日晚召开了学生和家长线上会议,明确活动要求,进行相关培训。

参观过程中,安全防护至关重要。1月16日和19日,学校两次与孤岛采油厂团委和采油管理七区负责同志一起规划参观路线,并实地"踩点",管控交通安全,确保室内外活动全程安全有序。

学习与思考是相辅相成的。活动方案下发后,东营与上海两地的老师分别组织学生搜集有关参观活动的资料,并提出疑问,以激发学生探究积极性,为研学实践做好知识储备和心理准备。

多方协同是活动开展的保障。学校除了与企业沟通,还利用微信接龙、共享文档等形式,与学生、家长对接,最终确定了活动参与人员,设计了分组乘坐交通工具的名单等。活动前一天,胜利孤岛一小和上海六师二附小两校的活动负责人在网络会议室调试连线设备,确保第二天活动中网络通信畅通,设备操作熟练。

二、活动实施阶段

2024年1月20日8:30,胜利孤岛一小"芦芽儿"低碳环保社团成员来到孤岛采油厂采油管理七区,与远在一千公里以外的上海六师二附小的同学通过云端连线的方式,进行"大手拉小手"绿色企业参观活动(见图2)。

图2 上海市第六师范学校第二附属小学与山东省东营市胜利孤岛第一小学师生线上联动打招呼

在这次活动中,首先由油田技术人员介绍了单位生产基本情况和光伏建设情况,接着上海六师二附小与胜利孤岛一小的同学们轮流提问,采油管理七区工程主任现场一一作答。

学生们提出的问题包括"三次采油技术包括哪些?""石油会用完吗?""光伏发电是否污染环境?""光伏发电会对气候变化产生影响吗?"涉及石油生产、光伏新能源、生物多样性、气候变化等方面,有广度,有深度,体现出学生强烈的探究意识、生态环保意识。

科普学习与问答环节结束后,大家参观了采油管理七区安全生产指挥中心,通过大屏幕实时观测到井场现场的景象,并通过对话了解到:2017年以来,胜利油田探索实施了标准化设计、模块化建设、标准化采购、信息化提升的"四化"建设(见图3)。"四化"系统不仅实现了生产源头实时采集数据,还实现了远程调参、开关井等智能操作。以前巡查一口井需要30分钟,现在巡查完管理区300余口油水井仅需要30分钟,工作效率得到了有效提升。

图3 工作人员介绍"四化"系统

离开安全生产指挥中心,同学们继续前往采油管理七区井工厂现场。油田技术人员现场教学,引领大家认识游梁式抽油机的构造和工作原理,强调了安全防范知识,介绍了井场一角矗立的一大片太阳能光伏板,让同学们了解可再生能源如何发电。

在离开采油管理七区前,胜利孤岛一小的同学们将自己精心准备的小礼物,献给油田工人叔叔阿姨,表达心中的感恩之情,同时送上诚挚的新春祝福。孤岛采油厂团委赠送给同学们抽油机拼装模型,希望大家在动手操作中,不断思考,不断学习,成长为明天的国之栋梁(见图4)。

这次参观活动对于上海、东营两地少年来说,不仅增长了见识,了解了光伏新能源,还感受到了油田工人劳动的辛苦。最后师生、家长、石油工人一起面对镜头,大声喊出活动口号:"黄河口长江口可持续发展,低碳生活从我做起!"

三、成果反馈阶段

两地少年云端联动,参观油田绿色企业后,教师布置了成果展示任务。六师二附小的17名同学制作了精美的学习小报,图文并茂地记录了当天学到的知识,以及各自最感兴趣的内

图 4　东营市胜利孤岛第一小学学生在井场展示手中的抽油机模型

容。胜利孤岛一小的 16 名"芦芽儿"低碳环保社团成员将抽油机模型拼装好,涂上颜色,然后安上电池,小抽油机就开始"工作"了,同学们爱不释手。

为真实了解学校和企业对本次协同开展气候变化教育实践活动的评价,胜利孤岛一小设计了《上海—东营两地联动,"大手拉小手"绿色企业参观活动反馈调查问卷》。问卷分学生卷和成人卷两种,学生卷共有 8 道题,成人卷共有 7 道题;其中主观题均为 4 道,包括指出该活动的特别之处、印象深刻之处,以及参加活动的收获、对类似活动的建议等。调查问卷分发给两所学校的老师、家长和学生,以及胜利油田孤岛采油厂的管理人员和技术人员。共收回成人卷 25 份,学生卷 33 份,回收率 100%,问卷有效率 100%。此外,六师二附小还提供了 3 份家长撰写的活动反馈单。参与者的积极反馈为科学分析本次活动得失奠定了基础。

本次绿色企业参观活动中,学生展现出了良好的思考力、学习力、实践力和综合解决问题力等多方面能力。根据活动参与度及学习成果提交情况,胜利孤岛一小评选出"低碳环保小达人"7 人,"低碳环保小卫士"9 人。另外,根据学生学习、思考、提问情况,评选出"应对气候变化探究奖"6 名。根据抽油机模型组装及活动手册提交情况,评选出"应对气候变化行动奖"5 名。

成效与反思

一、活动成效

通过《上海—东营两地联动,"大手拉小手"绿色企业参观活动反馈调查问卷》分析,可以反馈出以下活动成效。

其一，数字化学习成为最大亮点。活动充分利用数字化资源，借助数字化平台开展互动交流、问卷调查，提高学习效益。"如果没有数字化的加持，这种场景几乎是难以想象的；正是因为教育数字化的发展、更多样化的主体参与，带来了更丰富、更复杂的教育过程，教育效益更是成倍地放大。"[①]

如问及"你以前参加过这样的活动吗"，学生的回答很统一：没有。成人卷中，只有 1 人（占 4%）参加过。当问及"你认为这次活动最特别的是什么"，有超过 1/3 的被调查者提到"连线"或"联动"。部分学生卷的回答摘录如下：

王同学：跨越了地域，就像实地参观一样，身临其境地边看边学，还能及时交流互动。

徐同学：最特别的是跟远在上海的同学一起学习了更多的知识。

部分成人卷的回答摘录如下：

赵家长：通过视频通话将两地的孩子拉近，让上海的孩子也能拿到关于绿色企业的第一手资料！

陆老师：两地可以联动，学校深入企业，促进学生多方面发展。

其二，光伏新能源成为关注热点。学生卷第 5 题是"你认为这次绿色企业参观活动，最特别的是什么？"调查发现，同学们对于光伏发电、石油开采等方面的知识非常感兴趣。"光伏"成为此问题答案的高频词。在上海学生的学习小报中也可以看到，每个孩子都画出了光伏发电的样子，说明这一点对孩子们确实很有吸引力。

其三，互动提问留下了记忆点。调查发现，活动中善于思考提问的学生容易从群体中脱颖而出。学生卷第 6 题"你印象最深的同学是谁？"结果显示，胜利孤岛一小的郭启瑶同学、六师二附小的景梓萌同学给大家留下了深刻印象，因为她们积极勇敢、开朗大方。可见，两地学生在共学互学中，互为榜样，共同提升。

其四，参观企业激发了兴趣点。调查结果显示（见图 5），本次两地联动参观石油企业，学生、家长和老师对企业生产运行模式、远程监控智能管理、光伏新能源运用等非常感兴趣，印象深刻，从下面的反馈中可见一斑：

上海老师：最特别的是企业与学校联手，这在上海比较少见，学生对企业文化比较感兴趣，以前只从父母描述中略知一二。

企业技术人员：该活动理论结合实际，把从教室学到的东西拓展到生活的方方面面。

上海家长：印象最深的是施工现场比较荒凉，工作环境比较艰苦，石油工人的子女也都很体谅父母的辛苦。

---

① 李家成.数字化为协同育人提质增速[N].中国教育报，2024-03-24(04).

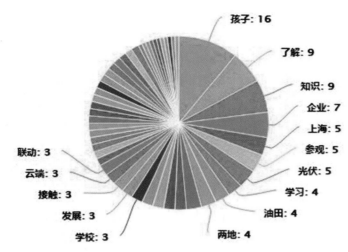

图 5 《上海—东营两地联动,"大手拉小手"绿色企业参观活动反馈调查问卷》(成人卷)
第 5 题"你印象深刻的是什么"结果统计

其五,激发了企业气候责任感。油田企业通过开放生产指挥中心和井场,接待师生家长参观,向公众展示了油田企业为应对气候变化所进行的生产方式转型,便于学生理解"发展可再生能源,采用技术创新实现节能减排"等治理方式。

企业人员反馈:

(1)作为企业,第一次与学校合作这样的活动,让孩子们能近距离接触到石油及新能源发展,也增加了企业的知名度,提升了社会责任感。

(2)为迎接学生参观,采油厂前期做足各项准备,制定安全预案,参观中向"小油娃"展示油田基本生产运行情况,让他们了解爸爸妈妈的工作环境,知道油田为了适应气候变化,在低碳环保方面做出的努力。

二、反思与跟进

一是要强化责任担当。气候变化是跨越国界的全球性挑战。气候变化教育应该面向多主体,多主题开展研究,需要学校、家庭、社会、政府、企业等协同推进,增强"人类命运共同体"意识,认识到点滴努力与系统更新之间存在关系,展望当下"做"与"不做"对未来的影响,从而采取低碳环保措施,为早日实现"双碳"目标努力。

二是要凸显继承发扬。胜利油田孤岛采油厂积极响应环保号召,承担社会责任,通过可再生能源发电、提高能源效率、推进能源数字化等措施,实现企业可持续发展。以前"小油娃"虽然生长在油城,但很少有机会到油田一线深入了解生产情况,感受父母工作辛苦。这次参观活动,以及活动后动手制作抽油机模型,让"小油娃"们了解了石油开采过程和父母工作环境,引发了孩子们如何继承和发扬石油精神、增强社会责任感的思考。

三是要提质共学互学。两地学生在共学中,呈现出不同的学习方式,比如上海的学生活动前广泛搜集资料,活动中大胆提问并做好记录,活动后完成学习小报巩固所学。东营的学生活动前制作感恩礼物,活动后动手制作模型具象感知。两校表示今后将互学互促,多层面深入交流。

今后,学校将以气候变化教育为抓手,结合新能源、生态环境、国防科技、农业生产、低碳生活、绿色城市、绿色校园等方面,深入开展多学科融合的项目化学习,促进学生将所学的气候变化知识渗透、体现在日常生活中。学校也将借助数字媒体技术,拓展活动内容,创新活动形式,促进校家社协同育人实践的可持续发展。

(赵嫔婷,东营市胜利孤岛第一小学副校长)

家庭绿色制冷原理探究及操作实践

研究背景

"绿色光年"是一家以"让公众快乐地践行可持续发展理念"为使命的公益组织。作为一家可持续发展教育机构,绿色光年通过招募并开展大学生导师师资培训,与中小学校合作开展课题研究,共同设计和打磨课程,持续向中小学生传播可持续发展教育理念。

2022年3月,南极康格冰盖从南极大陆断裂的新闻,引起了我们的关注。这个面积相当于纽约和罗马面积的巨大冰川将整体消失。科学家表示,由于从澳大利亚来的热浪席卷了整个南极大陆,整个南极所有科考站观测到的温度升高的数据全部比历史上平均高出40℃,这也是科学家们观测到南极有人类记录以来最高的升温记录。

在2023年夏天,全国都持续高温的情况下,几乎每家每户都要打开空调降温,这会引起热岛效应,让城市变得更加炎热,从而需要更多的能源来降温。那么要度过这个夏天,我们得消耗掉多少电呢?我们又如何用更低碳更省电的方式减少空调使用,或者更绿色低碳地让家里降温,同时还能省钱呢?带着这些疑问,"绿色光年"组建了由来自上海、杭州、苏州和无锡的13个家庭组成的团队,开启家庭绿色制冷原理的研究。

项目设计

项目设计(见图1)呈现以下特征:

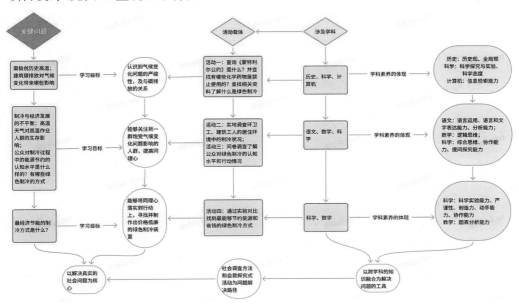

图1　项目设计思路

一是以解决真实的社会问题为核心目标。绿色制冷对减缓气候变化，提高高温作业人员福祉具有重要意义，也真实存在于我们每个人的身边。我们以解决实际生活中的家庭绿色制冷问题为目标，让学生在掌握制冷知识、节能技术的基础上，深入探索和研究，找到一种既经济又节能的绿色制冷方式。这一过程中，学生需要综合运用多学科知识，如物理学、化学、环境科学等，通过实地调查、问卷采访、实验研究等一系列活动，真正了解社会需求，提高自己的社会责任感。

二是以社会调查和自我探究式的活动为问题解决路径。我们鼓励学生走出课堂，走进现实社会，通过实地调查、观察、访谈等方式，了解家庭绿色制冷方式在现实生活中的应用情况，收集一手资料，为后续的研究提供实证基础。同时，我们也鼓励学生通过自我探究，如设计实验、进行数据分析等，深入研究制冷技术的原理，探索节能减排的途径。本项目充分体现以学生为主体，回应学生对知识学习的多元化需求。

三是以跨学科的知识融合为问题解决工具。在研究过程中，我们鼓励学生将不同学科的知识进行整合，如从环境科学的角度了解制冷剂，从经济学的角度分析各种制冷方案的成本效益，从工程学的角度优化制冷装置的设计。通过这种跨学科的知识融合，学生可以更深入地理解问题，找到更有效的解决方案。

项目实施

一、引出驱动性问题

我们首先思考驱动性问题有哪些特征，围绕生活性、开放性、多学科性、焦点性四大特征，提出了与项目研究对应的问题：如何设计出一款经济实用的节能降温装置？

但是，在这一核心的驱动性问题背后，有很多引导式的背景问题，我们又采用了"4W1H提问法"，分别是"是什么，为什么，在哪里，谁来做，怎么做"。这个方法也是学生需要掌握的重要的方法，例如本研究提出了以下问题：

（1）最近几年，你所感受到的高温天气是怎样的？为什么会出现这样的变化？

（2）你在什么地方经历过或看到过哪些极端天气？

（3）你认为这些极端天气给你的生活带来了什么影响？还可能会给哪些人群带来什么影响？

（4）你认为为什么会发生这些极端天气？

（5）你认为可以如何减少极端天气的发生？你的家庭是如何做的？

……

基于这些递进的问题给学生以思考铺垫，让一个全球性问题回归到家庭层面，从宏观到微观，真正核心的驱动性问题就顺其自然地出现了。

二、项目发展阶段

（一）第一阶段：案头调研

项目组先进行文献研究和学习，查阅了中国生态环境部官网、中国环境监测总站官网、百度等网站,收集并整理了蒙特利尔议定书的内容及发展过程,各种制冷设备的制冷原理和方法,绿色制冷剂的发展历程和特点。

(二)第二阶段:问卷调研

2023年7月23—29日,通过线上调查问卷,共计回收517份反馈"关于家庭的制冷方式"问卷调查表。同时,教师指导项目组成员从认识图表,理解和分析图表数据代表的含义,得出了以下结论(见表1):

其一,517名反馈者的家里安装空调率95.36%,每家有3~5台空调的占70%;37.14%的反馈者没关注过能耗,16.63%的反馈者认为制冷设备对环保影响小或没影响,17.80%的人在为了环保是否购买更高效的制冷设备时,选择了不购买或不确定。

其二,关于省电节能知识:共询问了7个节能知识,其中"使用空调时温度设置:制冷定高1℃,制热定低2℃,均可省电10%",仅有47.97%的反馈者表示知道,是知晓人数最少的节能知识。

表1 问卷调查分析情况

问题	分析
1.关注到的气候变化	反馈者主要关注到的气候变化是"每一年越来越热的"占85%,"自然灾害变多的"占64%。这说明气温的异常引起越来越多人的关注,进一步突出了气候变化问题的急迫性
2.家庭中主要使用的制冷方式	反馈者家庭中主要使用的制冷方式有:空调和电扇。空调占比高达95%,电扇占69%。说明空调的普及率非常高,对地球的臭氧破坏也更大,需要我们去研究绿色制冷的方式
3.家里制冷设备的数量	超过50%的家庭有3~4台制冷设备,仅有2%的家庭没有制冷设备
4.家中第一台空调是几级能耗	32%的反馈者选择一级能耗,一级能耗是国际先进水平,最省电,能耗最低。但有37%的反馈者没关注过。如果能够倡议生产厂家都售卖能耗更低的空调可能会有所帮助
5.家里第二、三台空调是几级能耗	29%以上的反馈者,家里的第二、三台空调选择的是一级能耗,有36%左右的反馈者没有关注过空调能耗。问题(5)和(6)都说明了一个问题:普及能耗知识仍然是非常必要的
6.先关窗户后开空调是否正确	64%的反馈者认为应先关窗户后开空调,其中有部分反馈者认为应先开窗户,开启空调5~10分钟之后,确保热污浊物排出室外,再关上门窗
7.使用空调时长	26%的反馈者是全天开空调,74%的人会适当调整关空调

（续表）

问题	分　析
8. 您是否知道：使用空调时温度设置：制冷定高1℃，制热定低2℃，均可省电10%？	52%的反馈者不知道这个节能小知识。说明节能知识的科普还没有做到位
9. 您是否知道：空调开机时，设置高冷/高热，以最快达到适宜温度时，改中、低风，既减少能耗，又降低噪声？	30%的反馈者知晓这个节能小知识
10. 您是否知道：使用空调房间最好使用厚质窗帘，以减少空气散失？	20%的反馈者并不知晓这个节能小知识
11. 您是否知道：使用冰箱时，冷藏物品不要放得太密，留下空隙让冷空气循环，减少压缩机运转次数？	22%的反馈者不知晓这个节能小知识
12. 您是否知道：使用空调制冷功能，搭配使用"除湿"功能，不仅省电，也能降低空调对人体的伤害？	近35%的反馈者不知晓这个节能小知识
13. 您是否知道：在打开冰箱前，应先想好要拿取的东西？	12%的反馈者不知晓这个节能小知识
14. 您是否知道：温度不太高时，可用风扇代替空调？	近9%的反馈者不知晓这个节能小知识
15. 您知道哪些新能源制冷装备？	有50%的反馈者没有听说过新能源制冷装备
16. 您对室内制冷方式发展有哪些期望？	环保节能、绿色、能源、低能耗、物理降温……

（三）第三阶段：实地访谈

为了解人体对于气候变化的感受，项目组成员对环卫工人、建筑工人、保洁阿姨和外卖小哥进行了访谈（见图2）。访谈结果显示，受访者对于绿色制冷设备的了解不多。但是无论在户外还是在室内的工作人员，都对于气候变化有着比较敏锐的感知。外卖小哥认为："我们每天风吹日晒，一点点天气的变化对我们工作的影响都挺大的。"保洁阿姨认为："现在夏天比往年更加炎热。"项目组成员在访谈时向受访对象宣讲绿色制冷技术的使用，受访对象表示未来应该多普及这一设备的应用。

（四）第四阶段：试验研究

项目组经过案头调研，发现日常生活中存在9种常见的制冷方式。组员通过对窗帘、水、冰块、冰晶和瓶子空调随机组合，对比室内外温度，研究最佳制冷方式（见表2、图3）。

图 2　小研究员们访谈户外工作人员

表 2　小研究员对 9 种制冷方式进行比对的研究结果

制冷方式	室内外最高温度温差/℃	室内外温差/℃
无窗帘遮挡	4.5	0.5
窗帘遮挡	4.9	1.3
窗帘遮挡＋水	7.0	1.2
窗帘遮挡＋冰块	7.1	1.8
窗帘遮挡＋冰晶	7.1	3.2
窗帘遮挡＋瓶子空调	7.8	2.2
窗帘遮挡＋瓶子空调＋冰块	6.7	1.8
窗帘遮挡＋瓶子空调＋冰晶	4.8	1.8
窗帘遮挡＋瓶子空调＋冰块＋冰晶	8.4	2.5

图 3　小研究员进行制冷实验

> **小研究员对于实验结果的原因分析**
>
> 第一，有窗帘遮挡的比无窗帘遮挡的降温幅度大。
>
> 第二，"瓶子空调"+水：通过风扇的蒸发作用来降低室内温度，这种方法的效果取决于室内湿度和水的温度。其可以在室内提供小幅度的降温效果。但是随着水的蒸发，冷却效果会逐渐减弱。同时由于水的蒸发会释放水分到室内空气中，可能导致相对湿度增加。
>
> 第三，"瓶子空调"+冰块：可以起到短期降温的效果。冰块的低温可以在短时间内使周围空气温度下降。但冰块会逐渐融化，冷却效果随之减弱。冰块融化后也会导致室内湿度增加。
>
> 第四，"瓶子空调"+冰晶：冰晶是一种可重复使用的冷却材料，空调扇加冰晶在较长时间内可以提供持续的冷却效果。与冰块类似，冰晶融化后也会导致湿度增加。
>
> 第五，"瓶子空调"：该方法的降温效果不太明显。如果所采用的瓶子空调的面积足够大，外部的风再大一些，这个降温效果可能会更好一点。
>
> 第六，若出现室内最高温度低于室外最高气温，室内最低温度高于室外最低气温，原因在于未做通风。

研究发现，制冷效果会受多种因素影响，如环境温度、相对湿度、空间大小和冷却介质的质量等。但通常来说，如果室内外的温差较大，表明制冷方式相对高效，能够迅速降低室内温度。反之，如果室内外的最高温和最低温之间的差异很小，可能是因为制冷方式的效率不高或者室内绝热性能较差。

室内外温差的变化也反映了材料绝热性能和维持温度稳定性的能力。在相同环境下，带有冰晶的空调扇目前看来制冷效果最强，同时功率不是很大，用电量也比较少。因此，项目组建议大众在日常生活中使用带有冰晶的空调扇。

（五）第五阶段：模型设想

基于实践结果，项目组成员展开想象，设计了使用空气压缩机进行降温的热泵技术、房顶隔热技术以及引入自然冷源三大技术进行降温。

第一,热泵技术。该技术的原理是使用压缩机将绿色制冷剂压缩成高压状态,随后通过换热器散发出热量,使其变成低压状态。接下来,风机将外部的空气吸入机内,经过集水槽中的水蒸气变成冷凝水,并通过隔热板与制冷剂进行交换,使空气得以降温。最后,冷空气再次经过风机,通过送风口向室内吹送。整个过程中,绿色制冷剂不断地循环流动,不会对环境产生污染。同时,集水槽通过循环利用减少了对自来水资源的浪费。因此,该模型是一种具有前瞻性、低碳环保、高效节能的未来绿色制冷模型(见图4)。

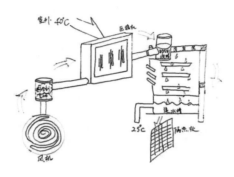

图 4　热泵技术原理

第二,房顶隔热技术。如图 5 所示,该技术的原理是在房顶建造一个游泳池,当风吹过泳池时,其温度会降低,而后继续流向楼栋的每一层房间。每层楼都有窗户和窗帘,确保风能够进入房间。房间内布置一个瓶子空调、一个冰晶空调扇,一个冰块箱和两扇窗户,从而实现降温。这一模型相比老式制冷设备更加环保,能够利用自然资源营造凉爽的感觉。同时,小研究员们也对于室内绿色制冷进行了设想,在房间内布置一台空调扇和一台普通风扇,两者同时使用,达到快速降温又节省功率的目的(见图6)。

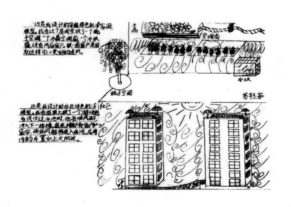

图 5　房顶隔热系统原理

图 6　左—房顶隔热系统俯视图、右—室内绿色制冷设想

第三，自然冷源。冷源，又称制冷剂。制冷剂是在制冷系统中不断循环并通过其本身的状态变化来实现制冷的工作物质。常见的制冷剂有丙烷、乙烯、氨等化学物质，会造成较多碳排放，加重温室效应。相比之下，自然冷源更加环保。小研究员据此设计了一个方案（见图7）。首先，选择来自社区或社区附近的自然冷源（比如池塘里的水），将水泵放入冷源连接水管，借助水泵将水压上来，进入水管。而后，使用升压泵增加系统压力，克服空调换热器的压力损失。再用水管连接空调主机的入口处和出口处，空调内部有换热器，换热器内部有密集的薄铁片，增加换热面积提高换热效率。最后，将换热器排出来的水投放到一个特意为冷水空调设置的水池里（冷水会自动下沉与更热的水分层）。

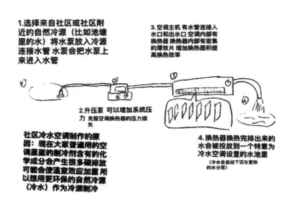

图7　自然冷源系统

成效与反思

一、初步成效

一是创新了气候变化教育的路径。气候变化教育不仅仅是引导学生关注能源变革，也应该引导他们关注那些因为气候变化而首先受到影响的人群。这是一个必要却容易被忽略的出发点——为了全人类的生存和发展，而不是为了解决问题而解决问题。本项目的学生在进行实地调查了解工人们的居住环境后，对于调研的场景感到难忘，他们的同理心被激发出来，对气候变化引发的问题就更加具有关怀意识。

二是激发了青少年的创新思维和动手能力。学生采用新的方法、技术和理论来探索绿色制冷方式，增加可持续应用。在模型设想过程中，充分考虑到运用自然冷源、物理隔热等方式，发展更加绿色环保的制冷方式，从而进一步降低能耗。经由本项研究，学生对气候变化的问题有了更深刻的理解，如对碳排放、温室效应的探讨等。同时，他们也对绿色制冷技术的长期应用和推广提出了设想，例如如何长期应用在家庭中。这个过程让学生对气候变化的成因、影响和解决方案有了更深入的理解。他们的创新思维和动手的能力都得到了提高。

二、反思

本项目通过带领学生开展家用制冷设备研究和社会实践,培养他们的跨学科思维、动手能力和气候素养,在创新气候变化教育路径上进行了积极的探索。但是由于部分知识的专业性较强、不容易学习和传播,在学习方法和学习方式上有待加强,需要聘请专业的人才担任绿色制冷原理探究的导师,并运用趣味性更强的方式向公众传播绿色制冷的重要性。

未来,项目组将继续完善项目设计,拓宽实施范围,将绿色制冷理念推广至公众视野,为应对全球气候变化持续贡献力量。

(谢楠楠,绿色光年项目副秘书长;

倪菡,绿色光年项目专员)

新能源企业助力气候变化教育的经验

研究背景

气候变化是关系人类命运共同体的全球性问题,也是21世纪人类可持续发展面临的严峻挑战。应对全球气候变化,保护生态环境,实现"双碳"目标,是全人类的共同使命和责任,同时也是各行各业推动自身高质量发展的内在要求。而光伏作为清洁能源的典型代表,正是应对气候变化、推动绿色发展、引领低碳未来的重要领域,尤其分布式光伏是发展绿色城镇、促进人与自然和谐共生的关键力量,能够为人类应对气候变化做出重要贡献,成为促进绿色低碳能源转型的中坚力量。

天合光能股份有限公司成立于1997年,是国内较早进军光伏领域的企业,现已发展成全球领先的光伏智慧能源整体解决方案提供商。"让太阳能造福千家万户"是天合光能不变的追求。20多年来,天合光能坚持践行绿色发展理念,用清洁能源守护绿水青山,一直是生态保护的积极参与者与坚定支持者,可持续发展也刻进了企业基因。企业还制定了2020年至2025年的可持续发展目标,包括碳排放管理、能源管理、水资源管理、废弃物管理等,均设置了具体指标。

天合光能关注"气候变化"已有十余年,董事长高纪凡先生认为,新能源企业有责任抓住应对气候变化的关键,致力于将能源的减碳作为零碳主力。因此,企业一方面抓住"光伏+"这一"十四五"规划的重点,努力探索"光伏+"新模式新业态、推动"光伏+"应用场景融合创新发展,以光伏之力应对气候变化,构建美好零碳新世界;另一方面基于少年儿童是祖国的未来,是中华民族的希望,从而创建"棵林环境保护协会",通过推出双碳教育公益项目等,借教育之力,发挥更多"天合力量",让守护生态的少年如星火燎原般,持续助力气候变化的减缓与应对。

项目设计

该项目立足天合光能这一新能源企业,结合气候变化教育的特质,充分利用"天合光能梦想与创新展示中心""天合田园生态农场"等场域资源,借助"棵林环境保护协会"的力量,并协同学校、社会多方,采取"走进去"和"迎出来"两种方式("走进去"即企业送课进校园,"迎出来"即学生前往企业相关基地进行研学与实践),从而持续开发"天合杯"双碳教育系列公益活动,如面向基础教育阶段学生的"少年硅谷"项目和面向高校学子的"新能源探索者"项目等。

一、项目目标

通过项目实施,以新能源企业之力,普及"光伏+"新能源知识,提升低碳生活理念的推广和传播力度,也为各类学生提供更多研学体验、社会实践的机会,在真实情境中激发他们对新能源研究的兴趣,促进广大青少年群体成为发展我国绿色低碳事业,应对气候变化的中坚力量。

二、项目推进规划

项目主要分三个阶段实施:第一阶段注重散点灵活的环保教育活动的实施,第二阶段主要是定点持续的双碳教育合作项目的实施,第三阶段聚焦气候变化教育的校企合作。

项目实施

一、第一阶段:散点、灵活的环保教育活动

一直以来,"天合光能"及其旗下的"天合富家"都在积极践行企业 ESG 责任。2011 年,现任天合光能旗下天合富家董事长高海纯携数名优秀的青年环保家共同发起、创办了非营利性环境保护组织——常州市新北区棵林环境保护公益协会,简称"棵林"(Co-link)。"棵林"的价值观是"LOVE":领导力+其他+远见+活力,口号是"你我心连心,独木变成林,改变心灵,保护地球"。"棵林"就代表着环境保护的爱心细化成一棵棵小树,让一棵棵小树连成一片森林,让片片森林守护地球。如今,"棵林"在美国和德国也建立了分社,三国成员已有 2 000 多名,在不断壮大着守护生态的力量。

"棵林"自成立后,不但借助天合光能的力量,还与有则集团、君合公司等企业建立紧密的合作关系,向 20 余家企业提供可持续发展咨询业务,包括提供碳排放、碳足迹、污水治理、精益生产等方面的建议;还有意识地与学校展开联动,最早以江苏省常州高级中学、常州市第一中学、常州市正衡中学、常州市新华实验小学等为试点,通过环保宣传、环保体验等公益活动来倡导健康绿色的低碳生活方式。早期,棵林与诸多非营利性组织、政府机构、学校合作发起并开展的环保教育活动较为灵活,其中包括"环保画展义卖""生态爱心实践""与来访的美国学者一起植树""公益环保滇藏骑行"等。这些活动虽然是点状、单次的,但是却焕发着环保教育的勃勃生机。就以"公益环保滇藏骑行"为例(见图 1),参与活动的 10 余名中学生,历时 33 天,翻越 9 座高海拔大山,穿越三江并流,沿尼洋河而上,重游茶马古道,沿途举办了 15 场环保宣导活动,得到民众的认可与赞许,还被 76 家媒体报道和转载报道,很好地辐射了环保教育行动影响力。

从天合光能生长出来的"棵林",肩负着绿色环保的责任,自带可持续发展的基因。多年的环保教育活动实施中,它作为链接学校、企业和社会组织的纽带和爱的桥梁,以自身"保卫环境"的决心,以心唤心,棵树成林,点亮了环保教育之星火,竭力照亮着青少年的绿色成长之路。

图 1　天合光能依托"棵林"早期开展的"滇藏骑行"环保教育活动纪念品

二、第二阶段：定点、持续的双碳教育合作项目

"棵林"在环保教育活动中的尝试，为天合光能助力气候变化教育奠定了厚实的基础。2023年，随着气候变化教育成为全球关注热点，身为GDP"亿万之城""新能源之都"的新能源企业，天合光能意识到科教兴国、人才强国的重要性。于是，天合光能借助"棵林"，打造了"双碳教育进校园"项目，签订了一批合作校，分"携手'童'行"和"走进高校"两条线，以长程系列的设计开展校企合作的双碳教育。

（一）双碳教育项目走进高校："新能源探索者"

为了推动党和国家"双碳"目标实现，让高校青年正确认识和把握碳达峰、碳中和的理论知识，树立新时代绿色低碳理念，助力气候变化教育，2023年11月12日，"新能源探索者——2023双碳教育进高校系列活动"在河海大学常州新校区启动。本项目采取"慈善＋教育＋双碳普及"的创新形式，有效整合高校和各类社会资源，凝聚校企力量，通过新能源企业研学实践、融入零碳生活展示、新能源公益课堂、第一届低碳生活短视频大赛、第一届风力与光伏发电模型搭建大赛等活动，为大学生们提供丰富的就业创新实践平台，吸引更多学生参与，成为宣传大使、行动明星，增强他们对新能源领域的双创意识和能力，助力"双碳"目标实现和人才梯队打造。

2024年3月15日下午，"新能源探索者——双碳教育进高校系列活动"总结会在河海大学江宁校区举行（见图2），会上还进行了"第一届风力与光伏发电模型搭建大赛"和"第一届低碳生活短视频大赛"表彰会，一等奖获奖团队代表陆雯发表获奖感言，表示："作为一名河海广电人，我将继续用镜头记录生活，用影像传递环保理念。我知道一个人的力量小，但大家一起做，力量就大了。我也希望我们都能行动起来，一起享受低碳生活。"天合光能也以此肯定和鼓励大学生们参与到项目中，带动更多人走出校园、走入企业，在校企合作创造的平台中提高解决实际问题的能力和创新能力，促进项目的阶段闭环和螺旋上升。

图2　天合光能依托"棵林"开展"新能源探索者"双碳教育进高校活动

"新能源探索者"项目在河海大学的成功实施,让大学生们在实践和创新平台中,看到了新能源应对全球气候变化的积极作用,对新能源、新未来有了更加深刻的理解和认识。

（二）双碳教育项目携手"童"行:"少年硅谷"

为了加强中小学和校外实践基地的合作,以学生发展、活动互联、资源共享等有效途径助力气候变化教育,2023年9月25日,"天合杯"双碳教育携手童行项目在常州市新北区龙虎塘第二实验小学启动(见图3)。天合公益基金会、天合富家能源股份有限公司与龙虎塘二小合作,共同推进"少年硅谷"项目建设,旨在利用课后服务、节假日,以研究性学习的方式定期组织学生走进新能源,了解新能源,培养研究新能源的兴趣。

图3　天合光能依托"棵林"开展"少年硅谷"双碳教育携手"童"行活动

自"少年硅谷"项目推动以来,天合光能携手学校,聚焦新能源与气候变化教育,开展了系列"走进去,迎出来"活动:研发人员走进去,环保公益课开讲——天合光能新产品和技术研发部人员为学生们讲主题为"应对气候变化,'碳'寻美好生活"的主题公益课,让学生认识光伏发电、风力发电、水力发电等不同的清洁能源及其作用,并指导绿色低碳的生活方式,在"大手牵小手"中共"碳"绿色发展;少儿玩伴迎出来,环保玩创向未来——学生组成少儿玩伴团,带着"光伏板的原理""光伏如何在农场应用"等研究主题,分批次前往"天合光能梦想与创新展示中心""天合田园生态农场"等天合基地进行校外实践体验,研究光伏板等新能源设施,感受"光伏+"的融创魅力,对新能源知识增强了解,切实激发学生的研究兴趣。

"少年硅谷"项目在常州市新北区龙虎塘第二实验小学的有效推进,让天合光能看到了小学生对于新能源的好奇心与探究力,为校企合作打开了新思路。

三、第三阶段:聚焦气候变化教育的校企合作

2023年12月22日,华东师范大学基础教育改革与发展研究所研究员、上海终身教育研究院执行副院长李家成教授和常州市新北区龙虎塘实验小学顾惠芬校长走进天合光能展厅参观,并与天合富家相关人员就可持续发展教育、校企合作中如何发挥家委会作用进行了交流,对共同开展气候变化教育达成了共识。

2024年寒假,一场主题为"走进清洁能源,创构绿色未来"的"龙龘龘富家杯"全国科创画大赛在校企合作中火热进行。活动鼓励学生与家长、伙伴一起去周边的清洁能源企业、基地看一看、访一访,了解了解、研究研究,然后用画笔和想象力勾勒出心中的未来世界。本次活动收到来自全国的学生作品70余份,我们在第二届"学校家庭社会协同开展气候变化教育"探讨会上进行了颁奖表彰(见图4)。活动不但推动了学生关注清洁能源、抵御全球气候变暖,也为天合光能后期的产品研发提供了儿童视角和别样思路,更促成了天合光能与龙虎塘实验小学气候变化教育合作行动的持续开展。

2024年初,校企双方在合作商议会上(见图5),进一步讨论通过了四类合作:生态校园建设——引进气候、气象观测设备,打造生态秘境,完成图书馆及校园各类屋顶的光伏发电建设,打造空中绿色庭院等;特色课程开发——增设新能源与建模整合的校本课程,进行清洁能源知识普及,组织新能源模型设计大赛等;特色活动开展——组织"绿色生态玩伴团",校家社企多方卷入,常态合作开展气候变化教育主题实践活动;促成成果传播——借助气候变化教育研讨会、学术成果发表等,多渠道传播合作成果。

聚焦气候变化教育的校企合作还在探索阶段,相信在不断丰富学校气候变化教育资源,打造面向全体学生的沉浸式新质教育体验的过程中,校企会从合作走向合一,创生出巨大的气候变化教育共育力和助推力。

图4 "龙龘龘富家杯"全国科创画大赛颁奖典礼　　图5 天合光能与龙虎塘实验小学校企合作商议会

成效与反思

新能源天生就具有改善全球气候变暖的能力。天合光能作为新能源企业，以创建无碳的新能源世界为愿景，责无旁贷地承担起以光伏智慧促进全球可持续绿色发展的责任。从关注"气候变化"到助力"气候变化教育"，天合光能从校企合作视角，也形成了一定的经验，并收获了可视化的成效。

从项目实践可见，最大的成效就是天合光能在合作模式上能与时俱进，使得项目更系统，有深度，体现可持续性。从最初点状、灵活的环保教育活动到定点、持续的双碳教育合作项目，再到聚焦气候变化教育的校企合作，天合光能从间接到直接，不断优化参与气候变化教育的模式，过程中签订了一批合作院校（单位），如河海大学、常州市新北区龙虎塘实验小学等，也开发了一系列切实促进气候变化教育的项目。这一过程中，参与人群和参与面不断扩大，每一次活动都受到了社会各界的关注，被不同平台宣传报道，从而带动、辐射出"二次"影响力，不断提升着气候变化教育的积极作用。天合光能始终热心于低碳环保事业、关心全球气候变化问题，在持续、系统、深入地介入气候变化教育，共建企业和社会可持续性高质量发展的未来。

当然，反观项目发展的全过程，我们也看到了更大的发展与优化空间。2024年，天合光能将在"天合杯双碳教育项目"的基础上持续探索校园模式，全新开展合作项目，汇聚产教资源，制定教学评价标准，开发与双碳进校园相关的一系列核心公益课程与创意实践项目，与校方合作共建低碳校园，加大低碳生活的理念推广和传播力度，为广大学子提供更多社会实践机会，也为我们的新能源项目在全国范围内的复制与迁移提供优秀的样板案例。同时，凝聚校企力量，全力打造校企育人"双重主体"，学生学徒"双重身份"，在人才培养、技术创新、就业创业、社会服务、文化传承、气候变化教育等方面进行深度探索，助力区域的高质量发展，共创一个自然向好的美好未来。

（吴钰涵，天合富家能源股份有限公司品牌主管、常州棵林环境保护公益协会联合创始人兼副理事长；潘虹，常州市新北区龙虎塘实验小学学生发展中心主任）

学在自然:"共同世界教育学"理念下的小学气候变化教育实践研究

研究背景

习近平生态文明思想提出了一套相对完善的生态文明思想体系,形成了面向绿色发展的四大核心理念。① 基于新自然观的当代教育改革,需要直面社会新转型,走向依教育所是而行、达自然而然之境,开创教育与自然内在关联的新阶段。②《气候变化教育指导纲要(试行)》指出"教育是应对气候变化的一个核心举措,通过影响价值观、思维方式、行为和生活选择,帮助学习者形成应对气候变化所需要的素养"。③

浙江省武义县壶山小学新城校区2022级3班以"共同世界教育学"④为指导,强调"共生共长"的重要性与必要性,反对自然与学校的分离,关注生命真实成长;致力于打通学校家庭社区自然四大育人场,融合区域资源、多学科力量,以开放的过程,创生"多维课堂+生态体验+气候益行"的实践模式,让每一个学生学会与自然共生共长,构建良好的教育生态。

项目设计

一、项目目标

通过"学在自然"项目的设计与实施,促成地方气候教育资源的整合,促成家校社力量的协同,构建完整的、连续的学习时空,扩展师生多主体参与的价值体系和认知格局,从个体的"小我"扩展到地球共同体的"大我",⑤推进生态文明素养有效提升。

二、项目规划

"学在自然"项目立足生命整体性,整合了学校、社区、家庭、自然的资源,确定了目标系统、内容系统、环节系统,构建了系列活动(见图1)。在此过程中,关注具体活动的价值开发,关注活动时间、内容、要求的衔接,充分考虑活动顺利推进的可能性。⑥

---

① 央视网.习近平生态文明思想是开放与发展的新思想[EB/OL].(2018-06-08)[2023-10-30]. https://news.cctv.com/2018/06/02/ARTIG5f4qcVSHsxKt6FEKdOv180602.shtml.
② 叶澜.溯源开来:寻回现代教育丢失的自然之维——《回归突破:"生命·实践"教育学论纲》续研究之二(下编)[J].中国教育科学(中英文),2020(2):4-29.
③ 气候变化教育指导纲要(试行),华东师范大学上海终身教育研究院,2024.
④ "共同世界教育学"融合了多种文化资源中具有生态意义的本体论和宇宙观,是一种基于生态协调与恢复的教育模式,其主要教学方式是教师与儿童一起在田间、山头、街道、城市花园和垃圾站等生活环境中散步,与万物相遇,进而促使思维方式从"主客二分"和"人类中心"到"万物相联"和"共同生成"的转变。(游韵,余沐凌.在"人类世"危机中构建"共同世界教育学"——联合国教科文组织《学会与世界共同生成:为了未来生存的教育》评述[J].华东师范大学学报(教育科学版),2023(12).)
⑤ 史舒琳.中国风景园林行业应对气候变化和支持双碳目标的现状、需求与策略[J].中国园林,2023,39(3):34-39.
⑥ 陶广璐.联合国教科文组织:以"全学校"路径推动气候变化教育[J].上海教育,2023(24):57-59.

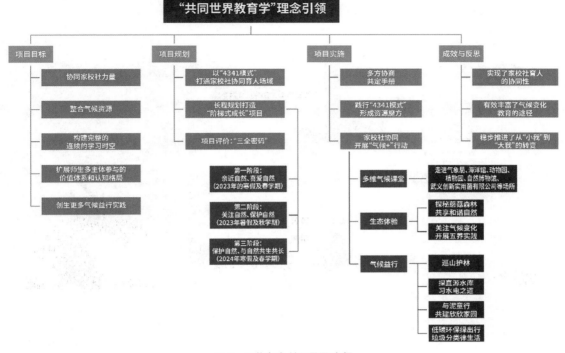

图 1 "学在自然"项目路径

(一)以"4341 模式"打通家校社协同育人场域

联合国教科文组织呼吁以"全学校"方式推进气候变化教育,并建立社区合作关系支撑教学。① 气候资源开发不局限于自然,是融合在校园生活、家庭生活、社区生活之中的,因此我们以"4341 模式",即以坚持学生主体、坚持生命整体、坚持五育融合、坚持协同育人的"4"坚持;校社联动、校企结合、家校携手的"3"联动;打通家庭、学校、社区、自然"4"大场域,努力实现"我与自然共生长"这"1"个目标。

(二)长程规划打造"阶梯式成长"项目

第一阶段是 2023 年的寒假及春学期,项目以"魔力菌菇 大展宏'兔'"为主题,努力协同学校家庭社区力量,打通学校家庭社区自然四大育人场域,让学生在真实的生态体验中感受到菌菇世界的奇妙。

第二阶段是在 2023 年暑假及秋学期,项目组聚焦"带着菌菇去旅行"主题实践,共同策划"水养""气养""食养""药养""体养"实践,编制《探秘菌菇世界 共创和谐自然》手册,鼓励学生走进自然,开展"生态体验""气候益行"等"气候+"实践,促进生态文明素养有效提升。

第三阶段是 2024 年寒假及春学期,以菌菇生长环境为切入点,联动武义县萤乡社区、武

---

① 李家成,王晓丽,李晓文.学生发展与教育指导纲要[M].福州:福建教育出版社,2016:379.

义县王古村、学校欣欣农场,编制《龙行龘龘 欣欣家园 与"泥"童行 寻龙记》,共同开展知土、护土、春植行动、共绘春日蓝图等一系列活动,探索土壤奥秘,形成护土意识,扩展价值体系和认知格局,从个体的"小我"扩展到地球共同体的"大我"(见图2)。

图 2 "学在自然"项目手册封面

如此,从亲近自然,到喜爱自然,再到萌生保护自然、与自然共生共长,开展生态益行行动,我们的项目实现了"阶梯式成长"。

三、项目评价

项目以"三全密码"为导向,以全学科、全时空、全主体组织发展性评价;构建"成长共同体",将教师、学生、家长三者有机结合,开展多元评价;从情感态度、能力、行为品质、特长等多维度进行评价;从即时性、过程性、总结性三个时间段开展评价。

项目实施

一、多方协商,共定手册

每期项目都以"师生互动、家校社交流、问卷调查、协商主题"的方式收集、整理金点子。在尊重学生主体意愿的基础上,班主任及学科教师、学生代表、塔山社区工作人员、大河源村村长、家委会代表齐坐一堂,共同召开"项目讨论会",主要讨论资源挖掘、参与人员、开展方式、开展时间、开展地点、价值意义、活动手册制作等内容,使活动融合儿童趣味、教师智慧、家长力量、社区支持,吸引儿童积极参与、家长主动投入、社区积极配合。报名当日每一组的报名人数都在20人以上,班级群中出现"抢人"热潮,各项活动的热度都居高不下。

二、践行"4341模式",形成资源魔方

项目组践行"4341模式",开展"场馆育人""先锋讲坛""社群建设"三大行动,形成多力驱动、多环交融的"全员、全程、全方位"育人格局。即以家长讲坛为活动阵地，以"学在自然"项目为切口，联合塔山社区、县气象局、创新食用菌有限公司、县反背林场等教育场馆、企业，充分发挥各主体、各场所的育人价值，打造立体资源魔方。如苏宸爸爸是气象站工作者，他便带领学生参观气象站；糖宝妈妈是教师，她便带领学生手工制作生态瓶等；荟荟外公是护林员，他便带领学生走进反背林场……

三、学校家庭社区协同，开展"气候+"行动

（一）多维气候课堂

除了利用学校课堂教学资源，项目组还充分挖掘校外气候教育资源，联动家校社力量，带领学生走进气象局、海洋馆、动物园、植物园、自然博物馆、研学基地、武义创新实用菌有限公司等场所，开发具有班级特色的多维气候课堂，让学生在更丰富、更真实、更开放、更具情境性的环境中习得气候知识，提升生态素养。如在沐沐妈妈的带领下，"气象小队"走进气象局（见图3），在气象局工作人员的讲解中，队员们近距离观看称重雨量器、闪电定位仪、蒸发传感器等气象设备，了解气象设备的功能及原理，知道了室外观测风向、地温、气温、雨量等的操作方法，探索了气象的奥秘，明白了气象数据在天气预报中的作用和意义。又如，参观自然博物馆后，学生们感叹"地球上的每个物种都值得被爱！"

图3 "气象小队"走进武义县气象局

（二）生态体验

气候教育强调"参与"或"体验"的学习方式，尊重学生的主体地位。① 于是，多方代表努力挖掘自然、社区资源，共同设计了系列生态体验活动。

子项目一：探秘蘑菇森林，共享和谐自然

从蘑菇森林出发，开展一系列活动了解菌菇及相关自然环境、气候；了解菌菇形态结构、学会辨认食用菌菇和毒蘑菇；探秘微观世界，了解菌菇生长环境；制作微观生态瓶，了解大自

---

① 戴剑.进阶式气候变化跨学科主题学习活动设计与实践[J].地理教学，2023(4):28-31.

然生态系统等。

子项目二：关注气候变化，开展五养实践

《中国应对气候变化的政策与行动 2022 年度报告》提出要开展"绿水青山就是金山银山"示范县创建。① 浙江省武义县践行这一理念，在绿色经济蓬勃发展的同时，环境保护和生态优势方面表现突出，如武义县履坦镇坛头村喜获"全球人居环境村落范例"大奖。"学在自然"项目以"环境、气候、爱护、保护、实践"为关键词，以武义县"五养"文旅特色为基础，开发水养、药养、食养、气养、体养系列活动，关注生态保护，共享绿色生活（见表1）。

表1 "五养"资源挖掘

五养	气候变化教育资源
气养	武义三面环山，绿意盎然，森林覆盖率达74%，是"全球绿色城市""中国天然氧吧"。武丽线附近就有省级示范基地毛竹园，茂林修竹，郁郁葱葱，清新的空气让人心旷神怡、神清气爽。下叶山休闲服务站挖掘当地"竹"这一特色产业，采用仿竹长廊、竹亭等形式，突出"气养"主题
水养	温泉是大自然赐予武义的珍贵礼物。武义温泉量大质优、温度适宜，富含偏硅酸等20多种对人体有益的矿物质和微量元素，有润肤养颜、祛疾保健的功效，誉称"浙江第一，华东一流"。此外，武义还有源口水库、溪里水库、直源水库等众多水库，有双源口水力发电站、横山水力发电站、直源电站等多个发电站
体养	养生之道，在于动静结合。"体养"主题仿古文化墙位于大公山岔口，是武义大型赛车场入口。在赛车场，不同于传统养生的闲适悠然，赛车爱好者可以尽情体验速度与激情。不只是赛车场，风景秀美的武丽线本身就是一条绝佳的骑行绿道
食养	武义有机茶认证面积、产量全省第一。茶叶、中药材、食用菌三大产业被认定为浙江省示范性全产业链，入选数全省第一。武丽线巧妙地在弯道挡墙上面设计了"食养"大型浮雕景观，展示各类高山蔬菜和有机养殖的禽畜，使人一目了然。特色农产品正是通过农村公路走出了大山，走进了千家万户。沿线大田乡盛产蓝莓，已经连续举办了好几届蓝莓节，带动了农旅融合发展
药养	寿仙谷有机国药基地是中国中药协会"药学科普示范中药基地""全国青少年农业科普示范基地"、浙江省中医药文化养生旅游示范基地，为了还原野外纯净无污染的环境，寿仙谷有机国药养生园的灵芝种植在仿野生有机栽培智能化大棚中，温度湿度受到严格监控，让灵芝宝宝在最适宜的环境里成长。此外，碗铺村拥有百里药材基地，主打"八味养生、四季花海"，种植有种类繁多的中草药

通过"五养"实践，学生们更了解家乡人民在生态文明建设中的协同贡献，从而积极参与其中。如，"气养武义"是了解武义的森林覆盖率达74%，是知晓森林的作用，是走进反背林场宣传森林防火；"水养武义"是感受到"每一度电都来之不易"，是"清理垃圾守护母亲河"的行动，是在世界水日走进社区宣传节约用水；"体养武义"是在壶山举行定向寻宝赛，观察沿

---

① 生态环境部.中国应对气候变化的政策与行动 2022 年度报告[R].2022:20.

途的动物与植物;"食养武义"是种植、观察、采摘、烹饪菌菇、茶叶、中草药;"药养武义"是了解灵芝养殖技术,萌生出让"科技与传统并存、自然与社会共生"的生态文明意识。

(三)气候益行

随着项目的推进,学生将"人与自然和谐共生"的理念付诸实践,开始关注人类面临的气候问题,开展了巡山护林、探秘水电站、与泥童行等"气候益行"实践。

巡山护林:在探秘菌菇世界的过程中,沈沐荟和洪妍熙同学了解到菌类是参与林地残留物分解的重要成员,就在妈妈的帮助下,组织了"巡山护林,守护绿水青山"的活动。进山途中,护林员引导队员们发现灵芝、蘑菇,介绍山林涵养水源、调节气候等作用……队员们直观感受到了菌菇喜爱的生长环境,也明白了大自然与我们的生存休戚与共的道理,表示一定将"呵护一草一木,守护绿水青山"记于心践于行。下山后,队员们和护林员一起挂护树牌、宣传森林防火知识(见图4)。

图4 挂护树牌宣传森林防火知识

"探直源水库,习水电之道":在知行爸爸的带领下,队员们顺着溪流而上,发现清泉源头;沿着溪流而下,看到蓄水水库。在"直源电站"工作人员的带领下,队员们参观电站,了解水力发电。

宁宁同学说:"学习到了发电原理,原来家里的电是这样来的,我们要节约用水用电。"

知行同学说:"我要从身边的小事做起,让垃圾不落地,希望大家都能慢慢成为一名爱护地球的小小环保者。"

若菌同学说:"我们和大自然共同生活在地球家园里,我们要保护好自然。"

"与'泥'童行,共建欣欣家园":随着活动深化,项目组以菌菇生长环境为切入点,探索土壤秘密。学校家庭社区协同开展"泥之源、泥之用、泥之语、泥之旅、泥之美"系列"与'泥'童行"活动(见图5)。如苏宸爸爸带领队员了解土壤作用、子涵妈妈带领队员们观察土壤中的生物、佳怡妈妈带领队员吟诵土壤美文、乐甯妈妈带领队员们到婺州窑制作陶瓷艺术品、昕妮妈妈带领队员开展春植行动……在多方力量的引导下,队员们通过观察、比较、操作、实验等方法,发现、分析和解决问题,呼吁大家保护土壤,为携手共建可持续发展的生态环境作出贡献。

图5 "与'泥'童行"活动设计

"低碳环保绿出行,垃圾分类律生活":班主任巩老师一直鼓励学生从点滴小事做起,践行绿色文化。通过"气候+"系列实践,步行上学的学生日益增多,他们用实际行动来宣传"低碳环保绿色出行"的理念。此外,学生们步行回家的途中也会随手捡起垃圾并分类处理,成为校内校外有担当的环保小卫士。

成效与反思

"学在自然"项目的成效主要体现在以下三方面。

一是实现了学校家庭社区育人的协同性。项目以"4341模式"广泛开展"多维气候课堂""你好,寒假(暑假)"等实践,每一个活动都有教师引领、家长参与、社区力量支持,实现学校家庭社区育人的协同性。项目推进过程中,获得了家长在精神上、行动上100%的支持(见图6)。每一位家长都是一位社会工作者,有着自己的社会关系网,每一张关系网都蕴藏着丰厚的资源,家庭与家庭在不同的活动中,也形成了高质量的交往关系、可共享的立体资源魔方。在这样的协同育人新格局中,教师与时俱进,格局开阔;家长自觉投入,互学共进;社区主动参与,协同发展;儿童舒展,胜任未来……

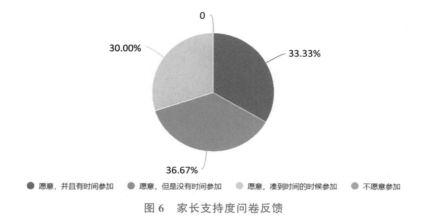

图 6　家长支持度问卷反馈

二是有效丰富了气候变化教育的途径。气候变化教育不应拘泥于书本,更不是呆板的说教。在"共同世界教育学"理念引领下,学生和教师、家长、社区工作者、企业工作人员共同在田野、森林、水电站、气象站、博物馆、社区、夜市、菌菇大棚等场所,通过"听、读、查、问、做、尝、说、写、画、唱、玩"等丰富多元的方式开展了系列体验式、互动式的"气候+"活动,让其在真实的、完整的、连续的生活体验中学习和成长。此时,孩子们的心、脑、手是融为一体的,他们以照片、视频、书法、绘画、手工等形式呈现各自的学习成果。这极大丰富了其气候知识,进而促使思维方式从"主客二分""人类中心"到"万物相联""共同生成"转变,①帮助他们更好地应对气候变化所带来的挑战,真正实现了为未来而教。

三是稳步推进了从"小我"到"大我"的转变。通过多主体的互动,学生有了新的体验、发现、创造。通过活动后的反思、对话,借助父母、老师、同伴的反馈,学生对自我成长形成了更清晰、更整体、更客观的认识。他们从一开始对自然感兴趣,到深入了解自然,再到萌生"人与自然和谐共生"的理念,提高对气候变化的认识和关注,从而促进学生开展"气候益行"系列实践……这是从"小我"到"大我"的价值提升与格局拓展,实现了自我觉醒和发展,也促进了地球共同体的合力构建。

通过回顾,我们也发现了进一步努力的方向:气候变化教育的资源开发不仅要有广度更要有深度;参与群体要由点及面,多方卷入。后续,项目组将在全面整合资源的基础上进行深度开发,形成多个项目化实践,营造全民参与的氛围。

(巩淑青,浙江省武义县壶山小学新城校区教师;
马骏,浙江省武义县实验小学教师)

---

① 游韵,余沐凌.在"人类世"危机中构建"共同世界教育学"——联合国教科文组织《学会与世界共同生成:为了未来生存的教育》评述[J].华东师范大学学报(教育科学版),2023(12).

长江探秘系列之船舶探索行

研究背景

长江作为亚洲最长的河流之一，承载了中国几千年的文明和发展，它不仅为中国提供了丰富的水资源，支持着众多人口的生活和城市的工业生产，同时也孕育了众多的文化名城和历史古迹，是中国人民的骄傲和文化自信的象征。

但在全球气候变化不断加剧的背景下，流域内面临着气候变化所带来的多种挑战。在国外，气候变化教育已经积累了一定的经验，但是在国内，目前还相对处于起步阶段。根据2024年2月上海终身教育研究院颁布的《气候变化教育指导纲要》目标中提到的"增强学习者的人类命运共同体、人与自然的地球生命共同体意识"可以感知，在中国开展气候变化教育亦是刻不容缓。

上海地处长江的入海口，位于长江三角洲的冲积平原，地处黄金水道，有着得天独厚的地理优势。上海市第六师范学校第二附属小学（简称六师二附小）背靠长江的二级支流——黄浦江，两者相距仅一公里左右，是浦江两岸的交通要道。本次活动将目光聚焦于长江上的船舶，以亲子家庭组合为主，引导学生发现船舶与气候变化之间的联系。

与此同时，本活动开展跨区域多模态联动，与中国的第二大河——黄河的入海口区域的小学、中职校协同开展可持续教育，通过资源共享，数字化联动教研的方式，实现同课异构。长江口—黄河口的联动也将成为中国首个跨区域联动开展气候变化教育的案例，希望本项目能够为上海以及中国乃至世界提供气候变化教育的新经验。

项目设计

本次的长江探秘系列选择了学生比较感兴趣的船舶作为切入点，一是借用社会资源和多方力量，带领学生探究气候变化与船舶之间的密切联系，提高学生对气候变化的认知水平，提升学生的行动能力，引导学生了解船舶运输对环境的影响，并能将所学的气候变化知识渗透在日常生活中。二是加强学校家庭社区之间的合作，形成良好的教育联动机制。通过学校通信、家长群等渠道推广宣传，向更多家长和学生宣传活动的意义和重要性。

通过这样的设计，促进长期稳定的家校社合作机制的建立，为今后开展类似的气候变化教育活动奠定基础；从案例中生成新经验并不断加以完善，提升教育活动的质量和影响力。

项目实施

一、启动仪式

2024年寒假期间,通过腾讯会议的方式,邀请了学校二年级部分班级的家长和孩子共同开启了活动(见图1)。课堂上气氛活跃,老师向同学们介绍了有关的基础地理知识,同学们手拿简易版的中国地图,在父母的陪同下,共同开启了船舶探索之旅的第一站。

图1　船舶探索营线上启动仪式

二、发现之旅——探索船舶历史与文化

中国航海博物馆(浦东新区申港大道197号)临近滴水湖,其承载着丰富的历史与文化价值,展示着航海发展的演变历程和海洋文化的丰富内涵,因此能够为学生提供深入了解海洋知识和航海技术的机会。另外,它是一个跨学科学习的理想场所,涉及地理、历史、科学等多个学科,能够促进学生的综合素养和学科交叉融合。它还反映了人类对海洋的探索精神和对自然环境的尊重,有助于培养学生的人文情怀和社会责任感。本次的一日研学全程以学生为主导,学生的主动发展性极高。

(一)研学前的头脑风暴

本次研学有别于传统的教师提前规划安排带领学生开展活动的方式,而是从第一站起,全程都让学生作为主人。1月27日,十几组亲子家庭在六师二附小西校区会议室开展了一次头脑风暴——活动当天来回的交通方式、公交换乘、目标设定以及参观路线,都由学生以小组为单位,家长在一旁充当"百度员"协助完成策划(见图2)。大家作出了"绿色出行"的统一决定,做好了充分的准备与行程规划。

图 2　学生活动线下策划

（二）研学中的互动体验

1月28日航海博物馆的一日研学路线由小组成员共同讨论制定。行程中，所有的活动以小组为单位，哪怕遇到意见相左、不能统一的情况，小组成员也都选择服从大局，没有一人脱离小组单独行动。事先选定的小岗位"文明提醒员""小组长""小助手"等各司其职，发挥了良好的作用。当天的活动过程井然有序，同学之间互帮互助，彰显了成员间的团队合作精神。活动结束后，很多家长和学生纷纷表示了对活动的肯定与喜爱。

朱思潼说："今天我来到航海博物馆，令我印象最深刻的是摆在大厅的大福船，福船里有两艘小船，我还看到船舱里有四张小床，我知道这几张小床是给远航的人过夜准备的。"

汤梦菡说："参观了航海博物馆，我印象最深的是大明混一图，这也是镇馆之宝之一，以前都是用汉文标注的，到了清朝，变成了用满文标注，这也是我们今天看到的大明混一图。"

李可轩说："我今天到了航海博物馆，我了解了不同船头的破冰船，因为船头很尖，所以可以破冰。"

周怡晨："我今天来到航海博物馆学到了很多知识，其中我最爱的是帆船，只要有风吹过，帆就会迎起，帆的另外一个作用是改变方向。"

朱思潼妈妈："今天也是我第一次来到航海博物馆，非常高兴跟着小朋友一起来到这个非常美丽的地方看到了一些跟航海相关的物件，包括大福船、破冰船、船模展览，都让我叹为观止。平常我们都是爸爸妈妈带孩子去哪里玩，今天非常荣幸，是孩子带我们去不同的地方，跟着他们去看世界，非常感谢六师二附小，非常感谢范老师。"

周怡晨妈妈说："感谢六师二附小今天组织二2班和二3班的亲子家庭来到航海博物馆，今天小朋友的收获很多，大人跟着小朋友一起也学了很多知识。在寒假中，学校里能组织这样的活动我们都觉得非常好,第一,这是一个跨班的学习交流,孩子们在学习过程中认识了新朋友。第二,是给孩子拓展了课外的知识,能让孩子开阔眼界学习航海历史知识。第三,是在航海博物馆我们一起聚餐、一起学习、一起交流,气氛非常好,在此也感谢学校和老师,希望以后还能有这样的活动!"

(三)研学后的反思总结

一天的研学活动带给亲子家庭满满的收获。3月13日,在学校二年级范围内举行了船舶探索营总结交流会,由两组亲子代表将他们寒假中在航海博物馆的所见所闻与收获分享给在场的同学、将气候变化治理的学习观念介绍给大家。也有不少同学在活动结束以后对气候变化教育有了新的认识,和父母一起继续去长江流域沿线游学,调查走访周边省市,探寻气候变化对船舶航运的影响。

三、智慧之谈——专家分享与解读

洋泾社区学校的顾校长给学生讲述关于长江口船舶以及黄浦江上的船舶故事,丰富学生的知识储备。每次专家专题讲座都会选择一个与气候变化、环境保护、长江生态等相关的主题热点,如:气候变化的科学原理、长江流域生态系统的特点、气候变化对长江流域水资源的影响等。专家在讲座中深入浅出地介绍相关主题的基本概念、最新研究进展以及实践经验,过程中还设置了丰富的互动环节,学生可以就自己感兴趣的问题和专家进行交流和讨论,除了理念知识的传授外,专家还就如何应对气候变化、保护长江生态环境等方面提供实践指导和建议。

四、创意之乐——船模手工艺制作

本次活动中选取了一些历史时期具有代表性的不同材质的船只,比如木质龙舟、积木福船、塑料救生艇等,在教师的指导下,学生完成船模的搭建,旨在培养学生动手能力和创造力。

在活动过程中,学生通过教师的指导示范,使用简单的材料和工具制作船模,包括船体、桅杆、船帆等部件,他们也可以根据自己的想法对船模的外观涂色装饰。这不仅让学生的动手能力得到了提升,还发扬了互帮互助的精神。

校内的制作活动结束了,但是学生的制作热情并未减弱。回家后,学生又根据自己的喜好,选择了特色船舶进行了亲子制作,其中有历史悠久的古代船模,也有当代船只,更有现代化的新能源动力船舶。船舶种类繁多,为了展示学生的制作成果,我们还举办了小型的校内船模展览会,将学生制作的成果以及船舶知识进行展示分享(见图3)。

这项活动,让学生通过亲手制作船舶深入了解了航运业的重要性,增强了他们对船舶航运领域的兴趣,并且提高了学生对气候变化的警觉性,意识到气候变化可能对航海业带来的挑战,从而增强了对环境问题的关注和责任感。

图 3　校级船模展览会

五、艺术之美——绘画比赛激发灵感

在前期了解船舶和气候变化之间的密切关联后,学生们对于船体的结构和材料、船只类型的选择、船舶动力系统的优化、船舶航行路线的调整都有了新的认识,每个学生都对心中的理想船舶有了新的期待和畅想。基于此,我们又开展了"画一画未来船舶"活动。本活动旨在让学生设计属于自己的未来船舶,提示如下:

你心目中的理想船只是什么样的? 它的动力形式是什么? 是大块头还是小家伙? 有何特殊之处能够俘获船员和乘客的芳心? 你能看到的如今这些船只起初都始于这些问题。帆船驶过海面、潜艇潜入深海、探险船远征异域,都有可能因为人们脑海中闪过的一点想法而有所改变。

六、联动之行——跨区域多模态学习体验

在寒假航海博物馆的一日研学活动之后,学生们对于船舶的认识有所增进,家长和孩子的兴趣也被激活,关注视野也从上海放大到了全国,由此想到了我国的第二条大河——黄河,在好奇心的驱使下,学生们从许多维度提出了想要了解的问题。微信群内,老师还针对学生的提问进行了搜集和整理。

后续,船舶探索之行即将组织一场线上线下的互联活动,推动和黄河口东营的学生开展跨区域的联动学习。

成效与反思

一、项目成效

通过这一阶段的项目学习,孩子和家长之间涌现出共同学习、共生共长的蓬勃状态,家庭亲子关系更加紧密,大家对气候变化教育也有了更深刻的理解。

(一) 身体力行,倡导可持续出行

寒假期间,亲子家庭组队开展的中国航海博物馆一日研学活动,以三个"全程"为主。一是全程的绿色出行:学校和航海博物馆来回近 150 公里,路途遥远,换乘不便。地铁+公交+步行,仅车程将近 5 个小时。二是全程的学生为主:不论是出行路线设计,注意事项的确

定,活动内容安排,都是孩子们现场讨论生成,父母充当了"百度员"。三是全程的学习互助:从前期策划到现场实施、再到活动小结,学生、家长、老师一直处在一个相互学习、相互鼓励、相互成就的良好状态中,没有老师事先提供的详尽活动方案和行程计划,一切均在讨论中生成,在行动中微调,家长和老师之间的默契,学生和学生之间的共学,诸多美好应运而生。

（二）探索船舶新能源方式,形成环保意识与行动力

项目以船为切入点,引导学生关注了船舶航运、能源、排放与气候变化之间的关联。气候变化教育在学生们的心中种下了种子。目前已经看到了孩子积极参加、家长热心投入的状态,后续也将继续通过老师的引领,教会学生如何采取个人可行的行动来应对气候变化,形成更加积极的环保意识和行动力,为建设绿色、低碳的未来社会做出积极贡献。

（三）关注家庭教育与可持续发展,促经验分享与前瞻意识

学校不仅仅是为家庭提供了一个项目主题,更是为家长和孩子搭建了一个共同探索、共同成长的桥梁。这一过程中,家长和孩子相互帮助、相互学习,是一种宝贵的成长体验。孩子们会通过自己的主动学习,向家长展示他们的成长和进步,家长则为之提供成人经验及智力支持。

在活动过程中,家长可以看到孩子们在课本学习之外的另一面,在线下活动策划以及出行时,孩子们展现出的智慧、勇气、吃苦耐劳,不仅让家长感到欣慰,也让他们更加信任孩子的能力,从而能更好地引导和支持他们的成长。

学校家庭社区协同开展气候变化教育之长江探秘系列之船舶探索行,虽然已经取得了一定的成效,但也存在一些需要反思和改进的地方。

二、反思重建

（一）加强多方参与和社区行动

建立稳固的学校家庭社区合作机制是保证教育活动顺利开展的关键。可以通过扩大家长委员会和家长志愿者团队,加强家校之间的沟通和协作;定期举办家长会议或座谈会,让家长了解学校的教育理念和活动安排;利用洋泾社区资源,邀请社区组织和专业机构参与教育活动,丰富教育内容。

（二）加强教育资源的联动

六师二附小位于滨江沿线的洋泾社区,是六师附小教育集团的成员校。洋泾有着丰富的人文教育资源,周边有环境学校、船舶研究所、社区学校等,距离我校都在一公里左右。将目光放大至全国,我校和山东东营长江入海口的小学之前也有过在线联动,由于两地独特的地理位置——分别位于长江口和黄河口,在后续项目推进中,需要加大联动的力度,将多方资源形成有效链接,扩大气候变化教育的影响力。

（三）发挥宣传的关键作用

学校将更好地利用新媒体平台,如微信公众号、网站、社交媒体等,传播环保知识和相关活动信息,扩大教育效应的覆盖面,吸引更多的学生和家长参与到气候变化教育中。

通过持续的反思和改进，本项目的效果将会得到进一步提升，为培养更多关注气候变化、具有责任感的未来公民奠定坚实基础。

（范漪猗，上海市第六师范学校第二附属小学教师）

学校家庭社区协同，融创"双碳"教育新"气象"

研究背景

小学生是未来公民，提升其应对气候变化挑战的意识与能力，尝试为气候变化问题制定出一定的应对策略，是教育的责任与担当。在小学教材中，各学科已然包含或渗透了与气候变化相关的内容。但我们也清晰地看到小学气候变化教育呈现出的问题：一是知识散点，各学科都有一点儿，不成体系。二是关系割裂，所学知识没有创造价值，主要表现为：各学科各自为政，学科与学科间相互割裂、缺乏综合；课堂知识与社会生活相互割裂、缺乏关联；学期学习与假期生活割裂、缺乏融通；知识学习与解决问题相互割裂、缺乏融创。

对学习者的生活有意义的知识才可能具有长久的生命力。因此，学生"应对气候变化"行动，是形成全社会"命运共同体"的纽带，是教育创造生活价值的载体，更是推动"综合融通"素养的魔晶石，还是培养综合性人才的加速器。以学校为主导，协同家庭、社会共同应对气候变化，不仅能以全局之力减缓气候变化带来的影响，更能助力培养学生的亲社会力、领导力、学习力等综合素养。

项目设计

气候变化是人类当前必须面对的既成事实，但气候变化教育绝不仅仅是学校的责任，学校与家庭、社会通力合作，才能寻找到科学的良方。我们一方面将气候变化相关的内容融入校园生活，采用任务驱动、项目式、跨学科学习等方式，指导学生从学科视角进行探究；另一方面，我们根据小学生年段特点，梳理了小学六年各学科中与"气候变化"相关的内容，制定了"气候变化教育"的年段目标、内容体系以及评价机制，依托家庭和社会力量，开展多样的跨学科活动；同时，借助学校以"互学共玩，联结生活、互学共长"为特点的假日亲子玩伴团激发创造性，开展系列化、长程化、多样化的假日研学活动（见图1）。

总之，我们的"气候变化教育"坚持目标导向、问题导向、实践导向，坚持校家社多主体"共育互育、共学互学、共生共长"，追求在综合融通的学习过程中成事，在具身认知及实践锻炼中成人。

项目实施

一、制定"接天气"的协同行动目标

为了对"全球气候变化"的实际形势有较为全面的理解，学校加入了全国气候变化教育研究联盟，项目组成员仔细研读了《巴黎协定》《国家适应气候变化战略2035》《中国应对气候

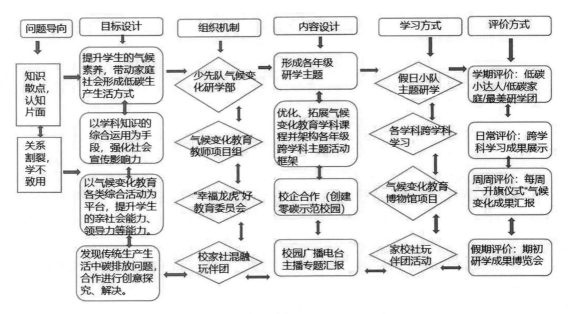

图 1 龙小"学校家庭社区合作开展气候变化教育"流程

变化的政策与行动 2023 年度报告》等资料文件,明确了"减缓和适应是应对气候变化的两大策略,缺一不可"。围绕"绿色生活""绿色消费""绿色生产"三方面的要求,根据学生年龄特点,我们以跨学科学习的方式,聚焦培养学生的社会责任感,为此,我们制定了不同年段的不同目标(见表 1)。

表 1 龙虎塘实验小学各年段"气候变化教育"目标

低年段	中年段	高年段
渗透"亲近自然、守护动植物多样性"的意识	着重研究"绿色消费、节约资源、低碳生活"	侧重了解"新能源之都"的"新能源企业"
以亲子玩伴、代际学习等方式,激发儿童乐于走进自然、参与"祖孙种植""认领绿树""绿色出行"	通过自主招募伙伴,了解"气候变化"给人类带来的影响,并利用周边的场馆实地体验、宣传	加强对新能源以及新质生产力的认识与了解,广泛宣传,积极运用
培养学生乐群、乐表达、乐动手实践的能力	培养学生发现问题、组织活动、协调解决、形成成果的组织力和创造力	培养学生的项目探究能力,尝试小发明和小改造

二、构建"聚众力"的学校家庭社区合作组织

高效的组织有汇聚和放大力量的作用,能提高效率。为此,我们从街区、学校以及学生层面构建了不同效能的组织。

区域"气候变化教育"委员会：为了推动气候变化教育微系统形成，提升社会教育力，多元汇聚资源，我校经多方论证，建立了街道层面融学生、家长、教师、社区人员以及市区领导和生态环境局领导、公益组织、新能源企业代表等多方代表加入的"气候变化教育委员会"，具有明确的工作目标、组织架构、职责分配。依托微信群及定期座谈会，委员会得以日常化运行，大大提高了学校家庭社区在气候变化教育方面互动的效能。

学校"气候变化教育"项目组：在国家课程的各学科中都有与"气候变化与环境保护"相关的内容，为了有效实施国家课程中相关内容并延伸、拓展、转化为实践，我们招募了有兴趣在这方面探索的各年级、各学科老师，组成了项目组核心成员，定期交流与研讨，引领本学科研究的深入与持续。而全体班主任则是项目组的主要成员，引领各自班级的导师团，开展各类班本化气候教育活动。

红领巾"气候变化"研学部：利用少先队阵地，分成"知识宣传研学队""绿色用电研学队""家庭减碳研学队""绿色出行研学队""早期采用者研学队"五个小分队，负责指导和组织大队委、中队委开展相关活动。

"学校家庭社区混融玩伴团"：为了让"应对气候变化行动"成为学生假期生活的重要组成部分，学校以团团联动的方式，通过自主报名在各年级招募了一批"气候变化探究小团长"，借助他们的力量，再选择自己心仪的小主题，招募志同道合的伙伴开展研学活动。"星星之火，可以燎原"，在一个个小团长的带动下，开展了丰富多彩的"低碳"学习与实践活动。

三、设计"接地气"的跨学科主题课程

中小学国家课程中学科较多，在不增加学生额外负担的前提下，我们以落实国家课程为基础，多学科联合，开展"跨学科"活动，并依托"班队会""综合实践活动"开发聚焦"气候变化"的班本课程。

（一）挖掘国家课程资源，做实"跨学科"活动

我们充分利用新课程标准中"各学科有10%的课时用于跨学科主题活动"的要求，筛选出适合的内容，开展"气候变化教育"活动。以语文学科为例，三年级的课文《花钟》介绍了不同时令开放的各种植物以及习作《为植物做名片》，基于此，教师与学生共同设计了跨学科主题活动"多姿多彩的花儿"；课余，走进校园图书馆和市区图书馆，寻找《植物百科全书》等介绍植物的书籍阅读；美术课上，观察校园和小区里的植物，选取1~2种为它们"画像"，再配上一段话介绍，办一个校园"画展"；劳动课上，与同学一起播下花种，做"小园丁"；双休日，与祖辈一起打造"一米阳台花花世界"，并在家长的帮助下，与祖辈在校园种下一片"月季花田"，增加碳汇。

（二）链接校级年级特色活动，开发个性化"班本课程"

校园活动是丰富多彩的，很多活动的设计如果能科学融入"气候变化教育"，不仅不会是简单地做加法、增负担，反而能起到"1+1＞2"的效果。如四年级组的特色文化是"书香阅读"，在世界地球日"遇见"校园读书节活动中，班主任们设计"我们的气变行动"课程：读"气变"、说"气变"、见行动，发动学生到图书馆寻找与"气候变化"相关的书，阅读后制作了好书

推荐卡,并向全校同学推荐;以小队为单位,寻找应对气候变化的策略,尝试实践后做小队汇报。在这样的过程中,学生到"气象局""自然灾害体验馆"等地进行参观与体验,利用已掌握的知识书写倡议书,向市民宣传"气候变化"倡议"低碳出行""绿色消费"。

四、建设"可持续"的假日玩伴团活动

学校的"玩伴团活动"需要家长和教师基于儿童立场,以学生兴趣爱好为驱动,借助社会资源并充分发挥家长和教师的教育合力为活动作保障与引领。活动中,亲子采用探索性学习、掌握式学习等方式实现共学互学。探索性学习是探索未知,而掌握式学习是运用已知。学生更热衷于掌握式学习,把已经学过的东西变成第二天性。① 这更有助于学生知行合一,综合运用所学知识,解决问题,锻炼能力。

(一)做好顶层设计

为了让学生的"应对气候变化"行动有的放矢、言之有物,学校聚焦于"气候变化我宣传""双碳计划我践行""节能永续化日常""增加碳汇增绿色""校企合作探新能"五大行动,制定了假日玩伴团研学的各年级主题(见表2),学生可以据此发现生活中的相关问题,寻找解决策略。

表2 龙虎塘实验小学各年级"气候变化教育"玩伴研学活动系列主题

年级	假日玩伴研学主题	年级	假日玩伴研学主题
一	我们是变废为宝小能手	四	我们是节约资源小标兵
二	我们是知行合一小园丁	五	我们是生态园林小主人
三	我们是绿色出行小先锋	六	我们是"新新能源"小创客

(二)建构活动模式

虽然每个年级的学生在相应的课程指导下,探究的内容皆有侧重,但是为了便于学生在家长的指导下有效开展活动,设计活动流程,学校以任务驱动,提出过程中需共同完成的"六个一":策划一个方案、发布一张招募令、拍摄一套活动照片、设计一张评价表、制作一个物化成果(心得、作品、创意等)、获取一套章(纪念章和素养章)。

(三)建设资源智库

气候变化的学习场域,不仅是课堂,更多的应是在家庭生活、社会生活场景中。为了让学生在广阔的自然天地、社会空间中学习,我们借助"气候变化教育委员会"形成了初步的学习资源库(见表3)。

---

① 艾莉森·高普尼克.园丁与木匠[M].刘家杰、赵昱鲲,译.杭州:浙江人民出版社,2019:188.

表 3 气候变化研学资源库(部分场馆资源)

图书馆类	企业类	参观体验类		公园类	田园类
常州图书馆	天合光能	常州博物馆	常州气象馆	紫荆公园	一米阳光田园
秋白书苑	森萨塔	安能自然灾害体验馆	常金米业陈列馆	东坡公园	佳农探趣
社区图书室	永琪自行车	小龙人安全体验馆	常州垃圾处理厂	圩墩公园	西夏墅艺术稻田
大众书局	温康纳公司	三江口公园文明实践基地	防震减灾科普教育基地	森林公园	常州九红生态园

五、丰盈"慧传播"的成果展评机制

（一）可视化：以纪念章和素养章的多元评价激发学生活动参与热情

一类是参加活动即可获得的"气候变化小领袖"纪念章。另一类是根据活动表现获得的素养章。为了保证活动的质量,凸显亲子共学的价值,学校协同家长、学生和社会人士共同制定了各年级的活动评价标准(见表 4),在每次活动结束前,进行多元评价:同伴互评、亲子互评。这样的双向评价,使亲子共长、家庭共学的民主氛围得到进一步激发。

表 4 "气候变化"研学活动素养章争章标准

一级指标	二级指标
互学共长	(1)主动学习气候变化相关知识,感受到学习新知的快乐,形成较强的学习力。 (2)在探究活动中能尊重、理解他人,善于发现、学习他人之长。 (3)善于运用知识,利用资源,与家长、玩伴一起,有效解决生活与学习问题。
勇于探索	(1)有好奇心,对周围环境中气候变化相关现象充满兴趣。 (2)能根据质疑的内容与同伴进行主动探究,形成共同的研究成果。 (3)自信、乐观,具有战胜困难的顽强意志和能力。
敢于担当	(1)遵守规则,培养初步的公民意识和命运共同体意识。 (2)主动参与气候变化探究与宣传活动,初步形成亲社会的情感和能力。 (3)乐于承担玩伴团探究活动中的岗位与职责,增强团队合作力。

（二）可再创：在大型活动中搭建展示舞台扩大活动影响力

一是每年寒暑假生活中的玩伴团活动,将会经过再加工后在学校期初博览展销会上现场展示与评比,日常双休日、小长假中的玩伴团活动也将以班级为单位在升旗仪式上进行展示和互动,再通过学校公众号向外辐射。这样的方式不仅让更多学生参与体验,还能提升创造力、合作力,激发成就感。二是打造了专属平台,利用校园广播专门设立了"龙娃说气候"栏目,并在校园公众号上开辟了"气候变化专栏",宣传学生的研学成果。

（三）可集聚：打造集成果展览与教育功能于一体的气候变化博物馆

气候变化博物馆包括两园一廊一中心。两园，即学校的半亩花田和与天合光能合作打造的"空中花园"，既能用太阳能屋顶发电，又是学生种植的花园；一廊，即一条生态秘境，拥有气象监测、气候学习、二十四节气探究等功能；一中心即玩伴中心，陈列学生玩伴团气候变化研学过程性成果，并能进行碳排放计算互动。

成效与反思

一、项目成效

一是营造了多主体关注气候变化的良好生态，形成了"命运共同体"实践力。寒假，我校开展的"气候变化"玩伴团活动占所有玩伴团活动的59%，各学科教师之间的合作也变得频繁，促进了教育力量的有效融合（见图2、图3）。学校、家庭、社区建立了友好合作关系，家长、社区、班主任、学科教师、学生的"命运共同体"的意识和情怀正在养成。通过学生自主、教师引领、家长保障、社会支持的好生态的营造，多主体的亲社会力都得到了提升。

图2　学生在玩伴中心介绍研学成果

图3　期初展销会上学生展示研学成果

二是促成了学以致用的融通型学习生态，提高了课外生活创造力。以重构综合性目标体系，着眼"气候变化"；驱动多元化活动方式，汇聚团队合力；实现资源互惠共享，提升校家社教育合力为路径，通过家校合作、校企合作、亲子互动等方式，多主体多元互动开展玩伴团活动，使学生感受到学以致用带来的价值感和成就感，课余生活更有自主性、丰富性，成为学期生活的延展、延伸和补充，也让家庭有效融入社会。

三是培育了校家社多主体的成长型思维，提升了学生的核心素养力。在研学活动发起、策划、招募、组织、评价、总结反思的过程中，学生在尝试中进步，在交往中自省，在挑战中自信，在合作中互律。积极参与项目活动的家长们，滋养了陪伴孩子共同成长的勇气和智慧，增强了自己的家庭教育者的角色意识，也使亲子关系更和谐。同时，参与此项目的教师理解了"学习必须把实践作为一个整体来考虑，并考虑到其中关系的多样性——既有共同体内部的关系，又有与整个世界的关系"[①]突破学科的困囿，走向跨学科、超学科的生活教育。

---

① J.莱夫·温格.情景学习:合法的边缘性参与[M].王文静,译.上海:华东师范大学出版社,2004:26.

四是搭建项目研学的多元平台,扩大了社会影响力。自项目实施以来,开展相关跨学科主题活动70余个、相关假日玩伴团活动300余个(见表5),开展全国气候变化教育联盟现场展示活动多次,被《中国教育报》、江苏省少先队、《常州晚报》、常州电视台等平台宣传20多次,教师在全国气候变化教育联盟研讨会上讲座分享10多次,受众约5万余人次。研究成果还被顾惠芬校长带到"第三届中日韩环境教师交流会暨中日韩环境教育和公众环境意识研讨会"交流并获得高度赞赏……在落实学校家庭社区协同育人方面、新时代背景下生态文明建设方面,形成了全域共育互育的教育新生态。

表5 "气候变化玩伴团"开展的活动(部分)

活动类别	活动内容	班级数	参与人数(概数)
节约能源,物尽其用	参观污水处理厂	12	300
	探索神秘水世界	5	80
	参观垃圾处理厂	15	260
	不让恐龙的今天,成为人类的明天	30	1000
	垃圾分类——我们的特别行动	20	1200
增加碳汇,人人行动	一起种多肉	12	200
	保护植物多样性	12	600
	我为植物做名片	12	500
	走进生态农场	5	50
	争做小农人	24	1200
	为校园植绿	36	200
	我与祖辈栽月季	12	100
变废为宝,创意无限	以废换绿公益集市	5	200
	以物换物爱心集市	20	500
	"瓶"空再造	5	200
	纸箱变身"大玩家"	12	1000
	布"袋"称王	5	50
校企合作,智慧科创	走进"天合光能"	13	90
	探究"新能源车"	2	90
	"森萨塔"互学共学	13	50
	共同打造太阳能空中花园	12	500
	龙鱻鱻杯清洁能源科创画大赛	15	100
	环保烟花秀	3	100

二、不足与改进

一是要加强气候变化教育实践中可视化成果的积累，促进经验推广。虽然学校有气候变化教育的顶层设计，活动丰富、形式多样，学生、家长、社区人士多主体的参与体验也较为深刻，但可视化的成果还是相对较少，一是项目组收集成果的意识不强，没有及时收集并分类存放与展示；二是文本经验较少，比如整理好的适合各年级阅读的气候变化书目表、校本化玩伴团活动手册、气候变化宣传文案（如诗歌、标语、快板词等），经验辐射方面还有较大空间。

二是要提升气候变化教育与学校品牌项目的融合度与效度，促进特色实践。学校多年来开展代际学习项目研究，而气候变化教育的主体恰恰是全社会全人类。因此我们已经尝试让祖辈、亲辈走进气候变化教育课堂，创设校内外多时空、多领域的气候变化教育大学堂，让多代互学共学过绿色生活。后续，将进一步做强代际协同开展气候变化教育等融合型学校特色实践。

三是要增强气候变化教育与社会协同育人的任务驱动，促进机制可持续。借助顶层设计，我们一方面依托国家课程中气候变化教育内容设计活动类"幸福作业"，加强常态实施。另一方面激发学生和家长的主动创造性开展校外玩伴团活动，增强动态生成性。同时，依托社区力量、校企合作，以"定向越野"的方式，在全国生态日、每月初等时间节点发起气候变化行动，不断催生新的学习与成果传播。

总之，在"双减"背景下，气候变化教育项目有效提升了家庭生活的质量和家长陪伴孩子的质量，变革了学习方式，促成了多元共学、内化生成、实践行动的"三合一"，以孩子为中心，依托家校社的合力创设了强大的学习场域，为全民追求更美好的绿色低碳生活打开了一条最佳路径。

（陈亚兰，常州市新北区龙虎塘实验小学副校长；
顾惠芬，常州市新北区龙虎塘实验小学校长）

对话专家

气候变化是当今世界全人类面临的共同挑战,是联合国可持续发展目标的重要关注点。气候变化教育,是当下一代人和下一代人能够理解、关注和应对气候变化的根本措施,也把教育对人本身的关注,扩展到人与自然这个更多的内容,更长的时间和更广阔的空间上。

《气候变化教育的中国行动(第一辑)》聚焦于这一非常重要和有挑战性的议题,通过案例汇编,涵盖理论研究、政策背景、项目设计、实施过程以及成效反思等气候变化教育的多个重要方面,体现了中国教育工作者在气候变化教育中的未雨绸缪的思想和扎实入微的实践。

该书第五篇围绕"学校家庭社区协同开展气候变化教育"的主题,从气候变化教育实施主体的角度,涉及校企合作、社区参与等多元化的教育模式,为读者提供了一个全面的气候变化教育视角。案例汇编展示了家庭、学校和社会三方面如何协同合作,共同推进气候变化教育。

各个案例中不仅强调了理论知识的学习,更重视将所学知识应用于实际问题的解决。本篇通过具体的项目实施案例,提供了气候变化教育实践操作的详细步骤和策略,具有很强的操作性和指导性。其项目设计的创新性,还体现了教育活动在促进社会参与和提升公众意识方面的实际效果。这种深入的案例分析为读者提供了宝贵的第一手资料,使其能够具体了解气候变化教育项目从构思到执行的每一个环节。"银芽气候课堂"将气候变化教育的受众从学生拓展到祖辈,通过祖孙共学模式,增强了年轻一代对气候变化的理解和应对能力,同时也让老年一代更新了环保知识,实现了知识的双向传递。"凤凰驿站"通过校企合作,将气候变化教育与自行车文化相结合,不仅提升了学生的环保意识,还增强了他们的实践能力和创新思维。"学在自然"通过家校社的共同努力,构建了一个完整的学习时空,让学生在真实的自然环境中学习和成长,加强了教育的社会性和实践性。

项目设计中的"双师"制教学模式,即学校教师与企业专家的合作,为学生提供了理论与实践相结合的学习机会。此外,本篇还提出了构建全方位分层次气候教育体系的策略,这种策略不仅关注知识的传授,更重视培养学生的跨学科思维和解决问题的能力,体现了教育创新的理念。

本篇的案例汇编在展示项目的实施成效之外,还包含了对项目实施过程的深入反思。案例中对教育活动的反思不局限于表面现象,而是深入到教育方法、内容和效果的评估层面。这种反思不仅仅是为了单个项目的改进,更是为了形成可推广的教育模式,形成一种"可持续"的教育范本。这种反思有助于不断优化项目设计,也能够为后续模仿、改进的老师们提供关键指引。

(张勇,华东师范大学环境科学系主任,教授;
徐倩华,华东师范大学环境科学系研究生)

附录　气候变化教育指导纲要(试行)

2024 年 2 月

教育是应对气候变化的重要力量,开展气候变化教育是提高公众对气候变化的认识和应对能力的重要途径,是强化气候行动的重要组成部分,是实现可持续发展的重要保障。

基于对国内外已有成果的梳理,尤其是基于华东师范大学上海终身教育研究院与各教育机构共同发起、组织推动的系列研究的经验总结,通过多次学术会议和学术研讨,特形成本指导纲要。

一、气候变化教育的理念与特征

(一)理念

气候变化是当前全球面临的重大问题之一,事关人类的生存与发展。应对气候变化是联合国系统的整体战略。

我国高度重视应对气候变化问题,以积极建设性的态度推动公平合理、合作共赢的全球气候治理,将应对气候变化纳入了国家发展战略,按照推动高质量发展的要求,统筹谋划经济社会发展和应对气候变化工作。

教育是应对气候变化的一个核心举措,通过影响人的价值观、思维方式、行为和生活选择,帮助学习者形成应对气候变化所需要的素养。

(二)特征

气候变化是跨越国界的全球性挑战。从具体可感知的气候变化,到在全球尺度、历史进程中认识气候变化,气候变化本身的复杂性,要求以多面向、跨学科思维来开展气候变化教育,体现综合性、实践性、丰富性。

(1)综合性:从学习内容看,气候变化教育兼有自然科学与人文社会科学的内涵;从学习过程看,气候变化教育要帮助学习者从多角度全面理解生态环境系统,掌握社会环境与生态环境及其内部各要素间的密切联系和相互作用;从学习途径看,气候变化教育可以通过跨学科的方式融入各科教学中,也可以单独设立气候变化专题教育课,还可以通过综合实践活动或依托家庭、企业、社区来开展,形成学校、家庭、社会协同开展气候变化教育的新格局。

(2)实践性:气候变化教育强调学习者在亲身体验中发现和创造知识;在解决现实气候变化问题的过程中发展创新能力及批判能力;在参与中增进交流与理解,形成正确的生态环境价值观;在实践中形成与自然环境和谐相处的健康生活习惯,增强积极参与促进可持续发展的行动意识。要避免单一学科的教育和割裂的教育方式,尤其要避免知识教学与生产生活实践的脱节。

(3)丰富性:气候变化教育的主体、对象、时间、空间、载体、方法多样,既有一般性的原理,也有不断生成的新策略;既要不断总结已有经验,也要不断面对全球气候变化的新问题、涌现而出的新方案。尤其是面向全民、凸显终身学习、重视不同教育主体、教育机构所开展的气候变化教育,更有无限的丰富性,体现教育的创造性。

二、气候变化教育目标与内容

(一)目标

气候变化教育需要在更新和发展学习者相应的价值观与思维方式,发展相应的知识与能力,注重对生活教育和行动能力的培养,促成生产、生活与参与治理方式的转型等方面,产生积极的效应。

1. 发展价值观与思维方式

(1)增强学习者的人类命运共同体、人与自然的地球生命共同体意识,形成尊重、协作、贡献于人类文明、人类发展、地球生命的价值取向;形成合理认识个人与人类关系的思维方式。

(2)增强知行合一、领导变革的价值取向,形成合理认识点滴努力与系统更新关系的思维方式。

(3)增强立足当下、主动作为的价值取向,形成合理认识当下与未来关系的思维方式。

2. 发展知识与能力

(1)发展应对气候变化的相关自然科学类知识。

(2)发展应对气候变化的相关人文社会科学类知识。

(3)发展学习者的领导力、实践力、综合解决问题力、终身学习力。

3. 形成在生产、生活和参与治理中的新习惯

(1)能将所学的气候变化知识与能力渗透、体现在日常生活中。

(2)能将所学的气候变化知识与能力渗透、体现在生产中。

(3)能将所学的气候变化知识与能力渗透、体现在社会治理、全球治理中。

(二)内容

气候变化教育的内容多元丰富。为利于全面理解、关心并应对气候变化的复杂性,可以根据具体学习人群而加以选择并系统化,但都服务于目标。如下内容可以作为基本的内容结构:

1. 气候变化的现状及紧迫性

(1)气候变化的概念与特征:理解《联合国气候变化框架公约》中所作定义:"指除在类似时期内所观测的气候的自然变异之外,由于直接或间接的人类活动改变了地球大气的组成而造成的气候变化。"

(2)气候变化现状与问题:气候变化对地理环境和人类活动产生深远影响,包括气温升高、极端天气事件增多、海平面上升、生态系统受损等。

2. 气候变化的多维度影响因素和应对

(1)科学之维:解释气候变化的科学原理和机制,推动科学素养的提升。

(2)文化之维:探讨不同文化对气候变化的认知和响应,促进跨文化理解。
(3)伦理之维:强调气候变化的伦理责任,培养全球公民意识和责任感。
(4)经济之维:解析气候变化对经济的影响,引导可持续发展的经济观念。
(5)政治之维:审视政治决策对气候政策的影响,保持高度的政治敏锐性并推动政治参与。

3. 气候变化的治理实践

(1)解读《联合国气候变化框架公约》和《巴黎协定》的内容和实施机制:分析全球合作的机制,以及本土创新和行动案例;强调减排行动和适应行动的关联性,强化监测评估的重要性。
(2)气候变化的应对与行动:应对气候变化的治理措施,包括减少温室气体排放、发展可再生能源、循环经济、保护和恢复森林等。

4. 生产方式的转型

(1)能源效益:研究可再生能源的技术和应用,提升能源利用效率,倡导可持续能源发展。
(2)温室气体排放管理:聚焦甲烷和其他非二氧化碳温室气体的控制和减排策略。
(3)循环经济和资源利用:推动生产方式向循环经济转型,提高资源利用效率。
(4)生物多样性保护:介绍保护生物多样性的重要性和方法,强调与气候变化的关系。

5. 生活方式的转型

(1)生活环境觉察:鼓励对环境的主动认知,提倡可持续生活方式。
(2)消费习惯更新:引导消费者选择环保产品,减少碳足迹。
(3)衣食住行的可持续选择:具体介绍在日常生活中的可持续选择和实践方法。

6. 文化传承与创新

(1)传承:探讨气候变化对传统文化的影响,强调文化保护传承的重要性。
(2)创新:鼓励社会参与应对气候变化,促进新兴文化与可持续发展实践的结合。

7. 科技的推动作用和数字工具的使用

(1)了解气候科学、气象技术和数据模型等科技手段在监测、预测和理解气候变化中的角色。
(2)探讨新兴科技如人工智能、大数据等在气候研究和解决方案中的创新应用。
(3)运用数字工具,如气候模拟软件、数据可视化工具,进行气候数据分析和展示。
(4)探讨技术创新对减排技术、清洁能源和可持续发展技术的推动作用。

8. 多方参与和社区行动

(1)社区行动认知:引导学习者认识社区中的气候挑战,了解气候变化对当地社区的影响。探索与社区组织、当地政府等合作,推动形成社区层面的气候变化应对方案,培养对社区环境的敏感性。
(2)可持续发展理念:推动学习者思考社区的可持续发展理念,明确社区在气候变化背景下的发展目标。介绍社区可持续性评估工具,帮助学习者了解社区可持续性发展与气候变化之间的关系。

(3)社区层面的气候教育:开展社区层面的气候教育活动,包括学校与社区合作,社区学校组织开展系列教育活动,举办气候变化主题活动、讲座、研学游等。

9.国际合作与全球视野

(1)理解不同国家和地区的气候变化挑战及应对方案,为学习者拓展全球视野提供具体案例和实践。

(2)开展国际合作,促进学习者的跨国合作学习,协同推动气候变化教育。

10.教育和宣传的关键作用

(1)强调教育在提高社会认知和推动气候行动中的关键作用。

(2)探讨媒体和社交媒体在宣传中的具体作用。

(三)面向全民的教育目标与内容结构表

附表1　面向全民的教育目标与内容结构

目标		参考内容要点							
一级	二级	幼儿园	小学	初中	高中	中高职	普通高校	社区学校	老年学校
1.形成正确的价值观与合理的思维方式	1.增强学习者的人类命运共同体意识,形成尊重、协作、贡献于人类文明、人类发展的价值取向;形成合理认识个人与人类关系的思维方式;具有实现人与自然的生命共同体和人类命运共同体相统一的价值认知								
	2.增强知行合一、领导变革的价值取向,形成合理认识点滴努力与系统更新关系的思维方式								
	3.增强立足当下、主动作为的价值取向,形成合理认识当下与未来关系的思维方式								

(续表)

目标		参考内容要点							
一级	二级	幼儿园	小学	初中	高中	中高职	普通高校	社区学校	老年学校
2.发展气候变化的知识与应对能力	1.发展应对气候变化的相关自然科学类知识								
	2.发展应对气候变化的相关人文社会科学类知识								
	3.发展学习者的领导力、实践力、综合解决问题力、终身学习力								
3.养成可持续发展的生产、生活和参与治理方式与习惯	1.能将所学的气候变化知识与能力渗透、体现在日常生活中								
	2.能将所学的气候变化知识与能力渗透、体现在生产中								
	3.能将所学的气候变化知识与能力渗透、体现在社会治理、全球治理中								

三、气候变化教育途径与方法

（一）气候变化教育的途径

（1）跨学科整合和项目化学习：教育工作者可以通过在教育机构内融通于各学科教学中，通过开发校本课程和组织项目化学习，通过发展校园文化，通过指导、推动基于家庭和社区的学习活动，促进学习者、家长和社区人士的深度学习、共同学习。

（2）亲子、代际学习与家庭社群学习：家长、祖辈可以通过支持基于家庭的学习，支持学习者的学校和社区学习，发展亲子、代际学习，开展家庭、社群学习等方式，促进不同背景和专业知识的人共同参与气候变化教育活动。

（3）社会资源的开放与合作：鼓励政府、企事业单位、私营部门、社会组织等开放教育资源，与家庭和学校合作，为工作人员、学生、家长、教师、社区居民等提供实践机会和专业知识支持。探索企业与学校的合作项目，如企业提供实地考察和工作坊，以增强学习者对气候变化的实际理解。

（4）数字化学习：重视数字化资源，指导数字化学习，借助数字化平台开展学习成果传播、互动交流，提高学习的效益。

（二）气候变化教育的方法

面向全民开展气候变化教育，倡导个体学习、群体学习、共学互学，提倡项目化学习、观察学习、体验式学习、探究式学习、课堂学习等多种方式，重视真实的生产与生活世界、书籍、视频、模拟软件等多类型学习资源。

教育过程强调针对学习者特征开展有针对性的教育。如针对青少年儿童，尤其强调对这一综合性、复杂性问题的觉察、理解，借助正规学习的力量，辅之以非正规、非正式学习，习得系统知识和技能，发挥项目化学习中的领导力，主动带动家长和社会人士；对于成年人，要强调与生产、生活和治理的高度结合，强调知行合一、家校社互动、亲子共学互学，鼓励参与气候变化教育和宣传活动；对于老年人，要强调人文传承，并鼓励老年学员加强科学学习和技术应用，鼓励代际学习，强调服务学习。

上述内容为通用性建议，教育者和学习者可以具体参考、采用、更新，发展起气候变化教育的智慧。

四、教育评价

(1)重视学习者素养的形成。通过对学习者参与学习前后的素养变化，形成评价结果。

(2)重视情境的创设与运用。在具体城乡发展、日常生活、生产与治理的情境中，对气候变化教育成效进行评估。

(3)重视数据的收集与分析。高度重视来自学习者、教育者和参与者的反馈，积累多类型的数据，强调数据支持下的评价结论形成。

(4)重视评价的主体与方式。充分发挥多主体的评价责任，如自我评价、同伴评价、教师评价、专家评价、社会评价等，并能共同推广、践行应对气候变化的可行性方案。

五、教育支持

(1)组织、支持教育工作者、家长、社会人士学习气候变化主题的专业知识，持续发展绿色技能，积累教育经验，提升开展可持续发展教育的智慧。

(2)为职前和在职教师专业发展培训建立课程框架，开发监测工具，将气候变化教育师资培养制度化。

(3)推动教育政策更新，在学习型社会建设背景下更新已有政策，尤其是鼓励成年人、老年人参与气候变化教育项目中。

(4)推动研究联盟组建，发动、鼓励教育者、气候专业人员、家长、社会人士等建立气候变化教育联盟，形成跨界、跨域的研究与实践共同体，创新协作、联动机制；既基于区域特点，开展重点领域的研究，有效突破；又通过互补互促，互学共研，整体提升气候变化教育的质量。

(5)设立专项，保障项目经费，鼓励研究，推动教育知识的生产、传播和转化。

(6)国际交流,建立或参与在线资源库建设,作为交流平台以分享我国在气候变化教育方面的优秀案例和实践,为实现全球可持续发展目标贡献中国力量。

执笔人：

华东师范大学　李家成

华东师范大学　朱丹蕾

江苏省常州市新北区龙虎塘实验小学　顾惠芬

华东师范大学　杨懿凝

广东省深圳市宝安中学(集团)实验学校　陈才英

上海市华东师范大学第五附属学校　黄婕

上海市浦东新区新场实验中学　施一凡

上海市崇明区教育学院　郭春飞

上海市南汇新城镇社区(老年)学校　李秋菊

山东省东营市胜利孤岛第一小学　赵嫔婷

山东省东营职业学院　王雅宁

2024 年 2 月 20 日

气候变化教育研究丛书
上海终身教育研究院　总主编

Climate Change Education: China in Action

Edited by
Jiacheng Li, Zhiping Yang, Huifen Gu

Volume 1

Climate Change Education Research Series
Chief Editor, Shanghai Municipal Institute for Lifelong Education
  Climate Change Education: China in Action (Volume 1)

Contributing Editors:
  Yazhen Wang, Qiaoling Wang, Ying Wang, Meng Wang, Yan Wang, Yijing Ping, Rong Fu, Chaoxia Long, Shuqing Gong, Min Zhu, Meifang Liu, Juan Sun, Tongtong Sun, Shuhan Li, Xingyue Li, Qiuju Li, Jiacheng Li, Lin Su, Yonghong Zhang, Jie Zhang, Yong Zhang, Jing Zhang, Yalan Chen, Caiying Chen, Baoyu Chen, Shuang Chen, Yuhan Wu, Jianfen Lin, Ziyi Yang, Zhiping Yang, Yining Yang, Xiaojuan Zhou, Yiyi Fan, Zelong Zhao, Jing Zhao, Pinting Zhao, Zhiwen Hu, Lin Jiang, Yao Yao, Yifan Shi, Weiwei Gu, Peipei Gu, Weibin Gu, Huifen Gu, Huanxin Qian, Qianhua Xu, Youchun Guo, Han Ni, Jie Huang, Xiaoyan Sheng, Yuyan Cheng, Yufang Zeng, Danyan Jiang, Nannan Xie, Meiqin Lan, Yan Cai, Xiaofeng Cai, Hong Pan, and Yishu Teng

Organizer:
  Shanghai Municipal Institute for Lifelong Education, East China Normal University

Supporting Organizations:
  Institute of Schooling Reform and Development of ECNU
  National Training Center for Secondary School Principals, Ministry of Education
  UNESCO Project on Education for Sustainable Development in China
  China Education for Sustainable Development Program Research Community
  Ecological Civilization and ESD Innovation Studio, Beijing Academy of Educational Sciences
  China Climate Change Education & Research Alliance
  China Intergenerational Learning Research Alliance

Also Supported by:
  Key Project of National Social Science Fund of China, "Research on Community Education System from the Perspective of Lifelong Learning for All" (AKA210019)

Preface

Climate change is an issue that we all can't ignore. However, education on climate change has not been lacking. Chinese educators have been consistently contributing to the field of climate change education.

In 2019, Shanghai joined UNESCO's Global Network of Learning Cities as one of the leading cities of the Education for Sustainable Development (ESD) Cluster. Against this backdrop, the Shanghai Municipal Institute for Lifelong Education (SMILE) started focusing on global climate change issues, exploring original local educational experience, supporting the implementation of the Sustainable Development Goals, and engaging in global dialogue and governance. Since 2023 in particular, SMILE has collaborated with relevant research institutes, universities, high-middle-primary schools, and other educational entities, and also linked up with enterprises, institutions and social organizations, to deeply engage in the practical exploration and theoretical innovation of climate change education. In pursuit of lifelong learning for all, from children to the elderly, SMILE has established a collaborative framework involving schools, families and communities to nurture the development of climate change education.

In 2023, this endeavor was presented for the first time to the national counterparts at the "Research on Students' Summer Vacation Life and Reconstruction of Life at the Beginning of the Semester" meeting, held on September 15, themed "Collaboration, Empowerment, and Creation". At this meeting, three case studies were shared with the attendees, including reporters Yalan Chen from Longhutang Experimental Primary School of Xinbei District, Changzhou City, Tang Ying from Shanghai Hongkou Experimental School, and Caiying Chen from Shenzhen Bao'an Secondary School (Group) Experimental School.

On December 12, 2023, the "1st Conference of Climate Change Education through a cooperative mechanism made up of schools, families and communities" was held at Qibao Mingqiang Primary School in Minhang District, Shanghai. The meeting was co-organized by the SMILE along with other units, with participants from all over the country, and a new pattern of multi-cooperation was formed, especially in cooperation with the local town government and the district meteorological bureau.

On March 1, 2024, the "2nd Conference on Climate Change Education through a cooperative mechanism made up of schools, families and communities" was held at Longhutang Experimental Primary School, Changzhou City, Jiangsu Province. This seminar convened an expanded array of government departments, research institutes and academic institutions, especially expert guidance from the Education and Publicity Centre of the Ministry of Ecology and Environment of the People's Republic of China. We collectively gained extensive experience from a large number of enterprises.

On April 26, 2024, the "Shanghai Climate Week 2024-Climate Change Education Forum" was held at East China Normal University. More case results were shared at the forum. With the aim of multi-cooperation and lifelong learning for all people from children to the elderly, and the collaborative development model of multi-sectoral and multi-principal co-operation, we established a clearer connection between theory and practice. In particular, the development of climate change education in the context of the development of universities for the elderly has achieved positive impact both nationally and internationally.

In mid-May of 2024, SMILE formally submitted a proposal to establish a "Climate Change Education Laboratory at East China Normal University". The launch of this multidisciplinary laboratory promotes the innovation of the climate change education model suitable in China, promotes the establishment of a China's independent knowledge system in this field, provides opportunities for domestic researchers and students to participate in international research, and also nurturing the development of a group of climate change education personnel with an global perspective.

The cases in this book come from educators in various provinces and cities in China, and committed to addressing international issues on climate change. Chinese education practitioners and researchers are confronting global issues with an open mindset, striving to connect climate change in daily life to participation in global climate governance, and communicating and sharing our ideas and practices more proactively.

The educational practices presented in this book reflect China's commitment. Chinese educators place special emphasis on lifelong learning for all, from children to the elderly, and have therefore expanded climate change education to encompass the entire population. Moreover, it has been implemented in a synergistic manner through a cooperative mechanism involving schools, families and communities, and climate change education has also extended to the family and society. This demonstrates the expansive educational perspective among Chinese educators.

Practices across the nation pass on Chinese wisdom. The rich traditional Chinese culture nourishes the lives of Chinese people, and the idea of ecological civilization directly contributes to the evolution of Chinese education. Through climate change education, Chinese wisdom can be more fully recognized, and also wisdom embedded in folklore will be embodied in concrete intergenerational learning of all kinds of people.

This phase of efforts also benefited the Chinese practices on climate change global understanding. Under the organization and promotion of the SMILE, the research results on climate change education in Shanghai have been released globally by the United Nations Educational, Scientific and Cultural Organization (UNESCO) as typical cases. SMILE has also proactively created various kinds of dialogues and exchanges, and provided pertinent educational materials, so that China's ideas, strategies, tactics and effectiveness on climate change education have been increasingly recognized and acknowledged by international community.

The publication of this book marks the culmination of these efforts. In the course of the publication process, our practices have been ongoing, international exchanges and cooperation have been intensified, the contribution to global climate governance through education continues to grow, climate change education, occurring in everyday activities including production, daily life, and governance, persists.

Jiacheng Li, Zhiping Yang, Huifen Gu
1st June 2024

Contents

Chapter 1 Climate Adaptation Oriented Campus Construction ············ 1
- Design of an Intelligent Lighting Control System for Classrooms in the Context of Energy Conservation ············ 2
- Construction of the Campus "Fish and Vegetable Symbiosis Hall" and Low-Carbon Education Practice ············ 10
- Build a Growing Climate Change Exploration Museum ············ 16
- The Construction Concept and Educational Practice of "Zero-Carbon Campus" ············ 27
- 🎓 Expert Review ············ 38

Chapter 2 Discipline-Based Teaching and Learning for Climate Change Education ············ 41
- Research on the Integration Practices of Primary Chinese Teaching and Ecological Education ············ 42
- Design of Teaching Activities for Primary School Chinese Language Unit Under the Influence of Ecological Education ············ 47
- Exploration of Promoting Low-Carbon Life Through Project-Based Learning in Primary School Mathematics ············ 53
- A Practice of Climate Change Education in English Teaching ············ 60
- Climate Change Education Based on Scientific Practice Activities: Taking "Measuring Precipitation" as an Example ············ 68
- Smart Learning, Wise Application, and Intelligent Innovation: Constructing an Integrated Model of Information Technology Education and Climate Change Education ············ 76
- Designing Teaching Activities for Morality and Law in Primary Schools Under the Context of Ecological Civilization Construction ············ 84

- Expert Review ········ 94

Chapter 3   Climate Change Education Based on Project-Based Learning ········ 97
- "Disfavored Shoes", Have You Been Disfavored? —Let the Idea of Recycling be Integrated into the Daily Life of Primary School Students ········ 98
- Research on the Recycling and Reuse of Milk Cartons ········ 104
- Implementing Climate Change Education Practice Based on STEM Integrated Activities ········ 111
- Design and Implementation of the Multidisciplinary Project "Regeneration of Coffee Grounds" ········ 119
- Exploring the Relationship Between Wampee Fruit and Climate to Revitalize the Countryside ········ 127
- Exploration of the Relationship Between Climate Change and Biodiversity from a Student Perspective ········ 135
- Design and Implementation of Project-Based Learning on "Harmony of Water and Air Nurtures the Symbiosis of All Things" ········ 143
- Exploring the Project-Based Learning of Science Popularization on "Climate Change and Environmental Protection" ········ 152
- Expert Review ········ 159

Chapter 4   Climate Change Education Based on Integrated Activities ········ 163
- Find the Yangtze River Estuary No. 2 Ancient Vessel and Promote the Yangtze Estuary's Vitality ········ 164
- The Past, Present and Future of Heavy Industrial Plant in the Yangtze Estuary ········ 174
- "Green Travel" Makes Carbon Emission Reduction Visible ········ 183
- Practices of Implementing Climate Change Education in Secondary Vocational Schools ········ 190
- Exploring Meteorological Proverbs, Investigating Climate Change ········ 199
- A New Paradigm for Learning About "Climate Change Education" Based on Embodied Practice ········ 207
- Expert Review ········ 218

Chapter 5 Collaborative Climate Change Education by Schools, Families, and Communities ·········· 221

- Design and Implementation of the Learning Pathways of the "Climate Change Classroom Targeted the Elderly and Children" ············· 222
- Green Innovation and Climate Change Education: Building a New Model of Lifelong Learning ················· 231
- School-Enterprise Collaboration Promotes Climate Change Education Practices on Campus ············· 242
- School-Enterprise Cooperation: Advocating "Green Productivity", Promoting Climate Change Education ················ 251
- Case Study on Student Collaboration in Green Enterprise Visits at Yellow River Estuary and Yangtze River Estuary ············· 263
- Investigation of the Principles of Green Refrigeration for Home Use and Practical Operation ················ 273
- Experiences of New Energy Companies in Supporting Climate Change Education ················ 284
- Learning in the Nature: A Practical Study of Climate Change Education in Primary Schools Under the Concept of Common Worlds Pedagogies ················ 292
- Yangtze River Ship Explorations ··············· 303
- School-Home-Community Collaboration Creates a New "Climate and Carbon" Education Paradigm ·············· 311
- 🎓 Expert Review ················ 322

Appendix Climate Change Education Guidelines (Draft) ············· 324

Chapter 1

Climate Adaptation Oriented Campus Construction

In the face of the growing challenge of global climate change, climate change education is imperative, and schools play an irreplaceable and critical role. Schools are the primary venue for exploring ways to reduce emissions in educational settings, the laboratories for enhancing climate adaptation and resilience, and the learning platform for fostering climate literate citizens.

The cases presented below are centered on the themes of "retrofitting energy-saving and emission reduction facilities," "improving the local environment of the campus," "building climate-themed venues," "building a zero-carbon campus," and so on. The cases are centered on the themes of "renovation of energy-saving and emission reduction facilities", "improvement of local environment", "construction of climate-themed venues", "construction of zero-carbon campus", etc. They include the design and construction of green buildings on campuses, adjustment and upgrading of lighting equipment, development of intelligent and energy-saving classrooms, construction of climate change education , teaching experiments, and thematic venues, and exploration of campus "carbon verification", energy audits and planning. It is especially worth mentioning that these actions are all participated in and experienced by teachers and students of the school, which demonstrates the courage, wisdom and commitment of educators to self-renewal and innovation in response to climate change.

Design of an Intelligent Lighting Control System for Classrooms in the Context of Energy Conservation

Research Background

In the context of current global climate change, energy conservation has become a growing important global mission. According to the *Paris Agreement*, "In the second half of this century, on the basis of fairness, and in the context of sustainable development and efforts to eradicate poverty, balance should be achieved between anthropogenic emissions by sources and removals by sinks of greenhouse gases."

At present, humanity urgently needs innovative solutions to reduce carbon emissions and enhance energy efficiency to mitigate the impacts of climate change. "Education serves as a core measure to address climate change, helping learners develop the necessary competencies by influencing values, ways of thinking, behaviors, and life choices." This case study explores the innovative application of emerging technologies in climate change research and solutions, focusing on how algorithms can be used to save energy.

Project Design

In many schools, homes, and offices, the energy consumption of indoor lighting systems accounts for a significant proportion. Nevertheless, traditional lighting systems lack intelligent control functions, leading to energy waste. This case, based on the lesson "The Smart Classroom in My Mind," starts from the "energy-saving pain points" discovered by students on campus. It explains how to guide students in designing an intelligent lighting control system and further optimizing and iterating.

1. Activity Objectives

(1) Help students integrate their life experience to understand the role of logical operations in intelligent lighting control systems.

(2) Stimulate students' interest in thinking, and develop their ability to analyze problems, propose hypotheses, and justify conclusions.

(3) Cultivate students' practical skills and their ability to solve real-world problems.

(4) Enhance students' awareness of building an energy-efficient campus.

2. Activity Process

(1) Introduction period: Teacher triggers students' thought processes through "energy-saving pain points" and introduces the teaching content of transforming a smart control classroom.

(2) Experimental exploration period: Teacher demonstrates Experiment A,

introducing the concept of "NOT operation."

(3)Communication and discussion period: Students conduct Experiments B and C, discussing and exploring the conditions affecting the switching of intelligent lighting.

(4) Expression and presentation period: Students share ideas for upgrading other equipment in the smart control classroom and engage in discussions on natural language descriptions and evaluation criteria.

3. Resources

(1)Experimental program: Before the activity, the teacher should design the experiment and prepare supporting materials for the students, including the experimental procedure and data recording sheet, thereby facilitating the smooth execution of the activity.

(2)Task clarification form: Teacher needs to provide a learning task sheet during the activity to help students clarify the requirements and steps to complete the tasks, guiding them to participate purposefully in the activity.

(3)Learning evaluation sheet: After the activity, teacher should ask students to fill it out to self-assess their performance and learning during the activity.

Project Implementation

1. Scenario Introduction: Students Leaving the Classroom with the Lights on

At the beginning of the class, the teacher presents a video showing students leaving the classroom with the lightson before their PE class—despite the classroom being empty, the lights remain on. This scenario makes a thought-provoking question: how can the classroom lighting be optimized to achieve intelligent control and save energy? This leads the core question: how would you design the smart control classroom in your mind?

2. Experiment Simulation: Students Understand the Basic Logical Operations "AND", "OR", and "NOT"

Before the class, the teacher created three programs based on the factors of "time," "light intensity," and "people indoors" (see Figure 1) to serve as experimental learning materials for the students. The three programs simulate different scenarios of intelligent lighting control: Experiment A: The light turns on if it is nighttime; otherwise, it turns off. Experiment B: The light turns on if the classroom is dim and there are people present; otherwise, it turns off. Experiment C: The light turns on if the classroom is dim or there are people present; otherwise, it turns off.

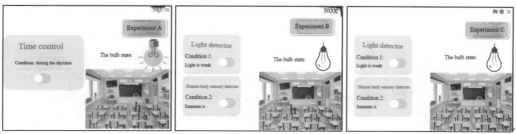

Figure 1. Schematic of Experiment A, Experiment B, and Experiment C programs

Firstly, through a combination of teacher-student Q&A and independent student discussions, students are helped to identify the potential factors that can affect the operation of smart lighting, namely whether there are people indoors, whether it is daytime, whether there is sufficient indoor lighting, and the sensors required to detect these factors (see Figure 2).

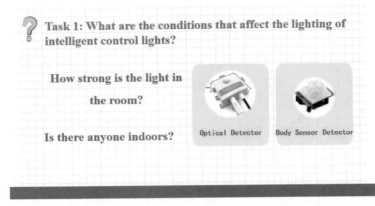

Figure 2. Identifying the potential factors affecting intelligent lighting and the required sensors

Next, students use the provided learning material "Smart Lighting Simulation Experiment Program." Under the teacher's guidance, they complete the three experiments and record the process and results in the table.

Experiment A: The teacher guides students to complete the steps of Experiment A. Students first independently control and find out whether the conditions are met. Then they should carefully observe the lights under different conditions and record the experiment process and results in the table. For Experiment A, the condition is: if the condition "daytime" is not met, the light remains on; if the condition "daytime" is met, the light turns off. Record the results as follows: define "1" for conditions met and "0" for conditions not met; in the light status column, record "0" for the light off and "1" for the light on (see Table 1).

Table 1. Experiment a record sheet

Condition: daytime	Lighting status
0	1
1	0

Experiment B: Students work with their deskmates to complete Experiment B. The light only turns on when both conditions, "low light" and "presence of people," are met.

Experiment C: Students work with their deskmates to complete Experiment C. The light turns on when either condition, "low light" or "presence of people" is met. The results of

Experiment B and Experiment C are recorded in Table 2 and Table 3, respectively.

Table 2. Experiment B record sheet

Condition 1: low light	Condition 2: presence of people	Lighting status
0	0	0
0	1	0
1	0	0
1	1	1

Table 3. Experiment C record sheet

Condition 1: low light	Condition 2: presence of people	Lighting status
0	0	0
0	1	1
1	0	1
1	1	1

Practical experience in the class showed, through manual operations and immersive experiences in these three experiments, students gained a deep understanding of the rules of the three basic logical operations and clarified methods for modifying the classroom lighting system. In Experiment A, they learned about the "NOT" operation, where the condition and the result are opposite. In Experiment B, they understood the "AND" operation, where the result is true only when both conditions are met simultaneously. In Experiment C, they mastered the "OR" operation, where the result is true as long as one condition is met.

3. Comparative Study: Students Consider from the Perspective of Energy Conservation

Students discuss with their engagement in discussions with their peers how to choose the appropriate solution for modifying the classroom lighting system from an energy-saving perspective. Through experimental simulations and drawing on their own life experience, students unanimously agree that the Experiment B solution (i.e., the "AND" operation: the light turns on if the classroom is not well-lit and there are people present; otherwise, it turns off) is more suitable for application in classroom lighting system modifications in a campus environment. This is because it can intelligently control the lighting based on the classroom's light conditions and human activities, thus more effectively conserving energy.

Utilizing digital learning methods to collaborate in selecting appropriate solutions, students not only enhance their understanding of energy conservation concepts but also develop teamwork skills and problem-solving abilities. Throughout the process of thinking and choosing the best solution, they continuously improve their logical thinking and innovation awareness.

4. Practice and Reinforcement: Students Use Different Logical Operators to Control Various Smart Devices

Students apply the new knowledge learned in this lesson—the three basic logical operations "AND", "OR", and "NOT"—to consider energy conservation. They independently think through and determine the appropriate logical operators based on the characteristics of the smart control devices and the task requirements. Then, students record these design ideas in a table, forming a design plan. Finally, students learn to use the completed table and the sentence structure provided by the teacher "The smart control device is _____ (name of the smart control device). When _____ (condition), it _____ (operation)," they endeavor to articulate the working principles and application scenarios of their designed smart control devices in words (see Figure 3).

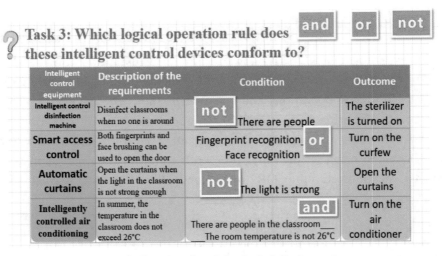

Figure 3. Practice of applying basic logical operators

5. Extended Application: Students Explore How Smart Control of Air Conditioning Maintains Constant Temperature

Students mimic the sentence structures provided in the previous section to describe, in natural language, the operational schemes for smart control devices in their envisioned smart control classroom using logical operators. From an energy-saving perspective, they try to determine which basic logical operations, or combinations of operations, should be used to control the given smart control devices.

In this process, students engage in deeper discussion. They question whether the operation step of "adjusting the air conditioning temperature" is specific enough—should it be adjusted to a higher or lower temperature? The teacher provides timely guidance, suggesting that specific conditions need to be considered for proper operation and that incorporating a logical operator "AND" or condition statements might be useful. Students then use the sentence structure to express their ideas and propose an adjustment scheme for the smart control air conditioning: if the smart control air conditioning sensor detects that there are people in the classroom and the indoor temperature is above 26°C, the temperature should be lowered; if the sensor detects that there are people in the classroom and the indoor temperature is below 26°C, the temperature should be raised. This concept is commendable but requires the use of condition statements. For the students, this is a new attempt and challenge, which stimulates their interest in learning about subsequent facility and equipment modification course content.

Effectiveness & Reflections

This case study is based on the issue of students forgetting to turn off the lights when leaving the classroom, which stimulates students to think about how to use the knowledge they have learned to transform the classroom lighting control system. In the experimental simulation period, students gained a deep understanding of logical operations through collaborative experiments and recordings. In the comparative exploration period, based on experiments and practice, they learned how to choose appropriate solutions from the perspective of energy conservation and emissions reduction, strengthening their awareness of electricity saving and environmental protection. In the practice and reinforcement period, through hands-on operations, they systematically organized the knowledge they acquired, enhancing their problem analysis, expression, and resolution skills. In the expansion and application period, students actively conceived other feasible smart control facility modification plans, using what they learned to design a smart control classroom, showcasing their exceptional imagination and innovation.

Transforming the classroom lighting control system is part of the school's smart energy conservation education. It not only enhances the students' learning experience but also significantly reduces the school's electricity consumption. However, to fully implement smart energy conservation education, further expansion from multiple aspects is still needed.

1. New Developments in the Discipline: Teachers Can Explore Systematic of Smart Energy Conservation Education in Upper Elementary Grades

As societal emphasis on sustainable development and environmental protection continues to grow, the systematic exploration of smart energy conservation education in upper elementary grades becomes particularly important. At this stage, students who have completed a semester of information technology studies possess a certain level of digital literacy and learning ability, making them ready for more systematic smart energy conservation education.

Schools can organize practical activities for upper elementary students, such as creating energy-saving technology projects or participating in smart energy-saving games, to foster awareness and skills in smart energy conservation through hands-on experience. Additionally, specialized courses or modules on smart energy conservation education can be introduced in upper elementary grades (see Figure 4) to achieve systematic teaching and management of energy-saving education.

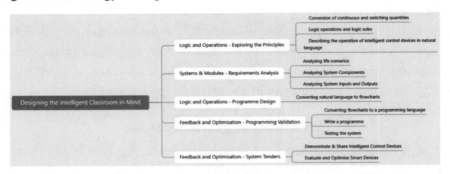

Figure 4. School-based unit design for intelligent energy education curriculum

2. Campus Digitization: Schools Can Apply Smart Energy Conservation Education in School Management

The campus is one of the key venues for smart energy conservation education. Building a climate-smart campus can facilitates the integration of smart energy conservation education into campus management. For example, digital tools can be used to establish a campus smart energy management platform, which monitors campus energy consumption in real time, provides smart energy-saving recommendations, and offers management suggestions, thus providing scientific basis for smart energy management in the digital age.

Jinshan District Qianjing Primary School in Shanghai has already planned such initiatives. For instance, they aim to implement smart management for campus water purifiers, enabling them to automatically turn on and off according to a schedule, thereby saving both electricity and water. They also plan to use smart management for the "small kitchen treasure" in restrooms, setting management strategies based on the daily routines of students and staff to conserve energy, extend the equipment's lifespan, and reduce the risk of electrical fires from regular use. Additionally, effective management and utilization of campus energy will be carried out to reduce energy consumption and carbon emissions.

3. Home-school-community Cooperation: The Society Should Promote Smart Energy Conservation Education in Families and Communities

Smart energy conservation education should not be confined to the campus but also requires effective and ongoing collaboration with families and communities. In the classroom, teacher guide students to envision and design their ideal smart control classroom, and also encourage them to discuss and plan ways to avoid energy wastage from household appliances with their parents at home. Collaboration with the surrounding community can integrate smart energy conservation education into community building and activities, such as inviting professionals to conduct lectures on ecological civilization in the school.

Systematic research of smart energy conservation education curricula, continuous advancement of campus smartification, and close cooperation among families, schools, and communities are three important exploration directions for smart energy conservation education at the elementary school level. We believe that through the collaboration and efforts of all parties, it is possible to enhance students' learning of smart energy conservation knowledge and cultivate their awareness of energy-saving practices, making a positive contribution to building resource-saving campuses and promoting global sustainable development.

(Yijing Ping, Teacher of Qianjing Primary School, Jinshan, Shanghai)

Construction of the Campus "Fish and Vegetable Symbiosis Hall" and Low-Carbon Education Practice

Research Background

Admist the growing environmental challenges, schools, as important institutions for cultivating future citizens, have the responsibility and obligation to lead in climate change education. The "Construction of the Campus' Fish and Vegetable Symbiosis Hall' and Low-carbon Education Practice" at Qianjing Primary School in Jinshan District, Shanghai, was thus conceived.

Environmental protection is closely linked to our daily lives. Cultivating students' environmental protection habits can not only reduce the negative impact on the environment but also contribute to building a sustainable ecological environment. The campus "Fish and Vegetable Symbiosis Hall" provides students with a platform to intuitively understand the connection between environmental protection and life. It also enables them to personally feel and experience the value and significance of environmental protection in practice, realizing that environmental protection is a global issue and is closely related to each of us. In addition, during this educational practice process, teachers constantly learn and explore new environmental protection education methods, thereby enhancing their professional quality in environmental protection education.

Project Design

1. Overall Design of the "Fish and Vegetable Symbiosis Hall"

The "Fish and Vegetable Symbiosis Hall" is located within the "Jingjing Vegetable Garden" section of the school (see Figure 1). A set of "fish and vegetable symbiosissystem" has been constructed in the hall, mainly consisting of the following parts:

(1) Infrastructure construction: This includes fish ponds, hydroponic cultivation beds, water pumps, and pipelines. Fish ponds are used for fish farming, hydroponic cultivation beds for growing plants, and water pumps and pipelines for pumpimg water from the fish ponds to the cultivation beds to achieve the recycling of water resources.

(2) Fish farming: We usually select fish species with strong adaptability and fast growth (such as goldfish) for farming. Through reasonable feeding and management, ensure the healthy growth of fish.

(3) Hydroponic cultivation: Utilize the nutrient-rich water in the fish ponds for

plant cultivation. Generally, plant varieties with short growth cycles and high yields are selected, such as lettuce, green vegetables, celery, etc.

Figure 1. "Fish and Vegetable Symbiosis Hall" of Qianjing Primary School in Jinshan District, Shanghai

In this system, the water in which fish are raised is rich in nutrients and can be used for plant irrigation after treatment. Plants absorb these nutrients, purify the water body, and provide a clean living environment for fish. This system not only achieves resource recycling, reduces the use of chemical fertilizers and pesticides, but also reduces environmental pollution and the consumption of water resources, achieving low-carbon and environmentally friendly agricultural production.

2. Design of Learning Objectives

(1) Knowledge acquirement: Students can know the causes and effects of climate change, in particular the relationship between agricultural activities and climate change through the construction and operation of the "Fish and Vegetable Symbiosis Hall". They can also know the principal of aquaponics.

(2) Habit formation: Students can develop energy-saving, water-saving and other environmental habits by participating in the daily management of the "Fish and Vegetable Symbiosis Hall". They can also develop interest in issues including environmental protection and sustainable development.

(3) Thinking cultivation: Students can better understand the interrelationships and impacts of the various components of the ecosystem by attending educational practice in "Fish and Vegetable Symbiosis Hall". Students can establish the concept of sustainable development and realize the importance of environmental protection and economic development.

Project Implementation

In the "Fish and Vegetable Symbiosis Hall", the school has carried out various low-carbon education activities, as listed below:

1. Theme Education Activity of "Energy Conservation and Emission Reduction"

In the "Fish and Vegetable Symbiosis Hall" project, students understand the scientific principles and practical applications of energy conservation andemission reduction through experiments and observations. Students observe that through efficient system design and meticulous management, the "Fish and Vegetable Symbiosis Hall" can significantly reduce energy consumption and carbon emissions while meeting production and living needs.

Students understand through personal operation of intelligent irrigation equipment that when the intelligent irrigation system is set to irrigate once an hour for 10 minutes each time, it can improve the utilization efficiency of water, ensure the accuracy and efficiency of the irrigation process, and achieve energy-saving effects. This allows students to truly feel the necessity of energy conservation and emission reduction, thereby taking energy-saving measures in daily life and reducing energy consumption. Furthermore, students also express their understanding and feelings about energy conservation and emission reduction and environmental protection through painting.

2. Theme Education Activity of "Recycling"

The water resource recycling system in the "Fish and Vegetable Symbiosis Hall" is an important educational resource. Under the guidance of teachers, students observe and participate in every link of rainwater collection, wastewater treatment, and recycling (see Figure 2). They learn the working principle of the rainwater collector for collecting rainwater and understand that the collected rainwater will be stored in the reservoir underground of the "Fish and Vegetable Symbiosis Hall" and eventually used for irrigating plants. By observing, they understand the entire process of how wastewater becomes reusable water resources after treatment and purification. Students also participate in the design and improvement of the recycling system, and find other links of the designed recycling system in the "Fish and Vegetable Symbiosis Hall" by communicating with parents or consulting materials.

Figure 2. Rainwater collector in the "Fish and Vegetable Symbiosis Hall"

3. Theme Education Activity of "Ecological Environmental Protection"

The "Fish and Vegetable Symbiosis Hall" provides students with a platform for observing and researching ecosystems. Here, students understand the composition and operation mechanism of the ecosystem, learn how to protect the ecological environment, and participate in a series of observation and documentation activities for ecosystem.

Students work in groups to regularly observe the growth status of fish and plants, as well as changes in water quality, etc., and record these observation results. The school also invited ecological and environmental protection experts to the "Fish and Vegetable Symbiosis Hall" on March 15, 2024, to give lectures and professional guidance (see Figure 3). The vivid case presentations and easy-to-understand explanations by the experts have stimulated students' interest in understanding the impact of human activities on the ecosystem. Through participating in the operation and management of the "Fish and Vegetable Symbiosis Hall", students have a deeper understanding of the importance and principles of ecological environmental protection and actively practice environmental protection in daily life.

Figure 3. Ecological civilization lectures conducted by the school

Effectiveness & Reflections

Following a semester of practical exploration, the campus "Fish and Vegetable Symbiosis Hall", with its unique charm, simulates the natural ecosystem and integrates aquaculture and hydroponic plants harmoniously. It not only realizes the recycling of resources but also maintains the balance of the local campus ecology and improves the overall quality of the campus environment. This project provides students with the opportunity to participate in the construction and management of the campus ecosystem, serving as a valuable platform for cultivating students' scientific literacy, practical skills, and innovative spirit. During the participation process, students deeply understand the low-carbon concept, understand the operation mechanism of the ecosystem, and realize the preciousness of resources and the importance of environmental protection.

During the project implementation process, students' problem-solving ability has been enhanced. Faced with the practical challenge of dirty pond water, students responded proactively, delving into the analysis of the issue, identifying causes, and seeking solutions. Ultimately, they successfully overcame this difficulty through innovative methods such as breeding black-shelled shrimp and using fishing nets to catch green algae (see Figure 4～Figure 6).

Figure 4. Students releasing black shell shrimp

Figure 5. Students using fishing nets to catch green algae

Figure 6. Students removing green algae

Furthermore, in the construction of the campus "Fish and Vegetable Symbiosis Hall", students' teamwork ability and communication ability have been enhanced. Within the team, each student plays an important role. They collaborate with each other, face challenges together, discuss and formulate solutions together, and deeply recognize their responsibilities and missions as global citizens.

In the implementation of this highly innovative and practical project, there are some areas for improvement. For instance, although students are highly motivated to participate in the project, some may feel confused and helpless during the practical process due to a lack of relevant professional knowledge and experience, making it difficult for them to continue. Therefore, teachers need to constantly improve themselves, deepen their learning, and strengthen systematic training and individualized guidance for students to explore more efficient solutions. For example: a cross-disciplinary cooperation platform can be constructed to promote the intersection of knowledge in multiple fields and the fusion of innovative thought; a step-by-step phased training strategy can be implemented to gradually consolidate students' professional foundation and improve their professional skills. At the same time, a project mentor system can be established to ensure that each student can receive individualized guidance and timely academic assistance that suits their needs. Through these comprehensive measures, we will strongly promote the construction and research of the "Fish and Vegetable Symbiosis Hall" to deeper levels, achieving more fruitful results.

(Zhiwen Hu, Teacher of Qianjing Primary School, Jinshan, Shanghai)

Build a Growing Climate Change Exploration Museum

Research Background

In 2023, Changzhou, Jiangsu Province, known as "City of New Energy", became one of the cities with GDP exceeding one trillion yuan. In the days when the good news was announced, various local media in Changzhou viedto report it, and the topic drew the attention and discussion of students: Why is it necessary to develop the new energy industry? What are the market prospects for the new energy industry? Based on this, the students discussed, investigated and learned about the close connection between the new energy industry and climate change.

In October 2023, after discussion and voting in the Red Scarf Council of Changzhou Hongmei Experimental Primary School, "Exploring Climate Change and Contributing Youth Power" was selected as the research theme for the 15th Science and Technology Festival. Students focused on "climate change" for research-based learning, and through interviews, observations, and experiments, they understood the impact of climate change on human life. At the closing ceremony of the Science and Technology Festival, our school established the first climate change education mentor team, inviting the chief forecaster of the municipal meteorological bureau, experts from the ecological environment bureau, and others to serve as mentors. At the same time, the school also launched a winter vacation climate change education research project—"Climate Change and Human Life" and deployed ten major winter vacation events such as "Parent-Child Visit Records," "Promotion Planning Team" and "Collection of Exhibits".

In November 2023, Changzhou Hongmei Experimental Primary School became a member of the Climate Change Education Research Alliance led by Professor Jiacheng Li of the Shanghai Municipal Institute for Lifelong Education, ECNU. In January 2024, *Opinions of the CPC Central Committee and the State Council on Comprehensively Promoting the Construction of a Beautiful China*, encouraging grassroots units such as communities and schools to carry out green, clean, and zero-carbon leading actions, and make full use of museums, exhibition halls, and science education centers to promote the vivid practices of building a beautiful China. The school immediately responded to the spirit of the document and continued to deepen the research on climate change education. In order to transform climate change education research from a short-term to a regular practice and to collect and exhibit the materials from students' research process, the school invested special funds to initiate the construction of an on-campus "Climate Change Exploration Museum". In January 2024, the Climate Change Exploration

Museum project was officially launched. In April, Changzhou Hongmei Experimental Primary School became the leading school for the science education museum curriculum project in primary and secondary schools in Tianning District.

Project Design

Centered around the research theme of climate change, Changzhou Hongmei Experimental Primary School, based on its existing research foundation, introduced resources from multiple parties to build a Climate Change Exploration Museum that integrates functions such as science popularization, experimental exploration, salon sharing, and interesting viewing. The museum will be integrated with the school's curriculum construction and community science education. The project design pursues the following goals: to build an immersive and exploratory new venue that suits local conditions; to develop a new museum-school integrated curriculum model; to establish a new team of scientific exploration mentors; and to construct a new advancing science popularization ecosystem with the times.

The Climate Change Exploration Museum provides teaching resources and space for schools to implement high-quality climate change education and scientific education. It is not merely a "display corner" for equipment, but an "inquiry laboratory"; not just a "material exhibition station," but a "hub of ideas"; not an "activity meeting room," but a "talent incubator"; and not a "classroom for teaching," but an "intergenerational growth area." Students can explore and research mysteries here, share research findings, hold scientific activities, and receive scientific enlightenment; teachers design interdisciplinary activities and develop museum-school courses here. Community residents visit the place on weekends, and volunteers carry out ecological civilization promotion activities here. This is a practice base for lifelong learners to jointly build a beautiful China.

Project Implementation

To plan the construction of the museum scientifically, Changzhou Hongmei Experimental Primary School visited new energy enterprises, went to meteorological bureaus and ecological environment bureaus for academic exchanges, and engaged in collaborative discussions with science associations and other units. At the same time, guided by the concept of "participatory learning," the school actively encouraged students to participate in the management, use, and the design of the "Climate Change Exploration Museum" activity space, making the school space richer and more diverse, with greater value for educating people.

1. Museum Design: Participatory Co-construction

1) Teacher participation, full integration of all subjects

The school established a museum-school project working group, led by the principal, and set up departments such as planning, curriculum, external publicity, activities, and logistics. Corresponding work systems were formulated to plan project advancement, organize specific activities, ensure facilities and equipment, allocate teacher resources, and ensure efficient implementation of the exploration museum construction.

Taking the curriculum department as an example, the heads of teaching and research groups and lesson preparation groups from various subjects took the lead in studying the literature on climate change education compiled by Professor Jiacheng Li and others. They sorted out the content related to climate change in the curriculum standards and textbooks of each subject, and listed a framework for the integration of climate change education into subjects across six years of primary school. At the same time, each grade determined the research theme of climate change education for the current semester, planned grade-level activities, prepared integrated lesson examples, designed cross-disciplinary assignments, made good use of the main classroom front, and developed extracurricular activity circles, closely combining subject teaching with practical activities to ensure the depth and breadth of research.

2) Student participation, full presence in all aspects

Starting from the perspective of children, the school observed and considered environmental construction from a child's viewpoint, upgrading and renovating campus venues to create a campus atmosphere filled with growth. We informed students about the construction plan for the exploration museum and provided information such as venue maps and data on proportions. We coordinated various subjects and organized activities like "I Design My Venue." When students become designers, builders, researchers, and users of the school's educational venues, their state of active learning is stimulated, and their sense of identification with the school is ignited.

Among the collected student suggestions, student Ziqian Wang proposed using the red plum blossom as a research vehicle for climate change. Because the Red Plum is not only the name of the school but also an expression of Chinese cultural imagery. Many Red Plum trees are planted on campus, making it convenient for students to observe and record the growth process. Following his proposal, the sixth-grade group initiated a phenological study of the red plum blossom. Students study the growth and glory of red plums and explore the changes in climate and weather for each year.

After serial discussions between teachers and students, the final draft of the exploration museum covers a indoor area of nearly 700 square meters, with five major spaces: a climate change science popularization area, a climate change exploration area, a climate change experimental area, a climate change teaching area, and an area for envisioning responses to climate change. The outdoor space of approximately 300 square meters includes a climate monitoring area, a plant cultivation area, and a stargazing area, meeting the needs of students for research and exploration in climate change. Through the integration of technology, art, and culture, the venue provides students with a multi-dimensional experience, and the content displayed in various thematic spaces demonstrates scientific, intellectuality, and humanity (see Figure 1, Figure 2).

Figure 1. Overhead view of the front of the Climate Change Exploration Museum

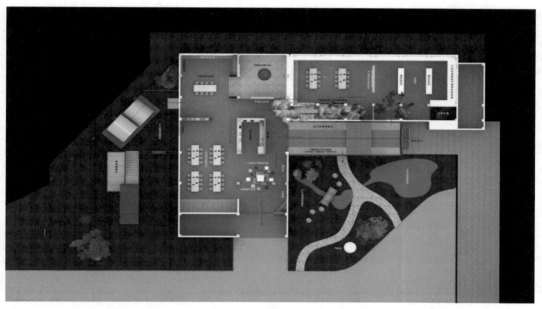

Figure 2. Top view directly above the Climate Change Exploration Museum

2. Curriculum Development: Cross-disciplinary Co-creation

To effectively develop and utilize the museum-school resources of the "Climate Change Exploration Museum" effectively, the school aims to cultivate students' scientific literacy and scientific interests, with a focus on the developmental rhythm of student learning. By improving curriculum elements and coordinating educational forces, the school has further extended and developed theme-based elective and experiential courses for climate change exploration, in addition to implementing the compulsory national curriculum. This has led to the construction of a three-in-one "Climate Change Exploration Museum" curriculum map, with compulsory courses as the main component, supplemented by elective courses and specialized experiential courses.

The project team has sorted through the compulsory textbooks at the primary school level and preliminarily structured a curriculum development plan that covers multiple angles of school life for teachers and students, integrating and interweaving cross-disciplinary and multi-disciplinary elements. The following "three-pronged" measures have been clarified:

1) High-quality implementation of "Daily" compulsory national curriculum clusters

By utilizing the organizational and activity mechanisms of the "Climate Change Exploration Museum," the school integrates the resources of various subjects in the national curriculum to form a national curriculum cluster on climate change education, exploring project-based learning and cross-disciplinary learning to enhance students' problem-solving abilities.

For instance, in the Science curriculum, we have outlined the key points related to the "Twenty-Four Solar Terms" in the first semester of sixth grade. This topic can be integrated with the Solar Term's songs in the Chinese language curriculum and the seasonal work in the labor education curriculum. Together, these elements can form a thematic activity task group centered around the "Twenty-Four Solar Terms" (see Table 1).

Table 1. Climate change education curriculum Resources (Science) discipline in Changzhou Hongmei Experimental Primary School

Grade/Book	Toipc	Objectives	Design
Grade1 Book 1	How much do you know about the weather?	Students can: • Know cloudy, sunny, rain, snow, wind and other weather phenomena, know the impact of weather changes.	After learning, students can: ①Talk about the weather conditions you know about. ②Tell me about the significant weather changesin Changzhou throughout the year.
Grade 1 Book 2	Observe the weather	• Under the guidance of the teacher, students can communicate the observation results of the weather through oral description, drawing and other means.	①Observe a weather phenomenon and try to describe it in words or draw pictures. ②Read some picture books related to weather and climate.

(Continued)

Grade/Book	Toipc	Objectives	Design
Grade 2 Book 1	Care about the weather	①Understand the meaning of weather symbols and understand common weather symbols. ②Describethe weather elements in a structured way based on reading the weather forecast. ③Talk about the impact of weather on people's production and life, plants and animals.	①Design your own weather symbols and understand common weather symbols. ②Watch theweather forecast, summarize the main points of the weather forecast, and try to be a "little announcer" to broadcast the weather. ③Talk about the impact of weather on people's production and life, animals andplants.
Grade 2 Book 2	Climate and animal	①Know that different animals are adapted to different climates. ②Can observe and summarize the physical characteristics and living habits of animals living in different climates. ③Know that animal survival is linked to climate change.	①Learn about different animals living in different climates and discuss why they can live in these climates. ②Observe and discuss the different abilities of different animals to adapt to climate change. ③Select an animal in a climate to make popular science cards for them and introduce their living habits.
Grade 3 Book 1	The winds of the Earth	①Know what causes the wind. ②Know what causes seasonal winds.	①Know the characteristics of cold air and hot air, and understand and explain how wind isformed through experiments. ②Team work to solve the problem: Why do typhoons happen in summer, and they always move from the sea to the land?
Grade 3 Book 2	Measure the weather data	①Skilled in temperature measurement. ②Cloud cover and rainfall will be measured. ③Know what wind direction and wind power are. ④Know what weather and climate are and the difference between the two.	①Learn to use a thermometer and measure the temperature of the day. ②Work in groups to measure cloud cover. ③Make a rain gauge to measure the amount of rain. ④Measure the wind direction and strength. ⑤Talk about what is weather, what is climate, can distinguish between weather phenomena and climate phenomena, say what is the difference between them.

(Continued)

Grade/Book	Toipc	Objectives	Design
Grade 4 Book 1	Climate and plant	①Know that different plants grow in different climates. ②Can understand the structural characteristics and ability of plants to adapt to different climates through observation. ③Grow a plant and learn that the growth of plants is related to the climate.	①Learn about plants that grow in different climates in different regions and discuss why they grow in this climate. ②Observe and discuss the different abilities of different plants to adapt to climate change. ③Plant a tulip, keep watching, and realize that climate is very important for plant growth.
Grade 4 Book 2	Climate and land surface	①Understand different landforms and the conditions under which they arise. ②By simulating the erosion effect of wind on the surface, I realized the uncanny workmanship of nature. ③Understand the global landscape changes caused by global climate change.	①Recognize different landforms and understand that different landforms are caused by different climatic conditions. ②Design simulation experiments to simulate the generation of wind erosion landforms. ③Be aware of the impact of climate change on the landscape, and understand the changes of the landscape caused by global warming.
Grade 5 Book 1	Water drop's journey	①Can describe the continuous circulation of water on Earth between land, ocean and atmosphere. ②Establish a preliminary awareness of the dynamic balance of the water cycle, and understand that the water on the earth is in a state of dynamic balance andconstant renewal. ③Be able to explain a series of consequences of global warming with knowledge of water cycle.	①Preliminary established the concept of water cycle by communicating "why the water in the sky never falls out". ②Use arrows and words todraw a water cycle diagram. ③Use the knowledge of water cycle to exchange and discuss why global warming will lead to glacier fusion and sea level rise. And to figure out what other natural disasters might happen next.

(Continued)

Grade/Book	Toipc	Objectives	Design
Grade 5 Book 2	Four seasons climate	①Summarize the climate characteristics of the four seasons and know that the formation of the four seasons is related to the earth's revolution around the sun. ②Know how the seasons affect some phenomena on Earth. ③Understand the causes of the four seasons through simulation experiments, and the causes of the regular natural phenomena related to them on Earth.	①Read the map, discuss and summarize the climate characteristics of the four seasons, and know that the temperature, precipitation, length of day and night in the four seasons are regular. ②Explore the relationship between the earth's tilt and direct and oblique radiation, and then explore the influence of direct and oblique radiation on temperature. ③Simulate the Earth's rotationand revolution through experiments to deepen understanding.
Grade 6 Book 1	History of climate understanding	①Understand the development history of meteorological observationtools and experience the importance of technology for meteorological observation. ②Know the history of human exploration of meteorology and have a simple understanding of the working principle of weather stations. ③Recognize the wisdom of the ancients by understanding and reciting the 24 solar Terms.	①Collect data, discuss and summarize the observation history of human meteorological observation and the iterative update of tools. ②Understand the weather station and how it works. ③Learn the 24 solar terms by heart, understand the meteorological connotation, know what customs different solar terms have, and understand the connotation and significance behind them.
Grade 6 Book 2	Ideal home	①Understand the impacts of global warming on human production, life, animals and plants. ②Realize the seriousness of the situation that global warming will bring irreversible consequences. ③Be responsible for the Earth and contribute to stopping global warming.	①Watch videos and pictures to understand a series of consequences caused by climate warming. ②Simulate sea level rise caused by global glacier melting through simulation experiments. Aware of the impact of global warming. ③Discuss what we can do to combat global warming and write a 200-word climate Change mitigation proposal.

2) Creative development of "Weekly" elective theme course clusters

During the weekly club activities and after-school service, the school offers a diverse range of theme-based elective courses at the exploration museum, catering to all students. The school strengthens guidance for science clubs and interest groups, encouraging and supporting interested students to engage in long-term, in-depth, and systematic scientific exploration and experimentation related to climate.

To accomplish these goals, the school has introduced high-quality external organizations. With the continuous influx of professional instructors from society and their orderly participation in after-school services, an increasing array of elective courses has been developed. These include "Interesting Chemical Elements," "Climate and Folk Customs" and "The Culinary Delights of the Twenty-Four Solar Terms". These courses are progressing towards the high-quality goals of being "open, personalized, and elective."

3) Sequential organization of "Monthly" museum-school course clusters

During periods for science, comprehensive practice, and labor, the school organizes at least one museum-school course per month. Based on the extension points in the textbooks and students' interests, teachers adopt a task-driven, teacher-guided, and student-led collaborative exploration approach for scientific thematic exploration and practical activities, developing museum-school specialized experiential activity clusters.

The school develops inwardly and expands outwardly, constructing museum-school course clusters and developing a "museum-school course circle." By making reasonable use of resources such as meteorological stations and science museums, the school collaborates with co-construction units to establish student education activity bases, expanding students' activity spaces and facilitating interconnection among multiple educational domains. Adopting a task-driven, teacher-guided, and student-explorative learning style, the school promotes participatory learning based on autonomous practice, turning society as a textbook and the world as a classroom. In the rich immersive context, students are guided to experience, reflect, and grow.

3. Mentor Cultivation: Diverse Co-growth

Through "three major actions," the school is committed to building a new team of mentors with strong scientific exploration capabilities.

(1) Master guidance, pointing the way forward. The school has specially invited distinguished experts in the field of climate research in China, as well as well-known trainers from the province, to serve as "research mentors". They provide regular training for the teacher team, guiding teachers to refine research topics and methods, offering valuable suggestions, and providing professional guidance to achieve high-quality development of the mentor team.

(2) Craftspeople mentoring, guiding specific practices. The school has assembled a group of professionals with expertise, talents, and passion, such as florists and

horticulturists, and building material experts. The school fully taps into the wisdom of these "craftspeople mentors", exploring research paths, optimizing research processes, and enabling the feasibility and growth of mentor practices.

(3)Young teachers' self-cultivation, guidance for growth. The school has established a young teacher growth camp, where peers collaborate and divide tasks. Through a goal-oriented, task-driven model, young teachers are encouraged to take charge of sub-projects, continuously refining their skills and achieving growth through project practice.

4. Science Popularization: Digital Sharing

The school extends an invitation to promote climate change education, striving to construct a new advancing ecology for science popularization with the times.

(1)Building a "Deschooling Climate Change Exploration Museum". The school has bridged the boundary between the campus and the community, sharing spatial resources for climate change education with residents. On weekends and holidays, community residents can visit the museum to participate in science popularization activities, promoting the deep integration of lifelong education for community residents and school education, creating a wall-less school and jointly building a future learning center.

(2)Building a "Digitalized Cloud-based Science Popularization Platform". The school endeavors to empower teaching and assignments with digital technologies in the exploration of climate change education, enriching the breadth, depth, and precision of teaching and learning, making teaching more effective. At the same time, the school is actively developing cloud-based science popularization activities. For example, our science teacher Ludan Zhang has launched a "Climate Change Education" video column, conducting research on themes related to around "Climate in Daily Life", leading students to analyze the causes of climate change, and carrying out simulated climate change experiments, simulating the process of global warming and sea level rise due to melting glacials. Through video channels, the concepts and practices of climate change education can radiate to a wider audience.

Effectiveness & Reflections

Through half a year of practice, we have initially achieved some results:

(1)Building a curriculum map: Comprehensive planning of the curriculum system for the Climate Change Exploration Museum. guided by the principle of the curriculum concept of "full participation and individual development", the school has improved curriculum elements, coordinated educational forces, and further extended and developed a three-in-one "Climate Change Exploration Museum" curriculum map of "compulsory courses as the main component, supplemented by elective courses and specialized experiential courses", based on the high-quality implementation of compulsory national science courses.

(2) Exploring pathways: Developing and innovating participatory museum-school

course teaching styles. Based on the concept of "participatory learning", we pay attention to the state of active learning participation of each student in museum-school course activities, fostering students' curiosity, imagination, and desire for exploration, guiding students to widely participate in exploratory practices, with a special focus on activity styles involving student participation in discovery, exploration, and creation.

(3) Enriching resources integration: Integrating multi-dimensional collaborative spatial domains of school-family-community. The school has forged partnerships with local sub-districts and surrounding communities, allowing students to go beyond the campus and into the community. At the same time, the school continuously conducts scenario-based and experiential scientific practice activities, making the school's exploration museum an effective vehicle for promoting ecological awareness among residents.

(4) Evaluation and guidance: implementing developmental processevaluation to stimulate individual participation. Grounded in subject core literacy indicators, and centered on students', the school conducts performance evaluation and summative evaluation of students' learning performance, and through value-added evaluation methods such as "Mei Hao" (The homophony of "happy") energy cards and competition for badges, stimulates students' enthusiasm and creativity.

(5) Implementation assurance: Exploring a new organizational management mechanism involving all subjects, all staff, and all processes. Centered around the theme of learning and practice in the exploration museum, the school has established a three-tier organizational structure: "school-level exploration museum-grade-level exploration area-class-level exploration camp". The school has organized internal teachers and introduced external experts, scholars, and industry leaders to form an "expert mentor team" to assist and guide various activities in the exploration museum, actively engaging in school-level, grade-level, and even class-level scientific theme learning and activities.

In the future, we will continue to deepen the venue construction, curriculum development, activitycreation and culture creation of the Climate Change Exploration Museum, and actively promote the implementation of climate change education.

(Yan Wang, Principal of Changzhou Hongmei Experimental Primary School, Jiangsu)

The Construction Concept and Educational Practice of "Zero-Carbon Campus"

Research Background

Students are the masters of the future society, and cultivating their environmental awareness and low-carbon lifestyle habits is of great significance for promoting the whole societyto achieve carbon peak and carbon neutrality. Compared to the construction of "green campus", "zero-carbon campus" pays more attention to the calculation of carbon emission sources, emphasizes the behavior of teachers and students and carbon neutrality, has a stronger sense of quantification, and also focuses more on participation. The construction of "zero-carbon campus" is a professional and complex systematic project that requires joint efforts and continuous exploration from all parties. This project examines the development of "zero-carbon campus" at The Fifth Affiliated School of East China Normal University as an example, presenting design ideas, implementation preparations, and related educational and teaching practices, which is a phased achievement review.

Project Design

1. Objectives

The construction of a "zero-carbon campus" at The Fifth Affiliated School of East China Normal University is divided into short-term and long-term goals. The short-term "green action" goal is to promote theimportance and practical methods of low-carbon life to students through classroom lectures, promotional brochures, and organizing environmental protection-themed activities both inside and outside the classroom, to practice low-carbon living and enhance the environmental awareness of all teachers and students in the school. At the same time, through the demonstration effect of the school, it will promote the joint efforts of school, family, and community to save energy and reduce emissions, and advocate a "zero-carbon life". The long-term "zero-carbon campus" goal is to offset the carbon dioxide or greenhouse gas emissions generated by the lives of teachers and students in the school through energy conservation, emission reduction, and afforestation, to achieve positive and negative offsets, reach relative "zero emission", and realize the goal of a "zero-carbon campus" in the school.

2. Schedule

Some researchers have pointed out that "some junior high schools can basically achieve zero-carbon operation, and the campus has the greatest potential to achieve the goal of zero-carbon operation. Optimizing the utilization of solar energy resources on campus has significant energy-saving and emission-reduction significance." The construction of a "zero-carbon campus" refers to the activities within the campus, such as integrating educational resources and optimizing curriculum settings, promoting low-carbon lifestyles through interactive practical experiences, implementing campus carbon verification, and constructing a zero-carbon campus energy system, to achieve self-balance between campus carbon emissions and absorption, cultivate students' awareness of environmental protection and low-carbon living habits, and achieve the goal of relative "zero emissions".

1) Integrate educational resources, optimize curriculum design, and strengthen the promotion of "zero-carbon campus" construction

The school conducts classroom lectures, promotional brochures, and organizes environmental protection-themed activities both inside and outside the classroom, such as the seventh-grade "Go to Qinshan Nuclear Power Station" walking course and the all-grade lunchtime climate change education of moral education course. It integrates educational resources to promote the importance and practical methods of "zero-carbon life" to teachers and students, allowing them to understand concepts such as carbon peak, carbon neutrality, zero-carbon, net zero emission, carbon sink, carbon footprint, and learn to calculate personal carbon footprint, campus carbon footprint, product carbon footprint, and other knowledge (see Table 1).

Table 1. Course design for the "Zero-carbon Life" Carnival in The Fifth Affiliated School of East China Normal University

Lesson progress	Course content
Lesson 1	The Origin and Development of Carbon Neutrality
Lesson 2	"Zero-carbon Campus" 1: Carbon Emissions Accounting and Carbon Reduction Design
Lesson 3	Experts say "Carbon": the Carbon Sink Effect of Wetlands and Coastal Wetlands
Lesson 4	"Zero-carbon Campus" 2: Optimizing Electrical Energy Facilities and Management
Lesson 5	"Green News" of the Week: Zero-carbon Life Science Popularization Week
Lesson 6	"Zero-carbon Campus" 3: Waste Sorting and Circular Economy
Lesson 7	"Bringing Away the Garbage: Campus Garbage Clean up Practice Activity"
Lesson 8	Starting from "Zero": Jointly Building "Zero-carbon Campus", "Zero-carbon Family", and "Zero-carbon Community"

2) Set up a low-carbon life exhibition area, organize environmental protection practice activities, and promote a low-carbon lifestyle

Set up a low-carbon life exhibition area to allow students to personally experience the practical methods and achievements of "zero-carbon life". Schools organize students to participate in environmental protection practice activities, such as energy conservation starts from me, garbage classification, green travel, waste utilization, plastic reduction action, etc., to cultivate low-carbon living habits. Encourage students participate in environmental knowledge contests, low-carbon handmade waste utilization, "Silent Spring" reading promotion and other activities to enhance environmental awareness and innovation ability.

3) Implement campus carbon audit

Accounting for carbon emissions is the first step towards achieving a "zero-carbon campus". "Looking at the exploration of 'zero-carbon' on domestic campuses, reducing campus energy consumption is an important energy-saving and emission-reduction measure. This is achieved through energy-saving project renovations, low-carbon system adjustments, technological path innovations, campus planning strategies, carbon sink value enhancement, green technology applications, renewable energy substitutions, and full participation in green education to achieve the goal of reducing carbon emissions on campus." This case mainly draws on the planning of a carbon neutral implementation path for a newly built medical college campus in Tianjin, by developing a campus carbon emission source accounting process, carbon emission reduction and carbon sink accounting management process, and consumer-side behavior emission reduction quantitative accounting process to achieve the construction of a zero-carbon campus.

Firstly, establish a campus carbon emission source accounting management process. The campus carbon emission sources are categorized into three categories: entities, products, and activities. Among them, entities include physical entities such as buildings, service points, and parking spaces for various purposes; products include water, electricity, fuel, paper, plastic, as well as the resulting garbage, etc. used daily; activities include services, express delivery, travel, meetings, recycling, etc.

Secondly, establish carbon emission reduction and carbon sink accounting management processes. Thecarbon emission reduction process requires more institutional campus community cultural construction and management mechanisms to ensure the effectiveness of actions. In accordance with the principles of reduction, substitution, recycling, and regeneration, the total carbon emission reduction is composed of different methods such as new energy substitution, green transportation, reducing plastic and paper packaging, and reducing waste disposal.

Thirdly, establish a quantified accounting process for emissionreduction through consumer behavior. Accounting for emission reduction through individual green behaviors is mainly based on the 3R principles of the circular economy, namely Reduce, Reuse, and Recycle. In addition, there are also alternative, sharing, remanufacturing, and leasing models. The emission reduction through green behaviors is calculated by comparing the carbon emission data reduced by adopting green behaviors with the corresponding baseline emission data of conventional behaviors, accounting for the reduction of carbon emissions due to behaviors such as reducing energy, electricity, additional packaging, transportation and distribution, or using renewable energy, or promoting recycling and other reduction measures.

4)Successfully established a "zero-carbon campus" energy system

The "zero-carbon campus" energy system mainly includes: achieving self-sufficiency in campus energy through the use of renewable energy sources such as solar and wind energy; constructing a campus-level energy carbon monitoring network to achieve transparency in green and low-carbon campus energy carbon data, establish a campus-wide carbon emission map, and create an energy and carbon network; establishing a carbon/energy asset management monitoring and evaluation system to classify/manage campus electricity consumption item by item, and finding the best solution through carbon/energy optimization; adopting energy-efficient architectural design, establishing belt farms, wall greening design, etc., to reduce campus energy consumption and carbon emissions.

3. Project Schedule

The "ECNUFS Zero-carbon Campus" development project includes two phases: the cultivation stage and the implementation stage.

1)Cultivation stage

The cultivation stage includes three aspects: "Formulating project plans," "Promotion" and "Training and guidance". "Formulating project plans" involves developing a detailed project plan before the project begins, including the project's goals, implementation methods, time schedules, and personnel assignments. "Promotion" refers to promoting the significance and goals of the project through channels such as campus broadcasts, bulletin boards, and online platforms to attract more students to participate in the project. "Training and guidance" involve organizing relevant training and guidance to enable students to understand the importance and practical methods of low-carbon living, and to cultivate environmental awareness and innovation capabilities.

2)Implementation stage

The project implementation phase is divided into five aspects: publicity and education, interactive experience, practical activities, competition activities, and summary evaluation. "Publicity and Education" mainly involves classroom lectures, promotionalbrochures, extracurricular activities with environmental themes, and other means to promote the importance and practical methods of low-carbon living. "Interactive Experience" includes setting up a low-carbon life exhibition area where students can personally experience the practical methods and results of low-carbon living, learn carbon verification knowledge and technology, and organize interactive experience activities. "Practical Activities" arrange for students to participate in low-carbon life practices such as garbage classification and green travel, cultivating students' low-carbon living habits. "Competition Activities" organize students to participate in environmental knowledge competitions, low-carbon handmade activities, and other activities to enhance students' environmental awareness and innovation ability. "Summary Evaluation" is to summarize and evaluate the implementation effect of the project after its completion, including participation evaluation, cognitive evaluation, action ability evaluation, and social influence evaluation, providing reference and experience for the development of subsequent activities.

4. Establishment of the Project Team

The "ECNUFS Zero Carbon Campus" development project is led by the school's principal Chao Qiu, vice principal Jiahong Liu, and Professor Dongqi Wang of the School of Geographical Sciences at East China Normal University. Five teachers with different disciplinary backgrounds and expertise in Chinese, Mathematics, Science, Moral and Legal Education, and Psychology are involved in the planning. In terms of project division of labor, each teacher not only leverages their own expertise but also actively explores interdisciplinary cooperation to jointly promote the smooth progress of climate change education curriculum development, interactive practical experience, campus carbon verification, and campus energy optimization.

Project Implementation

Founded in September 2023, The Fifth Affiliated School of East China Normal University is a new school. Under the planning and care of the Jiading District Government of Shanghai and the East China Normal University Education Group, the school uses green and environmentally friendly building materials. The inverted V-shaped roof design not only has unique aesthetic value but also offers potential for the use of clean energy in the future.

The construction of "ECNUFS Zero-carbon Campus" is still in its initial stages. Currently, a low-carbon and environmentally friendly campus framework has been preliminarily established, and some attempts and explorations have been made in activities such as climate change education during the winter vacation, the May Day holiday homework assignment, and green travel etc.

During the winter vacation in January 2024, students embarked on the "Aspiring Teen Journey" to comprehensively improve their climate literacy through activities in various fields such as sports, reading, art, Chinese, Mathematics, English, science, society, labor, etc.

The school has designed winter vacation school-based assignments for sixth-grade students, including "The First Week Full of Vitality", "The Second Week in a Different

World", "The Third Week of Spring Festival" and "The Fourth Week of Capturing the Beauty" to guide students to learn and experience through practice. Here are examples of the content for "The First Week Full of Vitality" and "The Third Week of Spring Festival".

1. Themed Task Section of "The First Week Full of Vitality"

The content is divided into two categories: mandatory tasks and optional tasks. The mandatory task is "Pay Attention to Climate Change · Young Pioneers in Action." The optional task is to watcha documentary *The Road to Zero Carbon*. The specific mandatory tasks are as follows:

(1) Visit the ecological base in the form of a team, and each student writes a visit note and reflection, approximately 600 words. Each team completes a news release, which requires rich graphics and texts, and can edit videos. The best ones will be selected for the school-level official account to post.

(2) Complete an investigative report centered around the broad theme of "climate change". Each squad is divided into five groups, with each group serving as a unit. With the collaboration of school, family and community, choose one of the following five directions, enter the community and enterprises, take part in social practice activities, investigate and write research reports, create PowerPoint presentations, and give a report and evaluation at the start of the school year.

①Energy efficiency: such as the principle of energy conservation and emission reduction using new energy vehicles, building insulation, process improvement, renewable energy, wind energy, solar energy, greenelectricity, etc.

② Carbon peak and carbon neutrality: such as policy background, relevant measures, forestry carbon sequestration, marine carbon sequestration, carbon footprint, carbon emissions, etc.

③Circular economy: such as garbage classification, wasteutilization, second-hand market, plastic reduction action, etc.

④Protecting biodiversity: such as understanding endangered species, the reasons and methods of protection, the relationship between climate change and protecting biodiversity, etc.

⑤"ECNUFS Zero-carbon Campus": With the purpose of building a zero-carbon campus, carry out surveys, practical research, interviews, and other forms of research on the main carbon emissions in the campus. Start with the development of new energy, low-carbon travel, and the use of new materials, and propose feasible carbon reduction actions.

After selecting a topic in the early stage, contact the science teacher for guidance to ensure the smooth progress of the project.

☆**Horizontal comparison**: Each class is divided into five groups to work together, conduct class group evaluations, and the winning groups were awarded the "Blue Rose Science Award".

☆**Vertical Evaluation**: Five groups from four classes working in the same direction formed a larger group for collaborative exploration. These five larger groups participated in the school-wide selection and the winning groups were awarded the "Blue Rose Science Medal".

Each class formed a team to visit, write newsletters, fill out activity records, and discuss the climate change theme inquiry report. Under the guidance of the teacher, the students formed a relatively complete well-illustrated news article, and made a courseware to report in each class.

Figure 1. Climate Change Education Event of "Hua Hong Rose Class" in January 2024

Figure 2. The "Huashen Magnolia Class" visited the Jiading Meteorological Science Popularization Museum in January 2024

Figure 3. The "Huamei Xiangyang Class" conducted a study tour at the Qinshan Nuclear Power Station in January 2024

Table 2. Eaglet Holiday Team Activity Record

Name	Hua Yue Wisteria Class	Team name	Huacai Young Eagle Team
Captain	Student Feng	Instructor	Teacher Ma
Time	January 28, 2024	Place	Shanghai Auto Expo Museum
Theme	Visit Shanghai Auto Expo Museum		
Duration	☐The whole day ☐Morning session ☑Afternoon session ☐_____ hours ☐Other _____		
Participants	Student Feng and others		
Absentees	Student Kong		
Social practice	☑Social Investigation, ☐Professional Experience, ☐Public Welfare Work, ☐Safety Training		
Activity content	Visit the Shanghai Auto Expo Museum and explore the human wisdom contained in automobiles.		
Activity harvest and experience	Today, our Young Eagle Holiday Team visited the Shanghai Automobile Museum together. After the visit, I had a particularly great harvest, and I would like to share it with you. 　　The Shanghai Automobile Museum is located on Anting Bo Yuan Road. The entire museum consists of five parts: the History Pavilion, the Technology Pavilion, the Brand Pavilion, vintage cars, and the Temporary Exhibition Pavilion. Before entering the museum, we took a group photo with the exclusive flag of the Young Eagle Holiday Team at the entrance. As soon as we entered, we could see the panoramic view of the Automobile Museum, which houses a variety of cars.		

(Continued)

Activity harvest and experience	First, we visited the History Pavilion. This pavilion shows the birth and development of automobiles, including Volkswagen, Cadillac, Lincoln, and more. Then, we went to the second floor. This floor presents the development of automobiles in the 20th century. In that era, many people couldn't afford cars, but people liked to take pictures in cars as souvenirs, including many celebrities. I like a silver Porsche produced in 1958, which was small and very beautiful, with a streamlined body and smooth lines. Then, we went to the third floor. Wow! We saw a car disassembled and hung on the ceiling, which showed how many parts a small car had. This taught me a lot and let me know that cars are first assembled with the chassis, then the body, then the seats, and finally the doors, to form a complete car. Finally, we went to the fourth floor, where there was a library about cars. My classmates and I picked out books about cars and read them with great interest. We reluctantly left the car museum at least at 4 p.m. and said goodbye to our classmates. Chinese Automobile Memory: Today, billions of cars of various shapes and colors are roaming on the China's Land. When we look back at the years four or five decades ago, cars were the dreams of millions of people. At that time, cars were deeply imprinted with time, and some have become symbols of an era, which will remain in our memory for a long time. We are now enjoying the convenience brought by the automobile civilization, but also bearing the troubles brought by the rapid development of the automobile society, such as car exhaust, road congestion, global warming, and so on. Facing these challenges, China has actively responded to the global call for environmental protection and vigorously developed new energy vehicles. The popularization of new energy vehicles not only helps to reduce environmental pollution and alleviate traffic pressure, but also promotes the transformation and upgrading of the automobile industry and is a key measure to achieve sustainable development. Nowadays, more and more new energy vehicles are running on the Chinese land, representing the direction of green, intelligent and future, adding a new chapter to China's automobile memory.

2. Reading Section Task for "The First Week Full of Vitality"

Reading works such as *Silent Spring*, *Robinson Crusoe* and *The Wonderful Adventures of Nils* to gain an overall understanding of the story's main plot, with a focus on certain scenes or the words and actions of key characters (see Table 3).

Table 3. ECNUFS Recommended Reading List on climate change education in "Aspiring Teen Journey"

Grade	Works	Author	Press
Grade 6	Silent Spring	Rachel Carson (American)	People's Education Press
	Robinson Crusoe	Daniel Defoe (British)	People's Literature Publishing House
	The Wonderful Adventures of Nils	Selma Lagerlöf (Sweden)	People's Literature Publishing House
	The Adventures of Tom Sawyer	Mark Twain (American)	People's Literature Publishing House
	Hi, I'm China	People's Daily Online, China Qinghai-Tibet Plateau Research Association, Planet Research Institute	CITIC Press

3. "The Third Week of Spring Festival"

Designing the "Staying up for the Spring Festival customs" must-do tasks, such as ①"I write the Spring Festival couplets": give them to family members and friends, with blessings accompanying; ②"I send SpringFestival greetings": edit sincere and thoughtful blessings to each of your primary school teacher, and send Spring Festival greetings to your relatives and friends, no less than 50 words, which can be expressed in both Chinese and English; ③"I do household chores": take the initiative to clean the home before Spring Festival's Eve, and keep the room tidy every day; ④"I cook my signature dishes": try to prepare your own signature dishes for the Spring Festival reunion dinner; ⑤"I design creatively": how to reduce atmospheric dust and not set off fireworks to celebrate the joy of the Spring Festival? The spirit and charm of the Year of the Loong① rely on your whimsical ideas to bloom with endless glory and love.

☆Complete the check-in task "Staying up for the Spring Festival customs" and share photos of each of the above five items. At the start of the school term, a "Social Loong Person" competition will be held.

Effectiveness & Reflections

The construction concept and educational practice of the "ECNUFS Zero Carbon Campus" consistently uphold the following principles:

Firstly, deep integration of multiple disciplines, which combines environmental education with subject teaching by integrating Chinese, Mathematics, Science, Moral and Legal Education, Geography, and other disciplines, to enhance students' environmental

---

① 2024 is "The Year of the Loong" on the Chinese lunar calendar.

awareness and overall quality through teaching.

Secondly, diversify the forms of activities, that is, adopt various forms of activities such as publicity and education, interactive experiences, practical activities, competition activities, reading recommendations, etc., to attract students' participation interest and cultivate their awareness of environmental protection and low-carbon living habits.

Thirdly, multi-scenario life practice, which focuses on the organization and implementation of practical activities, cultivates students' action ability and environmental awareness by guiding them to participate in low-carbon life practices such as garbage sorting and green travel, and promotes the development of low-carbon life throughout the school and even the whole society.

Fourthly, a multi-dimensional evaluation approach, which means using various evaluation methods to comprehensively assess the achievement of project goals. Through participation evaluation, cognitive evaluation, action ability evaluation, and social influence evaluation, the implementation effect of the project is comprehensively evaluated to ensure the effectiveness and sustainable development of the project.

In the implementation process, numerous challenges have also emerged. For example, how toeffectively divide and evaluate the work of each project team member, how to allocate reasonably within the limited class time, how to achieve the integration of the five educations in the zero-carbon campus construction project, and how to better mobilize the enthusiasm and initiative of students to participate in the design and planning of activities. The road ahead is not smooth. We will continue the construction of "ECNUFS Zero-carbon Campus" and explore new directions for the future of the earth's "carbon"!

(Jie Huang, Teacher of The Fifth Affiliated School of East China Normal University)

🎓 Expert Review

Climate change is among the most critical and pressing challenges to global sustainable development strategies. On June 5, 2024, the World Environment Day, United Nations Secretary-General António Guterres delivered a special address titled *A Moment of Truth* on climate action, urging the world to act and fulfil the *Paris Agreement* commitments, taking concrete action to address global climate change. The key to the solution lies within ourselves. Therefore, through climate change education, changing our perceptions, thinking, values, and etc. to drive, maintain and implement the greening of our surroundings is an important initiative to address global climate change. For the young, school serves as an accessible natural laboratory and a practice ground for learning climate change literacy. Building green and low-carbon schools is a direct reflection of the whole-institution approach advocated by UNESCO in the field of education for sustainable development.

Design of an Intelligent Lighting Control System for Classrooms in the Context of Energy Conservation fully embodies the practical implementation of subjects promoted by the most recent curriculum reform in basic education. This project takes real campus problems as the starting point and integrating the elementary programming tasks based on logic operations from the IT curriculum. Furthermore, this project leads students to complete the design of energy-saving intelligent control of classroom lighting in real task situations through demonstrations, group experiments, discussions and exercises, addressing the energy-saving challenges around them with scientific and technological method. The follow-up solutions and ideas can also be expanded to intelligent energy saving of additional equipment in schools, families and communities, having strong operability.

Construction of the Campus "Fish and Vegetable Symbiosis Hall" and Low-carbon Education Practice to emulate and learn from the emerging low carbon agriculture industry, and visualizes the relationship between agricultural activities and climate change through micro-ecological construction on campus. By facilitating students to participate in experimental observation, participatory design, data review and lectures, the project integrating energy saving and emission reduction, recycling, ecological and environmental protection education together, and at the same time effectively improves the quality of the campus' local natural ecological environment.

Build a Growing Climate Change Exploration Museum aspires to turn climate change education sporadic to ongoing, and to provide a physical base for ongoing re-exploration, re-communication, talent incubation and intergenerational learning by storing and displaying materials related to climate change education organized by the school at various times of the day, and ultimately forming school-wide of climate change education which engages the whole school, the whole discipline, and the whole society, guided by the principles of participatory learning.

The Construction Concept and Educational Practice of "Zero-carbon Campus" combines China carbon peaking and carbon neutrality requirement in the new era, focuses on the more realistic and professional theme of "zero-carbon", and aims to decarbonize the campus environment. With the goal of decarbonizing the campus environment, it aims to build a successful and realistic "Zero Carbon Campus" through the provision of thematic courses, the promotion of a low carbon lifestyle, the launching of activities and competitions, the implementation of "carbon footprint" assessments management in schools, as well as the overhaul of the campus energy infrastructure, as well as the enhancement of both physical infrastructure and digital systems.

The four cases strongly demonstrate the creative exploration and good development strategies of schools in the field of climate change education, which reflect the height, breadth, depth and dynamism of the construction of a green and low-carbon campus in the context of climate change education. We believe that authentic origins of the problem, full participation of teachers and students, skillful integration with subject curricula, full experience and substantive participation, realistic assessment of the degree of problem solving, and full linkage with social resources outside the school are common and effective experiences. We also recognize that the high degree of professionalism required by climate change education, and indeed by ESD itself, needs to remind schools, governments and societies that sustainability awareness and guaranteed professional excellence in the teaching force are particularly critical. This will help to guide and support students to move steadily and resolutely along the path of sustainable development in the long term, fostering a virtuous chain of development in the neighboring communities and enhance the robustness of the school education system in ESD for all.

(Min Zhu, Associate Professor of Department of Education, East China Normal University; Adjunct Researcher of Shanghai Municipal Institute for Lifelong Education)

Chapter 2

Discipline-Based Teaching and Learning for Climate Change Education

The comprehensive, practical and rich nature of climate change education means that all disciplines can and must play a role in its power. The acquisition of scientific knowledge on climate change, the enhancement of climate change leadership, practice, problem-solving and lifelong learning, and the cultivation of the concepts and values of harmonious coexistence between human beings and nature, the community of human destiny, and sustainable living and development all are contingent upon multidisciplinary engagement and cross-disciplinary integration.

The following cases show the integration of teaching and climate change education across Chinese, Mathematics, English, Science, Information Technology, Moral and Legal, and bring us possible ways and feasible strategies for the incorporation and creation of climate change education in the teaching of various subjects. This shows us that strengthening the integration of climate change education with subject education can enable different subjects to give full play to their unique advantages, provide students with a more comprehensive, systematic and profound learning experience, and realize the educational goal of collaborative education.

Research on the Integration Practices of Primary Chinese Teaching and Ecological Education

Research Background

Currently, ongoing climate change has negatively impacted natural ecosystem and human society. On 17-18 July 2023, President Xi Jinping pointed out in his speech at the National Conference on Ecological and Environmental Protection that China's economic and social development has entered a stage of high-quality development with accelerated greening and decarbonization, and that the construction of an ecological civilization is still in a critical period of superimposed pressures and heavy loads. Therefore, school education remains the most basic and important part of raising the awareness of ecological civilization among the entire population.

Primary Chinese class, as one of the important subjects in basic education, is both instrumental and humanistic, and is the foundation for learning other subjects. Integrating ecological education in Primary Chinese teaching, guiding students to establish correct ecological values, and improving their ecological civilization quality are necessary for students' growth and social development. Primary Chinese teaching materials encompass a wealth of ecological civilization education. The primary school Chinese teaching materials of the official version are rich in embodying the idea and content of ecological civilization education. They encompass topics related to such aspects as understanding nature, respecting nature, protecting nature, and harmonious coexistence between human beings and nature. It is an important basis for infiltrating ecological civilization education into language teaching and promoting the deep integration of language teaching and ecological civilizationeducation. Against the background of the global situation and trend of addressing climate change and ecological environmental protection, it is imperative for Primary Chinese language teachers to take into account the characteristics of ecological education in classroom teaching. And by relying fully on the teaching materials, it is possible to carry out ecological civilization education activities that are rich in content and diversified in form, so as to promote the formation of ecological civilization values among students.

Project Design

1. Selected Text

The object of this study is the text *What am I* in the first unit of the second grade of the Primary Chinese textbook of the official edition. The theme of the unit is "Secrets of

Nature". This unit vividly introduces the laws of change and scientific phenomena in nature from different perspectives, such as animals, weather changes, plants, etc., which is close to daily experience.

The text *What am I* employs the first-person narrative point of view. Featuring "I" as the main character, through self-reporting, it vividly and interestingly introduces the circulation phenomenon of water in nature. Water can become "vapor'", "clouds", "rain", "hail", "snow", etc.. The use of anthropomorphizes talks about the different places where water lives and about its mild or violent temperament. The article uses vivid images and innocent language to stimulate children's interest in learning. It aids the children to understand the laws of nature and its scientific truths and led them to get close to, love, observe and explore nature.

2. Research Objectives

(1) Guide students to understand the changes of water and cycle of water in nature to promote a better understanding of nature.

(2) Guide students to understand the significance of water to humans in both positive and negative ways, and cultivate their reverence for nature.

(3) Guide students to act on the value of water in practical ways, mitigate water disasters and enhance their ability to protect nature.

(4) Promote students to develop the concept of harmonious coexistence between human beings and nature.

3. Methodology

(1) Visual Demonstration. Teachers use illustrations, multimedia equipment and other channels to show the relevant statements in the text, bringing students a more intuitive and vivid experience, and helping students better understand the natural knowledge.

(2) Reading. Teachers can cultivate the habit of reading throughout the entire teaching process, in reading aloud, to cultivate students' ability to perceive the beauty of nature. In addition to in-class reading, teachers also recommend books about water to students, to expand reading outside the classroom, guiding students to a deeper understanding of natural ecological issues.

(3) Co-operation and Communication. Through face-to-face listening, speaking, reading, questioning, evaluating, and discussing between teachers and students, as well as among students, to enhance the critical thinking skills of students.

(4) Graphically Organizing. Students draw a comic strip with the title of "Water Changes" to show the different forms of water and describe them in words.

(5) Practical Experience. It can promote students' participation in environmental protection through a variety of practical activities.

4. Evaluation Design

This case primarily evaluates from three aspects: students' classroom exercises, recitation expression, and graphical representation. First, classroom exercises are considered an integral part of reading instruction, and timely evaluations are given immediately after the classroom practice sessions. Second, a star challenge is created (see Table 1), guiding students to make accurate self-assessments and evaluations of their recitation expressions. Third, students are guided to use graphical forms to demonstrate their understanding of the changes and cycles of water, and evaluations are provided based on these representations.

Table 1. *What Am I* recitation star challenge evaluation

Rank	Standard
☆	I can recite my favorite sentences correctly.
☆☆	I can recite my favorite sentences affectively.
☆☆☆	I can recite my favorite sentences affectively, and summarize the recitation in my own words.

Project Implementation

1. Texts-oriented Education About Water

The article *What Am I* focuses on introducing the natural transformation of "water" into "vapor" under the sun's heat, how this "vapor" rises into the sky and forms tiny dots that connect to create "clouds", and how these "clouds" turn into "rain", "hail", or "snow" when they cool, eventually returning to "water" and flowing into ponds, streams, rivers, and oceans. The core content revolves around the natural knowledge of water, such as the different states of water and the cyclical patterns it follows in nature.

To stimulate the students' interest in exploration, teachers first need to create a context to pique their curiosity. For instance, they might use a question like "Today, we have a mysterious friend visiting our classroom. Do you know who it is?" to encourage students to read the text carefully. Once the students are familiar with the text, teachers continue by asking questions that lead the students to discover the core content of the article on their own.

In the teaching process, teachers should guide students to express in their own words the five forms of water—"vapor," "clouds," "rain," "hail," and "snow"—and the conditions under which these transformations occur. Teachers should also encourage students to use their imagination to experience the different characteristics of rain, hail, and snow falling from the sky, to appreciate the wonder of water's changes, and to appreciate the beauty of nature.

To help students understand the knowledge about the water cycle, teachers guide them to read the third paragraph of the text, asking them to identify the various places water goes and to perceive the unique beauty of water in different natural environments. After class, teachers can ask students to review and organize their knowledge through recitation and by creating mind maps.

2. Skillfully Posing Questions to Cultivate Students' Reverence for Nature

The fourth paragraph of the text uses the technique of contrasting descriptions to illustrate the two sides of the dual aspects of water's nature. The teacher inquires "Do you think water is important? What would happen if there was no water?" Students can combine their own life experience and have a fullexchange. Finally, the teacher uses visual pictures to show the importance of water to human beings.

While water is indispensable for human survival, we must also recognize its potential to cause disasters. Teachers guide students to read the fourth paragraph freely, focusing on the pair of antonyms in the sentence "Sometimes I am gentle, sometimes I am fierce," to appreciate the dual nature of water. Then, by expanding on extracurricular materials, teachers introduce relevant natural disasters, such as the Indian Ocean tsunami on December 26, 2004, allowing students to deeply understand the workings of nature and the reverence humans must hold towards it.

3. Fostering Inquiry to Stimulate Awareness of Nature Conservation

Through the previous learning, students have learnt that "water" can bring benefits to human beings and cause harm. Based on the sentence "People attempt to harness me in every way so that I can do good things but not bad things", the teacher can further investigate this with the students. It is important to lead students to think about what can be done to make water more beneficial and avoid the harmful effects. Students can launch independent learning and enquiry after class by searching for information and asking adults for help. In addition, teachers can create a thematic sharing and exchange session. Students independently associated with life to share their own good ideas and how to start from their own lives, to be a small environmental protection champion. At the same time, the school can organize all kinds of environmental protection activities, such as "selecting the campus water-saving small guard", "selecting the campus environment beautician", "selecting the campus small gardener" and so on. Environmental protection behaviors are in practice, jointly building a green campus.

4. Cultivating Students to Live in Harmony with Nature

Natural water occurs in a variety of forms and has its own systematic cycle of change and characteristics and impacts. After students understand the above knowledge, the teacher may prompt reflection by, "What kind of relationship should be established between human beings and water, and nature?" The teacher can recommend extended reading materials to promote students' discernment. Relevant books include *Water Changes* by Wuzhang Xie, *Where Does Water Come From* by Korean writer Shin Dong-kyung, and *The Story of Water* by British writer Matthews. This enables students to further learn that human beings and the water resources on which they depend belong to a natural ecosystem, and that the survival and development of human beings are closely linked to the various parts of the ecosystem.

In this session, teachers need to design hands-on activities to visualize the ecosystem cycle. The activity "Big World in a Small Bottle" guides students to make an "ecological bottle" (see Table 2), allowing them to participate in the production process and observe the state of the ecological bottle. Students thus gain a profound understanding of the decisive role of the ecological environment for the survival of natural organisms, and realize that the sustainable development of mankind depends on the harmonious coexistence of man and nature.

Table 2. Production record of ecological bottles

Types		Name & Number	Design
Biological	Animals		
	Plants		
Non-biological			

Effectiveness & Reflections

This teaching case integrates knowledge, science and fun, and the content and methods take full account of students' practical life. Starting from the students' true sense of nature, on the one hand, it improves the reading ability and language and writing level. On the other hand, it also promotes students' agility, originality and critical thinking through contextual imagination, comparative analysis and inductive judgement. Students exercise the ability to analyze and solve problems by integrating and applying knowledge, thinking and methods from different disciplines. It is helpful to develop their awareness of nature and aesthetic sensibilities to promote the formation of a natural outlook of harmonious coexistence between human beings and nature.

The ultimate goal of ecological education is to motivate students to take the initiative to concern ecological environment and climate change issues. Consequently, students can effectively protect the environment, become responsible citizens of the earth and achieve sustainable development of human society. The follow-up research process shall strengthen the creation of experiential activities and bridge the gap between inside and outside the classroom, so that students may immerse in nature more frequently and take practical actions to protect the nature.

(Yazhen Wang, Teacher of Tianjin Economic-Technological Development Area International School)

Design of Teaching Activities for Primary School Chinese Language Unit Under the Influence of Ecological Education

Research Background

The *Compulsory Education Chinese Language Curriculum Standards* (2022 Edition) underscores that students should actively observe and perceive life. The curriculum resource citation also proposes to "maximizing the educational function of curriculum resources, optimize teaching and learning activities, the use of course resources should be aimed at promoting the development of students' core literacy, and exploring multi-angle excavation of educational value." So the Chinese Language curriculum plays an important role in cultivating students' outlook on life and values, and has significant advantages in climate change and ecological environment education.

This educational case attempts to create a unit-level integrated design for the sixth unit of the second volume of the department-compiled version of the Chinese Language textbook for the first grade. The theme of this unit is "Summer." The case attempts to create engaging real-life content for students, while improving their core literacy ability of the Chinese Language, taking into consideration of the characteristics of the learning task group. The case allows students to experience the beautiful scenery and wonderful implications of summer, cultivating students' aesthetic interest in loving nature, exploring scientific interest in nature, and fostering an ecological civilization literacy of harmonious coexistence between humans and nature.

Project Design

1. Content Selection

Ecological civilization literacy includes five key qualities: "ecological sentiment, survival skills, ecological wisdom, ecological aesthetics, and ecological practice." In Chinese Language textbooks, content related to ecological civilization education is integrated into each stage. In the lower grades of primary school, topics are primarily related to meteorological and biological phenomena. Diverse content includes "Four Seasons Poems," short essays, and illustrations in textbooks, which are key media for cultivating students' ecological civilization literacy.

This unit includes three texts: *Two Ancient Poems*, *Round Lotus Leaves*, and *It's Going to Rain*, as well as a "Chinese Language Garden" activity session. These texts depict the beauty of summer from different angles. This case is themed "Summer Travel," integrating "three tasks, six activities," encompassing character recognition and writing, text reading, and other contents. Teachers guide students to accumulate relevant knowledge and experience the joy of language learning through reading, writing, and performing. They also encourage students to appreciate the beauty of nature and cultivate their aesthetic sensibilities.

2. Objectives

(1) Cultivate students' awareness of observing and loving nature and enhance their aesthetic ability.

(2) Guide students to understand meteorological knowledge and seek natural laws and appreciate the wisdom of nature.

(3) Encourage students to become advocates for the environment and enhance their ability to participate in ecological protection activities.

3. Methods

The case involves the following methodologies:

Text study: Guide students to obtain relevant ecological knowledge through reading and analyzing texts;

Environmental perception experience: Lead students to observe nature and appreciate the role of a beautiful natural environment for humans;

Practical activities: Organize activities such as "creating bright classrooms" and "building green campuses" to actively practice environmental protection;

Display and sharing: Present students' observations and understandings of nature through viewpoint sharing, work display, program performance, and other mediums.

4. Assessment

The teacher used various metrics to measure and display the improvement of students' ecological civilization literacy, including the fluency and richness of students' expressions in class, their generative works, and their participation in environmental initiatives. Specific evaluation criteria are shown in Table 1:

Table 1. Assessment of thematic activities

Activities	Assessment criteria	Assessment methods		
		Student self assessment	Student peer assessment	Teacher-led assessment
Summer photography exhibition	Willing to share observations of summer with others.	☆☆☆	☆☆☆	☆☆☆
	Can listen attentively when others share and can speak in complete sentences when speaking.	☆☆☆	☆☆☆	☆☆☆
	Interested in summer and nature, with their own methods of observation.	☆☆☆	☆☆☆	☆☆☆

(Continued)

Activities	Assessment criteria	Assessment methods		
		Student self assessment	Student peer assessment	Teacher-led assessment
Studying the text *It's Going to Rain*	Can read the text correctly, fluently, and with emotion, understanding meteorological characteristics, and appreciating the beauty of nature.	☆☆☆	☆☆☆	☆☆☆
	Can perform to show the impact of climate change on small animals, understanding the laws of nature.	☆☆☆	☆☆☆	☆☆☆
	Can listen attentively when others speak and can speak in complete sentences when speaking.	☆☆☆	☆☆☆	☆☆☆
Learning weather proverbs	Willing to communicate with others to collect weather proverbs; happy to share weather proverbs with others.	☆☆☆	☆☆☆	☆☆☆
	Interested in the weather and nature, able to apply learned meteorological knowledge to life.	☆☆☆	☆☆☆	☆☆☆
Interviewing elders	Willing to communicate with others about climate and natural conditions.	☆☆☆	☆☆☆	☆☆☆
	Can express in their own words the phenomena related to climate change, expressing their feelings about climate change.	☆☆☆	☆☆☆	☆☆☆
	Can listen attentively when others speak and can speak in complete sentences when speaking.	☆☆☆	☆☆☆	☆☆☆
Environmental activity "Creating a Bright Classroom"	Can actively participate in activities to create a green classroom.	☆☆☆	☆☆☆	☆☆☆
	Willing to participates in the care of plants and understand the knowledge of care.	☆☆☆	☆☆☆	☆☆☆
	The cared-for green plants grow lushly.	☆☆☆	☆☆☆	☆☆☆
	Can apply environmental actions to other campus environments.	☆☆☆	☆☆☆	☆☆☆

Project Implementation

1. Carrying Out "Summer Photography Exhibition" and "Summer Evening Party" to Cultivate Students' Awareness of Observation and Being Close to Nature

The unit is themed "Summer," depicting the seasonal characteristics of summer from different angles. Studentswere given the opportunity to discover the characteristics of summer and feel its beauty in various activities, thus developing closeness and affection for nature.

(1)Summer photography exhibition: While the texts in the unit depict the beauty of summer from multiple angles, the students should understand the subject by going beyond the textbook and making real-life connections. The "Summer Photography Exhibition" is an activity that encouraged students to enter nature. First, the teacher set the question: "Have you seen the summer scenery in the textbook in life? Please take photos related to summer scenes and objects, and act as a 'Little Lotus Tour Guide' once." After class, students entered nature, observed life attentively, discovered and captured the beauty of summer with their eyes and cameras. Then, they displayed photos of summer-specific food, daily necessities, activities, and scenery (see Table 2) in the classroom. In the sharing session students talked about the joyful experience of summer so that their aesthetic interest, natural observation, and environmental protection awareness are enhanced.

Table 2. Content of students' photographs for the Summer Photography Exhibition

Content category	Content expression
Summer foods	Grapes, peaches, pears, slushies, shaved ice, milkshakes, cold noodles, plum juice etc.
Summer attire	Short sleeves, shorts, thin skirts, T-shirts, sweatshirts etc.
Summer activities	Swimming, water fights, rafting, playing in the sand at the beach etc.
Summer weather	Thunder, heavy rain, typhoons, scorching heat etc.
Summer insects	Mosquitoes, flies, dragonflies etc.
Summer flowers	Lotus, morning glory, sunflower, orchid, lily, cockscomb, water lily, crape myrtle etc.

(2)Summer evening party: In the form of a "party," the teacher created a joyful classroom where students expressed their appreciation of summer through reading nursery rhymes, singing children's songs, and reciting poetry. Students engaged in singing nursery rhymes *Chopping Mosquitoes*, talked about the joys and annoyances of summer; recited children's poems *Summer is a Child* together. At the story-sharing session, through the folk legend of *The Cowherd and the Weaver Girl*, the beautiful imagination of summer in the traditional China culture was displayed. The teacher guided students to enjoy the summer evening through star gazing and listening to the symphony of the wild.

2. Stimulating Students' Interest in Reading and Discovering Nature's Secrets by Creating the "Mysteries Before the Rain" Thematic Event

It's Going to Rain is a scientific fairy tale in this unitaddressing meteorological and biological phenomena before rain. In the teaching process, the teacher created a "Pre-Rain Exploration" themed activity to stimulate students' reading interest:

First, the teacher asked progressive questions to guide students to discover theatypical behavioral patterns of animals before the rain, such as: "Find which small animals are in the story?" "What are the swallows, small fish, and ants doing before the rain?" After students fully express themselves, the teacher guided students to explore the secrets before the rain: before the rain, swallows fly low, fish swim on the surface, and ants move to new habitats. Next, the teacher guided students through role-reading to understand the effects of humidity on these animals and appreciate the interconnections in the ecosystem. Finally, students participated in a role-playing reimagination of the animals to strengthen their memory of these natural phenomena.

For the "Chinese Language Garden," the case involves teachings of traditional Chinese proverbs about meteorology. The teacher guided students to read, understand and memorize these proverbs to expand their environmentalacumen. The unit furthermorefeatures the "Explore Meteorology" extracurricular activity, where students collected proverbs and folk wisdom from consulting family members and friends. Students then shared their findings through a sharing session.

3. Engaging Students in Environmental Conservation Initiatives Through Intergenerational Dialogues

To enhance students' awareness of climate change, the teacher designed an intergenerational communication activity "Little Reporters Interview Elders." Students communicated with their elders around the topic "Is summer the same now as it was in the past?" Through this activity, students learned that human activities have led to an increase in greenhouse gas emissions causing adverse climate effects.

To guide students to protect the environment through small attenable actions, the teacher designed the "Creating a Bright Classroom" activity, where students decorated the classroom with beautiful green plants, and composed caring guides for these plants to water them regularly.

Effectiveness & Reflections

Integrating ecological civilization education into Chinese Language teaching is a concrete action of implementing the new curriculum reform. Based on the concept of unit-level integrated design, the teacher created this teaching case to integrate textbook contents, stress the theme of the unit, and create activities that connect with the students' lives. Instead of traditional teaching methods, the teacher led students to go beyond the textbook and venture into nature, to make observations and to utilize the Chinese Language they learned to express their spontaneous love for nature.

Moreover, the case used the collaborative model of education involving school, family, and community. The multi-dimensional synergy allowed students to step out of the campus, into the community, and integrate into life, engaging in experiential learning across various settings.

Climate change education is a comprehensive and long-term endeavor, wherein Chinese Language teachersought to take an active role. They should continually enhance their awareness of ecological civilization and their professional competence and practical ability to integrate ecological education into their subject teaching. This enables students to establish a correct view of ecology from an early age, to have an awareness of environmental protection, to learn to understand the world, analyze problems, and solve problems with a scientific perspective, and to practice the harmonious unity of humans and nature.

(Jing Zhao, Teacher of Tianjin Economic-Technological Development Area International School)

Exploration of Promoting Low-Carbon Life Through Project-Based Learning in Primary School Mathematics

Research Background

Mathematics stems from everyday life, and learning math is inseparable from practical living. Adopting an energy-efficient and low-carbon lifestyle is a key trend in the future development of family life and represents a viable strategy to actively respond to climate change. As an elementary school Mathematics teacher, I am cognizant of the importance of integrating climate change education into Mathematics instruction. This case study is based on the "Recognition and Addition and Subtraction of Decimals" unit in the second semester of the fourth grade's Mathematics practice from the nine-year compulsory education curriculum in Shanghai. The investigation launches from the small practice assignment titled "Calculate Weekly Household Expenses". Significant variations were observed in the data submitted by students. To delve deeper into this phenomenon and to assist students in providing solutions for their families, the fourth-grade Mathematics teachers, after fully considering the mathematical content and the age characteristics of the students, decided to initiate a project-based learning theme called "A Little Housekeeper for Home Life". This aims to help students better understand and address the challenges faced in real life.

Project Design

"Engaging students with the real world not only deepen their understanding and mastery of concepts and cultivates their thinking skills but also guides them to respect nature and life, and understand what social responsibility means". This case study commences with addressing the issue of "How to Reasonably Purchase Daily Necessities", guiding students to take on the role of "Little Household Stewards". This role helps them experience the close connection between Mathematics and everyday life, appreciate the value and enjoyment of learning Mathematics, and develop computational skills and an awareness of application—core competencies in the subject area. Additionally, it enhances students' awareness and understanding of green consumption and low-carbon living, fostering competencies to address climate change. The specific objectives are as follows:

Students can

(1)Experience the extensive application of decimals in daily life and accurately use addition and subtraction of decimals to solve real-life problems, thereby developing computational abilities and application awareness.

(2) Engage in activities such as data surveys, observational comparisons, and comprehensive planning to enhance team collaboration, communication, and problem-solving skills.

(3) Appreciate the value and fun of learning, fostering concepts and awareness of green consumption and low-carbon lifestyles by taking on the role of "A Little Housekeeper" and exploring the issue of "How to Reasonably Purchase Daily Necessities".

Project Implementation

1. Identifying and Framing Problems Based on Real-Life Situations

1) Initiating conjectures

Teachers initially introduce students with the data tables from their submitted practical exercise on household expenses (as shown in Figure 1). The students are surprised to discover significant differences among households, with weekly expenses nearing 2,000 yuan for some, while others are around 200 yuan. This stark contrast arouses the students' intense curiosity. They actively engage in thinking, formulating hypotheses, and conducting their verification, ultimately concluding that the primary reason for lower household expenditures in the statistical table is the possession of ample stockpiles or the adoption of frugal and budget-conscious consumption habits; conversely, higher spending in some families results from a lack of budgetary management for daily expenses or tendencies towards impulsive and excessive spending.

Small Practice

Do you want to know how much money the family spends in a week? Visit you to be a little housekeeper. Ask your mum and dad every night and record each expense. Calculate how much money is spent every day and how much money is spent in total in a week.

	Monday	Tuesday	Wednesday	Thursday	Friday	Saturday	Sunday	Total
Cost (RMB)	265.9	176	374.64	241	695	123.4	104	1979.94
	Monday	Tuesday	Wednesday	Thursday	Friday	Saturday	Sunday	Total
Cost (RMB)	147.8	208.9	69	189.8	146	296.8	100	1258.3
	Monday	Tuesday	Wednesday	Thursday	Friday	Saturday	Sunday	Total
Cost (RMB)	0.00	125.78	0.00	0.00	75.64	0.00	0.00	201.42

Figure 1. Data from students' submissions for the practical assignment on the unit "Recognition and Addition and Subtraction of Decimals"

2) Provide scaffolding

Teachers facilitate a discussion by guiding students to observe a household expenditure log recorded for the month of April by a student's family in the class (as shown in Figure 2).

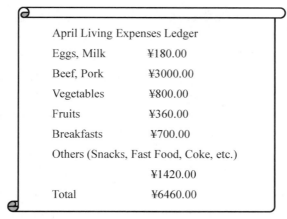

Figure 2. Handmade journal created by a student

Based on the students' observations, comparisons, hypothesis testing, and active discussions, the teacher identifies the essential question of the project as "How can we engage students' life experiences to systematically investigate, plan, and strategize, thereby developing computational skills and application awareness, and enhancing problem-solving abilities?" The driving question is: "During the period of home-based learning, how can you, as the family's 'Little Housekeeper', purchase daily necessities in a rational manner?"

2. Brainstorming: Analyzing and Solving Problems

The teacher initiates an online brainstorming session, where teachers and students collaboratively analyze and discuss the steps and key elements involved in solving problems. This helps students systematically create a shopping plan. Below is the discussion process and the four sub-questions that emerged during this process:

Step 1: Investigate what essential household items are missing at home.

Sub-question 1: What are the essential items for daily living?

Students enthusiastically raised their hands to speak and actively posted on an interactive message board, participating vigorously in the discussion. They eventually categorized the items into various groups: seasonings, dairy products, meats, fruits and vegetables, staples, household supplies, educational supplies, and medicines. Additionally, the students advocated for selecting products with environmentally friendly packaging and those locally produced to reduce carbon emissions during transportation.

Step 2: How to create a comprehensive shopping plan?

Students proposed essential components of a shopping plan from different perspectives, offering rational solutions for establishing a shopping plan to avoid overconsumption and waste. For instance, one approach is to list product names, ensuring that the listed items meet actual needs while advocating for a preference for environmentally friendly products; another is to check product prices, suggesting a comparison of prices across different channels.

Sub-question 2: How to check the prices of items?

Students collectively identified that the main methods for checking item prices include using online apps and visiting physical supermarkets. During this process, they discussed the importance of determining the quantity of goods needed before making purchases, avoiding overconsumption and reducing waste.

Sub-question 3: How much of each essential item should be purchased?

Students suggested that purchases should be made sensibly based on the characteristics of the items. For products with a long shelf life or those that can be preserved by freezing, it was advisable to buy in larger quantities to reduce the frequency of shopping trips and decrease carbon emissions. For perishable items with a short shelf life, such as vegetables and fruits, it was necessary to buy in appropriate amounts and choose seasonal foods whenever possible. Consideration should also be given to available storage space at home, such as sufficient refrigerator space, to prevent excessive purchases that could lead to food waste and overconsumption of resources. Moreover, finding suitable purchasing channels and encouraging low-carbon shopping methods should also be considered.

Sub-question 4: Where should essential items be purchased?

When it came to purchase essential items, students distinguished two primary methods: online and offline. For online purchases, this included placing orders through mobile apps or small programs. When buying online, it was advisable to combine multiple orders into one delivery to reduce the carbonemissions that came from separate deliveries. For offline purchases, this meant shopping at supermarkets. When making offline purchases, it was encouraged to walk or bike to the shopping destination. Taking advantage of bulk discounts through group purchases can also be considered. Ordering together with neighbors not only shares the discount but also reduces the carbon footprint of each household. During this process, it was important to estimate the total cost to ensure that the shopping plan did not exceed the family's budget.

Step 3: Evaluation, improvement, and presentation of results.

This project provided students with ample time and space, guiding them to complete tasks independently or in collaboration with classmates from the same community. Students conduct surveys, taking into account current family requirements and market dynamics, to develop a reasonable shopping plan (as shown in Figure 3). This ensured that the plan not only meets daily requirements but also stays within a controlled budget. Subsequently, students carried out the purchasing according to the planned strategy. They then received evaluations from parents and peer reviews. During this process, they documented meaningful stories or insights gained while acting as 'A Little Housekeeper'. The results and experiences from the project were showcased on the school's video channel, allowing the wider school community to recognize their efforts and learn from their experiences.

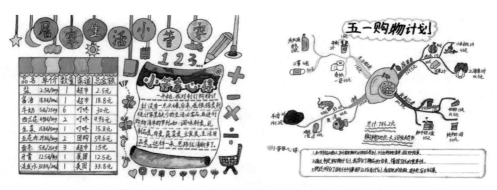

Figure 3. Family shopping plan developed by students

Effectiveness & Reflections

This learning activity has deepened students' understanding of "Recognition and Addition and Subtraction of Decimals" and enhanced their computational skills. Additionally, by advocating for active participation in green consumption, avoiding excessive hoarding, and reducing food waste, thereby increased the students and their families' identification with and adoption of a low-carbon lifestyle.

1. Developed Students' Computational Skills

In Mathematics education, using situational representation to characterize computational algorithms is one of the important strategies for developing students' computational skills. As seen in Figure 4, the student actively examined the expiration and shelf-life dates while making purchases, truly understanding the significance and gaining a profound understanding the real world. When calculating the total amount, the student accurately computed the total and figures out the discount amount based on the actual promotional activity, "Save 10 on every 100 spent". Furthermore, the teacher learned that the student, in pursuit of precise calculations, repeatedly checked the computations several times. This reflected a dedication to excellence and a scientific attitude of rigor and practicality.

Following the conclusion of the activity concluded, we conducted ongoing tracking of the performance of students in this grade level on the "Master calculation" section. We compiled statistics including the number of students achieving "Excellent," "Good," "Satisfactory," and "Needs Improvement" ratings, and continued monitoring until their graduation from elementary school. As illustrated in Table 1, by the time of graduation (6th grade), the overall average excellence rate exceeded 95%, with class 3 achieving an excellence rate of 100%. Overall, there is a clear upward trend in performance.

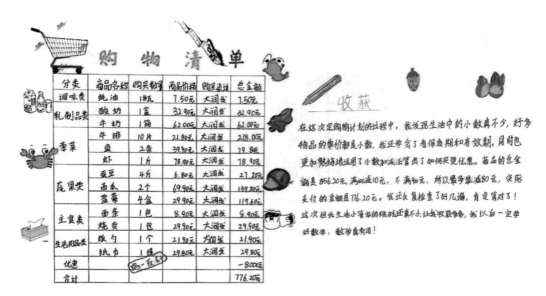

Figure 4. Family shopping plan created by the student and the written reflection

Table 1. Student performance statistics in the "Master calculation" section

Class	December 2021				June 2022				June 2023 (Graduation)			
	Master calculation				Master calculation				Master calculation			
	A/%	B/%	C/%	D/%	A/%	B/%	C/%	D/%	A/%	B/%	C/%	D/%
01	34	4	0	3	36	2	2	1	39	1	1	0
	82.93%	9.75%	0.00%	7.32%	87.80%	4.88%	4.88%	2.44%	95.12%	2.44%	2.44%	0.00%
02	29	6	4	1	32	7	0	1	38	1	0	1
	72.50%	15.00%	10.00%	2.50%	80.00%	17.50%	0.00%	2.50%	95.00%	2.50%	0.00%	2.50%
03	34	6	1	0	36	4	1	0	41	0	0	0
	82.93%	14.63%	2.44%	0.00%	87.80%	9.76%	2.44%	0.00%	100%	0.00%	0.00%	0.00%
04	33	5	2	1	36	4	0	1	38	2	0	1
	80.49%	12.20%	4.88%	2.43%	87.80%	9.76%	0.00%	2.44%	92.68%	4.88%	0.00%	2.44%
05	27	10	4	1	34	6	1	1	40	1	0	1
	64.29%	23.81%	9.52%	2.38%	80.95%	14.29%	2.38%	2.38%	95.24%	2.38%	0.00%	2.38%
06	29	7	3	2	37	2	1	1	40	0	1	0
	70.73%	17.07%	7.32%	4.88%	90.24%	4.88%	2.44%	2.44%	97.56%	0.00%	2.44%	0.00%
Total Sum	186	38	14	8	211	25	5	5	236	5	2	3
	75.61%	15.45%	5.69%	3.25%	85.77%	10.16%	2.03%	2.03%	95.93%	2.03%	0.81%	1.23%

Note: A—Excellent; B—Good; C—Satisfactory; D—Needs Improvement.

2. Enhanced Students' Awareness of Low-carbon Living

From the reflections submitted by students, it was evident that they had implemented a series of measures to promote sustainable consumption. Such measures encompassed avoiding the purchase of non-essential items, grouping orders with neighbors, sharing surplus goods with those in need, practicing rational consumption, and reducing food waste. These actions have reinforced the students' concept of green consumption and have enhanced their identification with alow-carbon lifestyle. The ongoing implementation of community group buying also demonstrates that people are gradually accepting and adopting a more sustainable and low-carbon way.

3. Helped Students from Good Reflection Habits

During the learning process, students fully leveraged their own life experience based on driving questions, personally experiencing data survey, overall planning, budgeting, and ultimately solving the problem of purchasing materials for their families. With these experiences, I believe that students will be able to investigate, plan, make plans, reflect on iterations in an orderly manner, and continuously improve their problem-solving abilities in their subsequent endeavors. In addition, these learning experiences have greatly promoted students' attention to daily life and reflection on family life, enhanced their understanding of low-carbon life, and transmitted this impact to families and communities, producing a greater radiation effect.

Of course, in the specific implementation process of this case, due to the fact that online discussions were primarily used, the number of real-time interactions between teachers and students, and among students, was reduced compared to traditional classrooms. To compensate for this shortcoming, it is suggested that students form groups freely to discuss at the group level, and then organize collective reporting and communication to increase the participation and interactivity of all students. In addition, in order to deepen students' understanding, we can collect and organize actual cases fromdifferent student families, and applying these for in-depth analysis and discussion, further guiding students to make rational decisions to avoid excessive consumption and waste, allowing them to more deeply understand the significance of low-carbon living.

(Juan Sun, Teacher of Qianjing Primary School, Jinshan, Shanghai)

A Practice of Climate Change Education in English Teaching

Research Background

As a vital force in combating climate change, climate change education can influence people's ways of thinking and choices of behavior. *Compulsory Education English Curriculum Standards* (2022 *Edition*) (hereinafter referred to as "New English Curriculum Standards") proposes that English courses should focus on core competencies, enabling students to gradually form the correct values, essential qualities, and key abilities needed by society through course learning, and to enhance students' awareness of "a community with a shared future for mankind" and the sense of social responsibility. Students can understand the significance of "loving and respecting nature, and living in harmony with nature" from the perspective of "people and nature." In addition, *Compulsory Education Curriculum Program* (2022 *Edition*) (hereinafter referred to as "New Curriculum Program") also advocates for interdisciplinary thematic learning should meet the design of no less than 10% of class hours. Teachers should guide students to understand the connections between different subjects in English learning and combine real-life situations to solve practical problems through interdisciplinary thematic learning.

Therefore, climate change education is consistent with the goals of living in harmony with nature in New English Curriculum Standards and the approaches of interdisciplinary thematic learning in New Curriculum Program. This case integrates climate change education into English teaching for middle school students, creates interdisciplinary thematic learning activities, advocates students to connect with real-life situations, pay attention to social hot spots, respond to climate change appropriately, and cultivate awareness to love nature and ability to protect environment.

Project Design

The text of this case, *The tiger lives in Asia*, is selected from Module 6 Unit 2 in *English* (*Grade 7 Volume 1*), focusing on five animals including elephants, pandas, zebras, tigers, and monkeys from aspects such as appearance, food, habitat, and living habit. The content is categorized under the theme category of "human and nature", with the theme group being "natural ecology", and the sub-theme being "loving and respecting nature, and living in harmony with nature".

In conjunction with the contentof the text, the requirements of the curriculum plan and standards, and the goals of climate change education, teachers encourage students to undertake the learning task of "making a brochure for protecting Siberian tigers", refer to the information on relevant websites such as the World Wildlife Fund, and combine real-life situations to carry out interdisciplinary thematic learning activities in groups.

1. Topic

Making aBrochure for Protecting Siberian Tigers.

2. Contents

Students are invited to participate in an activity held by the World Wildlife Fund (WWF) for protecting wild tigers. They need to make a brochure with the theme of protecting Siberian tigers, which explains the knowledge related to Siberian tigers, and introduces the ways of protecting them. The brochure also publicizes the importance of protecting Siberian tigers and maintaining biodiversity.

3. Involved Subjects

English: Write in English about the introduction to Siberian tigers, measures of protecting Siberian tigers, and the significance of protecting Siberian tigers for human survival and development.

Art: Draw pictures for the cover and inside pages of the brochure.

Geography: Understand the geographical distribution, habitat environment, and other relevant geographical knowledge of Siberian tigers.

Biology: Understand the morphological characteristics, living habits, and other biological knowledge of Siberian tigers.

4. Learning Objectives

(1) Through reading thetext and related materials, students sort out and integrate key information, accumulate English expressions and materials for introducing the morphological characteristics, living habits, geographical distribution, and habitat environment of Siberian tigers, understand knowledge related to ecosystem and natural science, and the relationship between biodiversity and climate change.

(2) Through consulting books, the Internet, and other related materials, students engage in inquiry-based learning, think about measures to protect Siberian tigers from multiple aspects and angles, actively discover, analyze, and solve problems, cultivate awareness and thinking of living in harmony with nature, and cultivate a commitment of protecting animals and the Earth in daily life.

(3) Through group cooperation to make a brochure for protecting Siberian tigers, students delve into the current situation of wild animals represented by Siberian tigers, deeply understand the close relationship between animals and climate, and between human and nature, and cultivate the awareness of "a community with a shared future for mankind".

(4) Through displaying the brochure for protecting Siberian tigers, students publicize

the importance and urgency of protecting wild animals to people around them, forming a good situation where everyone protects wild animals, maintains biodiversity, and jointly responds to climate change.

5. Assessment

"New English Curriculum Standards" points out that teaching evaluation should be throughout the whole process of English teaching and learning. In this learning activity, teachers adopt a combination of teacher evaluation, self-evaluation, and peer evaluation to improve learning effectiveness and implement the integration of "teaching-learning-evaluating".

1) Teacher-led assessment

Based on the teaching objectives, teachers assess the students' answers to questions, understand the students' learning process and difficulties, and analyze the discrepancies between the students' current learning level and the teaching objectives. For example, teachers judge whether students have completed the teaching objectives based on the activity of "Retelling the text according to the mind map", provide targeted guidance and help, and give necessary encouragement and praise (see Figure 1).

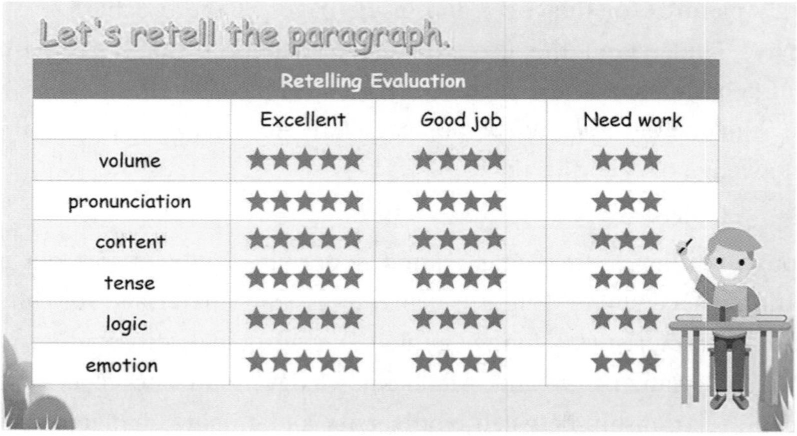

Retelling Evaluation			
	Excellent	Good job	Need work
volume	★★★★★	★★★★	★★★
pronunciation	★★★★★	★★★★	★★★
content	★★★★★	★★★★	★★★
tense	★★★★★	★★★★	★★★
logic	★★★★★	★★★★	★★★
emotion	★★★★★	★★★★	★★★

Figure 1. The Activity Evaluation Scale of "Retelling the Text According to the Mind Map"

2) Student self assessment

After answering questions, students carry out self assessment and peer assessment based on the activity evaluation scale given by the teacher (see Figure 2). In the evaluation process, students evaluate their own performance and actively engage in active self-reflection, which promotes self-supervised learning. In peer assessment, students play a full role as participants and collaborators in the activities, learn from each other's strengths and weaknesses, and promote self-growth.

Brochure Evaluation (评价)			
	Excellent	Good job	Need work
topic	★★★★★	★★★★	★★★
contents	★★★★★	★★★★	★★★
suggestions	★★★★★	★★★★	★★★
design	★★★★★	★★★★	★★★
colour	★★★★★	★★★★	★★★

Figure 2. The Activity Evaluation Scale of "Making a Brochure for Protecting Siberian Tigers"

Project Implementation

This interdisciplinary thematic learning activity is carried out in a combination of in-class teaching and post-class tasks, specifically divided into four parts: Theme Selection, Planning and Designing, Implementation and Summary (see Table 1). In the process of activity implementation, teachers break down large tasks into small tasks that students can perform, and students complete them through in-class learning and post-class inquiry. This method not only reflects the guiding role of teachers but also highlights the main learning role of students.

Table 1. The learning plan of "Making a Brochure for Protecting Siberian Tigers"

Learning tasks	Practical tasks	Period
Theme Selection (creating a learning situation; assigning tasks in each group)	Task 1: Students understand and analyze the thematic learning activity, connect with real-life situations and assign tasksaccording to students' strengths and wishes.	Period 1
Planning and Designing (improving design ideas, extracting and sorting out materials)	Task 2: Students read the text, extract and sort out relevant information, use mind maps to understand the structure of the text and specific details. They comprehensively introduce animals from aspects such as appearance, food, habitat, and living habit, which provides a knowledge base for protecting wild animals in subsequent tasks. Task 3: Through in-class discussions, students improve the design ideas.	
	Task 4: Students read relevant books, browse the World Wildlife Fund website and other Internet resources, search for supplementary materials, realize that the survival of wild animals is closely related to climate change, and seek protection measures.	Post-class task

(Continued)

Learning tasks	Practical tasks	Period
Implementation (Group cooperation)	Task 5: The group members cooperate to draw pictures for the brochure, and introduce Siberian tigers from aspects such as morphological characteristics, living habits, geographical distribution, and habitat environment, and complete the manuscript writing, therefore thinking about the impact of climate change on Siberian tigers, proposing measures toprotect Siberian tigers, and realizing the significance of protecting Siberian tigers for human development. Task 6: Display each group's first draft. Teachers and students respectively evaluate the first draft, and put forward suggestions for modification.	Period 2
Summary (Submission of works)	Task 7: Students combine the suggestions from teachers and students to modify the pictures and manuscripts and bind them into a brochure.	Post-class task

1. Creating Situation and Task Assignment

Initially, the teacher introduced the background and tasks of this learning activity to the students. Then, students formed groups of five, and they interpreted the learning tasks under the guidance of the teacher. In the process, students discussed autonomously, combining the brochures they have seen in life, and believed that the design of the brochure needed to be divided into two parts: first, in terms of the manuscript writing, the content should be concise and the structure clear, conveying the most important information to readers; second, it should be illustrated with vivid pictures, reflecting the theme and attracting readers. Afterwards, students were assigned different work according to their strengths and wishes: three students to write the manuscripts, and two students to draw pictures in the brochure.

2. Planning and Designing

Teachers used intensive and extensive reading strategies to help students extract and organize key information from the text, guiding them to integrate information from aspects such as appearance, food, habitat, and living habit. Students used mind maps to become acquainted with the structure of the text and retell details, which provided the framework and detailed contents for the introduction of Siberian tigers.

Initially, during the retelling of the text, students gained a preliminary understanding of five animals: elephants, pandas, zebras, tigers, and monkeys. Based on this, the teacher guided students to comprehensively understand these animals from aspects such as morphological characteristics, living habits, geographical distribution, and habitat environment. This provided ideas for writing the manuscripts about Siberian tigers in the brochure.

Throughin-class discussions, students further refined their design concepts and made assignments more specifically: one student to complete the material research and write geographical distribution and habitat environment of Siberian tigers; one student to complete the material research and write morphological characteristics and living habits of Siberian tigers; one student to complete the material research and write the current situation of Siberian tigers and measures for their protection; one student to draw the cover of the brochure; and one student to draw the illustrations inside the brochure. In the process of researching materials, students realized that due to climate change, the quality of Siberian tigers' habitats have changed, and the number of their prey has gradually decreased. Problems such as forest pests and diseases and grassland degradation caused by climate change also pose serious threats to the survival and reproduction of Siberian tigers. It is urgent to pay attention to climate change, protect Siberian tigers, and maintain biodiversity.

In addition, students also completed the review of relevant supplementary materials after class. Through learning tools such as reference books and the Internet, they learned about the geographical distribution, habitats environment, morphological characteristics, living habits, the current situation, and protection measures. They promptly carried out discussions, explore various approaches to protect Siberian tigers from multiple aspects and angles. In this process, students became aware of the serious situation of climate change and realized the urgency and importance of protecting wild animals and jointly addressing climate change.

3. Group Cooperation

According to thetask assignment, students completed the collection of materials, the writing of manuscripts, and the drawing of pictures. During the process, teachers provided necessary help and guidance. For example, during the collection of materials, geography and biology teachers answered professional questions; English teacher helped students in text translation; art teacher guided students in layout design. In the process of writing manuscripts, students cooperated with each other, gaining a comprehensive understanding, and cultivated the awareness of protecting animals and nature.

Subsequently, students in each group displayed the first draft in class and used the activity evaluation scale to carry out self-evaluation and peer evaluation. The teacher also gave the evaluation. Teachers and students discussed with each other and improved the brochure together, promoting learning in the course of evaluation.

4. Summary

Students combined the feedbackfrom classmates and teachers to refine the content of the brochures after class and submitted their works in a timely manner (see Figure 3~Figure 5).

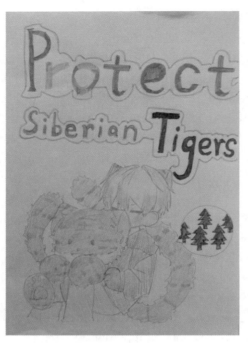

Figure 3. "Making a Brochure for Protecting Siberian Tigers" work display (cover)

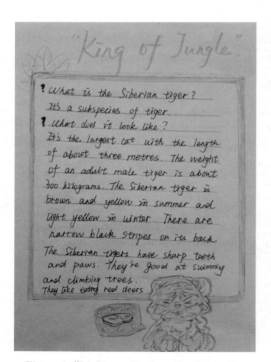

Figure 4. "Making a Brochure for Protecting Siberian Tigers" work display (inside page 1)

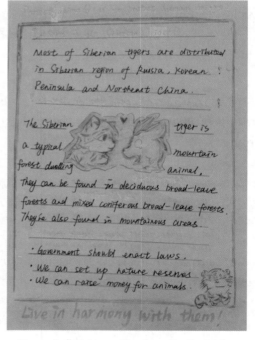

Figure 5. "Making a Brochure for Protecting Siberian Tigers" work display (inside page 2)

Effectiveness & Reflections

Climate change education is closely related to natural science. English learning similarly underscores the importance of the relationship between human and nature. Moreover, the interdisciplinary thematic learning in English is deeply compatible with the characteristics of climate change education in terms of its comprehensiveness, richness, and practicality. This case, through interdisciplinary thematic learning activities, creates a real-life situation about climate change, combining students' core competencies of language skills, learning ability, cultural awareness, and thinking quality. In the process of learning activities, students can feel the huge impact of climate change on animals and humans, and understand the significance of protecting wild animals and maintaining biodiversity.

However, climate change education is not only about imparting knowledge but also about cultivating students' competencies and abilities to actively respond to climate change. Therefore, when designing learning activities for different subjects, teachers should focus on the practical issues of climate change, call for students to experience personally, explore in practice, and solve problems by themselves. For example, teachers can guide students to study specific cases on climate change, let students discuss climate change issues from different angles, help them analyze the causes, impacts, and the response measures. Teachers can also encourage students to participate in climate change-related activities held by the community and allowing students to embody their roles in responding to climate change through practical actions, therefore cultivating their sense of social responsibility and promoting self-growth.

(Jie Zhang, Teacher of Tianjin Economic-Technological Development Area International School)

Climate Change Education Based on Scientific Practice Activities: Taking "Measuring Precipitation" as an Example

Research Background

Accurate and timely weather forecasts are of great significance for agriculture, daily life, and disaster prevention and refuge. Recording and sharing meteorological data plays an important role in studying the patterns of atmospheric phenomena, exploring the mysteries of nature, and accurately predicting the weather. Historically, people have consciously recorded meteorological information. Over the past 200 years, with the invention and improvement of meteorological instruments, meteorological data has become more accurate, and more meteorological stations have been established worldwide. An increasing number of meteorological professionals are continuously observing, recording, and sharing meteorological data. Based on this, more effective meteorological theories and analytical models have been established.

Precipitation is a fundamental characteristic of weather and an important data point recorded in the "weather calendar." One critical aspect of weather prediction is the observation of "water" conditions. Prior to this project, students at the Tianjin Economic-Technological Development Area International School, had already attempted to observe and judge rainfall through their senses. This case aims to guide students in grades 3-6 to understand the methods and ways meteorologists use to measure, record, and determine precipitation through the scientific practice activity "Cherishing the Water of Life, Valuing Water Resources." They will hand-make rain gauges and endeavor to utilize these self-made rain gauges to continuously observe and record precipitation.

Project Design

1. Activity Objectives

The main content of this activity is for students to engage in a one-month period as a survey cycle, during which they will use self-designed and made rain gauges to measure and statistics the rainfall and pH value of rainwater in their area. The specific objectives of the activity are outlined below: Firstly, To help students understand the scientific concept of "precipitation amount." Secondly, To teach students how to make simple rain gauges and learn how to use these gauges to measure precipitation. Thirdly, To enhance students' abilities in data organization, data analysis, and teamwork. Fourthly, To cultivate students' interest in exploring the mysteries of nature, guide them to understand the current situation and variation trend related to Earth's water resources through practical activities, and raise their awareness of ecological and environmental protection. Fifthly, By advocating and encouraging students to observe and record daily meteorological data like meteorologists, prompting them to participate in a long-term activity of weather observation, recording, and data analysis, to foster in students a consciousness of persistently engaging in scientific activities, and to cultivate careful and detailed observation habits as well as a persistent scientific research character.

2. Evaluation

1) Evaluation purposes and principles

The aim of this activity is to test the teaching effect, diagnose the teaching problem, provide feedback information, guide the teaching direction and regulate the process of scientific practice activities, so as to encourage the students to establish the scientific concept, form scientific thinking, willing to explore practice, excel in collaboration and sharing.

The evaluation of activities follows the principles of objectivity, scientific rigor, development and participation. The principle of objectivity is embodied in the fact that the evaluation process and results are based on facts and data, minimizing subjective judgments, independent of personal feelings, prejudices or positions, and ensuring the accuracy and reliability of the evaluation results. The scientific principle is embodied in following scientific methods, using reasonable evaluation standards and tools, and ensuring the logic and systematicness of the evaluation process. The principle of development is evident in the fact that evaluation focuses not only on current performance and results, but also on the long-term impact of activities and possibilities for improvement, encouraging participants to learn together and progress towards sustainable development of activities. The principle of participation is embodied in encouraging and promoting the active participation of students, improving the comprehensiveness and inclusiveness of evaluation, increasing the acceptability and possibility of implementation of evaluation results, and making evaluation of activities fairer and more effective.

2) Evaluation content

The evaluation content of this activity includes aspects such as "learning attitude," "learning outcomes," "learning ability," and "teamwork ability." The evaluation methods involve self-evaluation by students, peer evaluation by students, evaluation by parents, and evaluation by teachers (see Table 1). Outstanding participants will be awarded the "Young International Science and Technology Expert" certificate by the school.

Table 1. Comprehensive assessment form of students' study of scientific practice activity "Measurement of precipitation"

	Project assessment	Student self assessment	Student peer assessment	Parent-led assessment	Teacher-led assessment
Comprehensive assessment	1. Interest in activities, curiosity and spirit of inquiry				
	2. The mastery of knowledge				
	3. Experience of the process and method of the activity				
	4. The attitude and performance of cooperating with others and environment				
	Overview:				

Project Implementation

1. Preparation Stage: Understand the Basic Knowledge, Set Up Activity Group

In the preparation stage, the teacher spent three class hours to guide the students: to learn and understand the scientific knowledge about water, to become proficient in the utilization of the use of rain gauge, to understand the principle of rain gauge to measure precipitation, and through hands-on experiments, learn the basic knowledge and skills related to experimental operation; to make simple rain gauge, and learn to use simple rain gauge measuring precipitation.

According to the program of activities, students set up scientific practice groups in a free association mode before the summer vacation from 30 June to 3 July. Each student activity group prepares a set of tools, including maximum thermometer, minimum thermometer, rain gauge and precision pH test paper.

2. Implementation Stage: Making Rain Gauge, Measuring and Recording Precipitation

1) **Familiar with the relevant knowledge, lead to the core issues**

Teachers leverage science lessons to teach students scientific concepts and knowledge about water, such as the nature of water, the reserve and distribution of the Earth's water resources, the Earth's water cycle, the cause and harm of acid rain, the impact of human activities on the water cycle and water resources.

First, the teacher introduces to the students that precipitation is a basic feature of weather and a very important link in the Earth's water cycle; there are many forms of precipitation, rain, snow, hail and so on. Subsequently, the teacher asks the students to recall a recent rain event and ask, "Did it rain heavily or lightly?" "How did you judge the amount of precipitation?" The students were encouraged to recall where they could observe the amount of rain, such as the depth of puddles on the ground, put outside the container of how much rain. Coming up, the teacher asks the key question, "How do we know exactly how much rain is falling?" and shows the students how meteorologists use rain gauges to measure precipitation and grade precipitation based on how much it is falling, at the same time show meteorologists to measure precipitation data, instruments, etc.

2)Self-made rain gauge, simulated measurement experiment

The teacher instructs each activity group to prepare the necessary materials for the rain gauge: a method sheet, a thick-bottomed straight glass or plastic cup, scales, paper tape, cellophane tape, scissors, etc.. Students made simple rain gauges according to the steps explained by the teacher. When each group finished making their own rain gauge, to show the class. Teachers timely ask questions and organize discussions, such as "Does the size of the rain gauge aperture have an impact on the measurement?" "Can different sizes of rain gauge be used to measure precipitation?" and so on.

After the construction of the rain gauge, the teacher guides students to learn to check whether the rain gauge leaks, and to carry out a field simulation survey. Students go outside to select a suitable location, place the rain gauge, and simulate rainfall with a spray bottle. In this process, the students go through a process of collecting and measuring precipitation, and compared with the rainfall grade standard to judge the measured rainfall grade. When reading the rainfall data, the teacher reminded the students to pay attention to the relevant matters and discuss the reasons. For example, when moving the rain gauge, not to let the "Precipitation" in the rain gauge overflow, the rain gauge should be placed on a horizontal table; the line of sight should be parallel to the scale on the rain gauge.

Each member of the group then takes individual rain gauge readings in turn and reports the readings agreed upon after consultation within the group. Student representatives of each group recorded precipitation on "Our precipitation fill map". The teacher organizes a discussion: Why is the amount of precipitation different in each group? How to distinguish heavy rain, moderate rain, light rain and other different rainfall levels? Then the teacher show the climatologist's chart of rainfall grades based on how much precipitation there is, and the students compare it to make a judgment.

Next, the teacher guides the students in their extra-curricular activities of measuring and recording precipitation. The teacher and students work together to sort out the key considerations, such as determining the position of the rain gauges, and choosing as far as possible a more open spot with no shelter above or near it; Before measuring, students should fix the rain gauge to avoid being blown down by the wind; record the precipitation every 24 hours to determine the rainfall level; after each record, the water in the rain gauge should be poured clean to avoid affecting the next measurement andrecording.

3) Students on-site measurement, teachers timely guidance

The students choose one month during the summer vacation as a survey cycle, and measured and counted the rainfall and pH of the rainwater in their area using their own rain gauge and the rain gauge they bought. By comparing the measuring data of the self-made rain gauge and the purchased rain gauge, the causes of the errors are analyzed and the self-made rain gauge continuous refinement. Teachers provide remote guidance timely.

3. Summary, Show Stage: Analysis of Data, Show Results

After a month of scientific practice activities, students have gained a lot, and have also recorded the process and results of the activities in various forms. Students of each activity group analyze the recorded data, sum up the rule of weather change, and fill in the relevant data form, upload to the school designated mailbox. For the process records, the activities of the groups display the report with PPT, "Meipian App", text and other ways. In order to encourage students to participate in science and technology innovation activities actively and promote the spirit of loving science, learning science and using science, the school commended all the students who took part in the science and technology practice activities, and the award-winning works timely released to the school "Science Fly Book Group", displayed on the school's WeChat public account, students are encouraged.

Effectiveness & Reflections

In this activity, the teacher introduces the background knowledge of climate change, leads to the design practice of the self-made rain gauge, and leads the students to make the instrument and observe and record the precipitation data (see Figure 1, Figure 2). Through long-term observation, the students gain insights into the seasonal and interannual variations of precipitation.

Through case studies, students gain insights into the impacts of climate change in different parts of the world. Teacher issues timely initiatives to encourage students to think about how to reduce greenhouse gas emissions and propose strategies to address climate change. Figure 3-6 are graphs created by students base on recorded data during the activity.

Figure 1. The experimental apparatus prepared by the students

Figure 2. Primary records of students measuring rainfall and weather information

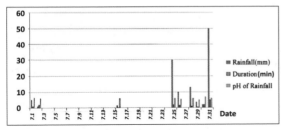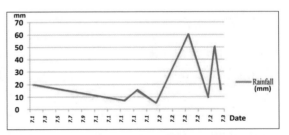

Figure 3. Column chart of rainfall recorded by students

Figure 4. Broken line chart of rainfall recorded by students

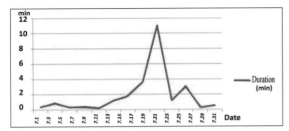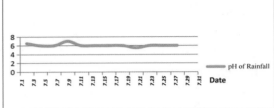

Figure 5. Student-drawn line plot of rainfall duration

Figure 6. Student-recorded line plot of pH value of rainwater

Some students wrote:

This summer vacation, My classmates Yu and I took part in the "Cherish the water of life" scientific practice activities. Our work was busy but meaningful. Through a month of measurement and statistics, In July, Hangzhou Road Tanggu District, Tianjin, under a total of 12 rain, each rain acid-base is different.

Acid rain does great harm to crops. It destroys soil components, reduces crop yields, kills fish and shrimp in lakes, corrodes buildings and industrial equipment, and destroys open-air cultural relics and monuments. Drinking groundwater polluted by acid is harmful to human body. We experienced an acid rain event here in July this year, and we should attach great importance to this issue. In addition, we feel more and more hot in summer, global warming has become an indisputable fact. Global warming will also have adverse effects on the circulation and utilization of water, thus endangering all mankind.

By participating in the survey, measurement and research activities, we not only learn a lot of knowledge, exercise the ability of scientific practice, but also know to cherish water resources, protect the environment. The Earth is the largest biosphere, is our common home. I love my family, we cherish our family!

Some students expressed emotion:

We know a lot of scientific principles in the course of the experiment, but also understand an important truth: to save water, cherish every drop of water!

From the data and research results submitted by students, although in the practice process there are many deficiencies for students, such as some observation, record data accuracy is not high, some students can not adhere to a month of measurement time. However, most of the students were able to deal with the difficulties positively and keep exploring the solutions to the problems.

In the process, they have learned simple methods of observing and recording weather, and have known that even if they were in the same district, even if they were just a highway away, the start time, duration, end time, and amount of rainfall may also be present different. Therefore, they know the reason and significance from multiple points collection of meteorological data. Students have also enhanced their ability to work in groups and to organize and analyze data, keenly aware of the importance of being rigorous, seek truth from facts and persistent in the conduct of scientific research. More importantly, they have developed the habit of actively observing weather changes, enhance their awareness of environmental information around, and understand the important role of water conservation in solving water crisis. They have developed the awareness of ecological environment protection.

At the same time, there are also some deficiencies worth reflection and improvement: one is that theory and practice are not closely combined. In the course of the activity, although the students participated in the practice, but the grasp of the relevant theoretical knowledge is not solid enough. Therefore, a series of effective measures should be taken to enhance the integration of theory and practice. For example, before the event, arrange lectures and seminars to ensure that students have the necessary theoretical knowledge; in the activities, design more tasks based on the actual situation, so that students can apply the theory in practice. The second is for students to improve the ability of autonomous learning. To this end, we can design more specific inquiry questions, encourage and guide students to think independently and solve problems, and constantly improve self-learning ability. In the future, we will continue to improve and enhance the effectiveness of the activities to promote the overall improvement of student literacy.

(Yonghong Zhang, Teacher of Tianjin Economic-Technological Development Area International School)

Smart Learning, Wise Application, and Intelligent Innovation: Constructing an Integrated Model of Information Technology Education and Climate Change Education

Research Background

China has made solemn commitments to achieve carbon peak by 2030 and carbon neutrality by 2060, demonstrating its determination and ambition to address climate change. As the masters of the future, children and adolescents' climate awareness and actions are critical to achieving these goals. With its unique characteristics of interdisciplinary integration, data-driven insights and technological innovation, the discipline of Information Technology provides rich resources and means for climate change education. With the help of information technology, students can have a more intuitive understanding of the scientific principles, impacts and coping strategies of climate change, as well as participate in the monitoring, analysis and response to climate change by means of information technology. This case endeavors to delve into the integrated practice of Information Technology subject teaching and climate change education, realizing the effective infiltration and in-depth promotion of climate change education through three dimensions: "Smart Learning," "Wise Application," and "Intelligent Innovation."

Project Design

1. Objectives

(1) Cultivate students' scientific cognition of climate change and understand the causes, impacts and mitigation of climate change.

(2) Improve students' ability of data analysis and scientific inquiry, and be able to analyze basic climate change data by means of information technology.

(3) Enhance students' awareness of environmental protection and social responsibility, and be able to actively participate in the action against climate change.

2. Content Design

The specific contents of this case are as follows: First, the learning activities of basic knowledge related to climate change. Related knowledge includes the meaning, causes, impacts of climate change and global measures to deal with climate change. Second, students' data analysis skills will be improved. This activity mainly guides students to use information technology tools (such as data processing software, programming languages, etc.) to analyze climate change data (such as temperature, rainfall, sea level rise, etc.). Third, interdisciplinary integration of learning activities. Combining the resources of

Information Technology, Science, Mathematics and other disciplines, students will be guided to assess the impact of climate change on the natural environment, ecosystems and social life. Fourth, climate change-related scientific research projects, environmental protection activities and community science popularization services.

3. Activity Approaches

(1) Interdisciplinary theme exploration. Climate change education involves multiple disciplines, which need to break disciplinary boundaries and deeply integrate Information Technology, Mathematics, Science and other disciplines. With the selection of teaching content, the application of teaching methods, the development of teaching resources, and the design of teaching evaluation as key links, a complete interdisciplinary theme project should be established to guide students to integrate interdisciplinary knowledge and solve problems.

(2) Experimental teaching. Utilizing information technology LABS or virtual LABS, organize students to conduct climate change data analysis and simulation experiments.

(3) Practical activities. Organize students to participate in practical activities such as field trips, social surveys, and research projects so that students can experience the impact of climate change and strategies to cope with it.

(4) Online learning. Make use of online learning platforms or social media to provide students with rich learning resources and communication opportunities to promote in-depth discussions on climate change issues.

4. Evaluation Design

Teachers' evaluation of students' learning outcomes in 'Knowledge mastery,' 'Skill application,' 'Attitude and action,' and 'Interdisciplinary integration' to achieve deep integration of information technology education with climate change education and enhance students' climate change awareness and action competence (see Table 1).

Table 1. Design table of learning evaluation for the integration of Information Technology and climate change education

Theme	Evaluation method	Evaluation contents	Evaluation level
Knowledge mastery	Examinations, assignments	Degree of mastery of fundamental knowledge on climate change	☆☆☆☆☆
Skill application	Project reports, data analysis reports	Ability to analyze climate change data using information technology tools	☆☆☆☆☆
Attitude & action	Observations, interviews	Attitude: Environmental awareness, social responsibility Action: Performance of actions to respond to climate change	☆☆☆☆☆

(Continued)

Theme	Evaluation method	Evaluation contents	Evaluation level
Interdisciplinary integration	Interdisciplinary projects or comprehensive practical activities	Ability to integrate knowledge and thinking across different disciplines	☆☆☆☆☆

Project Implementation

Climate change education is unique in its comprehensiveness, practicality, and richness, spanning multiple disciplinary fields. *The Curriculum Standards for Information Technology in Compulsory Education (2022 edition)* has built a complete and systematic content module of the whole learning section around six logical main lines: data, algorithm, network, information processing, information security and artificial intelligence. Each module has carefully designed interdisciplinary theme activities, which provides strong support and favorable conditions for the deep integration of climate change education in Information Technology teaching.

1. Smart Learning: Information Technology Contributes to Climate Change Education and Popular Science

Climate change intrinsically linked to daily life. Integrating climate change education into subject teaching aims to guide students to gradually enhance their awareness and action awareness of climate change in the process of solving practical problems by constructing scenes close to life.

As a treasure of farming culture, the 24 Solar Terms are the crystallization of the wisdom of our ancestors. In the section of "Online Learning and Life", teachers designed the activity of "24 Solar Terms Speaker", guiding students to collect and organize information related to solar terms online, and explore the wisdom contained in it through teamwork, to fully understand the impact of climate change, and finally to share the speech of the 24 Solar Terms with the help of mind mapping. The "clever" of this teaching method lies in stimulating students' interest, making knowledge close to reality, and guiding them torealize the importance of climate change and its close relationship with daily life.

In the process of delving into the 24 Solar Terms, the students asked, "Are the meteorological idioms handed down by the ancients scientific?" This core question, the teacher keenly captured this valuable classroom generation resources, and designed a comprehensive activity of "number" meteorological idioms. With "Are the meteorological idioms handed down by the ancients scientific?" as the core question of the unit, and break down the Chain of questions such as "How to verify?" "What data is needed?" "What tools are used?" "What conclusions are drawn?" Corresponding to the question chain, construct the activity string (see Figure 1). This problem-driven learning style can cultivate students' independent thinking ability and problem-solving ability, which is a kind of "clever" thinking in the teaching process.

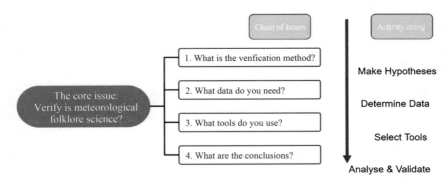

Figure 1. "Number" explaining the weather folklore "question-activity" design

For example, when assessing the accuracy of the saying "Major Heat and Minor Heat are not really hot, but the Beginning of Autumn and the Limit of Heat are the hottest", the teaching process is as follows: First, a hypothesis is proposed. The teacher assumes that the saying is scientific and prompts students to consider what data they would need to collect to prove the hypothesis. Next, the teacher guides students to think about using multiple sets of data for comparison to enhance reliability and validity. Then, students are led to choose appropriate tools to visually compare the different data sets. Finally, the data is analyzed and verified to reach a conclusion. As shown in Figure 2, except 2021, the data of Minor Heat and Start of Autumn solar terms in Shanghai in the past ten years are consistent with the rule of the idiom. The results obtained by the students are as follows: this proverb is scientific.

As for the saying "If it's not cold during the Great Cold, it will be not warm when the spring equinox comes next year", the students verify the data and find that the regularity is not obvious, or even does not conform to the rule.[①] As shown in Figure. 3, the students raise their confusion: Judging from the data, the proverb does not conform to the rule. Is the wisdom handed down by the ancients unscientific?

[①] The meaning of this saying is that if the Major Cold is not cold, the cold wave will be delayed, and the weather will not be very warm by the time of the next solar term, the Spring Equinox.

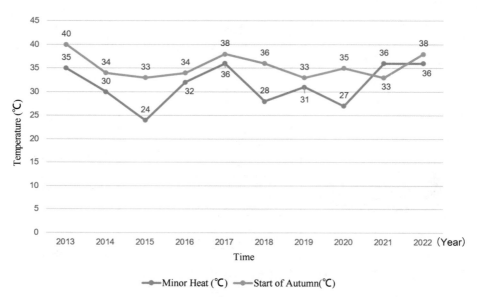

Figure 2. Comparison of the highest temperature on the day of Minor Heat and Start of Autumn in Shanghai in the past ten years

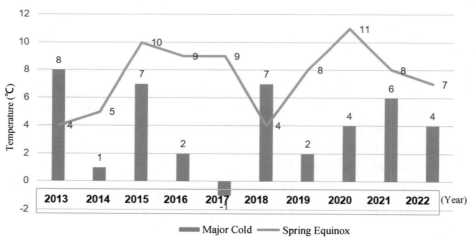

Figure 3. Comparison of temperature between the Major Cold and the Spring Equinox in Shanghai in the past ten years

In response, the teacher adeptly facilitates the students to delve deeper into the underlying reasons behind these idioms, especially the complex global climate change factors involved. Meteorological idioms are the results of wisdom accumulated by ancient people for many years, and they have strong local characteristics. Under the influence of global climate change, it is necessary to combine the contents of the idioms with scientific meteorological prediction, and dialectically apply the meteorological idioms to guide production and life by integrating the knowledge of various disciplines. Thus, students know how to verify the correctnessof meteorological idioms by collecting, collating and analyzing data. The pedagogical finesses is that it can cultivate students' scientific literacy, so that they can learn to look at problems from a scientific perspective, and at the same time enable them to have a deeper understanding of the laws and effects of climate change.

2. Wise Application: Intelligently Applying the Insights of Climate Change Explorations to Life

Climate change education prioritizes experiential learning to develop knowledge discovery, innovation and critical thinking. Under the guidance of the new curriculum standard, teachers have designed a project combining Information Technology and climate change education with students' life practice. Through engaging in practical experience, students are guided to not only gain knowledge and growth, but also apply the learning results intelligently to daily life, so as to maximize the learning value.

Taking "Homemade Weather Life Tips" as an example, teachers guide students to understand the importance of data in daily life: use via the campus meteorological science popularization data platform collect data, use technical tools to organize and visualize the data, make practical "Meteorological life tips", and share the research results in the whole school through multiple platforms such as campus TV station and financial media center. This process contributes to the dissemination of knowledge and wisdom sharing. At the same time, students can receive feedback and suggestions from others in the knowledge sharing, so as to further improve their own works. This circular process is cyclical "wisdom and application".

Through the "Weather Life Tips" project unit learning, students improve the use of data to speak, use data to solve problems awareness, and enhance the efficiency of data application. Some students independently explored the relationship between climate change and food, clothing, housing and transportation, combined with data visualization to make popular science exhibition boards, which were displayed and shared in the "MeteorologicalCarnival" organized by the school; Some students explored the scientific and technological topics of climate change and dressing, authoring a climate change research report *"Suggestions on the school Uniform Wearing of Primary School Students in Shanghai Considering Meteorological Factors"*, and promoted the 26° dress rule in the campus. In this way, "Wise Application" is also reflected in popularizing the results of climate change exploration to a wider group of people, improving their awareness and participation in climate change, so that more people can understand the impact of climate change and coping strategies, and forming a joint force of the whole society.

3. Intelligent Innovation: Wisdom-driven Innovation in Future Climate Change Education

In the journey of education towards addressing future climate change, it is far from sufficient to merely popularize knowledge. It is essential to cultivate students' problem-solving awareness from a young age and build their essential competencies for responding to climate change. To achieve this, relying on the unique "Space Capsule Mars Farm" learning space, the school's information technology society has innovatively carried out the "smart planting" project to build an "intelligent innovation experimental field" for students.

This project aims to simulate unnatural and even extreme weather conditions, intelligently manage the planting system through programming technology, in order to optimize the planting process and explore innovative strategies to combat future extreme climates. The project integrates multiple disciplines such as Information Technology, Natural Science and Labor Education, and selects a variety of plants such as pumpkin, cherry radish, petunia, dwarf tomato and strawberry as research objects to deeply analyze the key factors affecting plant growth.

At the project's inception, we asked a core question: "How to use Farmbot robot ecological box to simulate extreme environment and realize intelligent plant cultivation?" "And broke it down into multiple sub-questions, such as" Which seeds are suitable for growing in the capsule Eco box?" "Whatfactors affect plant growth in the space environment?" "How can the growth of plants be regulated by robots and artificial intelligence?" Etc., to stimulate students' strong desire to explore.

Subsequently, interdisciplinary teaching units are carefully designed according to the six elements of plant growth-light, temperature, humidity, water, gas and fertilizer, integrated with the core task of six-dimensional programming (see Figure 4). For example, in the study of light control, students learn how to adjust the height of the light panel according to the height of the plant; In the study of temperature control, students learn how to use the heating fan or fan reasonably according to the temperature of the eco-box; In the study of fertilizer control, students learned how to decide whether to make fertilizer according to the growth state of plants, and also learned how to use a professional soil detector to measure the current pH value of plants.

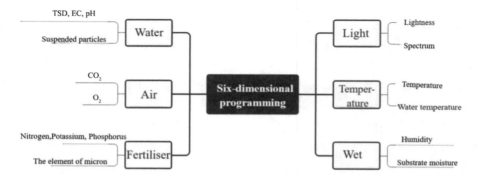

Figure 4. Six-dimensional programming diagram of the "Intelligent Planting" project

The micro-controlled ecological box in the school's "Space Capsule Mars Farm" learning space is endowed with various sensors to provide students with a real intelligent planting exploration environment, and students use robotics and artificial intelligence technology to program and control every factor of plant growth. This "trial and error" intelligent experimental field not only allows students to learn and explore in practice, but also provides valuable ideas andexperience for them to proactively address climate change issues in the future.

Effectiveness & Reflections

Nowadays, the role of Information Technology in climate change education is becoming increasingly prominent. The teaching mode of "smart learning, smart application and smart innovation" has markedly bolstered students' learning ability and problem-solving ability, as well as enhanced their environmental awareness. They began to pay more attention to the issue of climate change, and took the initiative to carry out relevant scientific and technological research around their daily life.

But climate change education is not something schools can do on their own. The close cooperation of the three parties (school-family-society) is the key to forming a joint force on climate change education. We need to work together to create a three-dimensional climate change education space for students, so that students can build closer ties with the real world, fully participate in and experience, and give full play to their imagination and creativity. Moving forward, we will continue to deepen the integration of Information Technology and climate change education, cultivate more responsible and action-driven future citizens, and contribute wisdom and strength to protecting the earth and green home.

(Yuyan Cheng, Tongtong Sun, Weibin Gu, Teachers of Mingqiang Primary School, Minhang, Shanghai)

Designing Teaching Activities for Morality and Law in Primary Schools Under the Context of Ecological Civilization Construction

Research Background

"Guide with morality, regulate with etiquette, and cultivate a sense of shame and discipline." The construction of ecological civilization requires strict adherence to ecological standards through regulations, and even more so, the cultivation of ecological civilization awareness through education. Integrating with ecological civilization education, promoting ecological civilization social practice is not only a new mission given by the times to moral and legal education, but also an inherent requirement of ecological civilization construction.

The *Compulsory Education Moral and Legal Curriculum Standards* (*2022 edition*) clearly identifies the development of students' core literacy as the fundamental direction of the curriculum, proposing to "help students form correct values, cultivate necessary character, enhance awareness of rules, develop social emotions, and improve key abilities." This case is based on students' perspectives and real-life experiences, guiding them to feel and explore nature, reconstruct the moral relationship between humans and nature, and transform "ecological civilization" a knowledge point that is accessible, perceptible, and in-depth from textbooks and classrooms.

Project Design

1. Objectives

"Man and Nature" is one of the three main themes of the primary school morality and law curriculum, alongside "Man and Self" and "Man and Society." Through analysis, we find that the Moral and Legal curriculum for grades one to five is designed based on the physical and mental development and moral growth needs of students in different stages of growth, guiding them to gradually form a harmonious coexistence between man and nature in a spiral manner. The content related to "ecological civilization" in the *Moral and Legal* textbook (5-4-education-system), is shown in Table 1.

Table 1. Content related to "ecological civilization" in *Moral and Legal* textbook (5-4-education-system)

Grade/Book	Unit	Text title
Grade 1 Book 2	Unit 2	Lesson 5: The Wind Blows Softly Lesson 6: Flowers and Grass Are So Beautiful Lesson 7: Lovely Animals Lesson 8: Nature, Thank You
Grade 2 Book 1	Unit 2	Lesson 8: Decorating Our Classroom
Grade 2 Book 1	Unit 4	Lesson 13: I Love the Mountains and Rivers of My Hometown Lesson 14: The Products of My Hometown Nurture Me Lesson 16: New Changes in My Hometown
Grade 2 Book 2	Unit 1	Lesson 4: Planting a Seed
Grade 2 Book 2	Unit 3	Lesson 9: The Story of a Tiny Drop of Water Lesson 10: Fresh Air is a Treasure Lesson 11: I Am a Piece of Paper Lesson 12: My Environmental Protection Partner
Grade 3 Book 1	Unit 2	Lesson 6: Let's Make Our School Better
Grade 4 Book 2	Unit 5	Lesson 13: Environmental Pollution We Know Lesson 14: Turning Waste into Treasure with Smart Methods Lesson 15: Low-carbon Living Every Day
Grade 5 Book 1	Unit 5	Lesson 15: The Earth—Our Home Lesson 16: Responding to Natural Disasters
High-grade in primary school Xi Jinping's New Era of Socialism with Chinese Characteristics for a New Generation (Students' Textbook)		Lesson 10: Lucid waters and lush mountains are invaluable assets.

Based on this, we further clarify the goals of the eco-civilization theme practical activities for each grade segment of the Moral and Legal Curriculum:

Phase 1 (Grades 1-2): Through participation in practical activities such as observation, planting or breeding, enable students to identify common plants and animals, understand their basic living habits, experience the closeness with nature, and learn to treat plants and animals with respect and care.

Phase 2 (Grades 3-4): Through participation in themed practical activities, students understand the close relationship between nature and human life, recognize that nature is our shared habitat, and understand the importance of protecting the environment, caring for animals, and conserving resources.

Phase 3 (Grade 5): Through participation in comprehensive projects or research, students analyze environmental issues using the knowledge they have learned, propose sustainable development solutions; instill values of sustainability, and recognize the importance of achieving harmonious coexistence between humans, society, and the environment.

2. Activity Content

This case study involves creating a series of extracurricular practical activities themed on "ecological civilization" to guide students in exploring in groups based on task lists after completing classroom learning.

The "Me and Nature" practice activities for Grade 1 students aim to let students be close to nature, feel the beauty of nature, and make enthusiasm the tone of the "ecological civilization" theme education. The "Me and My Hometown" practice activities for Grade 2 students enable students to further understand the place where they were born and grown up, establish a connection between their hometown and themselves through specific things, and provide a more concrete foothold for students' vast love for nature, making emotional understanding more genuine. The "Me and the Campus" practice activities for Grade 3 students are designed to stimulate students' sense of ownership of "I can also do something". The "Me and Sustainable Development" practice activities for Grade 4 students let "green environmental protection" enter students' real life, and promote their sense of responsibility and accountability with "成就感" (a sense of accomplishment). The "Me and Disasters" practice activities for Grade 5 students guide students to view the interdependence between humanity and nature and establish a correct view of sustainable ecological development.

3. Assessment

This activity evaluation will be structured through a grading system based on three dimensions: "interest in the activity", "activity capability", and "activity achievements". The evaluation will mainly combine students' self-evaluation, peers' evaluation, and teachers' evaluation.

Table 2. Comprehensive practical activity evaluation form

Name _____

Method		Student self assessment	Student peer assessment	Teacher-led assessment
Interest in the activity				
Activity capability	Collecting information			
	Communication and expression			
	Cooperative learning			
Task completion	Presentation and reporting			
	Activity achievements			

Project Implementation

1. Establish the Activity of "A Little Whale Making Friends with the Wind" to Guide Students in Perceiving Nature

The Jinshan District of Shanghai, situated in the North subtropical zone, where the monsoon prevails in Southeast Asia. The district has four distinct seasons, with a mild and humid climate. However, due to the alternating influence of cold and warm air, typhoons and rainstorms occasionally occur. Considering that first-grade students have strong curiosity and desire for knowledge, but may lack patience and self-discipline, in the practical activity of "A Little Whale Making Friends with the Wind," educators initially integrate classroom learning with the lesson "The Wind Blows Gently" to allow students to gain an understanding how to "capture" the footprints of the wind and guide students to observe through comprehensive practice (see Table 3). To heighten students' sensitivity to the "wind", teachers first guide them to think about the climate change behind the "wind" and let them record their feelings of the wind in different seasons and areas. Then, guide students visualize climate issues by drawing the appearance of the wind and the changes brought by the wind to their hometown. Finally, guide students to become plants in their hometown and create a dialogue with the wind to perceive the relationship between climate change and vegetation.

Table 3. "A Little Whale Makes Friends with the Wind" comprehensive activity task list

A Little Whale Makes Friends with the Wind		
Looking for the Wind Introduction: The wind is our closest companion, traveling through the four seasons and touching everything. _____, I go to _____ find the wind. I feel _____ _____.	**Understanding the Wind** Introduction: "Unravel the autumn leaves, scatter them across the sky, like a three-year-old's plaything; bring about the bloom of flowers in February." In Li Qiao's eyes, the wind carries away the autumn leaves and brings about the bloom of flowers in spring. What kind of changes will you bring to your hometown by finding the wind?	**Giving Thanks to the Wind** Introduction: Wind is not only our friend, but also a good friend of nature. Please choose a plant from your hometown and imagine what it would say to the wind. (　　) (wind)

2. Create the Activity of "A Little Whale Exploring Jinshan" to Promote Students' Attention to the Ecological Environment

Jinshan District boasts ten thousand acres of fertile fields, a fruit park, and the beautiful "Flowers Blooming on the Sea" ecological park, all of which are benefited from the mild climate environment. Combining with the study of the course "Local Products Nurture Me", the school creates the "Walking LOVE" activity, through which students can establish real emotional connections with the local ecology during their field visits as "A Little Whale Exploring His Hometown". In this activity, students visit their hometown based on their self-designed route, search for local products, and interview the wisdom of local people to comprehensively perceive the local ecological environment (see Table 4).

Table 4. "A Little Whale Explores Jinshan" practical activity task list

A Little Whale Explores Jinshan

Introduction: Jinshan District is located in the southwest of Shanghai, rich in natural resources and with a pleasant ecology. On a weekend, go with your parents to experience the tranquility of ancient towns, the sweet scent of orchards, the romance of fishing villages, and the innocence of the countryside. Use your feet to explore your hometown, use your eyes to record the unique local products that belong to you, and use sincere words to express your appreciation for the rich birthplace of these products.

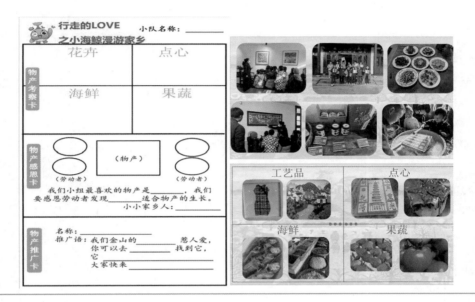

3. Initiate "A Little Whale Adds Color to Our Campus" Event to Encourage Students to Participate in Environmental Protection Actions

Jinshan District is almost the only place in Shanghai that has mountains, seas, islands, and domains, suitable for developing urban agriculture. The pleasant climate has given rise to many multifunction and unique beauty "rural complexes" such as Jinshan Fishing Village, Lvxiang Fruit Park, and Langxia Ecological Park. After studying the lesson "Making Our School a Better Place," students engaged in the comprehensive practice activity "A Little Whale Adds Color to the Campus," using the "Campus Survey Report" from the textbook to contemplate how to further improve and create a more comfortable school environment (see Table 5). The students also proposed specific and effective plans at the school's "Youth Representatives' Conference."

Table 5. "A Little Whale Brings Color to the Campus" comprehensive practical activity task list

A Little Whale Brings Color to the Campus			
A corner of the campus	Existing problems	Improvement measures	Subsequent maintenance

Students Proposal			
Proposer		Collective	
Co-signer			
Type of Proposal			
Name of Proposal			
Content of Proposal			
Suggestions for the Proposal			
Reply to the Proposal			
Remarks			

4. Establishing the "Little Whale Creative Moment" Activity to Shape Students' Aesthetic Awareness and Environmental Protection Concept

Following their study of the course "Turning Waste into Treasures with Smart Solutions", students integrated the "Little Ocean Whale Creative Moment" comprehensive practical activity into the school's "Jingjing Vegetable Garden" autonomous management through the activity of "Customizing Environmental-friendly Decorations for the Garden". This is an action to echo the environmental protection concept and care for the beauty of the garden.

Table 6. "A Little Whale Creative Moment" comprehensive practical activity task list

A Little Whale Creative Moment
Introduction: The planting season of "Jingjing Vegetable Garden" is coming soon! Please use your industrious hands and materials from daily life to create and decorate beautiful accessories for our garden. Let's make our garden colorful!
• Preparation Materials
• Production Process
• An Example

5. Develop the Activity of "A Little Whale's Disaster Avoidance Handbook" to Enhance Students' Disaster Avoidance Skills and Safety Awareness

Education on natural disasters is an important part of quality education, and knowledge and skills in disaster prevention and mitigation are essential practical abilities for students. Teachers, in combination with the study of "Responding to Natural Disasters", conduct practical activities through the "A Little Whale's Disaster Avoidance Handbook". Based on the observed problems in various fire and earthquake drills, students further plan escape routes for each class, draw "campus escape maps", establish a sense of crisis, and improve their ability to avoid risks. This "sense of crisis" is also transferred to family life, where students work together with their families to draw "family escape route maps" to comprehensively enhance their "safety consciousness" (see Table 7).

Table 7. Task list for "A Little Whale's Safety Manual" comprehensive practical activity

A Little Whale's Safety Manual
Introduction: Nature has given us endless treasures, but if we do not cherish it, Nature will also teach and punish us in its own way. What should we do in case of disasters?
• Campus Escape Route Map
• Home Escape Route Map

Effectiveness & Reflections

Moral and Legal Curriculum is a comprehensive activity course closely related to students' lives. The design of comprehensive practical activities is the key to closely connecting with students' actual life experiences. In the process of carrying out this comprehensive practical activity themed on "ecological civilization" has led to the recognition of three points of importance.

First is the importance of accurately grasping the concept of "ecological civilization". Teachers should continuously strengthen their study and research on ecological civilization education, carefully study the policies related to ecological civilization by the country and the party, comprehending their ideological essence, and integrate these ideas into teaching practice.

The second aspect is the importance of bringing "ecological civilization" into action. For primary school students, there is indeed a gap between "ecological civilization" and real life. It is imperative for educators to integrate living resources so that students can build a practical and profound concept of ecological civilization through "walking" in the social classroom and through real exploration.

The third aspect is the importance of promoting the construction of "ecological civilization" across all subjects. In 2019, China issued the document, pointing out the necessity of popularizing ecological civilization construction across all disciplines. In the *Opinions on Deepening Education and Teaching Reform and Improving the Quality of Compulsory Education*, it is clearly stated that ecological civilization should be one of the main contents of moral education. To cultivate students' ecological civilization quality, relying solely on the teaching of Moral and Legal Curriculum in primary schools is far from sufficient. More subject teachers need to be involved to collaboratively establishing a diverse and varied curriculum base for cultivating ecological civilization concepts and behavioral guidance.

(Yao Yao, Teacher of Qianjing Primary School, Jinshan, Shanghai)

🎓 Expert Review

Compulsory Education Curriculum Program and Curriculum Standards (2022 Edition) issued by the Ministry of Education clearly sets out the educational goals including "love nature, protect environment, love animals, cherish life, and establish the concepts of public health and ecological civilization, keep up with current affairs, love peace, respect and understand cultural diversity, and initially have an international outlook and a sense of the community of human destiny". The *General Senior High School Curriculum Program and Curriculum Standards for Language and Other Subjects* (2017 edition, revised in 2020), also emphasizes "ecological civilization and maritime rights and interests", as well as the cultivation goal of "respect nature, protect environment, and cultivate ecological civilization awareness". It can be seen that climate change education based on different disciplines has become an important part of the curriculum and teaching changes in basic education.

Climate change education grounded in various disciplines refers to the process of upgrading the content and quality of different disciplines with the concepts of ecological civilization and sustainable development, aiming at the iterative achieve iteratively of climate change literacy and different disciplines' core literacy, updating the teaching and learning methods with the principle of sustainability, and analyzing the climate change issues from the perspective of the different disciplines.

This chapter showcases excellent cases of the integrating teaching with climate change education practices in the disciplines of Chinese, Mathematics, English, Science, Information Technology and Moral and Legal, and presents three types of practice of climate change education based on different disciplines, which includes climate change education in different disciplines, large-unit teaching based on the theme of climate change, and discipline and climate change integrated practice activities. Its inspiration for the practice of integrating discipline specific content with climate change education lies in the following:

Firstly, it carefully analyzes the relevance of discipline curriculum standards to climate change education and grasping the overall characteristics of discipline integration is an important starting point for the integration of discipline and climate change education. For example, the *Compulsory Education Chinese Language Curriculum Standards* (2022 Edition) specifies the requirements in the [Reading and Appreciation] section, which are "care about nature, and express one's own opinions about people and things of interest". English curriculum standards emphasize the requirements of "national sentiment and cultural awareness within the context of a global community, as well as fostering critical and innovative thinking skills". The Physics Curriculum Standards put forward the requirement of "being close to nature, advocating science, being willing to think and practice, and having the desire and curiosity to explore nature. Also, students shall pay attention to the impact of science and technology on the natural environment, human life and social development, comply with scientific ethics, be aware of environmental protection and resource conservation, be able to contribute to the sustainable development of society within the limits of one's ability to achieve the great rejuvenation of the Chinese nation with a sense of responsibility".

Secondly, the integration of discipline and climate change education needs to grasp theintrinsic logic of each discipline. Building on the stance of the discipline, the iteration of the literacy goal, the optimization of the content integration, and progressively analyzing of the implementation path of climate change issues from the perspective of the discipline, it is necessary to realize the spiral progression in discipline literacy and the goal of climate change literacy education. For example, in the Chinese Curriculum case, it cultivates the students' awareness of nature and aesthetic ability, promotes the formation view of the harmonious coexistence of human beings and nature, and enhances the quality and realm of Chinese Curriculum adhere to the disciplinary literacy goals through the way of imagination on water cycle, comparative analysis, inductive judgement, and more.

Thirdly, the key to the design and implementation of the large-unit teaching of the climate change theme lies in the design and implementation of the iterative reconstruction of the climate change theme and the textbook unit, which helps students develop the habit of structured thinking, relevant meanings thinking, and the intrinsic drive of values. One challenge is extracting overarching concepts. Through the optimization of the content of the climate change theme and the disciplinary unit, broad concepts are identified to form a thinking framework for analyzing climate change issues from a disciplinary perspective, and meanings are generated between core knowledge, which then stimulate students' intrinsic drive to take action.

Fourthly, discipline and climate change integrated practice activities utilize fully 10 percent of the practice space to create a space for immersive experience and inquiry, and to enhance the integrated practice capacity of disciplinary perspectives in addressing climate change issues. For example, in this chapter, in Chinese Curriculum and Climate Change Education, English Curriculum and Climate Change Education, Mathematics Curriculum and Climate Change Education, Information Technology Curriculum and Climate Change Education, and Moral and Legal Curriculum and Climate Change Education, the vivid authenticity of the situation and immersive experience enrich the attainment of disciplinary literacy.

On the occasion of World Environment Day on June 5th, UNESCO launched the "Greening Education" initiative, calling for the enhancement of environmental education for young people so that they can better participate in addressing climate change and other pressing issues. In the context of China's modernization and the pursuit of a beautiful China in which "human beings live in harmony with nature", integrating climate change and sustainable development education throughout the education process and strengthening the cultivation of green and low carbon innovative talents represents the mission of the majority of teachers in contributing to the nation's strengthen.

(Qiaoling Wang, Secretary-General of the UNESCO China ESD Program Team and Deputy Director of the Institute for Lifelong Learning and Education for Sustainable Development, Beijing Academy of Educational Sciences)

Chapter 3

Climate Change Education Based on Project-Based Learning

In practice, we have found that climate change education is very suitable for the way of project-based learning, especially for the following three types of projects: lifestyle transformation, production mode improvement, and ecological and environmental protection. Project learning also brings together the strength of multiple disciplines in the previous article. Through the combination of subject teachers and the active communication of teachers responsible for the project, a comprehensive research state is formed.

The following cases are distributed in the above three types of projects, showing the process of Chinese educators planning and organizing project-based learning on this topic, and sharing the experience of students, parents and teachers in improving climate literacy in this process.

"Disfavored Shoes", Have You Been Disfavored?
—Let the Idea of Recycling be Integrated into the Daily Life of Primary School Students

Research Background

"Look, aren't my new shoes beautiful?"

"The color is really nice, and there are no shoelaces!" The other child showed a look of envy.

As soon as the class was over, I found that the children were looking at a pair of new and trendy sports shoes. The unique design of the shoes was that there were no traditional shoe laces, adhesive buckles, or magnetic drawstring buckles. Instead, there was a trendy button that could be pulled to loosen the shoe laces and pressed to rotate the button to tighten the shoe laces. The red and white shoes with Chinese characteristic were a festive, fashionable and convenient pair of Chinese style shoes that attracted everyone's attention. With old shoes replaced with new ones, what should we do with the old shoes?

With the rapid change of fashion trends, the number of old shoes is increasing imperceptibly. These "friends" that once accompanied us through spring, summer, fall and winter will become "disfavored shoes" once they lose their original appearance and function, and ultimately become a burden on the environment. According to statistics, up to 24 billion pairs of shoes are destroyed globally every year, with 90% of them either being land filled or incinerated. These discarded shoes are mainly composed of non-degradable materials such as rubber, plastic, and textiles, which take hundreds of years to decompose in the natural environment and release toxic substances during the process, causing pollution to soil and water sources. If incinerated, the resulting harmful gases increase emissions and cause pollution to the atmosphere. All of these factors will affect global climate change. If we can reasonably deal with these "disfavored shoes" and recycle them, it will not only reduce students' demand for new shoes but also reduce greenhouse gas emissions, alleviate global climate pressure, and contribute to a more sustainable circular economy.

In mid-September 2023, we conducted a survey on fifth-grade students of our school in this regard. The survey found that 48.51%~49.79% of the students tend to discard or store their shoes in the corner of the closet, while only about 30.64% of the students choose to reuse them. It can be seen that students' concept of recycling is still relatively weak. So, how can primary school students form the concept of recycling "disfavored shoes"?

Project Design

This student-centered project focuses on recycling and addressing climate change, based on the current situation of "disfavored shoes" recycling initiatives tailored to student. Through questionnaires, hands-on design and drawing, the project aims to enhance its interest and practicality. Its overall goal is to establish students' awareness of recycling, stimulate their competition and cooperation, and promote the concept of recycling in daily life. At the same time, it inspires students' potential and explores various new ways of recycling in life, making recycling a habit.

Project Implementation

1. Interesting Practice to Turn "Disfavored Shoes" into "Favored Shoes"

Through investigation, teachers found that many children like to use hand-painted graffiti, DIY handicraft decoration and other methods to transform "disfavored shoes" into their own favorite styles. Leveraging this insight, students pool their wisdom and turn the idea of "designing a unique and original unpopular shoe by themselves" into a reality.

1) Concept a figure and establish the consciousness of recycling

In October 2023, students started brainstorming. Drawing upon knowledge in various subjects, they used extracurricular magazines to find methods for designing composition and expanding their thinking. At the same time, they discussed and drew the design concept of the theme pattern that they liked (see Figure 1). They were extremely ingenious in elements, main colors, styles, etc. This small independent innovation design made the "disfavored shoes" become "favored shoes". These hands-on and personal participation experiences also sowed the seeds of recycling and caring for the environment in the hearts of students.

Figure 1. Design concept of "disfavored shoes"

2) **Paint a work and practice the concept of recycling.**

In late October 2023, the activities were in full swing, and the children's ideas for how to achieve economic and environmental protection as well as convenient recycling were endless.

"Teacher, I suddenly thought of it. To draw a pair of eco-friendly shoes, we can choose water-soluble pigments with strong solubility."

"Indeed. We should also use non-toxic and lead-free pigments or watercolor pens. Even if the shoes are thrown away, the pigments will not contaminate the environment."

"Besides the selection of paints, we can also match them by ourselves. Forexample, some students' shoes are not wearable, but their shoe laces are still good. We can match them to reuse and make them beautiful."

"This is a good idea. In addition, insoles can also be reused and matched into different shoes..."

The constantly inspired creative thinking and the consciousness of recycling enable students to come up with more and more creative ideas. Following approximately three weeks of design and practice, many pairs of "favorite shoes" that integrate aesthetic appearance and environmental protection concepts have finally been born (see Figure 2). They are changing their appearance and showing new life in the children's labor creation.

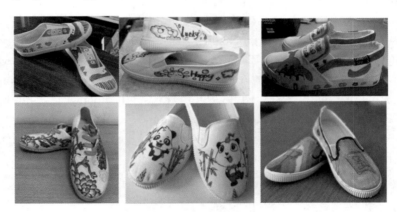

Figure 2. Reusable hand-painted shoes

3) **Make an exhibition to deepen the concept of recycling**

Through their own work and creation, students not only painted personalized and exclusive "favored shoes", but also held a model show to make "disfavored shoes" popular again (see Figure 3). In addition, students also created a personalized "disfavored shoes" showcase that integrates environmental protection, innovation, and aesthetics. With the continuous deepening of the event, students' concept of recycling has been continuously enhanced, and their desire for new shoes has also decreased.

2. Pioneering a New Path for Derivative Effects

Circulation is both the end and the beginning. Beyond academic pursuits, students not only discuss how to re-create neglected items and extend the service life of items, but

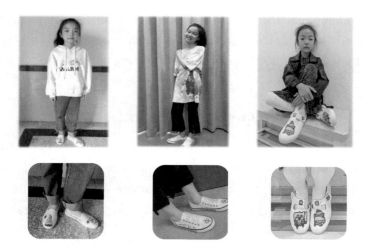

Figure 3. "Favorite shoes" model show

also subtly turn the awareness of recycling into a habit of life. The atmosphere of discussing the recycling of items in the class and campus is gradually becoming stronger.

"Beyond the realm of, besides reusable shoes, what else can be recycled to reduce greenhouse gas emissions and slow down climate change?" I asked during an extension class at the end of the "disfavored shoes" event.

A student replied, "I often see classmates tear unused draft paper into small pieces or into balls and throw them around, which is very wasteful. I think we can make a habit of cherishing paper. Don't tear paper randomly, and even if you've finished writing on it, don't throw it away. Recycle it and sell it together with domestic delivery boxes, newspapers, and other waste paper. This is also a way to reuse."

When other students heard that paper can be recycled, one of them immediately responded, "During online classes, when I printed exercises at home, my mother often printed on both sides for me, which also reduces paper waste and can be considered as recycling, right?" More hands were raised, and the thinking became more open. Some said that since their homes are close to school, they choose to walk or take the bus to reduce emissions from car exhausts. Some said that they need to break the habit of not turning off the lights to conserve electricity. Furthermore, students suggested recycle clothes that are no longer worn at home, turn "disfavored clothes" into "favored clothes", and create other new products. This not only improves innovation, but also protects the ecological environment.

With the continuous expansion of the use of thinking by teachers and students, our practical actions are constantly evolving. For example, teacher encourages students to attend lectures on local and societal environmental changes to bolster their environmental consciousness; teacher organizes book clubs focused on reading literature about the environment and climate; and teacher guides students in creating illustrations to raise awareness about climate change and the importance of recycling among community residents. Most encouragingly, an increasing number of families are getting involved, with more parents setting a positive example by practicing the principles of reuse and recycling.

Effectiveness & Reflections

1. Through Hands-on Practice, a Mindset for Circular Use is Fostered

Circular use is a philosophy, an attitude, and a way of life. It also reflects and affects the ability of every child to respond to climate change. Children in elementary school possess an innate curiosity for new things and have natural sensitivity and sympathy for environmental issues. However, due to the lack oftimely guidance, many children's environmental awareness is very weak: class napkins are everywhere, hand sanitizer in the toilet is misused to make bubbles, stacks of toilet paper are thrown away, and children have a fond of new clothes, toys, and so on. Through the "disfavored shoes" transformation activity, students understand the seriousness of environmental pollution and climate change issues and their impact on human life. They realize that everyone's environmental protection behavior is closely related to climate change and have a good sense of circular use.

In this process, we realized that interesting activities can greatly attract students' attention, stimulate their interests and curiosity, and improve their participation and enthusiasm. The "disfavored shoes" transformation activity is a combination of interest and practice, creating a positive and active atmosphere for students to feel happy while realizing that addressing climate change is a fun and meaningful action. When students are painting thematic patterns by themselves, they feel like "little designers"; when they become "little models", the feeling of being noticed and recognized bolster their self-assurance; when they practice the concept of circular reuse and continuously expand new channels on community platforms, the challenge and excitement enhance their ability to deal with complex problems and environments, as well as their psychological resilience.

2. Through Competition and Cooperation, Enhancing Team Cohesion and Collaboration Skills

Within this activity, both competition and cooperation play a very important role in education. In the process of concepting the figure, the competitive consciousness among students constantly stimulates their innovation, resulting in a number of unique and creative design drafts. This competitive consciousness enables them to better adjust their learning methods and strategies while discovering their inherent strengths and weaknesses. And every pair of "favored shoes" cannot be achieved without the reasonable division of labor and close cooperation of team members from design to realization. Thus, teamcohesion and collaboration ability can be improved.

3. With Diverse Stakeholder Engagement, the Concept of Circular Use Spreads Widespread

Students have experienced the significance of recycling in practice, while teachers and parents who participate in the project also reinforce their recycling consciousness.

Teachers should always permeate the concept of environmental protection in daily teaching, such as through Art, Music, Science and other courses, calling for the protection of the environment so that students can enhance their awareness and determination to protect the environment while being equipped with the idea of aesthetics. For example, in courses such as Chinese, Mathematics, English, Ethics and law, they can share stories about nature and environmental protection to inspire children to think about the feasibility of harmonious coexistence between human and nature.

The family is an important space to deepen the meaning of circular use. Parents' awareness of circular use profoundly affects their children's words and deeds, and parents' support also means support for their children's actions. Numerous parents are increasingly cognizant that they need to set a good example for their children through daily life when accompanying them in activities. For example, walking or cycling instead of driving, using reusable canvas bags instead of disposable plastic bags for shopping, and reusing a napkin many times by folding it. These small actions all affect children imperceptibly.

The development of this project not only allows students, parents, and teachers to have a deep understanding of the concept of reuse, but also provides a vivid practice for students to develop habits of reuse in their daily lives. Nonetheless, there are still some areas that need improvement in this project. For example, many students are enthusiastic about designing patterns and painting them by themselves, so most of the shoes they choose are white, which resulting in a predominant focus on monochromatic shoes. Although thisallows students to use their imagination in painting, it limits the reuse of multi-element objects in terms of matching. In project practice, we can more specifically encourage some students to engage in different elements such as cutting, pasting, and sewing to make major changes. Secondly, after the completion of the project, we can continue to follow up on the continuity of students' reuse concepts and the habit of practicality in their daily lives.

Circular use is a way of life, a social responsibility, and also a concept of sustainable development. Establishing the awareness of circular use and developing the habit of circular use will help students to understand and respond to climate change and improve their ability and quality to deal with it.

(Xiaoyan Shen, Teacher of Qianjing Primary School, Jinshan, Shanghai)

Research on the Recycling and Reuse of Milk Cartons

Research Background

In schools, students typically bring their own milk. Based on my observations, each class collects an entire trash bag of milk cartons daily. If a class on average produces 42 milk cartons per day, with the school having 36 classes, this results in 1,512 milk cartons daily. These cartons eventually end up in the trash.

I shared this observed phenomenon with the students in the class, and they conveyed considerable concern. Through further research, we learned that as living standards and consumption levels have risen, the number of people consuming milk has also increased. In 2022 alone, our country's dairy production reached 40.265 million tons, necessitating the use of billions of packaging boxes, most of which are buried or burned along with household waste. However, these milk cartons are predominantly composed of a large amount of non-degradable materials. Disposing of them as waste not only results in the squandering of resources and energy but also pollutes the environment and poses health risks to humans.

Everyone agrees that although small, the waste from discarded milk cartons cannot be overlooked. It is essential to raise public awareness about the hazards of improperly disposed milk cartons and encourage more people to join the environmental conservation effort. On October 10, 2023, the students proposed to undertake a study on the "Recycling and Reuse of Milk Cartons," aiming to maximize the true value of this waste.

Project Design

Before the class, teacher should research the materials, manufacturing methods, and reuse pathways of milk cartons. Based on this, the teacher guide students to:

Understand the types and materials of milk packaging boxes, realizing that despite their small size, milk cartons can cause significant environmental issues, teaching students to pay attention to environmental protection around them.

Use surveys, interviews, and field visits to deepen understanding of a project-based issue, enhancing students' interpersonal skills and teaching them to learn through cooperative learning.

Identify improper recycling and disposal methods for milk cartons, and based on their knowledge, propose reasonable optimization suggestions, establishing environmental protection and conservation awareness.

Promote research findings to draw more attention to the recycling and reuse of milk cartons and call for more people to become part of the environmental action.

Project Implementation

1. Initial Understanding of Milk Carton Types and Manufacturing Materials

From October 12th to 16th, 2023, students first conducted research on the sales of different packaged milk from the "Three Titans of China's Dairy Industry" — Bright Dairy, Mengniu, and Yili — on the online shopping platform Taobao's flagship store. They found that the sales of boxed milk from each brand far exceeded that of canned milk, and none of these flagship stores sold bagged milk. It seems that the demand for boxed milk is indeed very high.

Why, then, is the recycling rate for these small milk cartons so low? What non-degradable materials do they contain? Through online research, the students learned that the most commonly used types of milk packaging on the market are "Tetra Pak" and "gable top" cartons. Among them, "Tetra Pak" is divided into "Tetra Brick" and "Tetra Prism." Therefore, they conducted a comparative analysis of their structures and shelf lives.

The students discovered that these packaging types all use plastic and aluminum. In the packaging industry, such as milk cartons, there is a technical term called "composite paper packaging," which is a six-layer composite structure made from pulp, polyethylene plastic, aluminum, as well as printingink and coatings. What constitutes polyethylene plastic? By consulting a science teacher, everyone learned that its main component is polyethylene, which is derived from petroleum and is a petrochemical product. Aluminum, being a heavy metal, if buried, cannot decompose in the soil and can cause significant pollution damage to the land, livestock, and water sources.

As the research deepened, they discovered that there are already several technologies in our country specifically aimed at recycling and reusing such composite paper packaging. The more mature recycling technologies include hydro pulping technology, plastic-wood technology, color board technology, and aluminum-plastic separation technology. Simply put, these professional recycling and processing methods convert waste milk cartons into pulp, plastic particles, and aluminum powder, turning them into raw materials for other industries, thus achieving 100% recycling. However, such a composite structure is difficult to process as a single material. For us, lacking professional technological support, it is fundamentally impossible to meet the recycling requirements.

2. Student Survey on Local Handling Practices of Milk Cartons

On October 20th, 2023, the students decided to use people around them as the subjects of their investigation. They worked with their teachers to create a survey questionnaire, which they then distributed through their class WeChat group and Moments on WeChat.

From this survey, the students discovered:

(1) Considering the quantity of milk consumed daily by each household, the number of milk cartons generated is substantial. The waste from these cartons is significant and warrants our attention and research.

(2) Although most people are aware that carelessly discarding milk cartons impacts the environment, their understanding is superficial. There is a need to intensify educational efforts to make everyone truly understand the hazards of discarding milk cartons, thereby promoting the recycling and reuse of these cartons.

(3) Currently, the promotional methods for recycling and reusing milk cartons are varied and lack a cohesive strategy. Identifying which promotional strategies to use for different demographics and the platforms to leverage for advocacy are key aspects that need active exploration in our research.

To better understand the community's views on the reuse of milk cartons, with the help of teachers, an interview outline was also designed. Between October 25 and 27, 2023, they went into various communities to conduct more detailed interviews (see Figure 1), with the goal of thoroughly understand the situation regarding the reuse of milk cartons. They gathered the following information:

The demand for milk is high, and most people opt for boxed milk, with an average of 2 to 3 cartons generated per family. Furthermore, after consuming the milk, most people tend to throw the cartons away directly, with little to no consideration given to recycling or reusing these cartons.

Figure 1. Students conducting interviews in the community

3. Collective Wisdom: Crafting Innovative Recycling and Reuse Tips for Milk Cartons

In response to the issues highlighted by the survey, from November 6 to 15, 2023, the students conducted a comparative study on the treatment methods of milk cartons both domestically and internationally, gaining insights into:

(1) International practices: Leveraging online resources, the students identified several countries whose methods of handling milk cartons are noteworthy. For example, in Barcelona, Spain the Stora Enso Group utilizes advanced pyrolysis technology to sequentially decompose the fibers, plastic, and aluminum in old beverage cartons like milk cartons. In Japan, milk is commonly packaged in square paper cups; these cups are made from special high-quality paper, resulting in a higher recycling rate. Additionally, the outer packaging of milk cartons in Japan includes clear labels instructing how to dismantle the cartons.

(2) Domestic practices: Since 2002, enterprises in Zhejiang, Guangdong, Shandong, Beijing, and Fujian in China have emerged that use waste milk cartons and other paper-plastic-aluminum composite packaging materials for papermaking. In Taiwan, China, a "four-in-one" system is currently in place, which integrates the efforts of residential community residents, recyclers, local administrative departments, and a recycling fund to carry out resource recycling and waste reduction initiatives. At the 2010 Shanghai World Expo, recycled Tetra Pak packaging was transformed into over a thousand stylish, simple, and practical plaza seats placed in the event venues for visitors to rest. In 2019, with the formal implementation of waste sorting in Shanghai, Bright Dairy began a nationwide public welfare campaign to recycle milk cartons. At the 10th National Flower Expo in 2021, these collected milk cartons were turned into flower-viewing benches.

So, are these recycling and handling solutions worth learning from and adopting? The students conducted a comparison of the advantages and disadvantages and reached the following consensus:

Firstly, increase centralized recycling points for milk cartons. Currently, the disposal of milk cartons often involves throwing them away casually or mixing them with other household trash. If we could establish centralized recycling points like those in Japan, the processing would become much more convenient and efficient.

The students discovered that Bright Dairy has already launched a "Green Recycling Campaign for Milk Cartons," setting up over 3,000 collection sites nationwide. However, these are mostly located in first- and second-tier cities such as Shanghai, Nanjing, and Jinan, with fewer in third- and fourth-tier cities. The distances between sites are considerable, and there is a significant disparity between urban and rural areas. For example, the nearest recycling point to Changzhou is in Yixing. Therefore, the students suggest setting up recycling points in nearby residential areas or schools with high milk consumption to facilitate centralized disposal of waste milk cartons.

Secondly, optimize the design of milk packaging. The students researched several commonly used milk cartons and found that one product's packaging was printed with "Do not litter empty packages, keep the environment clean," while another fresh milk product noted "Recyclable material" and "Please clean, flatten, and deposit in recyclable container" post-consumption. Comparing the two, the latter places greater emphasis on the recycling and reuse of packaging. Instead of pursuing form over function, integrating educational elements into the design of milk cartons, such as the benefits of recycling and methods of milk carton recycling, could enhance people's awareness and inclination to recycle.

Thirdly, strengthen the promotion of recycling and reuse of milk carton packaging. Currently, there is a general lack of awareness about this topic. Only by making people fully aware of the hazards of indiscriminately discarding milk cartons and understanding the importance of recycling can they consciouslyadopt recycling measures. This necessitates intensifying promotional efforts and expanding the scope of these efforts. The students believe that promotion could start within classes, extend to grades, and eventually reach the entire school and the broader community.

4. Transforming Waste into Treasure: Revitalizing Milk Cartons

On November 24, 2023, after a thorough discussion, the students unanimously agreed that "turning waste into treasure" is the best approach to address the recycling and reuse of milk cartons. This led to the launch of the "Transform-a-Carton" series of activities.

Firstly, during class meetings, students learned and mastered the steps for cleaning discarded milk cartons, summarized into four simple steps: cutting open the used milk carton, washing it, letting it dry, and then disposing of it at the recycling station.

During this process, an issue arose: some students didn't drink all the milk, causing residue to spill when cutting the carton. Through online learning and practice, a nifty trick was discovered: after finishing the milk in a Tetra Pak, unfold the four corners of the carton and roll it from bottom to top, which helps push out the remaining liquid. This leftover liquid, once diluted with water, can even be used to water plants!

Secondly, students crafted recycling bins for milk cartons and placed them in the classroom's sanitary corner.

Thirdly, a system for claiming discarded milk cartons was established. After discussions, it was decided that based on the students' usual small roles and performances, teachers and group leaders would award "little red flowers." Collecting five of these flowers would allow a student to exchange them for a milk carton with the station manager, Senior Gao.

Fourthly, a "Transform-a-Carton" environmental art competition was initiated. Students could claim a milk carton based on the established claiming system and then redesign it (see Figure 2).

Figure 2. Students transforming old milk cartons into treasures

5. Promotion and Mobilization: Small Hands Pulling Big Hands to Build an Environmental World

From November 27 to December 22, 2023, the "Transform-a-Carton" campaign received enthusiastic responses, significantly boosting the students' confidence in getting more people to join the milk carton guardianship initiative.

On January 12, 2024, with the help of their teachers, the students engaged with the head of the school's "Red Scarf TV," using the platform to educate all students about the dangers of littering milk cartons. They disseminated proper milk carton recycling methods and called for more student participation, urging them to involve their families in the "Recycle Milk Cartons, Join Hands in Environmental Protection" campaign.

To further encourage people to join the milk carton recycling and environmental protection effort, the students also proposed to their local community property management to add more recycling points. They suggested implementing an exchange system where, for example, collecting 50 milk cartons could be exchanged for a box of fresh milk, motivating residents to actively participate.

Effectiveness & Reflections

This project on recycling old milk cartons has generated significant interestacross schools, homes, and communities.

Firstly, there has been an enhancement in students' overall competence. Every student was fully engaged in this research, contributing ideas and strategies for the smooth progression of the activities. The students' communication and collaboration skills have improved significantly. Moreover, through learning, the students have gained a deeper understanding of the hazards associated with discarded milk cartons, mastered the proper methods of cleaning them, and developed a greater awareness of how climate change fundamentally affects human survival and development. We need to be observant, think deeply, and use our own strengths to support new developments in environmental conservation.

Secondly, the recycling rate of used milk cartons has increased. Following the students' lead, many classes initiated a "Milk Carton Transformation" activity, which not only increased the recycling rate but also significantly reduced the workload of the cleaning staff. The transformation of waste milk cartons has become a beautiful feature within the campus.

Thirdly, the collaboration between home, school, and community has painted a united front in waste sorting. Through the "small hands pulling big hands" approach, we have called on more people to join the recycling and reuse of old milk cartons, enhancing public awareness of environmental conservation.

However, there are many areas where this project still requires improvement and deepening. For example, in the later stages of the research, students went into the community to spread knowledge about milk cartons and encouraged everyone to master the correct recycling methods. Yet, this initiative has remained mostly at the promotional stage without further creating activities that would deeply root the "rebirth journey" of milk cartons. Moreover, the project primarily focused on research and classroom practices, not making full use of surrounding resources. Moving forward, we intend to arrange field visits for students to professional milk carton recycling facilities to learn about advanced recycling technologies. Through school-corporate partnerships, weaim to launch more activities and contribute to education on climate change.

(Lin Su, Director of Student Affairs, Changzhou Economic Development Zone Primary School)

Implementing Climate Change Education Practice Based on STEM Integrated Activities

Research Background

Since the 18th National Congress of the Communist Party of China, China has incorporated the construction of ecological civilization into the overall layout of the "Five in One" initiative and placed it in a prominent position in overall work. The low-carbon pilot city in China, Shenzhen, Guangdong, has also achieved great success in soil and water conservation, sewage treatment, and water purification work, which has important reference significance for the coordinated governance of domestic water pollution and climate change mitigation, and contributing to moving towards a future "carbon neutral" society.

To promote harmonious coexistence between humans and nature, and promoting the construction of a global community of life, Bao'an Middle School (Group) Experimental School in Shenzhen, Guangdong Province (hereinafter referred to as "Bao'an Middle Experimental School") has become one of the initiators of the "National Climate Change Education Research Alliance" to actively promote collaborative education on climate change among schools, families, and society, discuss educational strategies to address the challenges of climate change, and seek the path of harmonious coexistence between humans and nature. In the current context of educational reform, how to effectively enhance science education has become a hot topic of work for frontline science and technology teachers. How to carry out climate change education practice based on STEM comprehensive activities has become an important task for the interdisciplinary team of climate education at Bao'an Middle Experimental School.

In order to establish the awareness of young people in coping with climate change, cultivate their scientific knowledge, methods, attitudes, spirit, andqualities in dealing with climate change issues correctly, and encourage more young people to participate in climate action, Bao'an Middle Experimental School is based on the characteristics of undergraduate innovation and the "1+3+5 Framework" climate change education program, namely: ① 1 Goal—establish a school with distinctive science and technology innovation features focused on climate change education. ② 3 Brands—development of school-based climate change readers, climate change learning month in October each year, and annual climate change education achievement exhibition in April. ③ 5 Dimensions—Society: introduce social education resources on climate change; School: build a climate change education science popularization base; Parents: establish a home-school education alliance for climate change; Teachers: develop school-based climate change education courses and readers; Students: encourage everyone to strive to become a climate change science popularization advocate. Facing elementary school students, Bao'an Middle Experimental School actively explores STEM planting and energy-saving and emission reduction project research based on biological gardens, agricultural fields, and greenhouse greenhouses; Targeting junior high school students, carry out technological inventions and project practice activities related to new energy, energy conservation and emission reduction, guide students to pay attention to the themes of soil and water conservation and water resource protection, practice meteorological climate and dual carbon theme projects, strengthen their understanding of climate change and ecological civilization, and participate in low-carbon environmental protection actions.

The distinctive science and technology education courses of Bao'an Middle Experimental School are divided into three series: science popularization courses, science action courses, and science superpower courses. Theinterdisciplinary team on climate change education has set up relevant project activities in all three dimensions and continuously transformed and optimized existing maker spaces. Among these, the interdisciplinary team on climate change education focuses on scientific research activities such as March Arbor Day, April Earth Day, May Labor Practice Month, June World Environment Day, and September and October energy conservation and emission reduction by setting up different projects. The information technology course introduces "future ecological maker education" that includes energy conservation and environmental protection, programming, 3D modeling, artificial intelligence, and other content. Climate change education related activities have also been integrated into subject expansion courses such as primary and secondary school clubs and literacy courses. For example, practical activities such as meteorological and climate science popularization and waste management have been added to club activities such as "Little Inventors" and "Little Experimenters" in primary schools; The "Science Box" club courses in junior high school have introduced courses in disciplines such as water technology, life sciences, engineering and technical sciences, and resources and environmental sciences. In competitive training courses, such as artificial intelligence exploration, AR/VR science fiction experiences, climate STEM projects, dual carbon projects, biodiversity projects, and whitelist competition training, a variety of climate change education projects and activities have also been added.

Project Design

STEM education advocates activities based on life experiment tasks, requiring students to create a series of physical models and explore, apply, and revise knowledge in fields such as science, engineering, and the environment during the production process. The materials used in STEM experimental activities are simple, but they allow students to experience the cognitive and operational approaches of scientists and engineering technicians during theactivities. Based on the STEM education philosophy, our school has launched the "Exploration of Water Quality Differences and Research and Development of Water Purifiers" project, promoting students to design and complete experiments by themselves, fully unleashing their innovative spirit, and closely linking the knowledge and skills they have learned with real-life situations, so that technological activities can flourish and bear fruit on campus, at home, and in society.

Table 1. Project design table for "Exploration of Water Quality Differences and Development of Water Purifiers"

Grades 3 – 9	Total class duration	6×60 minutes
Project Design Framework:		
Learning periods	Key points and difficult points	Methods
Class period 1	The concept of water, the causes and hazards of water pollution and the prevention and control measures, the indicators of water quality testing, the concept and identification of soft and hard water, and how to soften hard water.	1. The teacher explains vividly. 2. The teacher answers questions in a timely manner.
Class period 2	1. Calcium and magnesium detection reagents were used to identify soft and hard water. 2. Softening tap water by using three materials: water softening resin, activated carbon and molecular sieve, to explore and compare the softening effects of the three materials on tap water.	1. The teacher demonstrates and explains. 2. The teacher answers questions in a timely manner. 3. The teacher uses PPT demonstration.
Class period 3	The water quality test box is used to test different brands of drinking water commonly found on the market, mainly including the determination of residual chlorine content, solid dissolved matter content and whether it is mineral water.	1. The teacher demonstrates and explains. 2. The teacher answers questions in a timely manner.
Class period 4	Understand the principles of soil and water conservation and water purification, conduct product research on water purifiers on the market, understand the composition of water in nature, the necessary materials and principles required for impurities, and master the operation of filtration.	1. The teacher explains the principles of water purification. 2. The teacher introduces common water purifiers.

(Continued)

Class period 5	Under the guidance of the engineering design process of STEM education, the STEM experimental team carried out the project practice and product optimization of self-made simple water purifiers.	1. Students operate engineering project practice. 2. Students test and optimize water purification equipment.
Class period 6	1. Students prepare their presentations in advance, conduct project reports, and showcase their water purifier products. 2. Students improve their presentation and make their PPT slides clearer and more visualized.	1. Students debriefing. 2. The teacher guidance improvement.

Objectives:

(1) Class period 1

Be able to understand the concept of water, the causes and hazards of water pollution and the prevention and control measures, the indicators of water quality testing, the concept and identification of soft and hard water, and how to soften hard water.

(2) Class period 2

①By exploring whether different water samples are soft water, the calcium-magnesium detection reagent was used to identify soft and hard water.

②By exploring the softening method and effect of tap water, using three materials, soft water resin, activated carbon and molecular sieve, the tap water was softened, and the softening effect of the three materials on tap water was explored and compared.

③Students master the basic points of reporting PPT and do part of the presentation PPT.

(3) Class period 3

①Through the experiment of "exploring the difference in water quality of different brands of drinking water", students learn to use water quality testing boxes to test different brands of drinking water commonly found on the market, mainly including the determination of residual chlorine content, solid dissolved matter content and whether it is mineral water.

②Read the experimental data, record the experimental phenomena, analyze and summarize the experimental results.

③Students master the basic points of reporting PPT and do part of the presentation PPT.

(4) Class period 4

Students in the class form a STEM experiment group of 3-4 people, and use one or two pieces of paper to summarize the learning results of the group on the concepts or principles related to the experimental theme, as well as the literature consulted and cited, which should be accompanied by 2 pictures of the research content. The teacher should provide students with the path and method of data review, and organize students to hold a data review summary report meeting, so that students can learn to collect information, learn to ask questions, and learn to write data review reports

(Continued)

> (5) Class period 5
> By consulting the literature, we know the importance of soil and water conservation and artificial water purification, understand the principle of water purification, and conduct product research on water purifiers on the market. I know that the water purifiers commonly used in life mainly include pre-filter products, PP cotton filtration products, ultrafiltration membrane filtration products, reverse osmosis membranes, and nanofiltration membrane filtration products. And compare their roles, advantages, disadvantages. Carry out STEM R&D project activities for homemade simple water purifiers.
> (6) Class period 6
> ①Students make defense PPT in advance, make project reports and water purifier product displays.
> ②Through the guidance of teachers, students can realize and correct their own shortcomings, so that their report PPT is completer and more exquisite.

Project Implementation

1. Students Understand and Master the Basic Concepts of Water

In the inaugural lesson, the teacher explains the first scientific knowledge point-an overview of water, to enhance students' understanding of water. Firstly, the teacher introduces the issue of water pollution, guiding students to understand the hazards of water pollution and the importance of prevention and control work in addressing climate change. They also understand the concept of "prevention first, combined prevention and control" and the concept of comprehensive governance.

Moving forward, teacher and students will jointly explore the physical, chemical, and microbial indicators of water quality. On one hand, students make judgments about water quality based on their everyday experiences. This includes: ①Visual inspection: Observing water in a glass against light to detect any suspended particles; after three hours, checking for sediment at the bottom. ②Smelling: Smelling water from the tap to detect the presence of chlorine (bleach); a strong chlorine odor indicates excessive residual chlorine. ③Tasting: Drinking boiled water to check for a chlorine taste; a strong chlorine taste suggests excessive residual chlorine. ④Observation: Brewing tea with tap water and observing if the tea darkens overnight; darkening indicates high levels of iron and manganese. ⑤Drinking: Drinking plain water to detect any astringent taste, which suggests high water hardness. ⑥Inspection: Examining appliances like water heaters and kettles for scale buildup, which also indicates high water hardness. On the other hand, under teacher guidance, students use specialized equipment and reagents to test various types and brands of water quality, understanding key indicators such as residual chlorine, water acidity/alkalinity, zinc content, and total dissolved solids.

Finally, students use methods such as soap water, heating, and evaporation to distinguish between "soft water" and "hard water", and convert "hard water" into "soft water" through laboratory boiling and distillation.

2. Conduct Experimental Exploration to Understand Water Quality Differences

Students use one class hour to complete several experimental explorations under the guidance of the teacher. The first objective is to determine if different water samples are soft water; The second is to explore the softening methods and effects of tap water.

Building on this foundation, teachers set up real-life scenarios, stimulate students to think, and propose the question that needs to be studied—"Exploring the water quality differences of different brands of drinking water." Teachers and students jointly analyze the physical quantity of the experimental design-independent variable: different drinking water; Dependent variables: residual chlorine, conductivity (minerals), solid dissolved matter. Student project teams design experimental plans to explore the differences in water quality among different brands of drinking water.

3. Design an Experimental Plan for a "Self-made Water Purifier"

This implementation phase mainly consists of the following three aspects.

Firstly, the teacher explained the purpose of the experiment: Water scarcityhas become one of the most significant problems in the world today. We have previously learned concepts or principles such as filtration, adsorption, water purification, and water resource protection. We have acquired knowledge of the composition of water in nature, the necessary materials and principles for impurity removal, and have also mastered the operation of filtration. Accordingly, the primary objective of this experiment is to use our brains and hands to create a simple water purification device.

Secondly, students should master the literature review technique: form STEM experimental groups in groups of 3-4 people. Briefly introduce the learning achievements of our group on relevant concepts or principles in a short article, refer to and cite literature materials, and attach two pictures of the research content. During this process, the teacher provides students with a path and method for accessing information, organizes students to hold a summary report meeting, guides students to learn how to collect information, learn to ask questions, and learn to write a report on accessing information.

Thirdly, students design experimental exploration plans: in small groups, collaborate to study the principle, composition, and function of water purifiers, and propose a production plan for water purifiers. Conduct brainstorming to explore feasible solutions. Based on this, we will start making prototypes and evaluate them through experimental exploration, continuously improving them. Finally, present and exhibit.

4. Implement the Experiment of "Self-made Water Purifier" and Showcase the Experimental Results

Following the engineering design process, the STEM group embarked on a project to develop a simple homemade water purifier. The procedure was as follows: ①Research filtration and water purification knowledge; ②Brainstorm ideas on how to make a water purifier; ③ Create a design diagram for the water purifier; ④ Start building an environmentally friendly water purifier; ⑤ Consult relevant science books or other materials to test the model's accuracy and effectiveness, recording the comparative results; ⑥ Evaluate the model, analyzing the components and functions of each part of the water purifier; ⑦ Determine an improved design plan for the device; ⑧ Re-evaluate and assess the improved design; ⑨ Present the group's learning outcomes.

Finally, each group must submit the following outcomes within the timeframe specified by the teacher and present them as a group. The presentation should include: ① A 1-2 page paper summarizing the research findings on the principles of water purification, with cited references, and must include two images; ② A design diagram for an environmentally friendly water purifier; ③ A record, analysis, and explanation of the water purifier's performance testing; ④ A summary of the experimental tasks and practical process, which should cover the objectives of the tasks, a brief explanation of the testing process, and an interpretation of the results; ⑤ A "Self-Evaluation Scale" and "Reflection Record." The STEM project group is required to bring the homemade water purifier back to school, prepare it in the laboratory with the lower end connected to a beaker for collecting water; pour the wastewater into the water purifier along a glass rod for observation, and report on the water purification effectiveness, improvement strategies, and lessons learned (see Figure 1).

Figure 1. Environmental Protection Defender Homemade Water Purifier

Effectiveness & Reflections

The "Exploration of Water Quality Differences and Research and Development of Water Purifiers" project is based on the fields of engineering technology, natural sciences, and social sciences. Teachers guide students to carry out innovative research projects on how to improve water environment quality, strengthen water resource conservation and ecological protection, and enhance wastewater treatment capabilities. It can effectively implement the pertinent directives of the *Water Pollution Prevention and*

Control Action Plan issued by the State Council, which includes "strengthening publicity and education, and incorporating water resources, water environment protection, and water situation knowledge into the national education system". It has a positive promoting effect on improving the awareness and technological innovation ability of young people in ecological environment protection, guiding them to actively participate in water ecological environment protection and water conservation actions. In the next stage, we will guide students to conduct scientific and technological practice surveys on topics such as water resource conservation and protection, water ecological restoration, water environment status, and public awareness of water environment through the Chinese Youth Science Survey Experience Activity.

In the past six years, Bao'an Middle Experimental School has been promotingyouth action to address climate change. We have encouraged students to adopt different methods such as writing relevant articles or personal speeches, and have adopted low-carbon behavior education in schools, families, and communities to spread awareness and understanding of climate among young people; We also encouraged students to propose solutions, measurement models, or physical inventions targeting climate change and low-carbon through scientific inventions, programming, algorithm models, and other methods related to environmental protection. As the masters of the future world, teenagers are also an important driving force in addressing climate change. Climate change education for young people is also of utmost importance in sustainable development education. It not only helps young people understand and respond to the impact of climate crises, but also equips them with the knowledge, skills, values, and attitudes needed as agents of change.

[Caiying Chen, Chaoxia Long, Yufang Zeng, Lin Jiang, Xiaofeng Cai, Teachers of Bao'an Middle School (Group) Experimental School, Shenzhen, Guangdong; Jianfen Lin, Teacher of Shekou School, Nanshan, Shenzhen, Guangdong]

Design and Implementation of the Multidisciplinary Project "Regeneration of Coffee Grounds"

Research Background

Shanghai is the leading city in China in terms of coffee consumption. According to Meituan, up to March 2023, Shanghai has 8,530 cafes, with 3.45 cafes per 10,000 people. Estimates indicate that the annual per capita consumption of coffee in China is about 10-12 cups, and the annual per capita coffee consumption in Shanghai exceeds 20 cups, double that of the domestic average. According to this calculation, Shanghai produces an average of 15 tons of coffee grounds every day. How to use these coffee grounds to reduce the pressure of garbage disposal is not only a practical problem that needs to be faced, but also a research topic worth exploring.

Since 2018, Shanghai No. 1 High School has launched the exploration on how to cultivate the global competence of middle school students, and teachers of various disciplines in the school have jointly focused on "global issues" under the 17 overall sustainable development goals of the United Nations, such as renewable resources, responsible consumption, climate action, etc., and tried to explore thematic multidisciplinary integrated development projects, by setting up a "9 + 1" training workshop within their self-developed curriculum. Among them, the environmental protection workshop is composed of teachers of chemistry, biology and geography, which aims to focus on the establishment of students' awareness of environmental protection, the growth of real-life experience, the improvement of research skills and the enhancement of problem-solving abilities.

Project Design

1. Theoretical Basis and Overall Thinking

The design of the "Regeneration of Coffee Grounds" project follows the educational philosophy of science, technology, society, environment, and keeps to the idea of "Discover a problem→Dose a questions→Analyze the issue→Solve the problem→Elevate the finds" (see Figure 1).

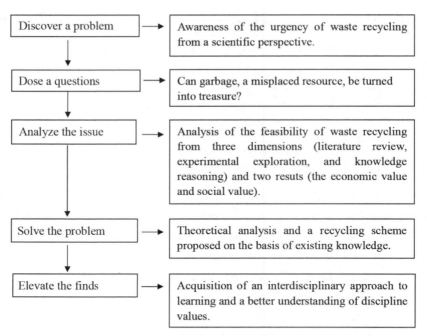

Figure 1. Design idea of "Regeneration of Coffee Grounds" project

2. Curriculum Theme, Schedule, and Content Organization

The "Regeneration of Coffee Grounds" project consists of several sub-courses. Taking the "Coffee Story" sub-course as an example, the course schedule is shown as Table 1:

Table 1. The course schedule of "The Story of Coffee".

Topic	Number of lessons	Activities
1. Classification and data evaluation of common coffeebrands commercially available	3	Purchasing common commercially available coffee beans in groups, classifying and evaluating them, collecting data, writing research reports and presenting reports.
2. Classification and recycling of coffee packaging	3	Sorting the packagings according to material classification, studying the recycling technology of different materials and testing them in practice.
3. The evolution of coffee-making and the practical production	2	Baristas from Starbucks are invited to the school to teach different coffee making methods, and students can actually experience the making of a variety of fancy coffees.

(Continued)

Topic	Number of lessons	Activities
4. Testing soil cultured by using coffee grounds	4	Experimental research on plant-nutrient soil from coffee grounds, including the measurement of coffee ground components, bacterial culture, and ratio planting experiments.
5. Production and testing of odor improver based on coffee grounds	2	Designingand making sachets to improve odors in refrigerators, shoe cabinets, classrooms, etc. by using the aroma of coffee grounds.
6. The origin of coffee andsome legends about coffee	2	Exploring the history of coffee and sharing interesting stories about coffee.

In particular, the order of the above topics is not fixed. Adjustment is allowed each semester according to different students' needs, including their motivation and participation. The length of the lesson is also an approximation, as some chapters are taught simultaneously with different student groups by relevant multidisciplinary teachers, it is difficult to accurately count.

Project Implementation

The theme "4. Testing soil cultured by using coffee grounds" is used as an example to illustrate how to achieve multidisciplinary integration in this project.

During this instructional period, teachers from three different disciplines: Chemistry, Biology, and Geography gave instructions in different laboratories. The class was divided into three groups, and each group took turns to do one of the three experimental projects. As shown in Figure 2, the first group of students carried out the chemical experiment A-DIS determination under the guidance of the chemistry teacher, the second group of students carried out the biological experiment B-microbial culture under the guidance of the biology teacher, and the third group of students conducted the geography experiment C-coffee grounds under the guidance of the geography teacher. Upon completion by each group students completed the experiments and gatheredcertain data, they moved on to the next lab.

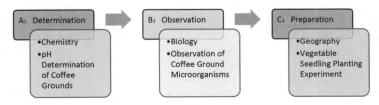

Figure 2. The task flow of the three disciplines of Chemistry, Biology and Geography

This arrangement not only ensures proper teacher guidance, but also improves the quality and efficiency of teaching. From the perspective of experimental operation, the experiment of bacterial culture and vegetable seedling planting often needs to last for several days or even weeks, and students need to go to the laboratory frequently to do some measuring and recording. From the perspective of student participation, in each class, each student group is capable of conducting all the experiments from multiple disciplines and record data respectively, which improves the efficiency. From the perspective of teaching, each lesson can be focused on the same project in different groups, which not only improves the guidance effectiveness but also reduces the burden.

Here are three examples of students' work sheets in this topic.

(1) Student Activity A: Garbage Sorting Game, pH Determination of Coffee and Coffee Grounds (see Figure 3, Chemistry Teacher's Guidance)

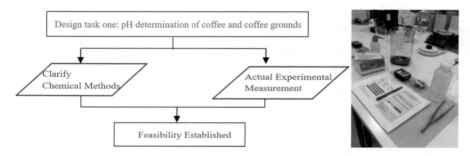

Figure 3. Determination of the pH value of coffee grounds

Student Activity Record Sheet-A

【Activity A-1】What barbage category(categories) can coffee and coffee-related items be sorted into?

Can	Coffee Bag	Glass Bottle	Plastic Bottle	Plastic Stirrer	Wooden Stirrer
Coffee	Creamer, Sugar	Coffee Beans	Paper Bag	Paper Cup	Expired Coffee Powder

【Activity A-2】Method for determining the acidity and alkalinity of the solution _____

【Activity A-3】Are coffee grounds acidic or alkaline?

Coke	Soda water	Milk	Sugar	Freshly grounded coffee	Coffee grounds

Refer to Scheme 1-pH Test Strip Method:

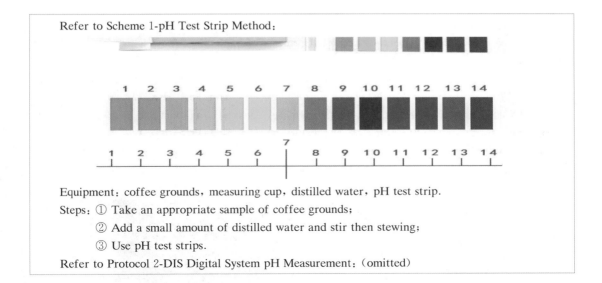

Equipment: coffee grounds, measuring cup, distilled water, pH test strip.
Steps: ① Take an appropriate sample of coffee grounds;
② Add a small amount of distilled water and stir then stewing;
③ Use pH test strips.
Refer to Protocol 2-DIS Digital System pH Measurement: (omitted)

(2)Student Activity B: Cultivation and Observation of Coffee Ground Microorganisms (see Figure 4, Biology Teacher's Guidance)

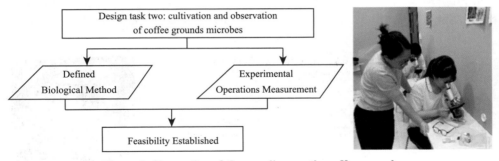

Figure 4. Observation of the mycelium on the coffee grounds

Student Activity Record Sheet-B

【Activity B-1】Agar plate making (preliminary preparation)

Two and a half teaspoonsof sugar, two and a half tablespoons of agar powder, two cups of low-salt beef broth. Two glasses of water. Mix them up in a pot and heat the mixture to boiling. Pour it into a Petri dish and cool it to 60 degrees Celsius. Store it in the refrigerator for later use.

【Activity B-2】Vaccination

Inoculate a small amount of coffee grounds on agar plates with a cotton swab and place it in a warm and humid (20℃—30℃) environment (incubator) for incubation.

【Activity B-3】Making temporary slide

Take a glass slide, place a waterdrop in the center, pick a little hyphae produced by coffee grounds with a dissecting needle or tweezers and put it in the water droplet. Then, cover the coverslip, and make a temporary load, ready for observation with a microscope.

【Activity B-4】Digital microscope observation

① Place a mobile phone on the eyepiece of the microscope and adjust the brightness until a bright white circular field of view appears.

② Place the temporary slide on the stage and adjust the distance until the mycelial image appears in the view.

③ If you look closely at the tip of Penicillium hyphae, you can see a broom-like structure, and there are also clusters of spores on the branches. The mature spores are blue-green.

【Activity B-5】Take photos and draw pictures

Use a pencil to draw the hyphae and spores of the mold you see with a microscope on the work sheet.

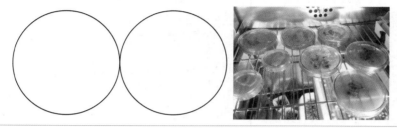

(3) Student Activity C: Preparation of Nutrient Soil with Coffee Grounds and Vegetable Shoots Planting Experiment (see Figure 5, Geography Teacher's Guidance)

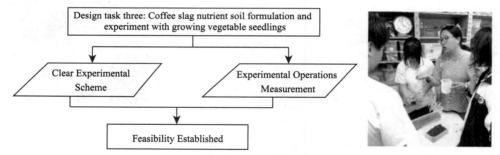

Figure 5. Instructing students to transplant vegetable shoots

Student Activity Record Sheet-C

【Activity C】Observation record of coffee grounds fertility experiment

	Pot 1 Coffee grounds, 100% Nutrient soil, 0	Pot 2 Coffee grounds, 75% Nutrient soil, 25%	Pot 3 Coffee grounds, 50% Nutrient soil, 50%	Pot 4 Coffee grounds, 25% Nutrient soil, 75%	Pot 5 Coffee grounds, 0 Nutrient soil, 100%
Number of plantings					

(Continued)

	Pot 1 Coffee grounds, 100% Nutrient soil, 0	Pot 2 Coffee grounds, 75% Nutrient soil, 25%	Pot 3 Coffee grounds, 50% Nutrient soil, 50%	Pot 4 Coffee grounds, 25% Nutrient soil, 75%	Pot 5 Coffee grounds, 0 Nutrient soil, 100%
Number of emergent seedlings					
Emergence rate					
Growing trends					
Grow__					
Grow__					
Grow__					
Grow__					
Grow__					
Grow__					
…					

Equipment: coffee grounds, seedling pots, nutrient soil, green vegetable seeds, watering cans, plastic wrap.

Steps: Soak the green vegetable seeds in water for 6-8 hours before the activity.

① In different color seedling pots, different proportions of coffee grounds and nutrient soil are loaded.

② After wateringthem thoroughly, put 5-6 green cabbage seeds in the seedling pots and cover them with thin layers of fine soil.

③ Use a watering can tomaintain appropriate moisture and cover the pots with plastic wrap.

④ Observe the germination and growth of seeds in different seedling pots, take pictures in time, and complete the experimental report.

Effectiveness & Reflections

This project starts from the real problemsin students' real life and encourages students' active participation, which is not only practical but also very interesting, arousing students' interest. The practice of such multidisciplinary integrated curriculum activities is also conducive to teachers to break he boundary of a single discipline and enhance the cooperation among teachers from different disciplines.

However, there are still some limitations in the implementation process. For example, only one class period (40 minutes) is assigned to this project per week, but the state of plants and microorganisms changes every day, which requires teachers and students to find time to measure and record them accordingly. Therefore, it is often difficult to meet high standards in terms of the frequency and accuracy of manual recording. Moreover, It is hoped that more technical devices will be available in the future to make up for the short of hands to keep records.

Are coffee grounds really garbage? Not really. Coffee grounds can alsopossess the potential to be repurposed into a new type of biofuel, a deodorizing and dehumidifying agent, and even a beautiful cup... The realization of these ideas requires the unremitting efforts of educators to cultivate more qualified futurists who pay attention to global issues, adhere to the concept of green development, and have climate change awareness.

(Weiwei Gu, Teacher of Shanghai No.1 High School)

Exploring the Relationship Between Wampee Fruit and Climate to to Revitalize the Countryside

Research Background

Zinan Primary School is located in a village in Chancheng District, Foshan City, Guangdong Province, and boasts rich local resources, prominently featuring the renowned Zidong Wampee. Since ancient times, households in Zidong Village have planted wampee trees in front of and behind their houses. Zidong Wampee is renowned for its large size, hairless skin, juicy pulp, small core, crisp skin, sweet taste, and refreshing flavor. This is an excellent entry point for climate change education. Commencing in February 2024, when the new semester begins, we will leverage local characteristics and activate resources to carry out comprehensive practical activities in classes, exploring the relationship between wampee and climate, and supporting rural revitalization. Specifically, we combine the activities with "Hello, Weekend," utilizing weekends for part of the exploration activities, embarking on the philosophy that "education knows no bounds, and growth never takes a holiday." During the initial stage of the comprehensive practical activities class, children discussed Zidong Wampee, and countless questions emerged in their minds. For example, "What special climate conditions contribute to the delicious taste of Zidong Wampee?" "Does global climate change affect the growth of wampee?" "How can we build a strong wampee brand to support rural revitalization?" With high enthusiasm, the children collaborated with teachers to develop an activity plan, and a series of climate change education activities with local characteristics are gradually unfolding.

Project Design

The overarching objective of this project is to gain an in-depth understanding and research on the relationship between wampee and climate, intuitively recognize the significant impact of climate change on plant growth, learn to care for the Earth, respect nature, cherish resources, form a concept of sustainable development, explore and utilize the industrial potential of wampee, and contribute to rural revitalization. Specific objectives are:

(1) Children will understand the relationship between wampee and climate through activities and gain scientific knowledge.

(2) Children will gain a deeper understanding of their hometown through promoting local culture through the wampee industry, enhancing their sense of belonging and

identity in participating in community-level governance.

(3) Children will strengthen their own and even the public's environmental awareness education via engaging, and increase their understanding of the relationship between the wampee industry and environmental protection.

(4) Children will work together with peers in activities to complete tasks such as visits, interviews, observations, and recording, fostering team spirit, communication skills, and problem-solving abilities.

The content design, implementation methods, and evaluation methods of this project are shown as Table 1.

Table 1. Contents, methods and evaluation of the project "Exploring the Wampee and Climate to Help Rural Revitalization"

Topic	Tasks	Implementation methods	Evaluation
Delicious wampee is a gift from the climate	1. Observe the growth of wampee. 2. Record the growth environment and climatic conditions of wampee. 3. Collect local climate data and find out the patterns related to the growth of wampee.	1. Conduct a field trip to the wampee plantation. 2. Interview the staff of the plantation. 3. Search for information in the library and analyze it.	Conduct self-evaluation, peer-evaluation, and give a thumb-up sticker for "participation, enthusiasm, and team spirit."
Climate change, wampee crisis	1. Understand the experience of wampee cultivation and specific examples of climate impacts. 2. Verify the specific impact of climate on the growth of wampee.	1. Visit Zidong Village to interview fruit farmers. 2. Conduct controlled experiments under the guidance of a science teacher.	Conduct self-evaluation, peer-evaluation, and give a thumb-up sticker for "participation, enthusiasm, and team spirit."
Protective action, guarding our homeland	Consider Measures to Address Climate Change and Protect Wampee.	1. Brainstorming. 2. Organize promotional activities such as designing promotional posters	Select the "Environmental Protection Ambassador".
Build a brand to boost revival	Consider Strategies to Build a Wampee Brand and Contribute to the Revitalization of Rural Areas.	1. Conduct field studies. 2. Interview the local community. 3. Summarize the golden ideas for rural revitalization and give a speech.	Select the "Outstanding Little Village Chief".

Project Implementation

1. Phase 1: The Delicious Wampee, Gifted by the Climate

At the beginning of 2024, students formed research groups and headed to the Zidong Wampee Plantation for field trips, where they observed the growth of wampees, recorded the growth environment and climatic conditions of the wampees(see Figure 1). The students carefully studied the materials related to wampee growth in the plantation and interviewed the staff there. Upon returning to school, the students utilized the information network in the library to collect local climate data, including temperature, humidity, light exposure, and rainfall, and analyzed these climate data to discern growth patterns associated with wampee growth.

Figure 1. Students observing at the Wampee Plantation

The students found that Zidong Village is located in the southern part of Guangdong, China, with a subtropical climate that is warm and humid. The annual average temperature is above 20℃, and the average temperature in winter is above 12℃. Zidong Village also enjoys ample sunlight, which helps wampee trees undergo photosynthesis, enhances the accumulation of nutrients in the tree bodies, expands the fruit-bearing area, and improves the quality and yield of the fruits.

At the same time, the students had an unexpected discovery. Through interviews, they learned that planting wampee trees is a tradition in Zidong. Many villagers plant wampee trees in front of their doors or in their yards if they build new houses. This is because the villagers believe that wampee trees have the effect of exorcising evil spirits, and they are insect-resistant, clean, and easy to care for.

Through this activity, the students intuitively felt the significant impact of climate on the growth of wampee. This helps them establish basic cognitive understanding of natural science and stimulates their interest in scientific exploration. At the same time, they collaborated with their peers to successfully complete tasks such as visits, interviews, observations, and recording, which not only cultivates their team spirit but also improves their communication skills and problem-solving abilities.

2. Phase 2: Climate Change, Wampee Crisis

Will climate change have an impact on the growth of wampees? To figure out the answer, students went to Zidong Village, where they interviewed agricultural experts and farmers with rich planting experience in Zidong, to learn about the experience of wampee planting and specific examples of the impact of climate (see Figure 2). Many of the farmers with rich planting experience are the grandparents of the students. Junlang's grandfather told us: Due to the severe pollution of ceramic factoriesin Nanzhuang Town before, the climate worsened, and the yield of wampee trees decreased. Then, as some ceramic factories in the town and the village were relocated, the environment gradually improved, and wampee trees began to bloom and bear fruit since then. Now, the harvest of wampee in the village is getting better year by year.

Figure 2. Students consulting with professional fruit farmer

In addition to conducting interviews and surveys, the students also consulted relevant books and materials on climate change and plants in the library. Under the guidance of science teachers, they designed experiments to simulate the growth of wampee under different climatic conditions. They designed control group experiments to test the effects of factors such as temperature changes, rainfall changes, and light changes on wampee growth.

After a series of studies, the students discovered that climate change would have multiple impacts on Zidong wampee. These impacts are directly related to the growth cycle, and ultimately the yield and quality of Zidong wampee.

Firstly, climate warming would lead to an extension of the growing season of Zidong wampee and an earlier flowering. High temperatures would alsoinhibit the photosynthesis of Zidong wampee, reducing its growth efficiency. Temperature fluctuations are prone to causing weak growth or even death of wampee. Secondly, flood disasters can cause the death of Zidong wampee. Too little rainfall would lead to drought, water deficit, and nutritional imbalance in the leaves and roots of Zidong wampee. Finally, the increase in carbon dioxide concentration caused by industrialization would also have serious impacts on the growth of wampee.

Climate warming not only affects the growth and adaptability of plants but also the balance of the ecosystem. By understanding the relationship between climate change and wampee growth, the students gained a deeper understanding of the importance of protecting the environment and reducing pollution.

3. Phase 3: Protective Action, Guarding Our Homeland

What can we do to protect delicious wampee? What can we do in the face of climate change? The students observed, reflected and discussed around the topic, and wrote down the measures discussed. They went to consult science teachers to verify whether they were feasible. They transformed the feasible golden ideas into promotional slogans or posters through their own ingenious hands, and posted them in schools, communities, and other places to call on people around them to work together to address climate change and protect the delicious wampee of their hometown.

The students suggested:

(1) Start with daily trifles, such as conserving water and electricity, remembering to turn off unnecessary lights and electrical appliances, and using energy-saving lamps instead of ordinary bulbs. This can save resources and reduce carbon emissions.

(2) Learn about garbage classification and recycling knowledge, and classify recyclable items to reduce overexploitation of natural resources and environmental pollution.

(3) Participate in more tree planting and afforestation activities to green the campus and surrounding environment. Trees can absorb carbon dioxide and release oxygen, helping to mitigate global warming.

(4) Promote the dangers of global warming and coping methods to family members, friends, and classmates, encouraging everyone to save energy, reduce waste generation, and adopt low-carbon travel modes such as cycling or walking.

This activity helped the students establish correct values. They have learned to care for the earth, respect nature, cherish resources, and formed a concept of sustainable development.

4. Phase 4: Building a Brand, Boosting Rural Revitalization

Zidong wampee tastes so delicious, how can we build a brand to boost rural revitalization? Our class conducted the "If I Were a Village Chief" activity, where I led students to conduct field research in Zidong Village as "small village chiefs" (see Figure 3), interviewed the people, collected opinions, and thought about how to build a wampee brand to boost rural revitalization. They proposed their own ideas and suggestions as "small village chiefs."

Figure 3. Students surveying public opinions

Here are the translations of the provided sentences:

The ideas they came up with are highly referential and surprising:

(1) Holding events such as a Wampee Fruit picking festival, where tourists can visit the plantations, experience the joy of picking, and taste the fresh yellow-skinned fruits. Organizing a Wampee Fruit cultural festival, including a photography contest and a tasting event to promote the popularity and reputation of the purple cave yellow-skinned fruits through various measures.

(2) Actively cooperating with e-commerce platforms to broaden sales channels and introduce yellow-skinned products to wider markets beyond rural areas.

(3) Developing deep-processing industries to process wampee fruits into juices, jams, and other products. Creating new culinary delights like wampee coffee, mille-feuille, and cakes to extend the industrial chain.

(4) Developing cultural and creative products based on yellow-skinned fruits, formulating effective marketing strategies and promotion plans. Utilizing social media, online and offline events, and collaborative promotions for publicity and marketing. Meanwhile, collaborating with relevant cultural institutions and tourist attractions to sell and promote yellow-skinned fruit cultural and creative products as specialty items.

In this event, we promote the integration of child-friendly concepts into rural planning and construction through a child's perspective. This allows children to understand the countryside, enhances their sense of belonging and identity in participating in community-level governance, encourages them to utilize their own strengths for participation and promotion, and contributes to the construction of their hometown and community, achieving common development between children and society.

Effectiveness & Reflections

Since February 2024, this project has achieved rich results in a short period of time. Through the joint efforts of teachers and students, a research manual titled "Exploring the Connection between Wampee Fruits and Climate to Support Rural Revitalization" has been compiled, which is distinctive and instructive. Additionally, one of the project leaders, Ms. Chen Baoyu, shared her experience in climate education projects at the "Shanghai Climate Week" education forum in April 2024.

Through practical activities, students gained a deeper understanding of wampee fruits, a local specialty crop, including their growth habits and relationship with climate. This intuitive learning method sparked students' interest more effectively than traditional classroom teaching, making knowledge more vivid and specific. Students improved their various practical skills, learned how to observe, interview, record, analyze data, conduct controlled experiments, and more. This fostered their practical abilities and scientific literacy. The activity deepened students' understanding and identification with rural culture. They realized that the countryside is not only their root but also an important place for their learning and growth. This cultural identity encourages them to contribute to rural revitalization. Students not only learned knowledge but also cultivated the consciousness and ability to contribute to rural revitalization. They can apply their learned knowledge and skills to practical life, contributing to the economic development, cultural inheritance, and ecological protection of rural areas. Through practical experience, students paid more attention to the natural environment and ecological protection, understanding the impact of climate change on crop growth. This fostered their environmental awareness and guided them to take environmental actions in daily life, inspiring others to embark on low-carbon and environmentally friendly actions.

The activity promoted cooperation and communication between schools, families, and communities. Parents actively participated in their children's practical activities, providing strong support and guidance. The community also provided students with practical venues and resources, forming a good educational synergy.

In April 2024, the project team conducted a questionnaire survey among students, and 94% of the students liked and identified with the project-based activity. Teachers also conducted related interviews, and received positive feedback from parents, elderly people, and staff members of the village committee.

A student's mother said, "This activity is fantastic. From the small classroom in school to the big classroom of society, the children's experiences have become richer, and they understand more about environmental protection. Now they even suggest green commuting for school and work." Luo's grandfather said, "I didn't expect that even at my age, I could still make a difference by sharing my years of experience with the students. I fully support this meaningful activity."

Ms. Luo, a staff member from the Zidong Village Committee, said, "The measures proposed by the students for building the wampee brand are very insightful and surprising. We will carefully consider and study them internally. Thank you, students, for actively participating in this activity and joining our ranks in revitalizing the countryside with your fresh perspectives and energy."

Of course, there are still areas for improvement in this project. This project has a long duration, often involving more work than evaluation. It is necessary to establish a more scientific and comprehensive evaluation mechanism to assess the effectiveness of the activities and students' learning outcomes. At the same time, attention should also be paid to cultivating students' self-evaluation and reflection abilities. sAfter the event, it is necessary to continuously follow up on students' growth and development. Regular follow-up visits and organizing related activities can maintain contact and interaction with students, encouraging them to apply their learned knowledge and skills to practical life. At the same time, attention should also be paid to the development of rural revitalization, providing students with more practical opportunities and resource support.

(Baoyu Chen, Teacher of Zinan Primary School, Nanzhuang, Chancheng, Foshan)

Exploration of the Relationship Between Climate Change and Biodiversity from a Student Perspective

Research Background

Kunming is located on the Yunnan-Guizhou Plateau, with a high altitude. It is experiences minimal influence from cold air from the north in winter and is influenced by the southwest monsoon from the Indian Ocean in summer, bringing abundant rainfall. This makes the city warm like spring all year round, earning it the nickname "Spring City."

However, in recent years, children have noticed that the weather in Kunming has become less pleasant. Some children said, "It's scorching hot in the summer, and we need fans in the classroom; it's freezing in the winter, and our parents bought new down jackets." Others observed that prolonged dry spells have caused the soil, which once nourished many plants, to crack, resulting in the wilting of many plants. These phenomena led the children to wonder: How does the persistent hot and dry climate affect species? Does climate change impact biodiversity, and vice versa? How can we respond to climate change? What can we do as elementary school students? With these questions in mind, the author designed a project-based study on the relationship between climate change and species.

Project Design

1. Objectives

The overarching objective of this project is to understand the relationship between climate change and biodiversity and to learn about daily life behaviors that respond to climate change.

Based on this general goal, we set specific sub-goals according to the capabilities of fourth and sixth grades students. The objectives for fourth grade activities are: ①Through methods such as consulting materials, consulting with family elders, field investigations, and organizing information, students canunderstand and clearly explain the relationship between Yunnan fruits and climate change. ② Through this project research, students can learn what they can do in daily life to mitigate the reduction of species caused by extreme weather, further raising their awareness of protecting biological resources and enhancing their sense of social responsibility for ecological protection.

The objectives for sixth grade activities are: ① Through understanding climate change knowledge, students can scientifically recognize the relationship between climate change and human activities, making the activities more scientifically grounded. ②Through sustained engagement in these activities, students can develop proactive environmental behaviors and encourage more people to incorporate environmental protection into their daily actions.

2. Content Design

According to the activity objectives, the author designed two thematic learning contents for fourth and sixth grades. The theme of fourth grade is "Exploring the Relationship Between Climate Change and Fruit Varieties," while the theme of sixth grade is "Exploring the Relationship Between Climate Change and Humans" (see Figure 1).

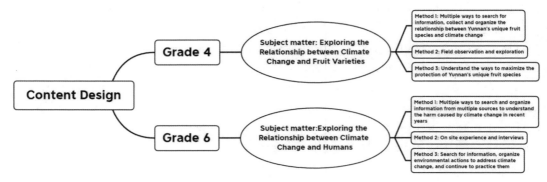

Figure 1. Content design of this project

The principles of activity design are:

Firstly, student Problem-Oriented: It aims to let students comprehensively use various methods to try to solve the problems in a project-based inquiry.

Secondly, integrating Regional Context: During the activity phase, students are guided to leverage their region's unique attributes as resources for correlated inquiry. This approach does not solely focus on climate change or biodiversity but investigates both due to their intrinsic connection.

Thirdly, basis of Daily Actions: After understanding the relationship between climate change and biodiversity, the content implementation will provide clear daily environmental action guidance for responding to climate change and protecting biodiversity. This effort will involve students, teachers, and parents from both grades to create an ecological circle.

Project Implementation

1. Project Initiation: Defining Activity Themes and Focus Issues

To better promote the project's implementation, it was integrated into the school's activities. The school's overall theme is "Collaboratively Creating an Ecological Circle."

On December 8, 2023, the author organized a project launch meeting for all class teachers in the school district, where they learned about climate change and climate change education. The teachers learned that the 28th session of the Conference of the Parties (COP28) to the United Nations Framework Convention on Climate Change was held in Dubai Expo City, UAE, from November 30 to December 12, 2023. On November 30, the opening day of COP28, a side event on "Mechanisms for Synergizing Biodiversity Protection and Climate Change Response" was held at the China Pavilion of COP28. This side event was guided by the Department of Ecology of the Ministry of Ecology and Environment of the People's Republic of China and jointly organized by the China Green Carbon Foundation, the China Environmental ProtectionFederation, the International Union for Conservation of Nature (IUCN), the Natural Resources Defense Council (NRDC), The Nature Conservancy (TNC), and WildAid. With this background knowledge, the class teachers discussed and actively researched countermeasures for the questions raised by the students. The meeting reached a consensus to conduct correlated inquiries on climate change and species with fourth and sixth grades as the main participants, focusing on "Fruit and Climate" for the fourth grade and "Humans and Climate" for the sixth grade.

2. Activity Preparation: Planning Activity Content and Designing Activity Plans

After the launch meeting, the class teachers of fourth and sixth grades led students to plan specific activity contents around the themes (see Figure 2). During this process, class teachers and parent committees intervened timely to guide the implementation of the activities. Figure 3 shows a project proposal for sixth grade.

Figure 2. The author led the students from three parallel classes of sixth grade to plan this activity

Figure 3. Project plan planned by sixth grade students

In the activity plan design process, we paid special attention to the differences in the development of abilities among students of different grades. For the fourth grade, the focus was on cultivating students' abilities to collect and organize information. In the sixth grade, the emphasis was on fostering students' environmental awareness and capabilities.

3. Publicity and Promotion: Encouraging More Participants

Through prior preparations, students became familiar with the goal design and activity content. Between December 28, 2023, and January 12, 2024, we introduced the project's related content and requirements to parents through parent meetings and school WeChat public account promotions, ensuring effective advancement of the activities using the school-family platform.

4. Implementation Guarantee: Field Investigations During Holidays

On January 12, 2024, students' winter vacation began as scheduled. Taking fourth grade as an example, some students returned to their hometowns, such as Dali and Lincang, to understand and record the local climate changes in recent years. Teachers required the observation period to be at least one week, including elements such as observation location, observation date and time, and observation methods. After the observation, students could draw a climate change line graph.

Subsequently, the students, accompanied by their parents, ventured into nature to observe the growth of seasonal fruits, visited fruit markets, orchards, and plant research institutes to inquire about the survival conditions of Yunnan fruits. During these investigations, students recorded their observations and research processes using photos and text. This activity lasted nearly a month.

5. Presentation and Exhibition: Diverse Methods and Carriers

From February 10 to 14, 2024, students presented their research findings in various formats, including Meipian (a Chinese social media platform), pop-up books, brochures, 3D models, and project research reports (see Figure 4).

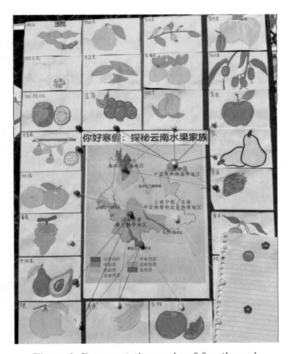

Figure 4. Representative works of fourth grade

6. Evaluation Design: Competency-oriented

After completing the first five stages of activities, a series of evaluation activities were organized. The evaluation of students' competencies focused on three dimensions: "Learning how to think," "Responsibility and commitment," and "Practical innovation." Specific evaluation criteria were differentiated for lower, middle, and upper-grade levels (see Table 1). The evaluation of students' works was conducted through school-level and grade-level exhibition and appraisal meetings (see Table 2 and Table 3).

Table 1. Student literacy evaluation of the project

Student literacy	Low grades (1-2)	Middle grade (3-4)	Upper grade (5-6)
Learning how to think	Under the guidance of peers, teachers and family members, I can basically know the current national or global frontier topics of "biodiversity and climate change".	Under the guidance of peers, teachers and family members, I can understand the current cutting-edge topics of "biodiversity and climate change" through the Internet, newspapers and other media.	They can independently understand the current frontier topics of "biodiversity and climate change" through the Internet, newspapers and other media.
Responsibility and commitment	1. Under the guidance of teachers, familymembers and peers, environmental protection-related behaviors (garbage sorting, restoring an area, not eating wild animals, not buying products without environmental protection recognition, etc.) 2. Keep order, love public property, can effectively save resources with the help of the family.	1. Will independently carry outenvironmental protection-related behaviors (garbage classification, restore an area, do not eat wild animals, do not buy products without environmental protection recognition, etc.). 2. Can abide by the public order, and form a preliminary social morality in the activities.	1. Will independently carry outenvironmental protection-related behaviors (garbage sorting, restoring an area, not eating wild animals, not buying products without environmental protection recognition, etc.). 2. Be able to actively participate in activities and encourage more groups to participate.
Practical innovation	1. With the help of teachers and family members, they can make simple oral planning andclarify their responsibilities. 2. Under the guidance of teachers and family members, I can conduct activities according to the requirements of the activities, and seek help from my family members and teachers when encountering difficulties.	1. Be able to independently carry out a relatively complete plan and form a personal plan. The elementsshould include: time, place and basic division of labor. 2. Be able to carry out activities according to the activity plan, and seek targeted help according to the actual situation in the process.	1. Be able to make complete planning independently and form an intra-group orclass general plan. 2. Be able to carry out activities according to the activity plan, and in the process, it will be flexibly adjusted according to the actual situation.

Table 2. Evaluation criteria for fourth-grade project works

Evaluation criterion	Student evaluation level
The work revolves around the theme.	☆☆☆
It is well-illustrated and complete in content.	☆☆☆
The work is creative.	☆☆☆

Table 3. Evaluation criteria for project works in Grade 6

Evaluation criterion	Student evaluation level
The work is closely related to the theme.	☆☆☆
Bind together in book form.	☆☆☆
Graphic, the activity process is strong and can continue to practice in life.	☆☆☆
The harvest in the work is real and specific.	☆☆☆

During the evaluation process, a fourth-grade teacher remarked with emotion: "This project-based activity has brought endless growth to the students. Listening to their explanations, it's as if we too have entered a vibrant fruit paradise with them. The delightful illustrations seem so vivid that one can almost smell the fragrance of the fruits. Through this activity, the fourth-grade students have led their parents in an attempt to create an ecosystem, jointly exploring the relationship between climate change and biodiversity, and striving to become advocates for ecological conservation." The school also promptly awarded children with high votes in the exhibition with the "Outstanding Work Award" and the title of "Activity Practice Expert," encouraging the children to continue to actively participate in activities and to inspire more people to join the ranks of ecological environment protection.

Effectiveness & Reflections

Following these activities, surveys were conducted among students, teachers, and parents to evaluate the project's effectiveness. A total of 874 questionnaires were distributed, with 654 valid responses. The results showed that nearly 95% of families highly recognized the initial research activities and expressed willingness to continue participating in subsequent activities. 98% of students felt that their activity content was diverse, and 85% of parents stated, "The activity has brought my child and me closer." 94% of teachers observed significant changes in students' ability to collect and integrate information and in their promotional skills due to the long-term biodiversity protection activities. Meanwhile, the students' project works showed a trend of increasing diversity.

Based on current research results, the author found that guiding students to pay long-term attention to the relationship between climate change and species and to continuously practice environmental protection actions cannot be achieved solely by schools and families. Therefore, in the upcoming research, we will seek support from relevant government departments and localenterprises for activity venues or educational support. We will also continue exploring from a biological research perspective, such as contacting the Kunming Institute of Botany, Chinese Academy of Sciences, the Institute of Zoology, Chinese Academy of Sciences, the Kunming Branch of the Chinese Academy of Sciences, and the Kunming Meteorological Bureau, to utilize geographical advantages for activities.

Climate change directly leads to the reduction of species, and the decrease in species inevitably causes breaks in the biological chain, which in turn threatens human survival. This project research and subsequent extensions aim to guide students in exploring the relationship between species and climate change more scientifically, practically, and sustainably. Focusing on climate change and biodiversity is a common topic and responsibility for all humanity.

(Xingyue Li, Director of Student Development Department, Xishan Cuizhi Yufu School, Kunming, Yunnan)

Design and Implementation of Project-Based Learning on "Harmony of Water and Air Nurtures the Symbiosis of All Things"

Research Background

Water is one of the most direct and important impact area of climate change, and the impact of climate change can be reflected through water. "World Meteorological Day," also known as "International Meteorological Day," is scheduled for March 23 every year. From 1961 to 2022, we have experienced 62 "World Meteorological Days," of which 9 themes are directly related to water. In 2023, it coincides with the 150th anniversary of the founding of the World Meteorological Organization. With the theme of "Weather, Climate and Water: Generations to the Future", the whole society is called on to enhance the understanding of the weather and climate system under the background of global warming. my country has always attached great importance to the rational development and utilization of water resources, and has listed water resources management as an important part of the national strategy.

In this context, a project-based learning activity with the theme of "Water and Gas Harmony: Nurturing Symbiosis of All Things" is carried out to cultivate students' environmental awareness, enhance their sense of social responsibility, take positive actions to effectively respond to climate change, and consciously practice the mission of ecological protection.

Project Design

1. Overall Process

Project-based learning is a learning method that uses projects as carriers. Regardless of the size of the project, it must follow the normal process of the project cycle. Based on the "four major stages and eight major templates" process provided by the Beijing Normal University project-based research team, the author designed this project-based learning process chart. Specifically, it includes four stages: project team establishment, planning and preparation, project implementation, and review and conclusion (see Figure 1). The four stages are repeated to form a continuously iterative and upgraded project-based learning system.

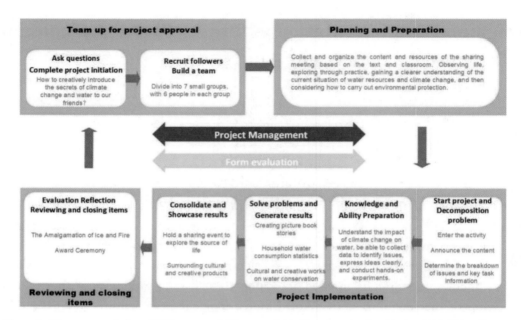

Figure 1. The process design of "Harmony of Water and Air Nurtures the Symbiosis of All Things" project

2. Target Design

Target design should be preferred (see Table 1).

Table 1. The student training objectives design of "Water and Gas Harmony, Nurturing Symbiosis of All Things" project

	Dimension	Ability performance
PBL core literacy goals	Have a sense of social responsibility and form correct values and reasonable thinking.	Be able to conduct simple sewage filtration experiments in a home environment to understand the water purification process; be able to understand the current status of water resources; be able to design and participate in water conservation activities to enhance social responsibility; develop relevant natural science knowledge to address climate change.
	Communicate and collaborate effectively. Develop learners' leadership, practical skills, and comprehensive problem-solving abilities.	Be good at listening and expressing, able to listen carefully to other people's opinions; able to achieve one's own goals through effective communication (notification, persuasion, inquiry, motivation, etc.); able to cooperate with others to achieve a goal together.

(Continued)

	Dimension	Ability performance
PBL core literacy goals	Actively explore and practice to develop sustainable production, lifestyle, and habits.	Be curious and proactive, be able to understand the principles of water circulation, be able to use scientific knowledge to explain experimental results, and understand the properties of water; maintain a passion for exploration and master exploration methods; be able to participate in scientific experiments on water, observe and record experimental results, apply knowledge in practice, verify ideas, and explore the world.
	Aesthetics.	Be able to express and create beauty through language, art and design, and be able to express emotions and knowledge about climate change or water through artworks.
	Use technology and develop information literacy.	Be able to use technology to scaffold your own learning and exploration; be able to apply information, communicate, and solve problems through information.

3. Content Design

Next, let's focus on content design (see Table 2).

Table 2. The content design of "Water and Air Are in Harmony and All Things Coexist"

Core driver issues	Overall mission	Final result
How to creatively introduce the secrets of climate and water to your friends?	Hold a sharing session to explore the source of life.	Sharing session and surrounding cultural and creative works.
Decomposing the driver problem	**Main tasks**	**Main products**
How much do you know about water, which is so familiar to you? What are the contents and resources needed to complete the sharing session?	Learn about climate and water: Rely on the text, appreciate the language, consult relevant information, understand the ever-changing nature of water, and introduce it in one or two sentences.	PPT and handwritten newspaper about the transformation of small water drops.
How to show the secrets of "Climate and Water" more interestingly?	Journey of exploring climate and water: Observe life, conduct practical research, understand people's understanding of water resources and usage habits through questionnaires, understand the important uses of water in life activities by recording household water consumption, have a clearer understanding of water pollution, and then think about how to purify water.	Water-saving slogans, Investigation reports, statistics, Scientific experiments on water, Three-line sayings on water-saving.

(Continued)

Decomposing the driver problem	Main tasks	Main products
How can we promote water resource protection and enhance people's awareness of ecological environment protection through sharing sessions? What can we do to let our sharing sessions have better promotion value?	Organize the information, prepare for the sharing session, and present your findings.	Improve the students' works and share drafts.

4. Assessment

The project evaluation focuses on three aspects: "Climate Change and Water Cycle Works", "Household Water Saving Survey Statistics and Water Saving Tips", and "Exploring the Source of Life Sharing Session". The evaluation objectives, evaluation evidence, evaluation methods and evaluation types are shown as Table 3.

Table 3. The evaluation design of "Water and Air Are in Harmony and All Things Grow Together" project

Contents	Objectives	Standards	Methods	Types
Works on climate change and the water cycle.	Actively read materials, be willing to think, gain a preliminary understanding of the relationship between "climate and water", and be able to introduce it to others in different ways; use the knowledge learned in a variety of situations to solve practical problems.	Design works that can clearly explain the relationship between climate change and water.	Evaluation rubric Self-evaluation and mutual evaluation.	Formative assessment Summative assessment.

(Continued)

Contents	Objectives	Standards	Methods	Types
Household water conservation surveystatistics and water conservation tips.	Design a questionnaire to understand people'sunderstanding of water resources and usage habits through the questionnaire, and put forward opinions and suggestions on improving the use of water resources through experimental methods. Find more ways to save water and protect water, and publicize how to reuse wastewater to cherish water resources by drawing handwritten newspapers and writing articles.	Practice and understand the principles ofscientific experiments. Promote and practice water conservation in daily life.	Evaluation rubric Self-evaluation andmutual evaluation.	Formative assessment Summative assessment.
Sharing Session on Exploring the Source of Life.	Enhance our understanding of climate and water, increase our attention to weather and climate, raise awareness of disaster prevention and mitigation, and take practical actions to protect our beautiful home.	Sharing session display Achievement exchange.	Evaluation rubric Self-evaluation and mutual evaluation.	Formative assessment Summative assessment.

Project Implementation

This project combines the school's "Seed Growth" featured course, explores project-based learning with "subject +" as the main feature, focuses on students' real life, carries out interdisciplinary learning, and improves children's ability to analyze and solve problems.

Facing the driving question "How to introduce the secrets of climate and water to friends creatively?", the core task is determined to be "holding asharing session to explore the source of life", and based on this, the problem and task are decomposed.

1. Subtask 1: Exploring Climate and Water

The results of this task are "Handwritten newspapers and presentations on climate and water". Through the study of the three texts "Who Am I", "Where is the Fog", and "Snow Child", students feel the magical changes of water. In addition, the group members read the picture book stories "Water Travel" and "Children Thinking about the World" together, watched scientific documentaries such as "Climate and Water" and "Pay Attention to Meteorology and Protect Water Resources, "and have a more comprehensive understanding of water. Based on this, the teacher guides students to use their imagination according to the content of the text and tell the travel story of the small water drop in the form of a story. Finally, they create their own works, draw the three states of water and the formation process of fog, snow, hail, etc., and introduce them in one or two sentences (see Figure 2).

Figure 2. Student work *Changes in Water*

2. Subtask 2: The Adventure of Exploring Climate and Water

The result of this task is "Family Water Saving Survey, Water Saving Tips". After learning about the role of water resources in our lives, the children went into life and explored the wonderful relationship between climate and water. The most impressive part was the experimental exploration. Under theguidance of science teachers and parents, the children carried out various interesting water experiments and wrote down what they saw and thought: Water Baby is a "magician" who can reverse the pattern drawn on paper; Water Baby is a "gymnast" who can "stand" on a small coin... Each small experiment allowed the children to see the characteristics of water: water refracts light, water has tension, "capillary phenomenon", water and oil are insoluble...

Facing the problem of weak awareness of water conservation, teachers and students jointly designed a questionnaire to understand people's understanding of climate and water. The survey results show that everyone generally understands climate and water, and knows that there are many ways to save water, but many people are not aware of the need to reuse water resources. In addition, by collecting, organizing and analyzing household water usage data, students can choose water-saving measures more specifically and establish good water-saving habits and green environmental protection concepts.

How to better protect water resources and improve water utilization in practice? The children learned about the principles of sewage filtration by watching the video and then started to experiment (see Figure 3).

Figure 3. Student work "Sewage Filtration Home Experiment"

3. Subtask 3: Improving the Quality and Effectiveness of Promotion Sharing Sessions

At the sharing session on exploring the source of life, the wonderful works (see Figure 4) showed the students' gains and enriched and sublimated their understanding of climate and water. At the end of the project, a summary and reflection activity were carried out based on the evaluation form, and various different types of awards were selected.

Figure 4. Student achievement report exhibition

Effectiveness & Reflections

This project has played a very good role in improving students' analytical thinking, innovative thinking, and practical thinking. For example, in terms of analytical thinking, after the experience of this project, many students develop a conscious thinking consciousness and have a positive thinking motivation when facing many related problems in life. The problem analysis ability is gradually improved. In terms of innovative thinking, throughout the entire project-based learning process, students use their favorite ways to discover, practice, and record their own processes and feelings. Teachers and parents also provide students with a lot of space for free imagination and self-creation, as well as scientific and applicable learning scaffolds, to help students gradually build a knowledge logic structure of themes and core issues, and promote the improvement of students' thinking literacy. In terms of practical thinking, this project has given students sufficient discussion space and practical experience in the analysis and exploration of issues such as "the relationship between climate change and water", "how to conduct household sewage filtration experiments", "how to obtain the current status of water resources", and "how to understand everyone's understanding of climate change and water", so that students' hands-on practical ability has been enhanced. At the same time, we also reviewed and reflected. The integration of this interdisciplinary project-based learning needs to be improved. For example, the analysis of interdisciplinary textbooks needs to be more in-depth, find the right points of fit, extract the common knowledge points, design project-based learning themes, and design problems driven by students' learning conditions according to the content settings, so as to achieve subject integration and optimization in curriculum development, and avoid the feeling of "patchwork" in interdisciplinary teaching.

A pool of clear water is the source of life and also the source of ecology. The prosperity and development of every person, every inch of land, and every country all rely on the blessings of climate and water. The "Harmony of Water and Air Fosters

Coexistence of All Things" project is ongoing and continuously improving. The green mountains and clear waters, and the growth of all things, require our joint care.

(Meng Wang, Teacher of Liangxiang Central Primary School, Fangshan, Beijing)

Exploring the Project-Based Learning of Science Popularization on "Climate Change and Environmental Protection"

Research Background

Facing the global challenge of climate change, governments and international organizations around the world are actively taking action to mitigate its impacts. Our country has constructively proposed a sustainable development strategy and a new concept of development, contributing Chinese wisdom to solving climate change issues, which are primarily manifested in energy crises, environmental changes, and natural disasters.

This issue, when extended to the field of education and teaching, becomes the responsibility of educators. As middle school teachers, we guide students through "Climate Change and Environmental Protection" science popularization project-based learning activities. We lead students to explore the causes, impacts, and countermeasures of climate change, enhancing middle school students' awareness of climate change and fostering their environmental consciousness and innovative abilities.

Project Design

The design of this project follows the principles of gradualism, integration and combination of theory and practice, and is divided into cognition, discovery, analysis, cooperative exploration, creation and other dimensions, as follows:

First, to study the law of temperature changes in China, let students understand the causes and effects of temperature changes, clarify the impact of temperature changes on the environment, and establish the social responsibility and scientific awareness of protecting the ecological environment.

Second, to analyze the impact of temperature changes on the environment. In particular, focusing on global warming, increasing extreme weather events, discussing the importance and measures of environmental protection, which can improve students' awareness and consciousness of environmental protection, and cultivate their sense of social responsibility andaction.

Third, to combine history, geography, science and other multidisciplinary knowledge to improve students' interdisciplinary learning ability and team spirit.

Project Implementation

1. Construct Study Groups and Assign Learning Tasks

In order to better guarantee the learning effect and the quality of project promotion, we divided the students into four groups, assigned the tasks, and each group received one research task (see Table 1).

Table 1. Task of "Climate Change and Environmental Protection"

Group and responsible content	Specific task
Group 1: Temperature changes in successive dynasties	Task 1: Review the literature of CNKI to understand the climate changes in the past dynasties.
	Task 2: Review ancient Chinese literature, poems, essays, notes and other records of climate.
	Task 3: Measure this year's temperature and check the local weather bureau since the 20th century. Bar chart of air temperature changes.
	Task 4 (Project results): Map the climate change of the past generations.
Group 2: The impact of climate change on the ecological environment	Task 1: Check the Internet and the CNKI to understand the impact of climate change on the natural environment.
	Task 2: Check the Internet and the CNKI to understand the impact of climate change on biodiversity.
	Task 3: Check the Internet and the CNKI to understand the impact of climate change on agricultural production.
	Task 4 (Project results): Hold a photo exhibition.
Group 3: How different countries deal with the environmental problems caused by climate change	Task 1: Check how the United Nations, the European Union and other international organizations to deal with the environmental crisis caused by climate change.
	Task 2: Check out how the developed countries in Europe respond to the environmental crisis caused by climate change.
	Task 3: Check out how the developing countries represented by China cope with the environmental crisis caused by climate change.
	Task 4 (Project achievement): Write an essay.
Group 4: Chinese wisdom and Chinese solutions	Task 1: Review the concepts of a community with a shared future for mankind.
	Task 2: Check out the concept of sustainable development.
	Task 3: Project achievement — Make publicity materials.

The first group's task is to investigate the temperature changes throughout Chinese history. The team members apply interdisciplinary knowledge including geography, history, statistical mathematics, English, and language arts to organize and analyze historical temperature data. By comparing temperature data across different historical periods, they explore the patterns and trends of temperature changes and their impacts on society and the natural environment.

The second group's task is to study the impacts of climate change on the ecological environment. Team members, drawing from biology, geography, and English, discuss how climate change affects biodiversity, ecosystem balance, and human living conditions. They will also conduct field investigations, experimental research, and literature reviews to deeply understand the specific mechanisms of climate change impacts on the ecological environment and propose strategies to address these issues.

The third group's task is to research how countries are addressing environmental issues brought about by climate change. Combining knowledge from English, ethics and governance, biology, geography, history, and other disciplines, they learn about the policies and practices of various countries in response to climate change. By comparing and analyzing the strategies and effects of different nations, they summarize experiences and lessons to provide references for China's response to climate change.

The fourth group's task is to propose a Chinese solution to the problem and create promotional materials. Team members, integrating knowledge from fine arts, ethics and governance, language arts, biology, geography, and other disciplines, propose solutions to China's climate and environmental issues. They also use creative and artistic methods to produce promotional materials, appealing to society's attention to environmental changes and enhancing public environmental awareness and participation through visual and linguistic impact.

Throughout the project's development and learning process, students are exposed to knowledge from multiple disciplines including history, geography, climate, and environmental science. They continuously improve their interdisciplinary research capabilities and deepen their understanding of environmental protection issues. Ultimately, they write research reports, present their findings, and engage in group discussions and exchanges, further deepening their understanding of the relationship between temperature changes and environmental protection.

2. Follow the Learning Path and Promote the Project Implementation

The completion process of each task can be roughly divided into seven stages. The first phase primarily involves teachers in designing and finalizing the research topics. From the second stage on, it is all completed by the teacher and students together. For the subtask "Temperature change and environmental protection in China over the centuries" the process mainly includes: "Develop a research plan"—"Imformation collection"—"Analysis and discussion"—"Report writing"—"Presentation and exchange"—"Reflection and improvement"etc. (see Figure 1).

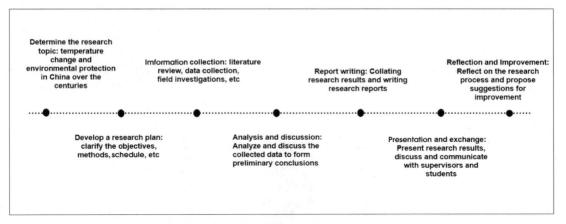

Figure 1. Path chart of the popular science project learning process of "Climate Change and Environmental Protection"

First of all, teachers give research supports, such as the path map of contemporary geography science research, common methods of natural science research and related papers, etc. Students make detailed and operational schedules according to clear research goals and methods. Then, under the guidance of teachers, students explore the problem of "climate change in China" by means of information and investigation, obtain the relevant data of climate change and explore the possible causes. Based on this, the students began to divide into groups to analyze and discuss the obtained data, use reasonable evidence to form their own preliminary conclusions, and sort out the research results and write research reports. After the formation of the group, each group basically completed the inquiry work at this time, and obtained more self-consistent research results and conclusions. Next, there is the presentation and communication. In this link, each group should take an appropriate way to exchange and show, and the teacher should participate in the consultant and guidance role, so as to obtain comprehensive and scientific research conclusions. Finally, students also need to reflect and improve the inversion phenomenon of some conclusions, and review the research process, research methods and experimental conclusions, so as to promote the continuous optimization of follow-up research.

Effectiveness & Reflections

Compared with the traditional learning form, project-based learning is a more open learning mode that directly refers to students' innovative thinking and practical ability. In this project, the students in each group presented their research results in different forms such as painting, audio, video and text, and also planned and organized practical activities such as venue inspection and garbage sorting.

Student designed a painting (see Figure 2). The main body of the picture is countless broken animal patterns, which can also be interpreted as a person divided into countless animals, which means the harmonious coexistence between man and nature. "Everything has a spirit, and it's not just humans who are considered the most spiritual beings among all things. In fact, there is no concept of a 'lord of all creatures.' What exists is an overabundance of anthropocentrism." Qinghan said. The reason for using the form of paintings is that she believes that patterns are more shocking than words for humans.

Figure 2. Work by student Qinghan Feng

On January 23, 2024, students came to the Garbage Classification Science Experience Museum in Nanshan to start an in-depth exploration of environmental protection and garbage classification (see Figure 3). In the experience center, the students encountered a warm-hearted and knowledgeable guide—a white-haired elder. He vividly introduced to the students the process of how garbage is produced, its harmful effects on the environment, the importance of waste sorting, and how waste can be turned into valuable fertilizers and energy through scientific treatment. His presentation deeply moved and impressed every student present.

Figure 3. Photo taken in Nanshan Energy and Ecological Park

This project learning has yielded a series of achievements (see Figure 4～Figure 7), presenting linkage, wide extension and openness. We arranged this project study in the winter vacation. During the vacation, students conducted in-depth research, analyzed data and sorted out materials, and made full preparations for the centralized report and presentation after the beginning of the semester. This kind of joint in-class and off-class methods, not only enables the students to better understand and master the knowledge, but also can cultivate their practical ability, innovative spirit and team spirit.

Figure 4. The "environmentally friendly" art works made by students

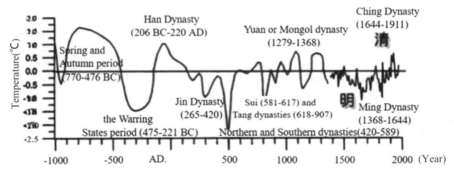

Figure 5. Winter temperature change map in the Central Plains of China drawn by group 2

Figure 6. "The impact of climate change on the environmental" PPT produced by group 3

(1) "People-centered" is the core essence of sustainable development

The basic content of the "people-centered" development philosophy is that development is for the people, development depends on the people, and the fruits of development are shared by the people.

The theoretical connotation of "people-centered" includes three aspects: First, the starting point of development is for the people. Development for the people is determined by the people's status as the main body of history. As the creators of history, the people are not only the main body of practice, but also the main body of value. Therefore, the target of development is not a few who are rich first, but the whole people. It also means emphasizing the dignity and worth of the human person, rather than blindly pursuing quantitative growth and rapid growth in GDP.

Second, the driving force of development can only come from the people. The people are the creators of social wealth and the decisive force for social change, and the driving force of social development can only come from the practical activities of the people. Development means breaking with the old and creating the new, and the sense of innovation can only be rooted in the practice of people's lives. Development is fraught with risks and challenges, which can only be solved by relying on the most enduring and greatest creativity of the people. Third, development belongs to the people, and the fruits are shared. People's sharing is the unity of sharing and comprehensive sharing by the whole people, and the unity of distributive justice and institutional justice.

Figure 7. The fourth group summarizes the Chinese wisdom and Chinese solutions on climate change and environmental protection issues

Overall, the project learning is a fruitful and meaningful exploration. Through in-depth research, teamwork and achievement demonstration, students not only increased their knowledge, exercised their ability, but also cultivated their sense of responsibility and mission. This student-centered and practice-oriented learning style improved students' interest and motivation in learning, and laid a solid foundation for their future study and life.

(Meifang Liu, Teacher of Foreign Language School of Shenzhen Bao'an Middle School; Zelong Zhao, Teacher of The Third Affiliated Middle School of East China Normal University)

🎓 Expert Review

President Xi Jinping pointed out that "building an ecological civilization is related to people's well-being and concerns the future of the nation". Since the 18th National Congress of the Communist Party of China (CPC), the CPC Central Committee has integrated the construction of ecological civilization into the overall plan for building socialism with Chinese characteristics, and the strategic significance of ecological civilization construction has become more prominent. President Xi Jinping stressed in the report of the 20th Party Congress that nature is the basic condition for human survival and development. Respecting, adapting to and protecting nature is an inherent requirement for the comprehensive construction of a modern socialist country. Therefore, we should take Xi Jinping's thought on ecological civilization as a fundamental guideline, firmly establish and practice the concept that lucid waters and lush mountains are invaluable assets, vigorously promote the construction of a beautiful China, and build a modernization in which human beings coexist harmoniously with nature. *The Action Plan for Enhancing Citizens' Awareness of Ecological Civilization (2021—2025) "Beautiful China, I am a practitioner"* specifies that ecological civilization education in schools should be vigorously promoted and incorporated into the national education system, with the aim of thoroughly studying and publicizing Xi Jinping's thought on ecological civilization, and guiding the whole society to firmly establish ecological civilization values and behaviors, and to promote ecological civilization. The aim is to study and publicize Xi Jinping's idea of ecological civilization in depth, guide society as a whole to firmly establish ecological civilization values and codes of conduct, and thus help to realize the Chinese dream of building a beautiful China and the great rejuvenation of the Chinese nation.

Schools serve as a crucial scenario for ecological civilization education, and education is one of the vital avenues to achieve the goals of ecological civilizationconstruction, playing a key fundamental role in ecological civilization construction. In 2021, UNESCO officially issued the outcome document of the ESD Roadmap 2030, which focuses on the contribution of education to the achievement of the SDGs, and advocates for the integration of Education for Sustainable Development (ESD) and the 17 Sustainable Development Goals (SDGs) into priority action areas including national policies, educational settings, capacity building of educators, empowerment of youth, local (community) action, etc., which includes Sustainable Development Goal 13 (SDG13)-Climate Action. Consequently, lifelong learning for all and learner-centered learning are reinforced, and countries are urged to use whole-institution models to promote ESD with the support of national policies.

The essence of project-based learning lies in employing real, challenging problems to develop students' lifelong learning skills such as creative thinking, critical thinking and collaborative abilities. Teachers should facilitate students to continuously explore and try to solve problems creatively, and eventually produce relevant results, so as to nurture students' sustainable learning ability and ecological civilization literacy in the process of trying to solve problems. The eight exemplary cases of climate change education presented in this chapter reflect the fact that climate change education is the main theme, encompassing multiple SDGs. The distinctive features of the project-based learning cases on ecological civilization and education for sustainable development are mainly reflected in the following aspects:

Firstly, it highlights interdisciplinary integration. Students need to integrate knowledge and skills from a number of disciplines, such as natural sciences, social sciences and humanities, in the learning process to thoroughly comprehend and address the issues of ecological civilization and sustainable development.

Secondly, it highlights the problem-oriented approach. Real and concrete problems or challenges, which are closely related to ecological civilization and sustainable development, are used as a starting point. Students need to conductinvestigations around these problems and deepen their understanding of the concepts of ecological civilization and sustainable development through the processes of data collection, problem analysis, and solution formulation.

Thirdly, it highlights the practical nature. The climate change project-based learning cases focus on practicality and operability, and motivate students to participate personally in practical activities, such as field trips, social surveys and experimental design. Through these practical activities, students can have a more intuitive experience of the current situation of the ecological environment and sustainable development, and enhance their understanding of the problems and their ability to solve them.

Fourthly, it highlights teamwork. Activities in the cases are conducted through a collaborative framework, in which students allocate responsibilities among themselves and work together to complete the task. This way of learning not only helps to cultivate students' teamwork ability, but also promotes communication and collision between students with different backgrounds and different perspectives, thus broadening their horizons and stimulating innovative thinking.

Fifthly, it highlights evaluation and reflection. In the case of climate change project-based learning, evaluation and reflection are shown to be an integral part of the process. Teachers and students need to self-evaluate and reflect on the learning experience, outcomes and co-operation, as well as receiving evaluation and feedback from their peers. This leads to deeper project-based learning and research in climate change education.

Looking ahead, climate change education should integrate education on ecological civilization and sustainable development throughout the entire educational process, and provide all-round human resources, intelligence and spiritual and cultural support for the construction of an ecological civilization. Education on ecological civilization and sustainable development will contribute to realizing a beautiful China and the 2030sustainable development goals, improve the quality of life for people globally and provide Chinese wisdom and Chinese solutions for global ecological governance and educational governance.

(Jing Zhang, Director of Ecological Civilization Education Research Office, Institute of Lifelong Learning and Education for Sustainable Development, Beijing Academy of Educational Sciences)

Chapter 4

Climate Change Education Based on Integrated Activities

In the school education scenario, integrated activities are different from subject teaching, but are rooted in Chinese classroom construction and student work. Many classroom teachers, grade leaders and others have accumulated rich experience in integrating climate change education as a prominent element in investigation and inquiry-based integrated activities, design and production-based integrated activities, cultural heritage-based integrated activities, and thematic study and research-based integrated activities.

The content presented below reflects the relevant explorations in this field and the practical wisdom of Chinese classroom teachers, grade-level leaders, parents and others.

Find the Yangtze River Estuary No. 2 Ancient Vessel and Promote the Yangtze Estuary's Vitality

Research Background

Climate change has had a profound impact on the Yangtze River Estuary. It is reported that in 2020, a large area of low-oxygen area appeared in the Yangtze River estuary in summer, and the high sea level raised the base water level of the storm, aggravating the disaster level. Tonggang Primary School in Shanghai Pudong New Area is located in Gaoqiao Town, close to the Yangtze River Estuary. Children accompany this hot land day and night. The impact of climate change on the Yangtze River Estuary will also indirectly affect the children's lives.

The news of the salvage and settlement of the No.2 ancient ship in Shanghai touches the heart of every student in Tonggang Primary School. On November 21, 2022, the Yangtze River Estuary No.2 Ancient Vessel was successfully salvaged out of the water, and more than 600 underwater cultural relics were stored and cleaned. On November 25, the Yangtze River Estuary No.2 Ancient Vessel came to the former site of Yangpu Shanghai Shipyard to "settle down", opening a new stage of cultural relics protection and archaeological excavation. Where the Yangtze River Estuary No.2 Ancient Vessel was found was at the mouth of the "golden waterway" of the Yangtze River and at the center of China's north and south coastline. Since ancient times, there have been countless underwater treasures and unsolved mysteries buried in this busy air route and in the complex waters.

Youth participation in global climate change governance is an important issue of global governance in the new era and an important development direction of global youth affairs. Based on this, it is urgent to pay attention to the past and present life of the Yangtze River Estuary No.2 Ancient Vessel and explore the sustainable development of the Yangtze River Estuary. The project "Find the Yangtze River Estuary No.2 Ancient Vessel, and Promote the Yangtze Estuary's Vitality" came into being.

Project Design

The project takes the phrase "The Yangtze River Estuary No.2 Ancient Vessel" as an example to explore independently the impact of climate change on the Yangtze River Estuary, which belongs to climate-related issues. Leaded by the third graders of Tonggang Primary School as the main body, together with the assistance from home, school and community union, it will explore the sustainable development of the Yangtze

Estuary through the ancient vessel and build a "a community with a shared future in climate for mankind".

The project aims to stimulate students' creativity and critical thinking, broaden their horizons, and cultivate their ecological civilization philosophy of respecting historical sites, conforming to and protect the nature. At the same time, it establishes a multi-functional education network with school as the main body, family as the pillar and community as the support, which can promote the communication between school, family and community, narrow the gap between each other, enhance mutual understanding, and deepen and efficient cooperation. It leads communities and schools to develop "a community with a shared future in climate for mankind", popularizing the knowledge of sustainable development in Yangtze River Estuary with practical actions.

Project Implementation

1. The First Search for the Yangtze Estuary No.2 Ancient Vessel

From January 20 to February 25, 2024, students responded proactively to this project. In the winter vacation, by searching for relevant information, drawing pictures of the ancient vessel and writing articles about it, students gained a deep understanding of the history and cultural value of the vessel, and also felt the unique vitality of the Yangtze River Estuary.

Firstly, checking and understanding relevant information. In the process of searching for relevant materials, students made full use of the library, the Internet and other resources to collect information about the historical background, construction technology and historical status of the Yangtze River Estuary No.2 Ancient Vessel. Through reading and researching, the students had a deeper insight of the vessel, and a clearer understanding of its important position and influence in history.

Secondly, knowing its magnificence through drawing the ancient vessel. Students also used their imagination and painting skills to paint the image of the vessel. These paintings not only showed the magnificence of the ancient vessel, but also reflected the students' love and inheritance of the ancient vessel culture. During their painting, the students further deepened their understanding of the structure and details of the vessel, while enhancing their perception and identification of the traditional culture.

Thirdly, writing articles about the ship to express feelings. The students wrote articles about the ancient vessel. These articles discussed its historical value, cultural significance and enlightenment and influence on modern society from different perspectives. Through writing, the students not only practiced their writing skills, but also deepened their understanding and thinking about the ancient ship culture.

Fourthly, having class meeting to make summaries and communications. In March 2024, the class organized a meeting to summarize and exchange the results of the project activities during the winter vacation. Students actively shared their experiences in searching for information, painting and writing. Through this meeting, the students not only showed their achievements, but also learned the knowledge and opinions from others. The class meeting further strengthened their teamwork spirit and cultural confidence.

2. Establishment of the "Facing the Future" Climate Change Team

On the basis of students' initial understanding of the various aspects of "the Yangtze River Estuary No.2 Ancient Vessel", the team of climate change in Tonggang Primary School called "Facing the Future" was formally established. Its establishment showed that students were dedicated to working jointly to explore the impact of climate change on the ancient vessel and create a sustainable future for the estuary.

About the question of how to seek the impact of climate change on this vessel, after the discussion of students and teachers, the team focused on the three aspects: "shipwreck mystery exploration", "cultural relics protection" and "the ancient vessel introduction". The following process evaluation as shown in Table 1 was taken as a criterion.

Table 1. Process evaluation of "Facing the Future" Climate Change Team

Evaluation dimension	Evaluation content	Student self assessment	Group assessment	Teacher-led assessment
Individual	To cooperate with the group, and actively participate in and finish teamwork			
	To discuss and communicate with members to analyze and solve problems			
	To think and listen, then express thoughts clearly and logically			
	To express his/her own views on global climate change			
Group	To design a complete report or initiative			
	To use the strength of the team to complete a series of climate change activities in a scientific and rigorous manner			
	To be a team player			
Self-suggestion:				
Suggestions from the group:				
Teacher-suggestion:				
Note: Student self assessment, group-assessment and teacher-led assessment are all graded as A (excellent), B (good), C (fair), D (poor).				

1) Exploring the mystery of the shipwreck

The "Facing the Future" team was divided into two groups to find the mystery of the sinking of the Yangtze River Estuary No.2 Ancient Vessel. Since there is no unified theory in the archaeological field about the cause of its shipwreck, the team speculated the mystery of the sinking by looking up ancient books, online information and documentaries, etc. The purpose is to make students make a historical and reasonable speculation.

According to the known official reports, the vessel is a merchant ship during the reign of Emperor Tongzhi (1862—1875) in the Qing Dynasty (1644—1911). It is also the largest and best-preserved sunken wooden sailboat in China's underwater archaeological findings, with a huge number of cultural relics on board.

The first group speculated that the cause of the shipwreck might be affected by the typhoon weather. They found a report titled "Why did the Yangtze River Estuary No.2 Ancient Vessel sink? East China Normal University estuarine coastal geology team to decipher" on the official website of East China Normal University, in which the Shanghai Chronicles recorded storm events during the Tongzhi period. The table 2 showed that Shanghai had experienced extreme weather events such as "hurricanes", "tsunamis" and "heavy rain". Although there is no direct evidence that the wreck was due to extreme weather, the existence of the *Shanghai Chronicles* makes this speculation possible.

Table 2. Record of Shanghai Chronicles

1864	the third year of Tongzhi Dynasty	Lunar June 10	Jiading county, Chuansha county, Xuhui county	Hurricane. Trees and houses including Xujiahui Public School were blown down, and tiles were blown around. Chuansha north and south gate suspension bridge flew into the city. Docked boats were drowned.
1866	the fifth year of Tongzhi Dynasty	The eighth day of the eighth lunar month	Chuansha county	Tsunami. After two or three times, it started to over.
1867	the sixth year of Tongzhi Dynasty	autumn		Hurricane, sea storm.
1868	the seventh year of Tongzhi Dynasty	Lunar May 29		Heavy wind and rain, trees and houses were blown down. Fields and houses were waterlogged several foot.

The second group speculated that the cause of the wreck is related to the characteristics of the wooden large junk. Although wooden large junks have superior performance in offshore navigation and large load capacity, they have large contract area with water and poor wave-breaking capacity. When encountering unexpected situation, itmay be too late to respond. The cause of its wreck was not officially determined, but the answer is not significant to this research. It focused on the process of thinking. Conjecture is not just imagining things out of thin air, but presenting facts and reasoning, and making reasonable assumptions by consulting relevant information. As a climate change team, it had to dig deeper, work harder, and think farther to make a potential connection to the ancient vessel in terms of climate change.

The New Jersey Student Learning Standards (NJSLS) 3rd Grade Interdisciplinary Teaching Example mentions that students can collaborate to design plans to solve real-world problems of extreme weather events (such as floods, heat waves, forest fires, droughts, and / or hurricanes). They can choose a weather event related to the local environment and do relevant research from multiple perspectives to determine the individuals and technologies in reducing weather-related risks. Students can then design plans to address the local weather threat posed by climate change and make recommendations to relevant stakeholders to promote real-world change.

Inspired by this, under the guidance, the children used their spare time to compare the ancient responses to extreme weather with today's practices, and reported in groups.

The Students' Reports

- In ancient times, large junk sailed on the vast sea, and when facing extreme weather such as stormy weather and huge waves, captains and crew could only rely on experience and wisdom to deal with it. They predicted and avoided possible dangers by observing astronomical phenomena, sea phenomena, and sensing natural regulations such as wind direction and water current.
- In the hull design, the ancient people also fully considered the ability to resist extreme weather. Large junks were usually built of sturdy and durable wood, with thick hull and flat bottoms to sail steadily in heavy wind and waves. Besides, vessels were also equipped with some emergency equipment, including lifeboats, life rafts, in case of emergency.
- However, despite the wisdom and courage of ancient people in dealing with extreme weather, their means and effects are still relatively simple and limited compared with modern technology.
- In modern society, we have a series of high-tech means such as advanced weatherforecasting system, satellite navigation technology and ship automation control system, etc., which have greatly improved the ability and accuracy to deal with extreme weather.
- During the design and construction process of modern ships, they not only pay attention to the stability and durability of the hull, but also give full consideration to environmental protection and energy saving. At the same time, modern ships are also equipped with a variety of advanced navigation and communication equipment, making navigation safer and more efficient. Moreover, in response to extreme weather, modern ships are also equipped with various advanced emergency equipment and technologies, such as automatic lifeboats, automatic fire extinguishing systems, etc., which greatly improve the self-rescue ability and survival rate in danger.

The students' reports were wonderful, and the slides were informative and vivid(see Figure 1). The teacher summarized on the basis of the children's display that this comparison not only showed the great changes in human technology and social progress, but also reminded us to cherish the convenience and safety brought by modern technology. And we should also know that human wisdom and courage are still our most valuable assets when facing extreme situations such as natural disasters.

Figure 1. Representatives of the group report on ancient and modern extreme weather

2) Protecting cultural relics

The impact of climate change on the salvage of the Yangtze River Estuary No.2 Ancient Vessel is a complex and severe problem. Due to the global warming, the sea level at the Yangtze River Estuary has been gradually rising, and environmental factors such as seawater temperature and salinity have also been hanging. These changes have brought many challenges to the salvage and preservation of ancient ship relics. The teacher led the students to think about the impact of climate change on thesalvage of cultural relics.

Firstly, the rise in sea level directly threatened the safety of the vessel. Its site was located in deep underwater, and the rise of sea level might lead to changes in the seabed topography around the site, which affects the structural stability of the vessel. In addition, the erosion of seawater will be intensified by rising sea levels, causing more serious damage to the wooden structure of the ancient vessel.

Secondly, changes of seawater temperature and salinity caused by climate change have an important impact on the preservation of cultural relics on the vessel. Rising water temperature can accelerate the decay of the remains, while changes in salinity may lead to increased corrosion of the hull material. All these factors further deteriorate the preservation status of cultural relics, and increase the difficulty and risk of salvage work.

To meet these challenges, the teacher and students have considered a series of measures to protect the relics. First of all, monitoring and research of the ancient ship site should be strengthened to grasp the preservation status and the changing trend of the site in time. Secondly, a reasonable salvage plan should be formulated, and advanced salvage technology and equipment should be adopted to ensure that the salvage work is carried out under the premise of protecting cultural relics. At the same time, the research and response to climate change should be strengthened, and measures should be taken to mitigate the effects of sea level rise and changes in the seawater environment on these relics.

In short, the impact of climate change on the salvage of cultural relics of the Yangtze River Estuary No.2 Ancient Vessel cannot be ignored. While protecting cultural relics, we need to actively respond to the challenges of climate change and contribute to the protection of the historical and cultural heritage in the Yangtze River Estuary.

Effectiveness & Reflection

A student-led project case, centered on this ancient ship, was explored and studied in depth. Under the guidance of the teachers, the students jointly made the climate change proposal of "Facing the Future" and the brochure of "Find the Yangtze River Estuary No.2 Ancient Vessel, and Promote the Yangtze Estuary's Vitality".

1. "Draw the Future" Climate Change Proposal

Climate change education is a long and continuous process, "Draw the Future" climate change team wrote an initiative to call on students, parents and society to pay attention to the Yangtze River Estuary No.2 Ancient Vessel and the Yangtze River Estuary environmental protection, and to contribute to the protection of our historical and cultural heritage and the maintenance of the ecological balance. They will jointly take the following actions:

(1) Strengthen the monitoring and assessment of the ecological environment of the Yangtze River Estuary, grasp the impact of climate change on the ecological environment of the Yangtze River Estuary, and provide scientific basis for the protection of ancient ships and archaeological excavation.

(2) Promote the formulation and implementation of ecological and environmental protection policies in the Yangtze River Estuary, strengthen ecological and environmental protection, and reduce the damage caused by human activities to the ecological environment in the Yangtze River Estuary.

(3) We will strengthen the protection and archaeological excavation of the No.2 Yangtze Estuary, adopt scientific methods and technical means to protect the ancient ship and its surrounding cultural relics, and leave precious historical and cultural heritage for future generations.

(4) Enhance public outreach and education to raise awareness and consciousness about climate change and cultural heritage protection among the general public, fostering a positive atmosphere across society for the collective concern over the Yangtze River Estuary No.2 ancient ship and the environmental protection of the Yangtze River Estuary.

This action not only reflects the students' deep understanding of environmental protection, but also shows their sense of responsibility to participate in the society and serve the society with practical actions.

2. "Find the Yangtze River Estuary No.2 Ancient Vessel, and Promote the Yangtze Estuary's Vitality" Brochure

The brochure "Find the Yangtze River Estuary No.2 Ancient Vessel, and Promote the Yangtze Estuary's Vitality" was compiled and organized by the teacher with materials provided by the students. The brochure aims to publicize the vessel in multiple dimensions and angles. There is a section on "Climate Change", which includes "Responding to Extreme Weather in Ancient Times Compared to Today" and "Reflecting on the Impact of Climate Change on Cultural Relics Salvage", so that students, parents and members of the public could see the efforts of the "Draw the Future" team on climate change.

The students comprehensively understood the historical background, cultural value and unique status of the vessel in the Yangtze River Estuary through reading materials, drawing and writing articles, and then designed a brochure after exchanging and summarizing the information. This not only practiced their data collection, painting and writing ability, but also makes them feel the broad and profound Chinese culture in personal practice, thus enhancing their cultural confidence and teamwork spirit (see Figure 2). With the process evaluation of "Facing the Future Climate Change Team" as the evaluation standard, the children will not blindly do various activities just for complete the tasks, but reflect in the activities and grow up in the activities.

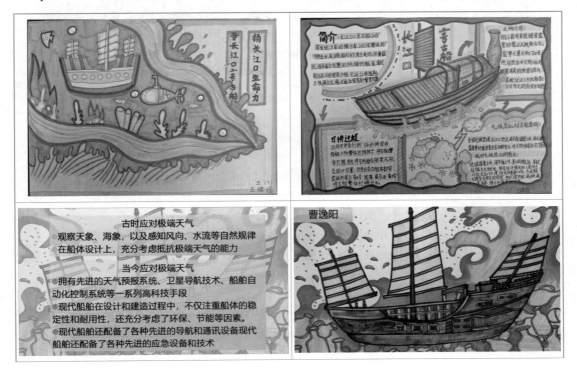

Figure 2. Brochure "Find the Yangtze River Estuary No.2 Ancient Vessel, and Promote the Yangtze Estuary's Vitality"

The project is not only an interdisciplinary educational practice, but also a cultivation of students' comprehensive development and their sense of social responsibility. The joint support of the school and the family provides a strong guarantee for the smooth progress of the project. Although there is no direct communication between the family and the school, teachers can sense the parents' recognition and emphasis on the project through the children's enthusiastic feedback. This tacit understanding and cooperation, for the growth of the children to create a more harmonious environment.

Of course, this project also faces many challenges. First, the professional ability of climate change education needs to be improved. In the process of infiltrating climate change education for students, teachers find that their professional ability needs to be enhanced, and they need to accumulate and supplement more cutting-edge knowledge in related aspects. Second, the propaganda and radiation range should be promoted. We can use the school's WeChat official account, short video and other communication means to innovate the publicity forms of communication, so that more people can know about climate change education and the historical value and cultural connotation of the second ancient ship in the Yangtze Estuary. Third, expanding public resources should be linked together. Linkage integration of the strength of the school, home and community, and jointly deepen the solid. In the future, we will carry out a series of activities with some schools located in the Yellow River Estuary to further promote the coordinated and sustainable development of the Yangtze River Estuary and the Yellow River Estuary.

(Youchun Guo, Teacher of Tonggang Primary School, Pudong, Shanghai)

The Past, Present and Future of Heavy Industrial Plant in the Yangtze Estuary

Research Background

Scientific research suggests that human activities, especially human activities since the industrial revolution, is primarily responsible for climate changes, most notably global warming.① Under the background of increasingly serious global climate changes, the impact of industrial development on the environment is an issue that cannot to be ignored. In this research, we study the heavy industry factories around Donggou Primary School, Pudong Northern Road, Gaohang Town, Pudong New District, Shanghai District. These factories flourished in the 1980s, making remarkable contributions to the regional economic development. On the other hand, the development was achieved at the cost of the localecological and climate changes.

To deepen the knowledge of the relationship between industrial development and climate changes, our school has launched a project, leading students to get a deeper understanding of the impact of industrial development on climate change.

Project Design

This project is based on the status of heavy industries located in the region same to Donggou Primary School. Grade-3 students are chosen as participants, who are expected to deepen the knowledge of the development journey, environmental pollution and green transformation of factories nearby the school through interviews, surveys, material collection and on-site factory visits. The project goal is to explore the impact of industrial development on climate changes.

Project Implementation

1. Pre-project: Discussions of Research Issue

On December 25, 2023, the teacher asked students before class began. "Do you know which heavy industry factories are nearby our school?" According to students' responses, the teacher asked more questions. For example, "How did these factories affect people in the past?" "What are these factories like now?" "What improvements

---

① In February 2008, Zalasiewicz and 21 other members of the Stratigraphy Commission of the Geological Society of London published a paper in *GSA Today*, arguing that human activities—particularly economic activities since the Industrial Revolution—have had a global impact on the climate and the environment. Zalasiewicz, J., Williams, M., Smith, A., et al. Are we now living in the Anthropocene? [J]. GSA Today, 2008, 18(2): 4-8.

can we make in the future to reduce the impact of industrial development on climate change?"

In the process of interaction, students tried to analyze the causes of the environmental pollution and the effects of the pollution on climate change. As the discussions deepened, the teacher brought forth a warm-up issue, that is, "How can we design a constant-temperature green plant to reduce the impact of industrial development on climate changes?" The teacher hoped that this warm-up issue would stimulate students' thinking.

2. Exploration: Formation of Implementation Plan

At this stage, the students conducted a series of research according to the key research issue. However, these Grade-3 research participants are still too young and needed to learn a lot of background knowledge. To facilitate the group learning, the teacher guided students to experience how to solve climate changes.

First of all, the students were divided into different project groups according to their research interests. Following that, the teacher organized the students to pinpoint the survey scope. Finally, representative factories nearby the school were selected as survey objects. Before January 2024, according to the requirements of the "Exploring Blessings" sub-activity in the "Winter Vacation Blessing Collection Activity" assigned by the school's cadet department, students conducted on-site visits, engaged in in-depth discussions with veteran factory workers, collected a large amount of first-hand information, and gained an understanding of the history, production status, and environmental impact of factories around the school. On March 1, 2024, in the classroom, students shared the data they had observed and collected over the past period in groups.

One group's investigation found that there used to be factories around the school that emitted an unpleasant odor that made it difficult for passersby to breathe. They speculated that this might be related to the raw materials used in the factories. Another group learned from asking their elders at home that the rivers around the factories used to be black and dirty, and that the smoke from the chimneys was continuously billowing, causing a noticeable increase in temperature when passing by the factories.

Through sharing, the students intuitively understood the past glory and changes of the factory. In order to make the following research more focused, the students were guided to think about what problems have been caused by the development of heavy industry, what problems need to be solved at the present stage, and what problems they may encounter in the future from the perspective of climate change. Each group discussed the key issues for further research according to the exchange.

3. Action: Knowledge Framework Construction

In this stage, students refine their knowledge construction of urban development, industrial development and climate change education through "experiential" activities.

Firstly, students learned the development process of Pudong. On March 22, 2024, the teacher organized students to visit the Pudong Exhibition Hal lto experience the magnificent development journey of Pudong since its adoption of the reform and opening-up over the past three decades. The students learned that the heavy industry factories nearby the school had once made remarkable contributions to the regional economic development. They were deeply aware that, as the successors of the new era, they should not only inherit the wisdom and experience passed down by the former generation, but also contribute their share to the future development through bold exploration and changes.

Secondly, the school actively communicates with the surrounding factories, enterprises and other partners, and jointly promotes the progress of the project through the activity of "sowing the seeds of innovation, illuminating the dreams in your heart". Taking SINOPEC Shanghai Research Institute of Petrochemical Technology as an example, on April 2, 2024, through field visits (see Figure 1), students gained an in-depth understanding of the actual situation of the development of petrochemical industry and its improvement and innovation, and the history of the development of the Shanghai Research Institute. From solving the problem of people's clothing at the beginning to focusing on the major national and industrial needs nowadays, the results of the research have been blossoming in a variety of fields. The simple explanation stimulated the students' interest in the petrochemical field, and also provided more practical suggestions for the students to design the "thermostatic" green factory for the future city.

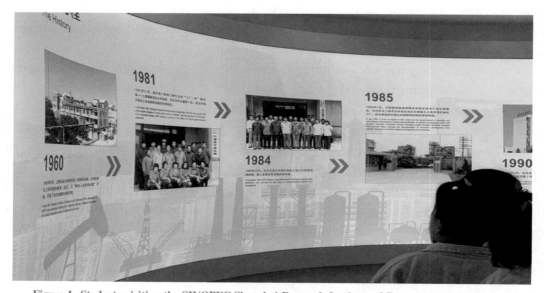

Figure 1. Students visiting the SINOPEC Shanghai Research Institute of Petrochemical Technology

Planting trees is one of the important measures to mitigate climate change. To actively respond to the call of "Lucid waters and lush mountains are invaluable assets", on March 12, 2023, students and teachers participated in the "ecological environment, building a beautiful Gaohang" forest chief system activity organized by Pujiang Academy in Gaohang Town(see Figure 2), contributing to the protection of green. Through this activity, the students learned that increasing the forest coverage rate can effectively reduce greenhouse gas emissions, regulate the temperature and other effects, all of which can help to slow down the trend of global warming, and create a more livable environment for the human survival and development.

Figure 2. Students participating in the forest chief system activities

After practicing and learning in this phase, the groups constructed their own understanding of the key issues in practice, and realized that industrial development is one of the important drivers of climate change, but there are also actions we can take to protect the environment and mitigate the effects of climate change.

4. Design: Forming of Project Works

In order to better display the project works in the next stage, April 15, 2024, each group needed to formulate the design scheme of the future urban "thermostatic" green factory according to the learning in the previous stage, and the group members will present the design works in the form of painting and composition(see Figure 3). Through the group discussion, they came up with the requirements for the design of future green factories: they needed to focus on technological innovation and the application of green production methods, and realize the coordinated development of industrial production and environmental protection through the clean energy and other measures. At the same time, some students also considered the integration of the factory and the surrounding environment, and put forward the idea of building an eco-friendly factory.

Figure 3. Students' paintings

5. Evaluation and Summary: Presenting Research Findings

The final outcome of this project consists of individual and team results. Team outcomes require groups to develop a design for a future urban "thermostatic" green factory. Individual outcomes require students to draw the design of the future factory and write an imaginary essay about the future factory.

The evaluation of this project is based on the combination of process evaluation and summative evaluation(see Table 1, Table 2), with process evaluation as the main focus and summative evaluation as the supplement, focusing on inter-group and intra-group evaluation of students. There will be success criteria for each stage of the assignment.

Table 1. Evaluation indicators of project activities

Evaluating indicator	Evaluation description
Teamwork	To understand the roles of yourself and your partners, set appropriate participation rules based on the content areas in which they specialize, and handle problems in communication appropriately.
Communication and sharing	To exchange ideas with peers or others in appropriate ways, with rich forms of expression, correct views, novel content and distinctive personality.

(Continued)

Evaluating indicator	Evaluation description
Creative expression	To analyze and demonstrate the existing information from different perspectives, and form different strategies, technical methods or problem solutions.
Information gathering	To reasonably analyze, process and handle information, to extract ideas from it, and to provide evidence to support the ideas.
Problem solving	Understand the constraints to solve the problem, set goals and take appropriate actions to complete the task, and continuously monitor, optimize or iterate results.

Table 2. Evaluation scale for teamwork performance in project activities

Evaluation scale			Evaluation by members outside the group	Evaluation by group members	Student self evaluation
Excellent	Average	Needs improvement			
Actively participates in team activities, and often speaks to provide advice.	Be able to participate in team activities with encouragement from others, less likely to speak.	Barely participates in team activities, does not speak up.			
Actively participates in the project practice activities, collects information and makes personal recommendations.	Participates in some project practice activities, collects information but is incomplete.	Does not participate in various information gathering or project practice activities.			
Participates in the production of project results, completes own tasks and assists peers.	Participates in making project results, completes his/her own task, but less helps others.	Does not initiate or participate in producing project results, does not complete his/her own tasks.			

Effectiveness & Reflections

By thoroughly exploring the close connection between heavy industry and climate change in the school's region, the project not only promotes the deepening of school education, but also makes students, parents and the society pay more attention to and understand the urgency of climate change.

1) Students have established the concept of climate changes and strengthened their awareness of environmental protection

Through project-based learning and practical activities, students have a deep understanding of the links, impacts and coping strategies between industrial development and climate change, and form a solid knowledge base on climate change. Students' understanding of environmental issues went from the surface to the deep, and they not only knew what they knew, but also knew why they knew it.

Various environmental activities, such as tree planting, garbage classification, energy saving and emission reduction, environmental protection publicity, etc. These activities make students experience the importance of environmental protection in practice, thus consciously enhancing their awareness of environmental protection. Students began to pay attention to the impact of their own behavior on the climate, and took the initiative to take environmental protection measures to reduce carbon emissions and realize a low-carbon life.

Realizing that they were a member of the global village, students actively participated in community activities to protect green activities, and promoted the green concept to more people, expanding the influence of education. With the depth of this project, students gradually developed good environmental behavior habits, such as saving water and electricity, reducing the use of disposables, promoting recycling and so on. The formation of these habits not only helps individuals to reduce the impact on the climate and the environment, but also has a positive impact on the people around them, forming a good upward atmosphere.

Here are some of the students' comments on their participation in the project:

Student Li: When I see the dark smoke billowing from the factory chimney, I always feel a little heavy in my heart. These emissions are one of the main culprits of global warming. If there is no improvement, our future living environment will be difficult to imagine.

Student Wang: My hometown is a highly industrialized city, where the sky is often covered with haze. I am well aware of the impact of industrial pollution on the climate. It has not only changed our air quality, but also caused serious damage to the ecosystem of the whole earth.

Student Liu: The impact of industrial development on climate change is all-round. It not only changes the composition of the atmosphere which leads to the intensification of the greenhouse effect, but also changes the surface morphology and intensifies the urban heat island effect. All these changes have a profound impact on our lives.Therefore, we must face up to this problem and take effective measures to deal with climate change.

Industrial development has a profound and extensive impact on climate change.

Through continuous education and guidance, it is hoped that students can keep their attention to and participation in climate change in their future study and life.

2) School-home-society connection for better educational effects

This project received strong parent participation, positive leadership from the school, and broad community support.

Parents have a deeper understanding of the impact of the industrial environment on climate change, through ndirect or direct participation in project activities. Some parents work in the factories around the school and participate together with their children through their narration and guidance to form a strong family support.

In addition to teaching about industrial development and climate change in the classroom, schools organize field trips, lectures and visits to stimulate students' interest and enthusiasm in learning about climate education.

The community provides venues, resources and other support for the project activities. At the same time, through these activities, it popularizes environmental protection knowledge among residents and raises the environmental awareness of the whole community through the project activities.

Here are some comments from parents about the impact of industrial development on climate change:

Student Wang: Industrial development is inevitable for the progress of the times, but we cannot ignore its negative impact on climate change. As a parent, I always pay attention to the health and future of my children. I hope the government can strengthen supervision and promote the development of green industries so that our children can breathe the fresh air and enjoy the blue sky and white clouds.

Student Zhang: I remember when I was a child, the sky was so blue and the water was so clear. As a parent, I hope I can educate my grandchildren to cherish the environment and protect our planet. At the same time, I also hope that the society can pay more attention to environmental education, so that more people will be aware of the impact of industrial development on climate change.

Student Liu: Industrial development is an indispensable part of modern society, but we can't let it be the main culprit of destroying the environment. As a parent, I will guide my children to pay attention to the issue of climate change and let them understand the impact of industrial development on the environment. At the same time, I will also set an example by reducing waste and pollution, and doing my part for the environment.

Through the linkage of school, home and society, resource sharing and collaborative parenting, the project activities have formed a good atmosphere, contributing to the fight against climate change and the protection of the Earth's home.

Inevitably, this project also faced many challenges. For example, as the project kept deepened, the students' knowledge of industrial development was not limited. Some factories were difficult to investigate on site because of their special conditions. Because of students' limited research perspectives and knowledge, their project outcomes were notcreative enough and their design lacked variety. Additionally, climate is a complex and abstract issue for a primary school teacher without a relevant academic background. Without the relevant research background, the teacher finds it hard to deliver the background knowledge to primary school students in a comprehensible language.

From the perspective of climate education, this project is not only an in-depth exploration of the relationship between industrial development and climate change, but also a vivid practice of climate education for students. Through the implementation of the project, we deeply realize the profound impact of climate change on human society and the natural environment, as well as the importance and urgency of addressing climate change.

From the perspective of climate education, this project is not only an in-depth exploration of the relationship between industrial development and climate change but also a vivid practice of climate education for students. It can inspire us to reflect on how to better integrate climate education content into education. This can be achieved by designing climate education activities that are more closely aligned with the local industrial development situation, as well as utilizing modern teaching methods and technologies such as virtual reality to help students gain a more intuitive understanding of the impacts of climate change.

To sum up, this project is not only a useful exploration of the relationship between industrial development and environmental protection, but also an important practice of climate education for students. Through the implementation and reflection of the project, we have strengthened our determination and confidence in promoting green industrial development and enhancing climate education. We will continue to deepen our research and practice in this field, and make a greater contribution to promoting climate education.

(Yan Cai, Teacher of Donggou Primary School, Pudong, Shanghai)

"Green Travel" Makes Carbon Emission Reduction Visible

Research Background

Global climate change has become a topic of widespread concern in today's society. It has led to widespread extreme weather events and has brought many negative impacts to human society. How to respond to climate change is one of the research topics of human science. The United Nations Educational, Scientific and Cultural Organization (UNESCO) defines climate change education as "learning to cope with risks, uncertainties, and rapid change," with the goal of building an understanding and capacity to respond to climate change and the impact of global warming on biodiversity.

China has long attached great importance to the issue of climate change and has been firmly committed to taking the ecologically-prioritized, green and low-carbon path of high-quality development. Carbon peak and carbon neutrality have been integrated into the overall development of the economy and society, with carbon reduction serving as the key strategic direction for ecological civilization construction. China is committed to building a modern society in harmony with nature. At present, it is of great significance to conduct "climate change education" for the youth. As the future masters of society, young people should strive to actively participate in responding to climate change and environmental protection, establish a concept of responding to climate change and environmental protection, cultivate environmental awareness, enhance social responsibility, and actively take measures to effectively respond to climate change.

Project Design

Under the guidance of A Distinguished Head Teacher from Changzhou, relying on the professional support of the Shanghai Municipal Institute for Lifelong Education, ECNU, and drawing on the strength of the research team, the author, serving as the head teacher of Class 4, Grade 4 in Changzhou, led several students in the class to form the "Green Miles" Playmate Team during the winter holidays and conduct project-based research and practice in an organized and planned manner.

The students had participated in or learned about some energy-saving and emission-reduction activities, but they often remained at the level of publicity and action without forming carbon reduction data, which indicates that the participants did not know the extent to which their behavior could contribute to energy saving and emission reduction. Therefore, the students, through discussion and research, formulated the "Climate Change" project-based learning activity plan (see Table 1). The students spontaneously formed a project-based learning research Playmate Team named "Green Miles".

Table 1. "Climate Change" project-based learning activity plan

Time	Method	Content	Exhibition
Jan. 6-7, 2024	Information search	Climate, climate change	Themed post, mind map
Jan. 13-21, 2024	Information search	Understanding the current status and impact of global climate change	Themed post, mind map, class exchange
Jan. 27-29, 2024	Information search	The causes of global climate warming and the composition of greenhouse gases	Themed post, mind map, class exchange
Jan. 30-Feb 18, 2024	Green travel	Learning on the way	Take screenshots, annotate, and upload the route map of your journey to the class group photo album
Feb. 19-25, 2024	Data processing	Classification of travel modes and mileage conversion	Data statistics table, statistical chart
Feb. 29, 2024	Final exhibition, classroom presentation	Learning from the results	Results report, experience sharing

Project Implementation

1. Learning Through Information Search

Students started by learning about climate change through online searches and reading popular science books. Then they knew global warming is mainly caused by human activities. It will cause a redistribution of global precipitation, melting of glaciers and permafrost, and rising sea levels, which will not only endanger the balance of natural ecological systems but also threaten human food supplies and living environments.

Based on this imformation, students drew climate change posters, studied the relationship between meteorology and climate, and compared ancient and modern meteorology to feel the impact of climate change on phenology. They also visited the "Trinasolar" Energy Science and Technology Museum to learn about solar power generation and its impact on reducing carbon emissions.

The students discovered through their study tours that the exhaust from gasoline-powered cars contains a lot of carbon dioxide. Each liter of gasoline burned produces about 2.4 kg of carbon dioxide, and the fuel consumption of gasoline-powered cars on city streets is much higher than on highways or in rural areas. Consequently, the large number of gasoline-powered cars in cities must lead to the worsening of the greenhouse effect. Therefore, the students spontaneously formed the "Green Miles" Playmate Team and decided to start with themselves and reduce the use of private gasoline cars to do something to reduce carbon dioxide emissions.

2. The Real Demand for Green Travel

In the previous "Green Travel" activities, the participants often used additional non-demand action routes, which did not have any actual "emission reduction" effect, but rather increased it. Taking this into consideration, they decided to use the real demand for transportation as the basis for the study, and to actually "reduce emissions" through the actions of ourselves and our families.

To truly reduce carbon emissions, we studied our actual travel needs and recorded our green trips in a class group photo album. During the winter vacation, all of the original data was statistically compiled by the team members (see Figure 1), and the team's green travel statistics table was created. After compiling all the data, The students found that our efforts resulted in over 4,000 kilometers of green travel, and 2% of the travel was "zero carbon" walking and cycling, 29% was private electric vehicle travel (including electric bicycles and electric cars), and 69% was public transportation (including buses, subways, and high-speed trains).

Figure 1. Green Travel Playmate Team

The students also found that by reducing gasoline consumption by approximately 295 liters, they could reduce carbon dioxide emissions by 709 kilograms. This is a truly inspiring result.

3. Multi-level Display of "Gaining New Fans"

In February 2024, leveraging the "Hello, Winter Vacation!" project, the class shared their research findings with the entire school. When other students saw the result of 709 kilograms of carbon dioxide reduction, they were amazed and exclaimed, "I never thought that green travel could really reduce so much carbon dioxide," and they expressed their desire to join in and contribute to combating climate change.

Changzhou Television also covered the event with an interview of the class, promoting the "class experience" to audiences across the city, encouraging more peers to join this "green team."

In March, the children shared their discoveries with their parents, and they engaged in in-depth discussions, achieving a collaboration between school, home, and community (see Figures 2 and 3).

Figure 2. Sharing their insights with the whole class

Figure 3. Parents engaging in heated discussions with their children

In April, Playmate Team (Grade 4, Class 4) shared their findings at school events and even participated in "Shanghai Climate Week," where the teacher and the students presented action experiences at a youth conference (see Figure 4). The teacher also participated in the round table discussion of the "Climate Resilience Community Forum" as the class guidance, sharing the synergistic effect of school-family-community cooperation in climate change education.

4. True Experience Makes Real Feeling

After that, students expressed their true feelings. They all said that such project activities were meaningful, and they improved their comprehensive abilities, enhanced their sense of social responsibility, and contributed their own efforts to addressing climate change.

Student Wang said: This winter vacation, our class organized a particularly meaningful activity—green travel. I started with cycling, walking, and taking the bus. Through this activity, I had a deeper understanding of the significance of green travel, and at the same time, I realized that green travel requires the joint participation and efforts of everyone. Only by working together can we make our city more beautiful and livable.

Figure 4. Youth Climate Action Science and Technology Innovation Presentation Event

Student Sun also expressed her thoughts. At the beginning of the holidays, she participated in the class activity of "Green Travel, Start with Me". First of all, green travel is a relatively environmentally friendly travel method, which saves energy, improves efficiency, reduces pollution, and is good for health and efficiency; at the same time, green travel is also a symbolic concept and a sustainable environmental concept. Therefore, choosing green transportation is not only choosing a healthy way of travel, but also choosing an active lifestyle.

Student Zhang believes that walking is a low-carbon travel method that protects the environment, exercises the body, and allows people to appreciate the scenery along the way. Start from myself and start from the people around me. Low-carbon travel, and do our part to protect the environment.

Student Guo also believes that green travel means not to block others and not be blocked by others. "When people stuck in traffic look at the endless line of cars ahead, I ride my beloved little bicycle, strolling through, that feeling is beyond words! I exercise myself and contribute a little to reducing pollution."

Student Huang also shared that he participated in a green travel activity during the winter vacation, starting with subway, walking, and taking the bus. Without reducing travel efficiency, it saves energy, improves efficiency, reduces pollution, and is beneficial to health. Start from myself and start from now. Make green travel the top choice for travel and make a contribution to environmental protection.

Student Liu said that she participated in a green travel activity during the winter vacation and deeply understood the importance of environmental protection. Choose public transportation and low-carbon travel to reduce carbon emissions and lighten the burden on the earth. The scenery along the way makes me cherish the beauty of nature even more. This activity made her realize that every small effort from each person can bring positive changes to the environment.

Many other students also expressed their happiness to contribute to protecting the environment by their own efforts, and they could consciously follow the 135-travel plan, which means walking within 1 kilometer, cycling within 3 kilometers, and taking public transport within 5 kilometers. They will take every possible step to walk, cycle, or take public transportation to set an example for low-carbon travel.

Parents also think that the green travel during the winter vacation made them experience the importance of environmental protection (see Figure 5). Every choice of green travel is a contribution to the earth. It is not just a slogan, but a lifestyle that needs us to practice andpersist in. Cycling, walking, and taking public transportation is not only good for reducing environmental pollution, but also good for our health. They call on more people to choose green travel and defend our earth with concrete actions!

Figure 5. Parents sharing their feelings and support for the activity

Effectiveness & Reflections

The best way to solve problems is to take actions. Not only do we need to know why and how we are taking actions, but we also need to assess the effect and significance of our actions in a real and scientific way. Using data to present the truth in front of everyone enhances the persuasiveness of the action effect.

Green travel may seem small individually but has significant collective impact in avoiding the effects of global warming. It avoids global warming, ice melting, and the extinction of many innocent lives.

The promotion of the class micro project has led the way, from one team to the involvement of an entire class, and finally to the sharing of experience and display of achievements to the whole school, community, city, and the country. While this project was successful overall, there are areas for improvement:

First, whether the students' green travel is "necessary travel" or whether there are travel data generated for the activity.

Second, there are many travel modes in the students' travel, including electric vehicles or electric bicycles. However, the calculation does not include greenhouse gas emissions from burning coal for electricity, which could be a direction for further research.

Third, this project activity is limited to reducing carbon emissions and does not focus on increasing carbon sinks. Future activities may extend to carbon sink enhancement.

(Rong Fu, Teacher of Longhutang Experimental Primary School, Xinbei, Changzhou, Jiangsu)

Practices of Implementing Climate Change Education in Secondary Vocational Schools

Research Background

The important task of secondary vocational education is to serve as the basic education for the system of cultivating applied talents. It has become a pressing issue for all secondary vocational schools to address how to raise the awareness among vocational students about the serious impacts of climate change on human economic and social development, human health, and natural ecosystems. In addition, it is crucial to explore how to integrate students' specialized fields of study into comprehensive climate change education activities within vocational education.

President Xi Jinping emphasized the "strategic importance of ecological protection" during his Northeast inspection tour, stressing that "ecology is a resource and ecology is productivity." In the winter of 2023, the Harbin Municipal Government, focusing on the attraction of unique environments, vigorously developed the ice and snow tourism industry. The cold climate is a prerequisite for the development of the tourism economy in Northeast China, and the occurrence of warm winters undoubtedly affects the development of winter tourism in this region. Raising young people's awareness of climate change and promoting environmental protection concepts are crucial to the sustainable development of Northeast China. As emphasized in the *Berlin Declaration on Education for Sustainable Development*, sustainable development education can cultivate learners' cognitive and non-cognitive skills, such as critical thinking, collaborative abilities, problem-solving capabilities, resilience, systematic and creative thinking in the face of complex situations and risks[①] For students majoring in nursing, it is essential to pay special attention to the impact of climate change on human health, thereby applying their professional knowledge to real-life situations, fostering noble professional ideals, serving people, and caring for lives.

In December 2023, Harbin Health School launched comprehensive climate change education activities in class 10, grade 2023, as a pilot class. Over a period of two months, students conducted a series of activities, including surveys, research, family interviews, and community lectures. A questionnaire survey showed that the project not only enriched students' science knowledge, but also significantly improved their cooperation and communication skills, with a satisfaction rate of 90%.

---

① Berlin Declaration on Education for Sustainable Development [EB/OL]. https://unesdoc.unesco.org/ark:/48223/pf0000381228.

Project Design

Taking into account the personalities, interests, and age characteristics of secondary vocational students, the teacher designed the climate change education as a student-centered, exploratory and comprehensive activity (see Figure 1).

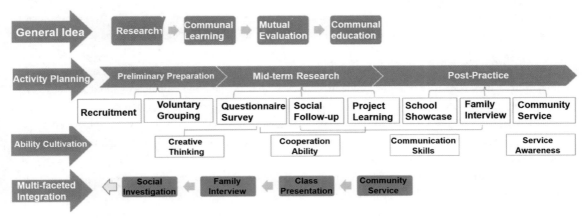

Figure 1. Comprehensive Activity Design Framework

Overall Objectives: To provide nursing students with environmental knowledge, values, and attitudes to adapt to the new climate realities through this comprehensive climate change education activity. By studying the impact of climate change on human health, students will use their professional knowledge and skills to serve the community and promote a noble professional ideal.

Specific Objectives: To enhance students' understanding of the current climate change situation in Northeast China and foster their environmental awareness. During the event, students will grasp methods of coping with extreme weather and improve their survival skills. They will also explore the relationship between climate characteristics and diseases, thereby enhancing their professional knowledge.

Project Implementation

1. Conducting Climate Change Education Research Activities

From November 7th to 12th, 2023, Class 10 Grade 1 of the nursing program was selected as the pilot class for activity recruitment and grouping. In conducting climate change education for secondary vocational students, the principles of being close to reality, professionalism, and daily life were adhered to. The theme of the comprehensive climate change education activity was determined as "Coping with Climate, Serving with Professionalism".

On November 5th, 2023, Northeast China was affected by a widespread heavy snow and rain storm, resulting in school closures and flight cancellations in multiple regions.

Teachers took this opportunity of the recent extreme weather experienced by students to initiate the climate change education by issuing a recruitment notice to all class members.

Climate Change Education Comprehensive Activity Recruitment Letter

Dear fellow students,

  We have just experienced a widespread heavy snow and rain storm in Northeast China. Over the past two days, we have had to suspend classes and stay at home as our studies and daily lives have been severely impacted by this extreme weather. The continuous heavy rainfall and snowfall have result in icy and snow-covered roads, causing vehicles to slip, flights and transportation to be suspended, and citizens' travel to be hindered. But how did such climate changes come about? And what do they affect human life? As secondary vocational students whi grew up and studied in Northeast China, do we truly understand the climate of our hometown? What changes have occurred in our hometown's climate in recent years? Let us join together in the "Climate Change Education" learning activity, and collaborate with peers from Shanghai, Jiangsu, Shandong, and other regions to discuss "matters of great importance" and contribute to the future of our planet.

  "Coping with Climate, Serving with Professionalism," fellow students, come and join the comprehensive climate change education activity now!

Harbin Health School
Class 10, Grade 1 Nursing Program
November 7, 2023

The principle of respecting students' wishes and encouraging their active participation is upheld in determining who will take part in the comprehensive climate change education activity. Any student who is willing to participate is welcome to join. The design of this comprehensive activity is grounded in promoting students' development, with emphasis on fostering their sense of participation and nurturing their abilities. Teachers have devised three activities: a questionnaire survey, social follow-up interviews, and project research, aiming to provide students with opportunities to showcase their strengths. Based on their individual interests, personalities, and abilities, students autonomously select their preferred comprehensive activity groups and negotiate to elect group leaders. Due to the large number of participants, the project research has been divided into four groups (see Table 1).

Table 1. Project recruitment and grouping

Group Name/Focus	Survey Questionnaire Group	Social Personnel Interview Group	Project Research Group 1	Project Research Group 2	Project Research Group 3	Project Research Group 4
Number of Participants	7	5	5	5	5	5

2. Conducting Exploration Activities on Climate Change Education

From November 13th to December 3rd, 2023, the implementation phase of this climate change education exploration activity was carried out.

Step 1: questionnaire surveys and social follow-up interviews were conducted. Teachers assigned the content for the questionnaire survey group, and students researched to identify 20 hot topics related to climate change. Through collective discussion, 10 issues closely linked to daily lives were determined. Students were encouraged to use online mini-programs to create questionnaires, distribute them, and analyze the survey results under guidance, culminating in the compilation of survey reports. Making use of their own networks, students actively shared the questionnaires to obtain more valuable survey data.

Step 2: teachers assigned the content for the social follow-up interview group. After collective discussion, students identified five key issues related to climate change (see Table 2). Prior to the interviews, they practiced with parents or peers, ensuring a balanced gender ratio among interviewees and covering various age groups. The interviews were recorded using audio and video equipment, and interview notes were compiled afterward.

Table 2. Outline of social follow-up questions

1. What changes have you noticed in the climate of Northeast China in recent years?
2. How did these climate changes affected your life?
3. What do you think are the human factors that influence climate change??
4. What measures should we take to address the issues arising from climate change?
5. Describe a scenario where you experienced extreme weather and how you coped with it.

Step 3: teachers and students discussed the survey results and interview transcripts together to determine the research themes for project-based learning. Group discussions ensued to design the project learning frameworks (see Table 3). Teachers emphasized inquiry-based learning and taught students how to retrieve literature and research materials, which fostered a sense of cooperation and teamwork. Adopting a feedback-interactive approach, teachers organized two interim group learning reports during the research process, enabling mutual learning and evaluation among groups. After each report, teachers provided timely feedback and suggestions, enriching and deepening the research content of each group.

Table 3. Project research framework

Theme	Research Framework	Presentation Method
Climate of Northeast China	Overview of the geographical environment Climate characteristics Climate change patterns Future development and impact predictions	A presentation using text, images, videos, and other materials.
Extreme Weather in Northeast China	Causes, hazards, and response methods for sandstorms, blizzards, floods, droughts, and extreme cold	
Climate Change Trends in Northeast China	Temperature trends over the past 50 years Factors influencing climate change Impacts on living environments Measures for climate protection	
Climate and Health in the North and South	Direct and indirect impacts of climate change on human health Strategies for adapting to climate change and protecting human health Climatic and dietary differences between northern and southern residents and related diseases Personal actions and lifestyle adjustments for maintaining health	

3. Launching Climate Change Education Publicity Activities

From December 4th to December 17th, 2023, the comprehensive activity outcomes were showcased in classes and the school.

Firstly, project learning groups shared their research findings with the entire class during a class meeting on climate change education, engaging in collaborative learning, mutual progress, and evaluation activities. After all presentations, classmates voted to select one group to represent the class and demonstrate their findings to the entire school. Following the class meeting, students were encouraged to create health and environmental posters that showcased their knowledge of climate change, extreme weather responses, and climate protection measures. Secondly, the social follow-up activity was extended by inviting all class members to participate in parent interviews, where they shared climate knowledge with their parents. Thirdly, on May 10th, 2024, the eve of Nurses' Day, ten students from the class visited the Zhaolin Community, using their nursing skills to offer volunteer services such as blood pressure and blood sugar measurements. In the process, they promoted knowledge about climate protection and health hygiene, and advised community residents to adjust their diets and lifestyles to maintain good health.

Climate Change Parent Interview Record Sheet
Interviewed Parent: _____
1. Do you think climate change education is important? Why?
2. What do you think is the major climate issue we are facing now? What measures should we take to deal with it?
3. Please describe a scene of extreme weather you encountered when you were young, and tell us how you coped with it at that time.
Today, I am a climate knowledge communicator: _____ To tackle climate change and protect the environment, we should do the following: 1. 2. 3. Parent Interview Time: Parent Interview Location:

Effectiveness & Reflections

The comprehensive implementation of regional climate change education activities has significantly contributed to raising the awareness of secondary vocational students towards climate change, fostering their collaboration skills, creative thinking, and inspiring them to take responsible actions, thereby nurturing them into health professionals with both moral integrity and outstanding technical skills.

This comprehensive activity, spanning from November 2023 to May 2024, was short in duration but yielded rich research results. Among the 45 students in Nursing Class 10, Grade 1, 32 participated, accounting for approximately 70% of the total class. During the project presentation and evaluation, the 13 students who did not participate in practical activities formed an evaluation committee with the group leaders to vote for the outstanding performance groups.

The activity achieved its goal of active participation and learning for all. Thequestionnaire survey team collected 193 online responses and completed a survey data analysis report, while the social follow-up team randomly interviewed 10 individuals and compiled videos, photos, and interview transcripts. These two sets of data provided a solid basis for the project learning group's topic selection.

Project Group 1 gained a comprehensive understanding of geographical knowledge by mastering the latitude and longitude, topography, and climatic characteristics and patterns of Northeast China.

Project Group 2 delved into extreme climate changes in Northeast China through data research. Historical records indicate that there were three major floods and three severe droughts between 1949 and 2007. Students used video materials to teach their peers about coping strategies for floods, droughts, severe cold, heavy snow, and sandstorms.

Project Group 3 examined climate change in Northeast China through literature review. Zhao Zongci etc. predicted a significant warming of over 3°C and possible increased precipitation in the region by the late 21st century based on 23 global climate models. Zeng Xiaofan etc. used the ECHAM5/MPI-OM model to conclude that the average temperature in the Songhua River Basin has continued to rise since 1980.

Project Group 4 investigated the impact of climate change on human health and found that high temperatures affect sleep quality, potentially leading to fatigue, anxiety, and depression. Climate warming may aggravate allergic symptoms and increase the risk of skin cancer. Differences in climate and diet between the north and south contribute to varying prevalence rates of cardiovascular diseases, hypertension, and diabetes. Based on their nursing expertise, students proposed strategies to adapt to climate change and protect human health, and raise public awareness of the health implications of climate change. A total of 25 climate change parent interviews were collected, and 10 students engaged in community volunteer service, distributing 30 copies of the "Climate Change and Health" leaflet.

At the end of the activity, the students expressed their enthusiasm and appreciation for the comprehensive climate education program.

Student Shi remarked, "During this period, our group studied the climate issues of Northeast China through WeChat videos. Even after our discussions, we continued to review our coursework together, and I even scored 100 on the weekly test."

Student Li shared, "I finally managed to create a questionnaire by using a mini-program. I realized the importance of rigor and a solid foundation in Chinese language knowledge. I'm grateful to my teacher for constantly helping me proofread the questionnaire."

Student Cao said, "I remember that day of the interview was particularly cold. At first, I encountered many rejections when interviewing strangers, but I didn't give up. I believe perseverance will surely get the job done!"

Student Ren expressed, "It was my first time participating in volunteer service. I felt so proud when I measured blood pressure for community residents and disseminated knowledge about 'Climate and Health.' I can use my professional knowledge and skills to help more people in the future."

Following the implementation of the comprehensive climate change inquiry activity, we also engaged in introspection:

Firstly, the design of comprehensive activities should be in line with learners' academic situations, continuously attend to their motivation, and stimulate their drive. In comprehensive activities of climate education, we design verious types of activities that enable learners to exert their strengths in multiple intelligences, so that they can perform learning tasks competently and excellently. Group learning in project-based learning fosters a relaxed, harmonious, and free learning environment, maximizing each student's autonomy and allowing individuality to flourish.

Secondly, it's necessary to establish the central role of learners, empowering them to engage in independent inquiry and apply what they've learned. Teachers recommend combining online and offline methods for research, such as libraries, bookstores, Baidu Scholar, and CNKI. Students voluntarily divide tasks within their weekend groups, conduct research, and mutually recommend websites for information retrieval. Students delve into tasks, constructing knowledge meaningfully and transforming it into transferable learning abilities. Through project-based learning, students' motivation and efficiency in academic courses improve significantly, which reflected in higher pass rates for project participants compared to their classmates in final exams.

Thirdly, boundaries between disciplines should be broken down with an integrated design across multiple subjects that combines cognitive learning with social growth. Problem-solving permeates the entire project-based learning process, as students continuously learn within a series of task frameworks, thinking and exploring like subject experts. Questions like "What is the future trend of climate change in Northeast China?", "What factors influence climate change?", "What measures are needed to protect the environment?" and "How can nursing expertise help prevent diseases caused by cold climates in Northeast residents?" are explored. Research-based, practical, and socialized learning fosters the transformation of fragmented multidisciplinary knowledge into a comprehensive force for exploring the world, fostering coherence and holistic understanding in learning and life.

Fourthly, specific requirements in comprehensive practical activities enhance students' comprehensive abilities. Submitting high-quality assignments on time requires necessitates reasonable task distribution within groups, which greatly enhances students' teamwork and communication skills. To better present research findings, students have mastered PPT creation, survey questionnaire applications, micro-video recording and editing, steadily improving their information technology skills and laying a solid foundation for adapting to the information society.

Lastly, most of the questionnaire responses on climate change in Northeast China come from Heilongjiang. Teachers and students should appropriately communicate andcollaborate with other vocational schools in the region to form a climate change education community, thereby improving survey reliability and validity and raising awareness among more vocational students about the importance of environmental protection. During project-based learning, unclear task distribution in some groups led to heavy workloads for group leaders. To prevent this, teachers should listen to group task reports to ensure full participation. The depth and breadth of the social follow-up questions on climate change need to be improved; next time, extensive collection of questions before screening and implementation can enhance students' deep thinking abilities.

(Shuang Chen, Lecturer of Harbin Health School)

Exploring Meteorological Proverbs, Investigating Climate Change

Research Background

The China Youth Daily pointed out that many teachers have found in their teaching practice that when they discuss global climate change with students, it often feels very distant to them. When studying climate change, students find it difficult to perceive its effects and often feel a sense of powerlessness. Therefore, they believe that climate change education should start at an early age. Wuyang Middle School in Wuyi County, Zhejiang Province, strongly agrees with this view. They have promoted "climate change education in schools" by conducting research on meteorological proverbs, with the aim of improving young people's awareness of climate change, scientific literacy, and humanistic feelings. Practice has shown that conducting surveys on meteorological proverbs among middle school students can bring them closer to climate change education.

Meteorological proverbs originated in China's ancient agricultural culture and were formed by the laboring people through long-term observation, summary, and accumulation during production activities. These proverbs have been passed down through generations and still shine with wisdom today, becoming a valuable asset in our response to climate change.

Wuyi County, under the jurisdiction of Jinhua City in Zhejiang Province, is located in central Zhejiang and has a subtropical monsoon climate with distinct seasons. It is a typical mountain city with a geographical pattern of "eight parts mountain, one part water, and one part field", and has abundant resources of mountains, water, forests, fields, lakes, and grasslands. Wuyi has been inhabited since the Neolithic era and is one of the typical representative areas of the ten-thousand-year-old "Shangshan Culture" circle. The local dialects in Wuyi include Wuyi dialect, Xuanping dialect, Yongkang dialect, She ethnic dialect, Fujian dialect, and Nanjin dialect. The complexity of Wuyi's local dialects brings richness and diversity to its folk proverbs.

"Climate change education in schools" is a new topic and task given by the times to contemporary primary and secondary basic education. In order to effectively improve students' understanding of climate change and their ability to participate in building a community with a shared future for mankind, organizing middle school students to start with the investigation and research of the numerous vivid local meteorological proverbs in their daily life is an important means to enhance their awareness of climate change.

Project Design

Climate refers to the average state of the atmosphere in a region over many years, primarily reflecting the area's characteristics of cold, heat, dryness, and humidity. It is a long-term stable weather phenomenon, which makes it difficult for students to directly perceive climate change. The meteorological station in Wuyi County has only been in operation for 63 years by 2024, and the weather data records only cover this period. Thus, the time frame available for reference is relatively short, making climate change less noticeable. Meteorological proverbs, on the other hand, come from China's ancient agricultural culture. They were formed by laboring people through long-term observation, summarization, and accumulation during production activities and were passed down through generations, covering a long period of time. Comparing meteorological proverbs with modern meteorological data can reveal traces of climate change. Based on this, a project was designed for students to investigate and verify meteorological proverbs to understand the occurrence of climate change, thereby bringing climate change closer to them.

Project Implementation

"The Study of Meteorological Proverbs" is a project-based learning process centered on schools and organized by families in collaboration with meteorological departments and scientific associations. It consists of two phases: activity preparation and activity practice. Initiated by Wuyang Middle School, this project involves seventh-grade students of the class in the investigation and verification of meteorological proverbs. Wuyang Middle School collaborates with the Meteorological Bureau and the Science Association to allow students participating in the project to engage in exploratory learning activities that involve investigating and uncovering local meteorological proverbs. Through this process, students gain a deep understanding of meteorological knowledge related to climate change, as well as learn essential survival and life skills(see Table 1).

Table 1. Implementation process and outputs of the "Exploring Meteorological Proverbs, Investigating Climate Change" project

Part 1: Investigate students' metacognition and train students in investigation techniques		
1. Investigating students' metacognition on climate change	Investigate the understanding of current global climate crises, meteorological proverbs, and local meteorological proverbs in Wuyi among students and their parents.	67% of students are aware that the world is currently facing a climate crisis; 51% know more than two meteorological proverbs; 94% are interested in learning about local meteorological proverbs from Wuyi; and 98% want to make a positive contribution to addressing global warming. Additionally, 100% of the parents support the implementation of this activity.

(Continued)

2. Popularize meteorological proverbs to enhance their understanding	Popularize the characteristics of climate change and meteorological proverbs to provide students with more information.	The teacher provides students with information about meteorological proverbs in the classroom.
3. Document and annotate the process of collecting meteorological proverbs	Label the application scenarios of meteorological proverbs based on the following elements: subject, time, space, event, background, psychological aspect, and functional. elements.	Create a form to record meteorological proverbs.
4. Issue an initiative, conduct promotion and planning	Print the initiative letter and draft the activity plan.	Distribute the documents and keep relevant copies in the classroom.
Part 2: Official implementation process of the activity		
1. Conduct the formal investigation based on the "Winter Break · Hello" activity.	During the winter break of 2024, the school distributed meteorological proverb survey forms and encouraged students to use written and video recordings to inquire about meteorological proverbs from their elders. The collected information was then shared on the WeChat group.	Collected 99 local meteorological proverbs.
2. Brainstorm in groups to determine which meteorological proverbs to investigate.	On February 22, 2024, students from Grade 7 Class 4 at Wuyang Middle School, as the main participants, were divided into different groups and selected group leaders. Each group discussed and identified the proverbs that needed to be verified and then visited the meteorological bureau for confirmation.	"地上暖,天上孵雪卵" — "Warm on the ground, the sky hatches snowflakes" "霜降未降,廿天稳当" — "If frost hasn't fallen, the next twenty days will be stable" "立春雨,一春雨" — "Spring rain on the first day of spring, rain throughout the spring" "八月半,乱打扮" — "Mid-August, weather in disarray"

(Continued)

3. Visit the meteorological bureau to review data, verify the proverbs, and write the research report.	On February 23, 2024, each group consulted with the meteorological department staff to verify the reliability of the proverbs. The supervising teachers encouraged them to think about the patterns revealed by the meteorological data.	Ten groups, ten investigation reports (presented in Meipian).
4. Conduct timely communication and reflection.	Students reflected on the gains and shortcomings of the research process.	Reflections from teachers, students, and parents
Part 3: Summarize the activity and share insights explore the enhancement of climate change awareness and skills from the perspectives of students, parents, and teachers		
Students	The activity enhanced our understanding of proverb culture and traditional wisdom; developed team collaboration and interpersonal skills; and improved planning, writing, and investigative abilities.	
Parent	I feeling very fortunate to be able to participate in the activity with my child.	
Teacher	Students are not only recipients of climate change education but also creators of the climate change education environment.	

1. Preparation of the Meteorological Proverbs Investigation

This phase includes four main steps: investigating students' metacognition of meteorological proverbs, training students in research techniques, launching initiatives, and conducting publicity and planning work. Firstly, the project team assessed students' knowledge of meteorological proverbs to understand their metacognition of the subject. Next, they provided education on the challenges posed by climate change and the knowledge of meteorological proverbs. This provided an opportunity for the project team to expand students' knowledge of meteorological proverbs. To facilitate the investigation, the project team also introduced methods for recording and annotating proverbs. After completing these preparations, the team printed brochures and an initiative letter, and then distributed them to students in the class to encourage their participation.

2. Formal Launch of the Meteorological Proverbs Investigation

This phase includes the following steps: During the winter break of 2024, meteorological proverb survey forms were distributed, and students conducted investigations by asking their elders about meteorological proverbs and making both written and video recordings. After the winter break, the students returned to school and were divided into groups. Each group selected one or two meteorological proverbs of interest for verification (see Figure 1). After coordinating with the meteorological bureau, the students visited the bureau after working hours to consult with on-duty staff for verification of the proverbs (see Figure 2). They conducted thorough investigations and wrote related reports. After completing the activity, the students reflected on the shortcomings of the research process.

Figure 1. On February 22, 2024, students selecting meteorological proverbs for verification on campus

Figure 2. On February 23, 2024, students visiting the meteorological bureau for proverb verification

3. Summary of the Meteorological Proverbs Investigation

This phase involves collecting feedback from students, head teachers, and parents after the investigation is completed to understand the gains and suggestions for improvement. Reflections from the participating students, parents, and teachers reveal the enhancement of awareness and skills related to climate change from all three perspectives.

Student Zhao: Participating in this activity made me realize how closely meteorology is related to our lives. Learning and mastering basic meteorological proverbs can help us better cope with climate change and reduce the impact of natural disasters. Studying and practicing meteorological proverbs deepened my understanding of the importance of meteorology for both the nation and personal life.

Student Shao: In this investigation of meteorological folk proverbs, we focused on collecting numerous vivid proverbs from our surroundings, reviewed some literature, and organized the information by combining excellent previous research with our own limited knowledge. This experience not only deepened my understanding of proverb culture but also revealed the rich knowledge and meteorological awareness of ancient laborers in China.

Student Zhang: As the group leader for this activity, I needed to coordinate with team members. At first, I was concerned that some might not participate after I sent out the information. Fortunately, I eventually found suitable team members. This activity taught me many things, including the importance of good interpersonal relationships. It reminded me that we should not forget to support and learn from friends during our regular studies.

Student Tao: We didn't plan our investigation before we started, which led us to consult data multiple times. Under the guidance of meteorological experts and teachers, we learned that a single set of data may be random and not represent a general pattern. This experience taught us the importance of planning before undertaking tasks and that persistence, like a scientist's, is crucial to obtain scientific conclusions.

Parent of Student Zhang: In this activity, I was no longer a passive observer but an active participant. I saw my daughter transition from receiving the task to successfully organizing the activity, which made me very happy. I also enjoyed learning meteorological knowledge with my daughter, experiencing climate change together, and enhancing our awareness of climate change. I am grateful for the supportive class environment that allowed us to grow and succeed together in this activity.

Parent of Student Ma: My child introduced me to meteorological proverbs for the first time during this activity. I helped with data collection and analysis and learned about the impacts of climate change on daily life, production, and economic development. This activity gave me with a deeper understanding of climate change education and significantly increased my awareness of its importance.

Head teacher: As a science teacher from rural Wuyi, I had heard many meteorological proverbs and taught some meteorological knowledge. However, linking these proverbs with climate change education was a new experience. The students' enthusiasm for researching meteorological proverbs, their dedication to group formation, their persistence in data collection, their humility in seeking expert advice, and their caution in verifying results were impressive. Students are not only recipients of climate change education but also creators of a rich educational environment. Everyone involved, including students, teachers, parents, and meteorological bureau staff, experienced a meaningful climate change education.

Effectiveness & Reflections

Through the implementation of the project, the expected outcomes in enhancing students' climate change skills, collaborating with families and communities in climate change education, and creating a model for climate change education have been achieved:

Firstly, enhancing Students' Climate Change Skills and Scientific Literacy: The project aimed to explore and apply local meteorological proverbs in climate change education. With the collaborative help of teachers and experts, students independently selected research targets, used real local meteorological proverbs to describe weather features and environmental phenology as learning resources, and extracted key materials. Through teamwork, they addressed an open-ended climate change problem. This project-based learning process cultivated students' abilities to handle challenges and uncertainties, such as acquiring knowledge on meteorology and climate change, planning investigations based on proverbs, managing project implementation, and enhancing group communication and cooperation. The process of forming groups, constructing hypotheses, designing investigation plans, collecting evidence, reflecting, and reconstructing led to conclusions and successful investigations. This not only improved students' scientific literacy but also provided real insights into data changes like temperature and precipitation, and taught skills and methods for dealing with climate change, resulting in tangible improvements in students' climate change skills and scientific literacy.

Secondly, collaborating with Schools, Families and Communities for Comprehensive and Lifelong Climate Change Education: The complexity of climate change requires a multifaceted approach to education. The climate proverb investigation at Wuyang Middle School established a new method for climate change education centered on the school, involving families and leveraging social forces like meteorological departments. During the proverb investigation, students educated their family members about climate change, and older generations shared local climate history, natural conditions, and cultural traditions. This intergenerational learning through proverbs integrated climate change education into various family age groups, highlighting its comprehensiveness and lifelong nature. Meteorological departments played an expert guiding role, facilitating multi-party participation in climate change education.

Thirdly, creating a New Model for Climate Change Education: The goal of climate change education is to mitigate and adapt to climate change. By tracing meteorological proverbs and comparing them with current weather representations, researchers can discover signs of climate change, thereby enhancing their awareness and focus on climate change. Wuyang Middle School developed a process for investigating meteorological proverbs, creating a valuable model for future work. The project distilled three practices essential for project-based climate change education: developing localized topics, creating standardized processes, and leveraging students' subjectivity. Localized topics refer to exploring valuable climate change education projects based on local characteristics. Standardized processes include activity preparation, implementation, and summary. It is crucial to emphasize students' subjectivity throughout, focusing on inspiring, guiding, and encouraging students to complete related investigations and promote autonomous learning.

The project has space for further improvement. After conducting the meteorological proverb investigation and research, students, teachers, and parents have acquired some climate change knowledge and developed a varied understanding of the importance of climate change education. Students also showed a strong interest in meteorological proverbs. However, students were limited to listening to elders and consulting meteorological bureau data, without experiencing climate change firsthand. Future investigations could explore the following three aspects: First, actively leading students into fields to experience the relationship between phenological changes and climate change. Second, guiding students in field observations and research to verify the relevance of proverbs through continuous recording of meteorological elements over time. Third, conducting in-depth studies of previously verified proverbs. Due to the regional and literary characteristics of meteorological proverbs, continuous tracking is needed to assess their applicability in contemporary and future societies.

(Xiaojuan Zhou, Teacher of Wuyang Middle School, Wuyi, Zhejiang)

A New Paradigm for Learning About "Climate Change Education" Based on Embodied Practice

Research Background

As human beings live in harmony with nature, teachers should take up the mission to strengthen climate change education and encourage students to gain a deeper understanding of climate change. Cultivating students with a sense of mission and responsibility is the only way to promote climate change education to a new height, and to jointly protect the clear water and lush mountains of the Earth. In the traditional educational patterns, school education tends to make students sit in front of a desk, face a computer or a textbook, spend a lot of time understanding the teacher's experience and abstract terminology, and memorize the knowledge and skills required for exams through mechanical memory. This "sedentary education" makes the teacher the "porter" of knowledge and students the "receiver" of knowledge. It is not only bad for the physical and mental health of students, but also for the absorption and transformation of knowledge.

Based on the theory of embodied cognition, this project advocates the integration of physical learning and climate change education, promoting active teaching and learning. It transforms professional contents that are difficult for students to understand into students' embodied experiences, thus promoting the full understanding and mastery of abstract knowledge.

Project Design

The project includes the stages of startup, planning, implementation and evaluation, which are divided into the steps of "group learning, information search and questionnaire research", "school-enterprise collaboration and program design", "enterprise visits, expert lectures and sailing simulation", "result formation, evaluation and reflection" and other steps. The overall idea is shown in the Figure 1.

1. Based on the Questionnaire Survey to Find Out the Learning Condition

Before this activity, teachers used a "questionnaire" to find out the learning condition. Specifically, before the visit, the teacher investigated whether middle school students knew about climate change. Do they know that their parents' work environment is related to climate change? Do they know the contribution of enterprises in the field of climate change? What are their thoughts or experiences after the visit to the Shanghai Merchant Design and Research Institute (SDARI). Are they willing to join in similar

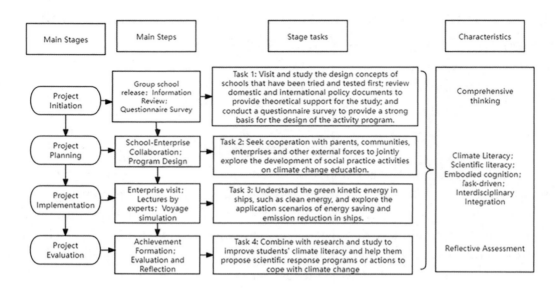

Figure 1. Overall Project Idea Framework

activities int the future?

The questionnaire enables the teacher to understand the real ideas and educational needs of students(see Figure 2), and at the same time provides information resources for reflection on the activities, which facilitates the improvement of the activity program and in-depth study.

2. Enhancing School-enterprise Linkage and Drawing up a Blueprint for Education Together

Adhering to the concept of "education is the collaboration between school, family and society", teachers try to promote "climate change education" through "school-enterprise collaboration". Shanghai Merchant Design and Research Institute (SDARI) has conducted in-depth research of new energy sources and vigorously developed green energy saving and intelligent ship technologies. The energy-saving devices such as SATR and fan ducts, reflect the institute's persistence in the pursuit of green and intelligent science and technology. On Jan 29th, 2024, 13 students and 3 teachers from Xinchang Experimental Middle School went to visit Shanghai Merchant Design and Research Institute (SDARI) together.

Firstly, they visited the exhibition hall and laboratories of the institute to learn about the development of the institute and achievements in scientific research, and to see ship models and energy-saving parts. Secondly, the experts of the institute were invited to carry out lectures on energy saving and emission reduction based on ship design, fuel selection, routing scheme, etc. Then hey interacted with the students. Finally, students were divided into groups. Under the guidance of the tutors, they were able to simulate

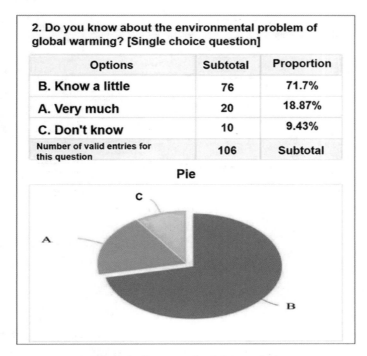

Figure 2. One example of data analysis

the sailing of the ships.

3. Focusing on Embodied Cognition to Experience Climate Change

The target group of this activity is the seventh grade students, whose logical thinking ability is at the initial stage and limited by the "confinement" of the traditional education "classroom" and lack of active learning awareness.

Through the study and practice in the Ship Research and Design Institute, the students were deeply involved in the learning about climate change, which stimulates their enthusiasm for learning and self-motivation. Through the learning framework based on "embodied cognition" (see Figure 3), the three-dimensional structure of "study and practice →cooperative inquiry→thinking enhancement" is constructed. The main steps are "raising questions, analyzing problems, and solving problems". The students may improve their climate literacy through a "thematic, interdisciplinary, and project-based" learning approach.

The specific objectives during the activity are: to understand the enterprise culture and improve scientific literacy; to listen to lectures given by experts and learn the measures taken by ship enterprises to cope with climate change; to broaden students' horizons and to stimulate innovative thinking. The communication with the experts of the Institute can help students understand the green development of the ship design industry more clearly and lay the first scientific foundation for coping with climate change.

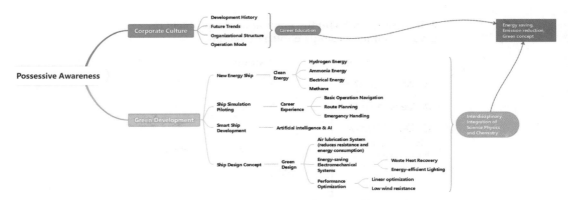

Figure 3. Learning framework based on embodied cognition

Project Implementation

1. Encounter with the Challenge of Exploring the Ship's Culture

After the students gained some understanding of climate change, the teachers took them to the Shanghai Ship Design Institute to explore the relationship between ships and climate in depth. In order to ensure the relevance and effectiveness of the study, the teachers defined the core question "Find green energy in enterprises" before the start of the activity and designed a program of the activity based on it.

In the design institute, students were guided by their tutors to engage in embodied learning and practical exploration in groups (see Figure 4). They first visited each exhibition area in order according to the exhibition layout of the institute. During the visit, the students carefully watched various ship models and physical products, recorded in detail the characteristics and concepts of different kinds of ships, and paid special attention to the specific application of green and energy-saving products on ships.

Figure 4. Exploring ship culture

When facing the concept of "Energy Transformation for Lower Carbon, Innovation in Design for Reducing Energy Consumption", the students seemed a bit confused. They said that although they generally understood the goal of "saving energy and reducing emission", they were not very clear about how to realize it and the technical details of it. In order to help the students better understand this design concept, the three tutors gave a brief explanation and guidance.

2. Dilemma Breaking and Enhancing Cognitive Skills

On the second floor of the Shanghai Merchant Design and Research Institute (SDARI), ship design experts systematically elaborated on the background of decarbonization, essential energy saving, green energy and smart development of the ship industry (see Figure 5). These contents largely involve knowledge from multiple fields, and the implementation of climate change education needs to be based on a multidisciplinary approach. After the lecture, the students actively participated in the discussion part and asked a series of questions, such as "What is the definition of clean energy?", "What are the gaseous components of white smoke emitted from ship chimneys?", "How are domestic wastewater and garbage on board treated? Are they directly discharged into the sea?" and so on. These questions fully reflect the students' curiosity and spirit of exploration in the process of embodied cognition.

Figure 5. Expert lecture "Exploring the Path of Ship Decarbonization and the Development Trend of Intelligent Ship"

In response to these questions, the ship design experts answered them one by one, taking into account the subject foundation of the seventh grade students. For example, the explanation of "clean energy" incorporated 9th grade chemistry knowledge, and the discussion of exhaust and wastewater treatment introduced relevant physics knowledge.

This interdisciplinary approach not only helped to solve students' cognitive confusion, but also promoted their integration and understanding of interdisciplinary knowledge. For questions that could not be answered in time, students were encouraged to record the questions, seek help from their tutors, ask their parents and search for information to find answers and solutions, which leads them to a broader and deeper understanding of global issues such as climate change and environmental protection.

In addition, the students also experienced the process of ship navigation in the Ship Navigation Experience Museum, operated the ship and made decisions during the navigation in person, so as to understand more intuitively the work content, work procedure, work environment, and the knowledge and skills required(see Figure 6). This kind of practical experience not only stimulates students' interest in the ship-related jobs, but also helps them to improve their self-efficacy in the operation. During the study, attention was always paid to the rigor and systematic nature of interdisciplinary learning to ensure that the information conveyed was accurate. Through this activity, the students not only broadened their cognitive horizons, but also managed to lay a solid foundation for their future studies and career development.

Figure 6. Ship sailing experience

3. Learning Together for Multi-dimensional Growth

After returning to school, the students were divided into groups according to their learning interests and difficulties, and solved problems related to "chemical formulas inclean energy" and "hull design and physics" by searching for information, learning to do experiments, seeking help from tutors and consulting parents.

In the evaluation stage(see Table 1), some groups demonstrated the sustainability of clean energy by creating a "future community power system". Some groups introduced new energy ships to the rest of the class through interactive communication. Most of the students' evaluation feedback forms mentioned that they "felt the importance of climate change education", "learned about the efforts of enterprises in green development", "I want to spread the awareness about climate change, too!".

Table 1. Evaluation form for cooperative group learning

Name of the group							
Group leader		Members					
What was gained from the event:							
What difficulties were encountered in the activities:							
Activity insights, inspiration for the future:							
Self-evaluation:							
Evaluation content	Division of labor (20 points)	Level of participation (20 points)	Help and learn from each other (20 points)	Sort (20 points)	Self-inquiry (20 points)	Total (100 points)	
Student A							
Student B							
Student C							
Student D							
Teacher-led assessment:							

From the students' self-evaluation and group evaluation, the dynamics of "embodied cognition" is fully demonstrated, which is manifested in the following ways: firstly, "climate change education" is upgraded from emotional cognition to physical cognition; secondly, the students are not limited to the study of "ship culture", but gain interdisciplinary knowledge and even knowledge within and across grades; thirdly, the students try to use different forms and methods to let more people join in climate change education.

4. Multiple Ways of Evaluation to Enhance Scientific Literacy

First, expand thinking and generate sparks of wisdom. Conventional ways of thinking and formalized modes of teaching often limit students' imagination and make it difficult for them to reach out to a wider horizon and think more deeply. Climate change education requires students not to be satisfied with superficial understandings and explanations, but to go deep into the nature of the problem and the connections between things. For example, in this activity, teachers not only want students to understand "clean energy", but also to integrate, develop and utilize it to solve the problem of climate change.

Student Peng expanded the boundaries of his thinking and engaged in scientific exploration. Interested in clean energy, he envisioned an efficient and environmentally friendly Future Power Generation System (see Figure 7). This system makes full use of "renewable energy" and simulates energy conversion and storage technologies to reduce dependence on traditional fossil fuels and alleviate the problem of energy shortage. Knowing the importance of new energy vehicles in reducing environmental pollution and promoting energy transformation, he also kept a watchful eye on the field of new energy vehicles. He completed the "Survey on Charging Facilities for New Energy Vehicles in Shanghai" and put forward useful ideas and references.

Figure 7. The model design of Future Power Generation System

Second, improve scientific literacy and achieve self-promotion. To cope with climate change requires students to explore and try in practice, be able to think independently, analyze deeply, and dare to put forward their own opinions and solutions to problems, not to be an armchair strategist.

Student Wang actively participated in various science and technology competitions and successfully developed the "Indoor and Outdoor Air Quality Monitoring System". This system can monitor indoor and outdoor air quality in real time, provide accurate and timely data support for people, help improve the living environment and improve the quality of life. By virtue of his wisdom, he won the "Golden Ring Award" in the

"Shanghai 15th Brain Olympics Innovative Learning Activity Innovation Competition". The article he wrote, "Improve Indoor Air, Build a Healthy Life", is of great significance to improve environmental awareness and healthy life(see Figure 8).

Figure 8. Scientific paper "Indoor and Outdoor Air Comparison System"

Third, adhere to the belief of green environmental protection and make it lead our life. In the communication and exhibition part, some students introduced new energy ships, elaborated on their important role in reducing pollution and protecting water ecology, and imagined the beautiful vision of the harmonious coexistence of science and technology and nature; some students skillfully used recyclable materials to make models, which not only carried out the concept of environmental protection, but also demonstrated the perfect combination of creativity and wisdom(see Figure 9). Through the activities, the students conveyed their love and pursuit of green life.

Figure 9. Modeling using old newspapers

In order to give students a clear idea of their performance, the teacher uses an effective evaluation that combines process and outcome, and to conduct self-assessment and peer assessment within the group, helping students to review their performance in the activity and look forward to future actions. It has become an inevitable choice of the times that teachers should motivate students to face the increasingly serious climate and environmental problems. Students must adhere to the awareness of environmental protection and sustainable development. Teachers, as the navigators of the future world, should shoulder this historical mission and promote the construction and development of a better China.

Effectiveness & Reflections

This project realizes the common growth of individuals and groups through collaborative innovation between schools and enterprises. During the program, students visit enterprises, learn the "green culture" of enterprises, and understand the development and application of "green technology". Students' innovative spirit, practical ability and scientific literacy have been improved..

In addition, during the implementation of the project, our school has carried out a series of campus construction under the theme of "ecological civilization". For example, we have designed a variety of learning areas by integrating the knowledge of climate change, and initially formed a green ecological campus pattern; carried out club courses such as plant appreciation and student guide, to integrate ecological civilization into curriculum teaching, so that students can experience nature and cultivate awareness of environmental protection while learning knowledge; promoted the concept of ecological civilization and the knowledge of environmental protection through the campus broadcasting, display boards, and the campus network; established a monitoring mechanism and set up an energy-saving and environmental protection monitoring group to regularly check and evaluate the campus' electricity and water consumption, garbage classification, etc.

Through study, social practice and school education on climate change, students realize that they are responsible for the earth and the future. They are willing to make more efforts to protect the environment, reduce waste, save energy and water resources, and actively participate in environmental protection activities such as "garbage classification" in their daily life. They also take the initiative to publicize the concept of environmental protection to their families and friends, and drive more people to participate in environmental protection. They gradually understand the relationship between "man and nature, society and self", and realize the urgency and importance of maintaining the ecological balance.

However, attention still needs to be paid to details such as student participation, organization and management of activities, and evaluation and feedback in the implementation process.

One of them is to clarify the cognitive goals. Embodied cognition encourages students to solve problems based on their own understanding and feelings in a particular learning environment. It is important to note that there are no universal, set-in-stone answers to these questions. For example, "How can I design a boat so that it is low carbon and environmentally friendly?" Instead, they are designed to stimulate creative thinking and problem-solving skills, encourage students to think about problems from multiple perspectives, and generate diverse and creative solutions.

In later assessment feedback, teachers found that some students expressed more admiration for the skills of ship designers and were fascinated by the coping skills of seafarers in the maritime environment. This interest helps to broaden students' horizons and may inspire them to explore related fields. But when students' attention is drawn to specific jobs such as ship design and navigation, these need to be more explicitly linked tothe context of climate change. For example, the importance of ship design and navigation technologies in reducing carbon emissions and improving energy efficiency could be emphasized, as well as the potential role of these technologies in facing the challenges of climate change. Teaching methods and content need to be continually adapted and optimized to ensure that students are able to develop a comprehensive and in-depth understanding of climate change issues and recognize the responsibilities and the roles of different professions in relation to this issue.

Second, integrate learning resources. In order to complete the worksheet, students can make full use of a variety of resources. These resources may include book literature, Internet resources, laboratory equipment, educational software, teacher guidance, peer discussion and so on. Cognition is not only the reception and processing of new information, but also the construction and reconstruction on the basis of existing knowledge and experience. During this activity, students have not yet studied the frontier knowledge of "clean energy", and the focus of the activity is more on "what is clean energy". If this activity is used as a pre-study, the deeper topic can be "how clean energy can be applied in daily life" and "the development prospect of clean energy". Therefore, it is crucial for students to collect information and build up the necessary knowledge before starting the study. These preparations facilitate the learning process by providing students with the necessary cognitive foundation so that they can find relevant information and solutions to problems more quickly and effectively.

(Yifan Shi, Teacher of Xinchang Experimental Middle School, Pudong, Shanghai)

🎓 Expert Review

Integrated activities serve as a crucial mechanism to develop students' comprehensive qualities in school education, which can enable them to transcend disciplinary silos, allow students to learn and grow through practice. In Chapter Four, cases cover different stages of primary, secondary and secondary vocational education, integrating climate change education through a diverse array of activities. For example, the investigation and inquiry integrated activities in Shanghai usually involve field trips and in-depth research, so that students can experience environmental problems first-hand and gain a deeper understanding of the impacts of climate change; the design and production integrated activities encourage students to use what they have learnt to design and produce eco-friendly products or programs that embody environmental sustainability; and proverbs related activities about climate and cultural heritage integrated activities combine the excellent traditional Chinese culture with environmental protection concepts, so that students can inherit the culture while at the same time learn the concepts of climate change and environmental protection; and the integrated activities of theme study focus on in-depth study of a specific theme, allowing students to learn about climate change and understand the scientific principles and impacts of climate change in a project-based exploration.

In general, climate change educationacross nations has different emphases, but they all emphasize students' participation and experience, as well as the integrated application of interdisciplinary knowledge. For example, climate change education in the United Kingdom focuses more on outdoor practice and hands-on experience. Schools organize outdoor teaching and learning activities, such as visits to wind farms and participation in tree planting activities, to enable students to experience the use of renewable energy and the impacts of climate change. Climate change education in the United States focuses more on scientific research and data analysis, and the integrated climate change practical activities at the basic education level will allow students to participate in climate change research projects, such as analyzing local climate data and participating in the construction of climate models, in cooperation with scientific research institutions. Climate change education in the Netherlands focuses on developing students' critical thinking and problem-solving skills, and schools often use simulation games and activities to make students understand the complexity and urgency of climate change, such as simulating the United Nations Climate Change Conference.

When designing climate-themed interdisciplinary activities, teachers need to consider multiple dimensions to guarantee that the activities are effective, interesting and educational. Compared to climate change education practices in other international countries, integrated activities in China have a unique advantage by integrating local culture, history and reality. For example, through projects such as an expedition to the Yangtze River Estuary No. 2 ancient ship and climate proverbs, students were able to connect climate changeto the local history and culture, and this integration significantly enhanced students' cultural understanding of the impacts of climate change and the transmission of traditional values. Cases in Chapter Four offer valuable insights. These experiences are as followings:

Firstly, to understand the characteristics of the geographical area in which the school is located: pre-analyzing the climatic characteristics of the area in which the school is located, such as temperature, precipitation, seasonal variations, etc.; considering local climate-related issues such as extreme weather events, sea level rise, ecosystem changes, etc.

Secondly, to design student-centered integrated activities: tailoring the activities to the age, personality and interests of students to ensure that they capture their attention and stimulate their curiosity. The activities are designed to take into account the cognitive and practical abilities of the students, ensuring that they are neither too easy nor too difficult.

Thirdly, to consider cross-disciplinary integration in the design of activities: through cross-disciplinary projects, students will be able to understand the multidimensional impacts of climate change from different perspectives; and climate change education will be integrated into the teaching and learning of different disciplines, e.g., geography, science, mathematics and arts curricula.

Fourthly, to strengthen the practical and exploratory nature of teaching methods: priority should be given to designing practical activities, such as weather observation, carbon footprint calculation, and community greening, so that students can learn by doing; and students should be encouraged to take the initiative to explore the issue of climate change by means of observation, experimentation and investigation.

Fifthly, to fully leverage the advantages of the cooperation mechanism for school-family-society activities: parents, community members and climate experts will be invited to participate in the activities to increase students' social interaction and learning opportunities; cooperation among students will be promoted, and their communication skills will be improved through group discussions and project reports.

Sixthly, to establish a systematic evaluation and feedback mechanism: local education administrations need to design an effective evaluation system to assess the learning outcomes and participation of students in activities; this will better ensure that the design of activities has a certain degree of adaptability and can be adjusted according to student feedback and the latest research on climate change, so that climate change education can be viewed as an ongoing process rather than a one-time event.

Seventhly, to pay attention to the emotion-driven and climate-action behavioral stimulation of integrated activities: the achievement of the SDGs needs to start with children, and in school education, teachers are required to encourage students to apply their learnings to practical actions, such as participating in environmental protection projects and promoting climate-friendly behaviors.

These integrated activities for climate education with unique creative features can integrate climate change education into daily teaching and learning, enable students to realize the importance of environmental protection through their own experience, and cultivate their environmental awareness and sense of responsibility. However, compared with the teacher-led integrated activitieson climate education with distinctive features and rich variety, the support of climate teaching and learning resources behind them is somewhat limited, which may lead to challenges in the sustainable implementation and systematic development of the integrated activities. These challenges may exist in the limitations of the coverage of integrated activities. Although integrated activities have achieved certain results in climate change education, the coverage of these activities is often limited due to resource and time constraints, which do not ensure that all students are able to participate in them; the lack of long-term tracking and evaluation of the effects of students' environmental protection behaviors and practices after the implementation of the integrated activities in many schools indirectly affects the climate-based integrated activities' continuous improvement and deepening; due to the lack of effective cooperation with external resources such as communities and enterprises, the content of these comprehensive activities may be out of touch with the actual environmental protection needs, resulting in a significant reduction in the effectiveness of education; and finally, the relative lack of professional knowledge and practical experience of many teachers in climate change education can limit the quality and effectiveness of comprehensive activities.

In order to promote climate change education more effectively through comprehensive activities, it is recommended that teacher training be strengthened, resource allocation be optimized, student participation be increased, and a long-term evaluation mechanism be established. Furthermore, interdisciplinary cooperation and international exchanges should be promoted, leveraging international expertise, in order to enhance the quality and impact of climate change education in a comprehensive manner.

(Yishu Teng, Associate Professor of Institute for Advanced Studies in Humanities and Social Sciences, Beijing Normal University)

Chapter 5

Collaborative Climate Change Education by Schools, Families, and Communities

The comprehensive nature of climate change education requires that the time and space of education be aligned with people's production and living spaces. It demands that the target audience of education be the entire population, and that the implementation of education be highly collaborative.

This chapter, includes various cases, such as intergenerational collaborative climate change education, corporate and social organization involvement in climate change education, and collaborative climate change education by schools, families, and communities. These cases highlight the multi-stakeholder and multi-institutional cooperation that characterizes the field and underscore the uniqueness of climate change education. The cases also come from different regions, and the practices are still evolving, indicating a growing space for exploration and innovation.

Design and Implementation of the Learning Pathways of the "Climate Change Classroom Targeted the Elderly and Children"

Research Background

The response to climate change is a holistic strategy of the United Nations. The article *UNESCO: Promoting Climate Change Education through a Whole School Approach* states that the climate crisis is a global reality. There is no solution without education. As climate change and its impacts accelerate, it is particularly important for young people to engage in climate change education, which is a key strategy to promote understanding and positive action on climate change among the younger generation. The classroom, as the primary venue for education, is uniquely positioned to facilitate this process. Grandparents, as an important force in society, can combine their experience and wisdom with their grandchildren's innovative spirit and learning ability to create a new educational model—the "Climate Change Classroom Targeted the Elderly and Children". The "Climate Change Classroom Targeted the Elderly and Children" is one of the key programs of the Silver Sprouts Alliance.[①]

Project Design

1. Objectives

Overall goal: Through the design and implementation of the "Climate Change Classroom Targeted the Elderly and Children" learning pathway, we will enhance the awareness and ability of grandchildren to respond to climate change, improve climate literacy, and promote the popularization of environmental education.

The specific objectives are:

(1) Using the "Climate Change Classroom Targeted the Elderly and Children" as a venue to learn about meteorology and enhancing learners' awareness of environmental protection.

(2) Enhancing the ability of the elderly and children to prevent and mitigate disasters and cope with climate change through multi-dimensional mutual learning and integration such as "class meeting + family learning".

(3) Tapping internal and external resources, carrying out thematic practices,

---

[①] The "Silver Sprout Alliance", created in 2021, is a brand developed by Wuyi County Cooked Creek Elementary School in cooperation with Wuyi University of the Elderly, which is committed to intergenerational learning through the introduction of high-quality resources for the elderly, and is a cooperative community centered on mutual learning and mutual assistance to promote cross-age communication and knowledge transmission.

promoting the practical transformation of theoretical learning, and enhancing the scientific spirit andclimate literacy of the elderly and children.

2. Methods

The design concept of the "Silver Sprout Climate Classroom" is comprehensive and innovative, encompassing intergenerational co-learning curricula, interactive teaching across generations, and community-based projects. The specific activities include grandparents guiding their grandchildren in watching educational videos, engaging in outdoor observations and experiments together, grandchildren teaching grandparents gamified learning, and both generations reading relevant books together. Practical projects are implemented collaboratively by grandparents and grandchildren, involving field trips, outreach, mock drills, and developing family climate action plans. The approach encourages active participation from students, teachers, parents, and community members, fostering a diverse and interactive learning community.

3. Development Phases

The first phase was from December 2023, when the project teachers participated in a special training on "Climate Change Education in Schools" organized by the Wuyi County Education Bureau. This marked the beginning of establishing a research team and designing the project.

The second phase spanned from January 2024 to the end of the winter vacation in February 2024. The focus was on conducting intergenerational co-learning activities, conducting classroom co-learning sessions centered on foundational knowledge and coping strategies, and engaging in interactive teaching where grandparents teach their grandchildren and vice versa.

The third phase began with the start of the spring semester in 2024 and continues to the present. It aims to explore internal and external school resources, promote the practical application of theoretical learning, and enhance the scientific spirit and climate literacy of participants through a variety of practical activities.

Project Implementation

The Berlin Declaration on Education for Sustainable Development emphasizes that Education for Sustainable Development (ESD) should be an essential element of the education system at all levels, with environmental and climate action as core curriculum content, and that ESD should be approached from a holistic perspective. The "Climate Change Classroom Targeted the Elderly and Children" is a creative and far-reaching educational program that requires careful planning and preparation, and the following is a detailed implementation process.

1. Teachers Participate in Training, Form Research Teams and Develop Implementation Plans

On December 22nd, 2023, "Decima-Winter Vacation, Hello!" Climate Change Educationin School and Wuyi County Primary and Secondary Schools Weather Folk Proverbs Investigation and Practice Research Project Seminar was held in Wuyi County Cooked Creek Primary School, and the author participated in this activity as the backbone teacher of "Silver Sprout Alliance". The project was supported and guided by Prof. Jiacheng Li, deputy director of the Shanghai Municipal Insititute for Lifelong Education of East China Normal University. Guoqiang Lei, a special researcher of the Lifelong Education Research Institute of East China Normal University and former director of the Wuyi County Textbook Institute, explained in detail *Treading the Snow and Searching for Proverbs to Welcome the Arrival of Spring* — *"Decima-Winter Vacation, Hello!" Climate Change Education Weather Folk Proverbs Investigation and Practice Research in Schools Action Program*.

Taking this training as a new starting point, the school convened a meeting about the leaders of both sides of the "Silver Sprout Alliance", and after discussion, we determined the learning objectives, learning contents and practical programs, and developed the program idea map as shown in Figure 1.

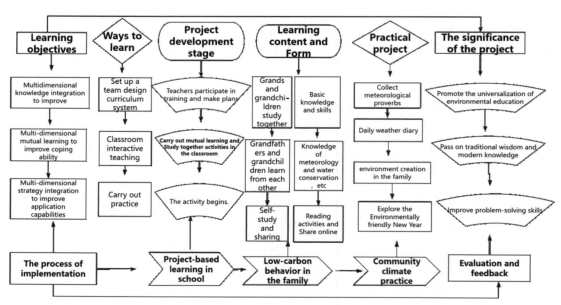

Figure 1. Schematic diagram of the design and implementation of the "Silver Sprout Climate Classroom" learning pathway

2. Development of Co-learning and Mutual Learning Activities

The function of education in responding to climate change is to guide students to actively adapt to climate change and actively participate in mitigating climate change by imparting climate-related knowledge, establishing correct values and enhancing coping skills.

In order to actively promote climate change education in schools and enhance the climate literacy of young people, the author has carried out a rich form of grandparent-grandchild co-learning and mutual learning activities on the platform of the "Silver Sprout Climate Classroom". The co-learning contents of the basic class include: the difference between climate and weather, greenhouse effect, carbon footprint, carbon cycle in nature, renewable energy and non-renewable energy, impacts of climate change, international agreements on climate change, etc.. The co-learning contents of the climate change strategy include: energy saving and emission reduction, transportation and travel, waste disposal, dietary choices of sustainable lifestyles, green shopping, water resource protection, science and impacts of climate change, exploring the local impacts of climate change, community engagement in action, community greening, environmental activism, and more.

In the "Climate Change Classroom Targeted the Elderly and Children", the elders participated in the classroom with their rich life experience as interpreters of meteorological knowledge and disseminators of local culture, sharing their life experiences and observations of environmental changes, while the younger generation brought in the latest scientific knowledge and action strategies on climate change. The elders told their grandchildren about their years of farming experience, how they predicted climate change by observing the sky, animal behavior, etc., and how they used traditional wisdom to adapt to these changes, sharing their experiences on how they chose drought-tolerant crops in drought years, and how they prepared for flooding when the rainy season came. The grandchildren, on the other hand, used their knowledge of modern science to introduce their the causes and effects of global warming to their grandparents and the measures to deal with it. They showed the impact of greenhouse gas emissions on the climate through charts and data, and explained how to slow down warming by reducing carbon emissions, planting trees, etc. They also shared some of the applications of modern science and technology in dealing with climate change, such as smart agriculture and weather forecasting(see Figure 2).

Figure 2. Grandparents and grandchildren learning together and learning from each other about climate knowledge site

3. Implementation of Practical Activities

The practical learning activities are designed to raise multi-generational awareness of climate change, create a sense of urgency, and improve the ability to respond to climate change and solve problems, around which the "Climate Change Classroom Targeted the Elderly and Children" has carried out a wealth of practical activities.

Climate classroom in folklore — collection of climate proverbs. Meteorological proverbs have distinctive local characteristics, carrying a wealth of folk wisdom and life experience, the complexity of Wuyi's local dialect has brought about the richness and diversity of folk proverbs, providing a large number of vivid and lively meteorological proverbs for us to carry out the "climate change education in school". I give full play to Wuyi meteorological proverbs and the advantages of a wide range of folk meteorological proverbs, to carry out "folk meteorological proverbs" collection activities. The children took the opportunity of New Year's greetings to crowdfund meteorological wisdom from their elders, and presented the results in a variety of ways: some in the form of drawings, some in the form of recorded videos, and some in the form of a beautiful story by parent-child cooperation. The class received a total of 102 proverbs.

Climate classroom in daily life — climate action diary punch card and family weekly record. The most important meaning of learning knowledge lies in transformation and application. Students embodied the climate change knowledge and ability they learned in their daily life and recorded their daily climate action practice in the form of diary punch cards. Each of the 45 students in the class insisted on drawing and writing a weather diary, and they wrote more than 900 weather diaries in total. Their actions have influenced one family and many touching stories have happened. On the basis of the diary cards, the students recorded the situation of their families for a week in the five aspects of "clothing, food, housing, transportation and observation", as shown in Table 1, and adjusted and improved the low-carbon strategies of their families by monitoring the trend of low-carbon behaviors of their family members.

Table 1. Climate change and low carbon behavior household observation

Time:		
School: Class: Name:		
Categorizing	Records of family members' behavior or use of objects	Quantity
Clothing	1. Total number of new clothes/pants/shoes purchased	
	2. Number of times clothes were washed in a washing machine	
Food	1. Number of times the whole family ordered takeaway	
	2. Quantity of various types of meat bought by the family (in kilograms)	
Housing	1. How many bags of garbage were thrown out (based on garbage bags being almost full)	
	2. Number of bags taken back from shopping trips	
	3. Number of bottles and paper cups generated at home	
	4. Number of hours the television was on	
	5. Number of hours you play with your cell phone/tablet by yourself	
	6. Induction cooktops/gas stoves/hours burned (in hours)	
Transportation	1. Number of taxi rides/drives in own car	
	2. Number of trips on public transportation (subway, bus)	
	3. Number of walks (count as 1 if you walk by yourself for more than 10 minutes)	
	4. Number of times I rode in an electric car or motorcycle	
Observation	1. Number of sunny days	
	2. Number of days of rain and snow	
	3. Extreme weather (unit: times)	

"Climate classroom" in the family—advance red packets for environmental creation. After learning about climate change in the classroom, the students went home to share with their grandparents their "golden ideas" for welcoming the Chinese New Year in a low-carbon manner. In the process of exchanging ideas with each other, the children unleashed their creativity into full play and came up with bold ideas. With the support of their parents, each of them was able to get an advance of 100 RMB (see Figure 3), a large sum of money that made them both excited and cautious, and all of them actively participated in the activity because they needed to be responsible for their own money. With the assistance of their teachers the children ordered canvas bags online, printed flyers, and planned with their teachers to make billboards at an advertising agency. Each of the participants acted diligently to protect the environment.

Figure 3. Students advance red packets to parents

"Silver Sprouts Climate Classroom" in the Community—"Carbon Quest" Green New Year Activity. On the afternoon of February 5, 2024, everyone gathered at the entrance of the supermarket, holding canvas bags and waving signs, while calling out and handing out bags and flyers: "Grandparents, uncles, aunts, come and get yours! These are canvas bags we customized with our own pocket money. Using our bags for shopping at the supermarket is more environmentally friendly." "Sort your trash with the 'pick-up' method, enjoy your meals with 'clean' plates, travel green when visiting for the New Year, and be frugal for a green New Year." The elders also patiently explained the content of the proposal to the citizens. Their actions and sincerity touched many people, and many citizens expressed their willingness to try a low-carbon lifestyle during the Spring Festival to show their support for environmental protection(see Figure 4).

Afterwards, the children brought their bags and followed their parents to the supermarket with the remaining 70 RMB. While practicing green shopping, they also distributed low-carbon action initiatives to the public, and their actions influenced many people who shopped at the supermarket.

Figure 4. Grandchildren and parents at the Green New Year event

Effectiveness & Reflections

Intergenerational learning about climate change is not only important for family and individual growth, but also has a positive impact on the sustainable development of society as a whole.

Firstly, it promotes the popularization of environmental protection concept and environmental education. The "Silver Sprout Climate Classroom" enhances people's awareness of environmental protection by explaining the impacts of climate change and response measures.. Especially for children, who are in the critical period of value formation, participating in such classroom activities will help them better understand the seriousness and urgency of the climate change issue, and thus build up a correct concept of environmental protection. Student Li lives with her grandmother, who told us that since joining the "Silver Sprout Climate Classroom", the whole family has become involved in low-carbon living, supervising and learning from each other, and has gained a lot of knowledge about energy-saving lighting, green traveling, water conservation, and reducing food waste. By participating in such classroom activities, several generations can gain a deeper understanding of these scientific knowledge and thus improve their scientific literacy. This is of great significance in promoting the popularization of environmental education and raising public awareness of environmental issues.

Secondly, it has conveyed traditional wisdom and modern knowledge across generations. As a unique form of education, the "Climate Change Classroom Targeted the Elderly and Children" is able to combine traditional wisdom and modern knowledge across generations. On the one hand, through the transmission of traditional wisdom, the older generation can help the younger generation to establish a harmonious relationship with nature and to understand the interdependence and mutual influence between humans and nature. On the other hand, the younger generation can also pass on modern scientific knowledge to the older generation. Modern scientific knowledge is key to addressing climate change, and the younger generation can convey the scientific principles, social knowledge, and coping strategies they have learned to their elders, helping them better understand and respond to the challenges brought by climate change, contributing to the sustainable development of the future. Through such classroom learning, grandparents and grandchildren not only enhance their understanding and trust in each other but also jointly contribute wisdom to address climate change.

Thirdly, creative thinking is stimulated and problem-solving skills are enhanced. In the faceof the global challenge of climate change, intergenerational co-learning can stimulate the innovative problem-solving thinking among family members. Silver Sprouts Climate Classroom involves multiple disciplines, and by integrating the knowledge and methods of different disciplines, this interdisciplinary way of thinking helps to break down stereotypes. "The Silver Sprouts Climate Classroom focuses not only on the transfer of knowledge, but also on the cultivation of problem-solving skills. Through hands-on activities, discussions and case studies, participants learn how to apply what they have learned to solve practical problems. This problem-oriented learning approach helps to develop participants' problem-solving skills.

The "Climate Change Classroom Targeted the Elderly and Children" has been actively exploring ways to improve students' climate literacy by imparting climate-related knowledge and developing their interdisciplinary thinking and adaptive capacity, as well as enhancing the ability of both generations to adapt to and cope with climate change, and promoting intergenerational communication and cooperation. However, as part of the knowledge is relatively boring to learn, there is room for improvement in teaching methods and approaches, and there is a need to use sufficient visual aids, examples and stories, and to design role-playing and mock conferences to help them better understand and apply their knowledge. There is a need to continue to focus on the motivation of learners and to strengthen the systematic nature of the program. In the future, we will continue to improve the design of the learning pathway and expand the scope of implementation, so as to promote the participation of more families and members of the community in climate action, and jointly contribute to the fight against global climate change.

(Meiqin Lan, Teacher of Shuxi Primary School, Wuyi, Zhejiang; Translator: Eryang Lan, Zidie He)

Green Innovation and Climate Change Education: Building a New Model of Lifelong Learning

Research Background

Firstly, policy incentives. Amidst the formidable challenges of global climate change, green innovation and climate change education have become essential strategies to address the environmental crisis. Recognizing its role as a key component of the China (Shanghai) Pilot Free Trade Zone, the Lin-gang Special Area is proactively aligning with national strategies to foster a green, low-carbon development model. This initiative is further supported by educational efforts aimed at cultivating a strong sense of environmental responsibility and the capacity for action among citizens. *The China (Shanghai) Pilot Free Trade Zone Lin-gang Special Area Ecological Field Green Low-Carbon Development Implementation Plan* calls for a comprehensive approach that guides public practice and encourages broad participation in ecological green initiatives. It emphasizes the importance of enhancing conservation awareness, ecological integrity, and a commitment to a green lifestyle, integrating the principles of green, low-carbon living into the conscious actions of the public. The plan also emphasizes the need to strengthen foundational research in low-carbon technologies and to expand scientific outreach and education to improve the public's understanding of energy conservation and low-carbon living. Through the use of both online and offline platforms, the initiative aims to effectively raise public awareness and comprehension of energy-saving and low-carbon principles, and foster a collective commitment to sustainable practices and environmental stewardship.

Secondly, community sustainable education project support. Nanhui New City University for Senior Citizens collaborate with theChina-UK Low Carbon College of Shanghai Jiao Tong University (LCC, SJTU) have embarked on an innovative community sustainable development education project in Lin-gang. Nanhui New City University for Senior Citizens, serving as a public welfare educational institution for the elderly, is distinguished by its focus on green technology innovation and low-carbon science popularization, and strives to establish an open and integrated platform for lifelong learning. Since its establishment in 2017, the China-UK Low Carbon College (LCC) has been dedicated to research in low-carbon technology and talent cultivation, promoting the dissemination and application of low-carbon scientific achievements. By integrating resources from universities and communities and conducting a variety of educational and popular science activities, the project aims to enhance residents' understanding of climate change and to encourage community members to actively engage in green, renewable, and low-carbon sustainable practices. It provides valuable learning resources and participation opportunities for residents, thereby contributing to the green development of the Lin-gang.

Thirdly, the need for intergenerational learning to address climate change. Climate change is underscored by the distinct perceptions and reactions to this global issue across different age groups. The elderly, being particularly vulnerable, require support in comprehending and adapting to the health risks associated with climate change, as emphasized in the *Lancet Countdown on Health and Climate Change 2022 China Report*, which notes the heightened mortality risks faced by those aged 65 and above. Middle-aged individuals, pivotal to family structures, need to increase their environmental consciousness and adeptly convey this knowledge to their offspring, with educational content ranging from climate change fundamentals to strategies for household energy conservation and environmental education. This generational transfer of knowledge can instill a green lifestyle within families, propelling broader societal engagement. Meanwhile, the youth, poised as future leaders and agents, are crucial to the climate change education movement. Their educational journey should prioritize the development of environmental awareness, scientific acumen, and innovation skills, equipping them to recognize the immediacy of climate change and to actively contribute to climate initiatives. Engaging in participatory learning and practical experiences, the youth can cultivate the competencies to tackle climate change challenges and embrace sustainable living and consumption patterns.

In summary, the goal is to construct a comprehensive educational system that spans from the family to society, fostering intergenerational and diverse co-learning and mutual learning across different strata. This approach is designed to achieve lifelong learning, thereby effectively addressing the challenges of climate change and promoting sustainable social development.[①]

Project Design

Goal: The goal is to implement activities through a cooperative mechanism made up of schools, families and communities. This effort aims to establish a distinctive brand of green innovation and climate change education with significant influence and appeal. By providing comprehensive and continuous educational services, the goal is to facilitate intergenerational learning, advance the construction of green and sustainable communities, and contribute to the development of a beautiful China and the global response to climate change.

[①] Cai, Wenjia, et al. The 2022 China report of the Lancet Countdown on health and climate change: leveraging climate actions for healthy ageing[J]. The Lancet. Public health, 2022, 12(7): e1073-e1090.

Specific objectives:

(1) Enhancing Public Environmental Awareness and Climate Change Perception: By implementing a variety of diverse and practical green innovation and climate education activities, individuals will be encouraged to gain a deeper understanding of the impacts of climate change on the environment and society. This approach aims to strengthen their recognition of green lifestyles and the principles of sustainable development.

(2) Cultivating Technological Innovation Skills and an Innovative Spirit: By integrating cutting-edge technology and sustainable development concepts, we aim to ignite individuals' interest and passion for technological innovation. This strategy is designed to enhance their technological literacy and innovative capabilities, to meet the challenges of a rapidly evolving society.

(3) Stimulating Enthusiasm and Responsibility for Green Actions: Through educational initiatives and community co-construction, individuals are guided to recognize the impact of their actions on the environment and climate change. This process aims to inspire a voluntary commitment to adopt green lifestyles and participate in environmental protection actions, fostering a desire and motivation for sustainable living.

Project Implementation

The project's curriculum is meticulously designed to comprehensively address the challenges of climate change and to foster sustainable societal development. It encompasses a diverse range of courses, including lectures, company visits, and practical courses, tailored to meet the needs of learners of different ages and backgrounds. The lecture course is taught by professional teachers from LCC, SJTU. The theme of lecture covers climate science, low-carbon technology, environmental protection skills and sustainable development concepts. The project firstly launched in 2023 and targeted key populations for young people, the middle-aged and the elderly by strengthen resource integration and cross-border learning cooperation through a series of unique educational activities to promote community environmental awareness and green lifestyle among different groups.

1. Series of Activities to Promote Climate Change Targeted for Young People and Families

In 2023, the project targeted youth and families in the community by launching a series of activities titled "Exploring Carbon Science, Heading Towards a Green Future—Series Activities of Addressing Climate Change and Promoting Low-Carbon Development in Communities" (see Table 1). These initiatives aimed to cultivate a spirit of technological innovation and environmental consciousness among young people and their families. Concurrently, the project sought to improve their scientific literacy and practical skills.

Table 1. Project schedule for 2023

Project Theme: "Exploring Carbon Science, Heading Towards a Green Future—Series Activities of Addressing Climate Change and Promoting Low-Carbon Development in Communities"(see Figure 1) Participants: 30 groups of families (A student in the second grade or above accompanied with one parent.)	
Theme	Schedule
Theme 1: Low-carbon Energy	Lecture: "Trends in the Development of Low-Carbon Energy in the Context of Climate Change". Company Visit: Visit Shanghai Lin-gang Hongbo New Energy Company, investigate the contribution of rooftop photovoltaic to climate change. Practical Course: DIY mini weather station model workshop.
Theme 2: Waste Recycling	Lecture: "Advancements in the Resource Recovery Technology of Electronic Waste". Company Visit: Visit Shanghai Household Waste Science Exhibition Hall(see Figure 2), understand the "past and present life" of garbage. Practical Course(see Figure 3): Make a desktop vacuum cleaner from recycled plastic bottles.

Figure 1. Group photo of "Exploring Carbon Science, Heading Towards a Green Future—Series Activities of Addressing Climate Change and Promoting Low-Carbon Development in Communities"

Figure 2. Company visit: participants visit Shanghai Household Waste Science Exhibition Hall

Figure 3. Practical course: group photo of participants make a desktop vacuum cleaner from recycled plastic bottles

Strategy and thinking:

(1) Combining hands-on practice with creativity: The curriculum emphasizes practical engagement where children transform discarded plastic bottles into useful tools, thereby enhancing their manual dexterity and creative thinking. The design of the courses focuses on demonstrating how to turn waste materials into valuable items, fostering an awareness of environmental protection and the concept of resource recycling.

(2) Family-oriented environmental learning: The program is designed to be family-inclusive and aims to enhance parent-child communication and collaboration, and to collectively strengthen the family's environmental consciousness. Through the reusing of discarded resources, children are awared that everyone can contribute to environmental protection through small daily actions. This approach helps them recognize their role and responsibility in addressing climate change.

(3) Popularizing scientific knowledge: Through the hands-on creation process,

children acquire scientific knowledge that enhances their scientific literacy and understanding of the importance of technological innovation in addressing climate change. By exploring the potential value of waste materials, their innovative spirit and problem-solving abilities are ignited. Through practical activities, children learn to addressenvironmental issues with innovative thinking, equipping them to become active participants and catalysts in the future efforts to address climate change.

2. Series of Activities to Initiate Green Science and Technology Education Targeting the Elderly Population

In March 2024, the project targeted the elderly population, focusing specifically on enhancing their environmental adaptability and quality of life. The initiative aims to strengthen their environmental awareness and social engagement, stimulate thought andinnovation, and anticipate their contributions to the nation's dual-carbon development strategy(see Table 2).

Table 2. Project schedule for 2024

Project Theme: "Smart Elderly—Special Program for Green Innovation and Climate Change Education" Participants: 50 learners are publicly recruited to the community for each activity	
Theme	Schedule
Theme 1: Low-carbon and Smart Transportation	Launch of the Project: Prologue Lecture: "The Key Technologies of Low-speed Unmanned Driving in the Field of Low-carbon Intelligent" taught by Xue Dong, Associate Professor of China-UK Low Carbon College, Shanghai Jiao Tong University(see Figure 4). Laboratory Visit: Visit unmanned driving laboratory of low-carbon college, and explore how technology promote climate change(see Figure 5).
Theme 2: Smart Life	Launch of the Project: Study Tour Company Research: Participants visit the Smart Manufacturing Research Institute at Shanghai Jiao Tong University to experience smart living and new technologies and breakthroughs in smart manufacturing, appreciate the aesthetics of industry, and understand the new developments in the Lin-gang Special Area and technological innovations through hands-on experiences with collaborative robots, heart-dog health monitoring, and other projects. This experience aims to showcase the conveniences that smart living brings to people. Practical Course: Participants will edit today's activities into a research study vlog, combining the theme of smart living and climate change education. The lead instructor will introduce the filming and production of short videos through theoretical explanations, practical operation skills, and on-site Q&A sessions. The course aims to enhance the digital literacy and digital learning capabilities of the elderly, help them bridge the "digital divide," and enjoy the benefits of digital development.

Chapter 5 Collaborative Climate Change Education by Schools, Families, and Communities | 237

(Continued)

Theme 3: Low-carbon Energy	Lecture: "Development Trend of Low Carbon Energy in the context of Climate Change" is taught by Zhendong Zhang, Associate professor of China-UK Low Carbon College, Shanghai Jiao Tong University. Promoting green development in an all-round way and adjusting economic structure and energy structure is the top priority. This course focuses on introducing the development path of energy technology, energy structure and energy management in China.
Theme 4: Low-carbon Monitoring	Lecture: "Radiation Heat exchange, Greenhouse Effect and Carbon Monitoring" is taught by Tao Ren, Associate professor of China-UK Low Carbon College, Shanghai Jiao Tong University. By introducing the relationship between solar radiation, earth radiation and the absorption of the atmosphere, and then talking about the formation of greenhouse effect, we advocate people to actively explore low-carbon technology and practice low-carbon life. Laboratory Visit: Remote sensing carbon dioxide monitoring laboratory platform visit

Figure 4. Associate professor Xue Dong from Shanghai Jiao Tong University give a lecture

Strategy and thinking:

(1) Professional and Practical Curriculum Design: The project promotes the establishment of a new paradigm for lifelong learning by conducting a series of activities under the theme of "Education on Green Science and Technology and Climate Change". These include "Science Popularization Joy Classroom," "Green Low-Carbon Exploration

Figure 5. The participants visit unmanned driving laboratory of Shanghai Jiao Tong University

Group," "Smart Life Technology Classroom," "Lin-gang Science and Technology Study Tour," "Green Living Practice Course," and "Elderly Wisdom Learning Group Development." The curriculum covers knowledge areas such as smart life, green low-carbon practices, and intelligent technology, with the goal of enhancing the scientific literacy, practical skills, and innovative awareness of the elderly.

(2) Interactive and Practical Educational Approaches: The curriculum is designed with a series of easy-to-understand lectures, company visits, hands-on activities, and other diverse engagements. By collecting and promptly responding to participants' feedback, teaching strategies are adjusted to enhance the adaptability and sense of participation among the elderly. This approach ensures that the educational content is both engaging and practical, effectively facilitating active involvement and knowledge absorption among senior learners.

3. Building a Comprehensive Hierarchical Climate Education System for Families

By the end of April 2024, the project aims to target families that include an elderly member, an adult, and a child to construct a comprehensive educational system that spans from the family unit to the broader society. This initiative is designed to foster intergenerational learning and mutual education for achieving the goal of lifelong learning. In doing so, it effectively meets the challenges posed by climate change and promotes the sustainable development of society(see Table 3).

Table 3. The Intergenerational Learning Project schedule in 2024

Project Theme: Green Technology Innovation and Climate Change Education Special Project Participants: Open Recruitment for 30 Families Comprising "One Elderly, One Adult, and One Child" to Participate Together in the Community

(Continued)

Theme	Schedule
Circular Economy	New chapter of intergenerational learning—Small Hands Holding Big Hands Lecture: "Evolution of E-Waste Recycling Technology" is taught by Jia Li, associate professor of China-UK Low Carbon College, Shanghai Jiao Tong University. By introducing the basic process of e-waste recycling, and the important role of e-waste recycling in accelerating the cycle of material and energy, creating economic vale and environmental value. Laboratory Visit: "Shanghai Jiao Tong University low-carbon Team Turns Garbage into Treasure". To see how super visual garbage sorting robot with the advantages of fast speed, good vision, cost saving to help garbage classification.
Astronomical Technology	Participants go to Shanghai Astronomy Museum to learn China's astronomical technology development by a vivid, interesting and intuitive scientific experiment content in a relaxed and humorous way, which also integrate stage performance elements, and spread science and scientific ideas. ① About "Liquid" (targeting all age): To learn density difference, Bernoulli's principle and liquid surface tension and other related principles. ② About "Vibration" (targeting all ages): Vibration of sound waves can be a melodious song or a sharp scalpel, depending on how to use. ③ About "Transmission" (targeting all ages): Participants go together into the space station to learn the space station thermal control system, vibration and noise reduction and heaven and earth communication principle, through science can understand the "guide". ④ About "Singing in Mars": By showing the power of the sound fire extinguisher, participants can learn the characteristics of the sound in Mars, and visualized the sound with laser. ⑤ "Tianwen" in chinese: With the experimental demonstration, participants can learn principles behind the "China Sky Eye", "Yutu Ⅱ", "Tiangong" space station Hall propeller and other major powers.

Strategy and thinking:

(1) Diverse Activities: Through a variety of formats such as lectures, interactive discussions, hands-on experimental operations, and competitive quizzes, to stimulate the learning interest and enthusiasm of participants.

(2) Community-wide Coverage: By leveraging community education initiatives, there is extensive dissemination of knowledge on climate change, mobilizing community members to participate collectively. This approach fosters a positive atmosphere of community-wide engagement and propels the widespread adoption of green lifestyles. Efforts are dedicated to promoting climate change education across different age groups, nurturing environmental awareness and action, and establishing a new paradigm for lifelong learning in green innovation and climate change education.

Effectiveness & Reflections

From October 2023 to April 2024, a total of 12 activities on climate change education were successfully held, attracting over 2000 participants and positively impacting the public's environmental awareness, technological innovation capabilities, and green action initiatives. The project, jointly conducted by the government, community, enterprises, and universities, has enriched the activities, innovated the model of sustainable community development education, and demonstrated the widespread attention and support from all sectors of society for climate change education and green innovation.

(1) Platform Construction: By integrating resources from the government, universities, and enterprises, we have meticulously designed a diverse range of climate change education courses and successfully established an information sharing platform to facilitate collaborative services. The establishment of this platform has broken the traditional model of single-source education, promoted the sharing of resources among various parties and provided more opportunities for community residents to participate in community education. Through multi-party collaboration, the project has achieved tangible results within the area. The collaborative learning involving school, family, and community has not only driven the coordinated development of green communities but also provided a solid foundation for future educational projects.

(2) Cognitive Enhancement: Through a series of social participation activities focusing on green innovation and climate change education, residents' sense of environmental responsibility is cultivated. The curriculum content is designed to be closely related to the daily lives and practical needs of the residents, thus stimulating the ecological sentiments of community dwellers and enhancing individual ecological literacy. This approach ensures that theoretical knowledge is effectively implemented in practical actions. The project has significantly improved participants' awareness of social development trends and the role of cutting-edge low-carbon technologies in national efforts to combat climate change.

(3) Behavior Change: By integrating climate change education with community education through innovative practice, participants have directly experienced the impact of low-carbon technology and gradually developed habits of adopting a low-carbon and environmentally friendly lifestyle. This research-oriented approach not only enhances participants' understanding on climate change but also leads to positive behavioral changes, injecting new vitality into the sustainable development of communities. Furthermore, by sharing their experiences and insights, participants contribute to social innovation and development.

This project has achieved significant results in enhancing public awareness of climate change, fostering technological innovation capabilities, and stimulating the enthusiasm for green actions. Through a cooperative mechanism made up of schools, families and communities, the project has achieved resource sharing and propelled the joint construction and diversified development of community education. The successful implementation of the project has not only been effective at the educational level but has also promoted the widespread adoption of environmental awareness and the practice of green lifestyles at the community level. It has provided a strong practical example and valuable experience for building a new paradigm of lifelong learning in the local area.

However, there are also space for reflection in the implementation of the project. First, the scope of the project needs to be expanded. Although the project has attracted many community residents to participate, there are still some groups who have not been able to join for various reasons. The project team shall focus on how to better attract and include a broader range of groups, especially those who are less familiar with or less interested in climate change education. Second, the project practice still needs continuous innovation: How to use existing resources more effectively and continuously explore new resources, or find more partners and financial support to achieve greater impact and coverage, is also important for us to continue to think about and strive for.

(Qiuju Li, Vice President of Nanhui New City University for Senior Citizens, Pudong, Shanghai; Ying Wang, Program Manager of Cooperation Exchange and Executive Training Office, China-UK Low Carbon College, Shanghai Jiao Tong University)

School-Enterprise Collaboration Promotes Climate Change Education Practices on Campus

Research Background

Since the 18th National Congress of the Communist Party of China, climate change education has shown a vigorous and encouraging development. President Xi Jinping has pointed out that the challenges posed to humanity by climate change are real, serious and lasting. Internationally, the latest advancements in climate change education have focused on four key areas: green curricula, green teachers, green skills, and green communities. Domestically, China's approach to climate change education features advanced concepts, coordinated policies, school initiatives, and integrated research and training, creating distinctive Chinese strategies and practices. Promoting a green development mode and lifestyles has been emphasized as a critical issue. How to cultivate and practice a healthy lifestyle has become a highly significant concern for both the education sector and broader society in China. Therefore, Qianjing Primary School strives to explore new approaches in conducting climate response education by capitalizing on regional characteristics.

1. School Development Needs: Seeking Innovative Campus Feature Projects

Qianjing Primary School, committed to meeting the educational needs of the future, continuously upholds an innovative spirit and explores education models that evolve with the times. To further broaden its educational horizons and deepen students' understanding of environmental protection, the school plans to launch a series of unique campus feature projects, inviting valuable suggestions and collaborative opportunities from all sectors of society.

Currently, the school is actively evaluating several project ideas, including the idea of expanding and enhancing the "Jingjing Vegetable Garden" into a comprehensive teaching module covering sustainable agricultural education; upgrading the "Little Whale News Media Center" to improve its professionalism in communicating about climate change and environmental protection. In Addition, teachers are considering introducing an interactive climate change simulation system to enable students to learn through first-hand experiences of various climate scenarios.

To implement these new projects, Qianjing Primary School is actively seeking cooperation with local companies, such as exploring green commuting options on campus with Phoenix Bicycles Co.; establishing joint ventures with Shanghai AILU Packaging Co., Ltd., and Jinshan Caojing Chemical Plant, and others. Through such collaborations, we aim to support environmental initiatives and provide students with practical experience platforms.

With the active participation and contributions of community members, businesses, and educators, not only new educational projects are developed by the school but through these feature programs, students will also gain knowledge and participate directly in environmental and social responsibility practices. We firmly believe that this will help cultivate a deep understanding of climate change and a long-term commitment to environmental protection among students.

2. Corporate Development Needs: Seeking Transformation for a Legacy Brand

Phoenix Bicycle is a well-known local brand in Jinshan District with over a hundred year of history. As global attention to climate change grows and the bicycle culture flourishes along with healthy living concepts, Phoenix Group believes it is vital to incorporate climate change education into educational collaborations. Consequently, the company has partnered with schools to establish "Phoenix Station" both on and off campus, organizing a series of activities centered around climate change education. These initiatives guide students to integrate environmental protection into their daily lives, such as cycling instead of driving to reduce carbon emissions and combating global warming. This move is believed to not only enhance students' physical fitness but also boost their innovation competence and sense of responsibility towards environmental protection, contributing to a greener future.

Project Design

The "Phoenix Station" project is dedicated to raising students' awareness of environmental protection, as well as enhancing their understanding of low-carbon transportation through learning about the history of bicycles and participating in cycling practice activities. The project employs diverse teaching methods, including project-based learning, lectures, practical experiences, and multimedia interaction, to cultivate students' scientific exploration, teamwork, and sense of community responsibility. Feedback from parents and businesses is leveraged to ensure the sustainable development of education.

1. Objectives

(1) Raise students' awareness of climate change and environmental protection by understanding the history of bicycles and their contribution to sustainable transportation.

(2) Cultivate students' practical skills and team spirit, improve their scientific exploration and problem-solving abilities, and strengthen their awareness of public engagement and community responsibility.

2. Contents

(1) Theoretical learning: The scientific basis of climate change and its impact on the

environment; the history of the bicycle, its role in urban transportation, and its future development.

(2) Practical activities: Beautification of helmets, bicycle assembly and maintenance workshops; design and participation in environmental cycling events, including route planning and safety guidance.

(3) Cultural education: The status and influence of the bicycle in different cultures; case studies of sustainable development, such as bicycle-friendly cities around the world.

3. Methods

Teaching methods: Teachers employ a "lecture + practice" model, disseminating theoretical knowledge through lectures and interactive discussions, and deepening understanding through hands-on operations and experiences. They also utilize multimedia and virtual reality tools to enhance the interactivity and fun of teaching.

Student participation: Students are divided into groups to complete different project tasks, such as bicycle maintenance and environmental cycling planning. During the activities, students take on different roles during the activities, such as project manager, recorder, etc.

Corporate participation: The school invite professional coaches from Phoenix Station to participate in course lectures and outdoor cycling activities. Teachers organize open classes to promote awareness of climate change.

4. Assessment

Process Evaluation: To ensure that students meet standards in both theoretical learning and practical operations, teachers should gather feedback from students on activities through questionnaires and group discussions, and promptly adjust our teaching plans accordingly. Additionally, by designing and distributing questionnaires, we collect feedback from parents to understand the long-term sustainable impact of the project.

Outcome Evaluation: A structured assessment through exams and practical skills tests evaluates students' knowledge acquisition and application abilities. This will assess the students' ability to integrate their learned knowledge with real-world problems, such as planning and executing a small-scale climate action project.

With such comprehensive and systematic planning, the "Phoenix Station" project aims to enhance students' awareness of climate change and response strategies, while strengthening their practical skills, creating a lasting motivation for environmental protection.

Project Implementation

1. Facility Construction and Resource Integration: Collaborative Contributions

Continuing the consistent approach of the "Qianjing Li" experiential courses, this project commenced with the development of physical venues centered on bicycle riding as an experiential medium. After several rounds of discussions, three experiential facilities were designed: "Jing · Phoenix Station," an outdoor cycling apparatus, and a Phoenix-themed classroom. Following consultations with experts, the school added a bicycle experience area, showcasing a variety of bicycles for students to explore andtry out. Additionally, an area was created around the Phoenix-themed classroom to display the history and culture of bicycles, allowing students to engage actively, have fun and learn.

Considering the practical circumstances of school, the Phoenix Group resolved the essential issues of funding, facilities, and equipment. The primary contribution from the school was providing the space, using land as "capital." This school-corporate collaboration project also received significant support from the district government and the Education Bureau. The "Phoenix Station" project was recognized as a major initiative for the year in Jinshan District and became a model case of implementing integrated education in the district, laying a solid foundation for both school and corporate parties to use bicycle cultural education as an opportunity to jointly promote education about climate change education(see Figure 1).

Figure 1. Off-campus Phoenix Station

2. Activity Implementation and Educational Practice: Joint Efforts

Firstly, to stimulate students' intrinsic motivation for learning. To spark an interest in climate change education, our school has frequently collaborated with Phoenix Group to organize various campus activities. These include fun cycling events, "Jing-Phoenix" bicycle model assembly activities(see Figure 2), "Old Brands, New Wonders" visits to historical brands in Shanghai, and festive winter funfair events to welcome the new year as part of the series of "Phoenix into Campus" activities. Phoenix Group has worked closely with us, actively participating in each event, sponsoring bicycles for the cycling activities, sending experts to guide the model assembly, inviting students to visit their enterprise, and arranging guides for the brand visits, thereby ensuring all school activities achieve desirable outcomes.

Figure 2. "Jing-Phoenix" Bicycle Model Assembly Activity

Secondly, we have strengthened a new model of collaborative education. In our practical activities, we explored the integration of the study of various subjects with the Phoenix cycling activities, incorporating low-carbon environmental awareness into our teaching process. The cycling competitions held by the Physical Education Department were strongly supported by the Phoenix Group, which not only provided bicycles but also sent professional technicians to explain techniques on "how to cycle faster," ensuring the successful execution of the event and effectively enhancing our students' cycling sportsskills(see Figure 3). In the Natural Science discipline, we designed an experimental activity involving a power-generating drum that uses the motion of bicycle wheels to generate electricity. This generated electricity can power electronic devices such as smartphones and tablets, playing a role in energy conservation and emission reduction.

3. Curriculum Development and Academic Exploration — Exchange and Cooperation

Curriculum is the blueprint for education. Therefore, as soon as the designated hubs were completed, we immediately began curriculum development without delay. In the design phase, both the school and the corporate side came together, each delegating personnel to form a writing team. After multiple research discussions, from name selection to structure, from content to steps, and from formats to teaching methods, every detail was meticulously refined to achieve fruitful results.

Figure 3. Campus Cycling

The curriculum was named "Phoenix Cycling" and became a distinctive course developed in collaboration between our school and Phoenix Group. Utilizing collaborative resources, the course offers experiential and participatory activities to create an interactive learning platform, which has been well-received by both students and parents. The "Phoenix Cycling" specialty course is implemented in three stages: enlightenment, advancement, and customization.

(1) Initiation Course: Students learn about the history and evolution of the "Phoenix" brand and types of bicycles, beginning with basic theoretical teaching. This course integrates the Phoenix Group's culture into the school culture, focusing on classroom learning together to reinforce the classroom's pivotal role.

(2) Advanced Course: This primarily involves skill-based learning, guiding students through the theoretical aspects of skills with activities like assembling bicycle model components and embellishing bicycle helmets as core content.

(3) Customized Course: In this stage, the Phoenix Group culture is incorporated into various aspects of the course through project-based integration of disciplines. This also fully leverages the benefits of integrated education, resulting in a pleasing situation where all subjects and families are actively involved. In the school project, Teachers Ms Sun and Ms Gong conducted two activities entitled "Endless Joy of Cycling: Exploring Project-Based Learning from an Integrated Educational Perspective" and "A Parent-Child Survey Activity on Low-Carbon Environmental."

Fifth-grade students tackled the driving question: "Have you learned to ride a bike before your age to ride on the road?" "Where do you ride? Please design a safe and healthy cycling plan." Throughout the process of exploring and solving real-life problems, they constantly raised questions, analyzed issues, and found solutions. After calculating cycling distance, time, speed, and the calories burned or BMI index, students planned out healthy and reasonable cycling plans. The submitted plans were organized into categories: sports cycling plans, sightseeing cycling route recommendations, and optimal routes for commuting to and from school. Each plan had its unique features, and different travel purposes led to different designs. Through this project-based learning, students not only exercised their bodies and experienced the joy of cycling, but they also reinforced their knowledge and skills in mathematics, appreciating its close connection to everyday life.

4. Teacher Training and Dual-instructor Model — Goal Consistency

Firstly, the implementation of the "dual-instructor" system was introduced. Considering that school teachers were unable to address some technical questions regarding bicycles, the dual-instructor model was proposed. In this setup, for each activity, Phoenix Group sends an expert, and the school selects a teacher to carry out the task. In practice, Phoenix Group and Qianjing Primary School work closely together to promote dual-instructor teaching. Under this model, both teachers participate in the teaching process where is in charge of teaching and managing students' learning, while the external instructor is in charge of answering questions and practical operations. Therefore the system works to enhance the quality of education and ensure the smooth progression of teaching activities(see Figure 4).

Figure 4. Off-campus Mentor Q & A Guidance

Secondly, the advantages of the dual-instructor system were fully leveraged. Under the guidance of both school and corporate instructors, using Phoenix brand bicycles as the activity platform, various classroom inquiry activities such as "Exploring the Centennial History of Phoenix," "My Family's 'Phoenix' Story," "Innovative Ideas for 'Phoenix'," and "A Century with Phoenix, Accompanying Us Along the Way" were conducted, along with extracurricular parent-child activities, small team activities, and squad activities. The professionals from Phoenix Group explained the history, development, and other technical aspects of Phoenix bicycles, which has greatly enhanced the effectiveness of the activities.

Teacher Hu from the natural sciences expertly conveyed theoretical knowledge about the history and classification of bicycles in China in a simple and clear manner. Meanwhile, the external instructors used their extensive practical experience to teach students safe cycling skills, achieving a harmonious integration of knowledge and skills. Additionally, the collaboration between schools and corporations meticulously planned a series of innovative activities including the "2024 Fun Garden Welcome Event," "Bicycle Helmet Creative Design Competition," and "National Trends Enter Schools, Phoenix Supports Growth." These activities not only greatly enhanced the students' practical abilities, innovative thinking, and teamwork skills but also injected new vitality and color into the campus cultural life, allowing each elementary school student to grow happily in a rich and diverse experience.

The joint school-corporate "dual-instructor" model can effectively enhance the effectiveness of activities and profoundly demonstrate the collaborative nature of education. Particularly significant is the cultural resonance between the two parties, which facilitates the better preservation and development of both campus culture and corporate culture.

Effectiveness & Reflections

Collaborating with Phoenix Group has enabled our school to effectively implement climate change education using bicycles as a medium, making it a standout feature of our school. During the project, students explored their personal needs and collected real data, which greatly fueled their enthusiasm and deepened their understanding of the necessityof green commuting. At the same time, we have also experienced the immense appeal of school-corporate collaboration in conducting climate change education. Looking ahead, we have several Reflection on future prospects.

1. Deepening Collaboration with Phoenix and Strengthening the "Phoenix Station" Brand

Firstly, further integration with technology. We should keep pace with technological advancements and enhance our projects with internet technology. By launching a joint school-corporate "Phoenix Internet Website", we will encourage more students, parents, workers, and community members to contribute to promoting "green commuting".

Secondly, we will refine our thematic focus: By emphasizing activities such as "Cycling Through the Weather," we aim to increase student interest andtry to expand collaborative learning among schools, families, and the community to expand from schools to society, thereby continuously broadening the impact of meteorological science popularization.

2. Spreading the Success of the Collaboration with Phoenix and Highlighting the "Climate Education" Brand

Firstly, we will further explore and utilize our existing venues. Within our distinctive "Qianjing Li" project, there are many untapped resources which can be developed for climate education. For instance, our school's "Jing Jing Vegetable Garden" naturally correlates with climate change, and Jinshan is a region where agriculture is highly developed with many companies engaged in vegetable cultivation and business. In line with school-corporate collaborations, we plan to extend outside the school to find suitable vegetable enterprises to develop our "Jing Jing Vegetable Garden" into another exemplary school-corporate collaboration.

Secondly, we will develop new experiential education venues to enrich our educational offerings. We will continue to establish new experiential venues focused on strengthening climate change education, such as envisioning the construction of "Qianjing Primary School Weather Station" for students to participate in, experience, and explore at any time.

Climate change education requires us to continually cultivate and explore with vigor. We hope our school can rely on "Phoenix" to spread its wings, attract more enterprises to join the climate education initiative, and collaboratively strive towards the realization of the ecological civilization education advocated by President Xi Jinping.

(Huanxin Qian, Principal of Qianjing Primary School, Jinshan, Shanghai; Danyan Jiang, Teacher of Qianjing Primary School, Jinshan, Shanghai)

School-Enterprise Cooperation: Advocating "Green Productivity", Promoting Climate Change Education

Research Background

Climate change is an ongoing phenomenon for which human activities have been identified as the primary cause. As early as 1992, the United Nations General Assembly adopted the *United Nations Framework Convention on Climate Change*, which explicitly mentioned human activities as a contributing factor: "'Climate change' refers to a change in climate that is attributable directly or indirectly to human activities that alter the composition of the global atmosphere and is in addition to natural climate variability observed over comparable time periods.[①]" Therefore, the focus of climate change education should be on teaching and changing human activities. Production and daily life, as the two main activities of humans, should both become significant topics for learning.

Climate change education should start from an early age, as primary schools play an irreplaceable foundational role. However, current primary school climate change education often focuses solely on students' daily life at school and home, neglecting production activities, which are perceived as "distant" from students. In reality, without changes in production methods and products, it is challenging for humanity to adapt to new lifestyle changes required by climate change. To nurture a generation that is conscious of and capable of participating in climate actions, primary education must strengthen the connection between current educational practices and future industry needs, and enhance the systematic nature of climate change education.

The Longhutang Experimental Primary School in Xinbei District, Changzhou (referred to as "Longxiao"), has developed a unique approach by focusing on school-home-community cooperation at the street level. Since the initiation of climate change research in May 2023, it has gradually formed a distinctive model of school-home-community cooperation. Longhutang Street, located in the Changzhou National High-Tech District, prioritizes the expansion, strengthening, and optimization of industries based on town-specific characteristics, aiming to forge a robust "industrial chain." This has led to the formation of a distinct industrial cluster characterized by "one specialized, one new, and one modern" industry. In this context, school-enterprise cooperation in climate change education has entered the school's research vision, focusing on the relationship between climate change and production activities, as well as the productivity

---

[①] United Nations. United Nations framwork convention on climate change[EB/OL].1992[2023-12-04].https://unfccc.int/resource/docs/convkp/conveng.pdf.

that drives their development, thereby clarifying two major research questions and scopes.

1. Green Productivity: The Weak Link in Climate Change Education that Needs Strengthening

President Xi Jinping has emphasized that "green development is the underlying tone of high-quality development, and the new form of productivity is inherently greenproductivity. We must accelerate the green transformation of development modes to support carbon peak and carbon neutrality." Green productivity, characterized by its environmental friendliness and sustainability, is a crucial factor that climate change education must research, advocate, and promote. Green production, along with green consumption and green living, forms the chain of actions people take to combat climate change. As the "input end" of this chain, green production plays a pivotal role and should be a key component of climate change education. However, while green consumption and green living receive considerable attention in current primary school climate change education, green production is often overlooked. This gap highlights not only a weakness in the existing research but also a significant potential for further study and exploration.

2. School-enterprise Collaboration: A Key Area for Developing Primary School Climate Change Education

In current research and practice, school-family-community collaboration is often simplified to just school-family cooperation, particularly in primary schools. While Longxiao has made substantial progress in fostering school-home-community partnerships, its primary external partners remain the community and government departments. Since the initiation of climate change education research, these partnerships have focused mainly on promoting and practicing green concepts, green consumption, and green lifestyles within families and communities. However, the educational value that enterprises and their production activities can offer has not been adequately explored.

Additionally, a search of the China National Knowledge Infrastructure (CNKI) for journal articles on topics such as "climate change education" and "collaboration" over the past five years reveals a trend shown in Table 1. This trend indicates an increasing interest and investment by researchers in climate change education. Nevertheless, there is a significant gap in research on collaboration in primary schools,, particularly in the dimension of school-enterprise partnerships. This gap highlights a critical area for innovative development.

Table 1. Trend in the number of publications related to "Climate Change Education" and "Collaboration" (2019-2023)

Year	Climate Change Education	Climate Change Education + Primary School	Climate Change Education + Collaboration	Climate Change Education + School-Enterprise Collaboration
2019	24	0	2	0
2020	19	0	0	0
2021	28	0	0	0
2022	38	2	0	0
2023	57	1	1	0

Project Design

Based on the research context and questions outlined above, this project focuses on school-enterprise cooperation. It adheres to the main idea of promoting interdisciplinary collaboration between the climate change industry and academia, following the principlesof co-construction, sharing, and mutual learning. The project aims to refine an integrated climate change education content system that encompasses both production and daily life. This approach seeks to explore pathways for implementing "whole institution" climate change education, while simultaneously influencing the families and communities of students and enterprise employees, as well as relevant departments, to awaken a broader sense of climate responsibility and continuously propagate a culture of sustainable development. This leads to the formation of the following research objectives and framework:

1. Research Objectives

Students can:

(1) Clarify the relationship between green productivity and climate change education, based on the anthropogenic causes of climate change, and construct a comprehensive climate change education framework integrating production and living elements.

(2) Explore deeply through university-enterprise collaboration to mutually promote the "whole school" and "whole enterprise" implementation of climate change education, and innovate paths and models for collaborative education involving schools, families, communities, and society.

(3) Develop green education partnerships among peers, teachers, parents, enterprise employees, community members, and others from a lifelong learning perspective. Through cooperative climate change education, foster integrated competencies across multiple stakeholders and promote the dissemination of sustainable development concepts.

2. Research Approach (see Figure 1)

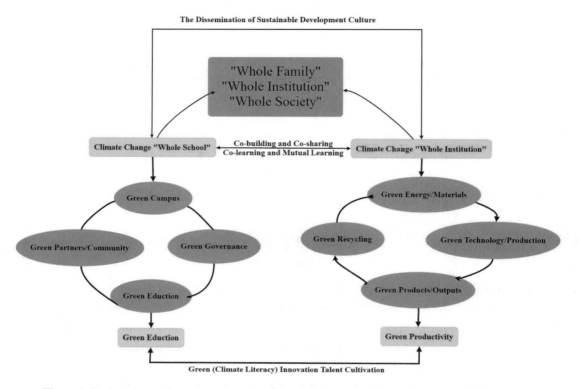

Figure 1. Design concept for university-enterprise collaboration in climate change education projects

Project Implementation

1. Research and Analysis: Focusing on Core Elements of Green Production and Establishing a School-enterprise Experimental Community

First, the project team gathered information through online data searches and consultations with local economic and technological development bureaus to understand the overall layout of enterprises in Longhutan, particularly the characteristics of high-tech companies. From this, we learned that by 2023, there are approximately 7,000 registered enterprises, with 103 high-tech companies currently in effect.

Next, based on the key elements of green production that adapt to climate change—energy or materials, technology, products, and circular economy—we identified relevant enterprises. These enterprises were classified into two categories: those that had previously collaborated with the school and those that were particularly representative in certain fields. We established cooperation intentions through connections and communication with parents, government leaders, and public welfare organizations.

Subsequently, project team members visited enterprises and engaged in in-depth discussions with enterprise management to gain a better understanding of their culture, teams, production, or product characteristics(see Figure 2).

Figure 2. Project team members visiting Trina Solar Limited and Wencan Machinery Manufacturing Co., Ltd.

On this basis, the school paired up with the first batch of climate change education enterprise communities and developed a comprehensive plan for the direction and contentof cooperative research (see Table 2).

Table 2. Climate change education university-enterprise cooperation communities and research directions

Research (Green production) elements	Cooperative enterprises	Climate functions
Green energy/ Materials	Trina Solar Limited / Trina Fuja Energy Co., Ltd., Polymers New Materials Co., Ltd.	Clean energy, decarbonization
Green technology	Wencan Machinery Manufacturing Co., Ltd.	Emission reduction, carbon sequestration
Green products	Sensata Technologies, Inc.	Low carbon, energy saving
Green recycling	Ruisai Environmental Technology Co., Ltd., Everbright Environmental Energy Co., Ltd.	Resource conservation, low-carbon recycling development

2. Creating a Model: Focusing on the Primary Goal of Carbon Neutrality and Exploring the Educationalization of the Energy Revolution

For more than a century, humans have relied on burning fossil fuels such as coal and oil to obtain energy for electricity and heating. However, the resulting greenhouse gases, such as carbon dioxide and methane, are major contributors to climate change and global warming. Thus, the goal of carbon neutrality primarily targets the energy revolution. Leveraging parental resources, the school initiated deep collaboration with the most representative new energy enterprise on the street, Trina Solar Limited, to explore a feasible basic model that can be further promoted.

Trina Solar has already formed a photovoltaic industry cluster. Considering the cognitive characteristics of primary school students, we further focused on Trina Fuja Energy Co., Ltd., a subsidiary primarily developing household photovoltaic products (hereinafter referred to as "Trina Fuja"). The collaboration has now reached its fourth phase.

Phase 1: building collaboration vision. In early 2024, the school project team visited and studied at "Trina Fuja Energy Co., Ltd." and invited the company management team for discussions and workshops at the school, facilitating the participation of various departments such as the street office, ecological environment bureau, parent committees, and university experts (see Figure 3). Both the school and the enterprise integrated their cultural pursuits and formed a collaborative vision for climate change education titled "Transmitting Smart Energy Intelligence for Zero-Carbon Production, Cultivating a Green Future of Zero-Carbon Living." Based on their respective practical resources and experiences, they preliminary agreed on three "One" action plans: short-term cooperation on an activity, mid-term collaboration to construct a facility, and long-term partnership to establish a model.

Figure 3. Multi-party participation in university-enterprise visits, discussions, and collaborative research

Phase 2: creating classic activities. During the winter vacation of 2024, the school and "Trina Fuja Energy Co., Ltd." jointly organized the "Approaching Clean Energy, Creating a Green Future—Long Dada (which means happiness in Chinese) Cup National Science and Innovation Painting Competition." This event encouraged students to engage in intergenerational learning within families (with a focus on linking with families of enterprise employees), participate in peer group activities for experiential learning, both online and offline. Students explored the relationship between clean energy and climate change, gained understanding of solar energy products, and designed new energy products based on the needs and experiences of family Lunar New Year celebrations. The outcomes were showcased in the form of science and innovation paintings. On March 1st, during the "Second National Symposium on Climate Change Education through a cooperative mechanism made up of schools, families and communities" hosted by the school, an award ceremony was held for the event, where the company awarded solar products to the winning students. Subsequently, the results of the activities were disseminated through platforms of both the school and the company, using methods such as short videos and presentations to reach a wider audience(see Figure 4, Figure 5).

Figure 4. Peer group study and award ceremony in the Long Dada Cup National Science and Innovation Painting Competition

Phase 3: curating a series of collaborations. The successful implementation and impact of the National Science and Innovation Painting Competition have accumulated experience and strengthened confidence in the collaboration between the school and "Trina Fuja Energy Co., Ltd." The company recommended the school to Trina Solar Limited. On April 9, 2024, leaders from Trina Solar's Public Welfare Foundation and Trina Fuja's brand management visited the school. They discussed long-term sustainable cooperation with the core project team members and specifically discussed this year's cooperation plan (see Figure 6). After several refinements and discussions, they formulated the "Photovoltaic + Dual-Carbon Education—2024 Climate Change Education Cooperation Action (Trina Corporation & Longhutang Experimental School)" plan. This plan focuses on developing a series of collaborative activities including site construction, curriculum development, activity implementation, and outcome dissemination(see Figure 7). Trina Fuja's technical personnel also visited the school promptly to conduct surveys and design photovoltaic project plans for the construction of climate change education site.

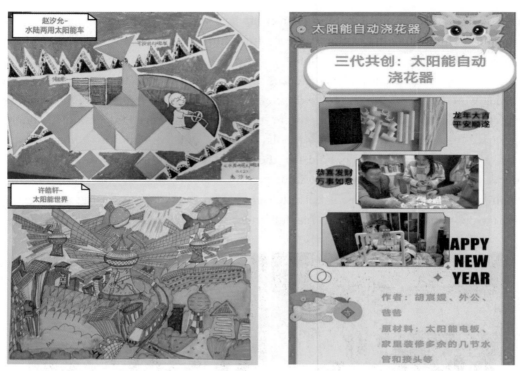

Figure 5. Selected works from the Long Dada Cup National Science and Innovation Painting Competition

Figure 6. University-enterprise discussions on annual cooperation plans

Theme Module	Specific Content	Implementation Method	Time
Base Construction	Ecological Secret Construction	Introduction of climate and meteorological observation equipment	April-August
	Campus Photovoltaic Construction	Installation of photovoltaic power generation systems on the roofs of the library and various campus buildings, creating a green rooftop garden.	
Course Development	In collaboration with the school's modeling club, develop the course "Smart Home Design and Production" to popularize clean energy knowledge, promote Trina Solar's concepts and products, and innovate new products.	1. School-based course development 2. Competition model design	June-August
Activity Implementation	Organize a Green Ecology Buddy Group, involving multiple parties such as family, school, community, and enterprise, to regularly cooperate on thematic practical activities for climate change education.	1. Visit the Kerlin Dual Carbon Labor Base 2. Initial experience in the new energy profession	April-June
Outcome Dissemination	Disseminate cooperation outcomes through various channels	1. Participate in or host climate change education seminars (domestic/international)	April 26: Shanghai Climate Week Climate Change Education Forum September: Hello Summer Holiday
		2. Publish academic results (papers, cases, books)	September - December

Figure 7. Partial excerpts from the "Photovoltaic ＋ Dual-Carbon Education — 2024 Climate Change Education Cooperation Action" proposal and campus photovoltaic project proposal

Phase 4: reviewing cooperative experience. Leveraging the "2024 Shanghai Climate Week" held from April 25-27, the school facilitated teachers, students, and management personnel from "Trina Fuja Energy Co., Ltd." to collectively review and summarize their cooperation experience. At the climate change education forum during Climate Week, Principal Gu Huifen and Jiang Ning, Public Relations and Brand Manager of Trina Fuja Energy Co., Ltd., shared the collaboration outcomes from the perspectives of the school and the company. Teacher Fu Rong and student teams also presented their collaborative research and practices at the "Climate Resilient Community Forum" and the "Youth Climate Action Science and Innovation Works Presentation." (see Figure 8). Xinhua News Agency published a special report titled "Changzhou: Schools and Enterprises Attend 'Climate Week' Together to Create' New Experiences'."

Figure 8. Teachers, students, and management personnel from "Trina Fuja Energy Co., Ltd." disseminating the school-enterprise cooperation experience at the "2024 Shanghai Climate Week"

3. Multi-"Ring" Practices: Focusing on the Integrated Fusion of Production and Life, Strengthening the Popularization of Climate Change Education

Through cooperation with Tianhe Fu Jia Energy Co., Ltd., the initial realization of "energy revolution" in "climate change education" has been achieved. Based on this experience, the school has comprehensively reviewed its preliminary cooperation with other enterprises, including Wenkangna Mechanical Manufacturing Co., Ltd., Sensata Technology Co., Ltd., and Evergreen Environmental Energy Co., Ltd. Each of them focuses on key aspects of green production such as "green technology for emissions reduction and accumulation," "low-carbon energy-saving green products," and "green circular classification and regeneration," expanding the core values of educational themes.

In the process of practice, the project team is committed to breaking down barriers between green production and green living (including green consumption), and constructing an integrated "chain" of climate change education, primarily through three major action strategies:

(1) Bilateral conduction between schools and enterprises: Teachers, students, and parents visit enterprises through activities such as buddy-group research. Meanwhile, enterprise personnel visit schools through forms like public service. Students and parents advocate for "green lifestyle" to enterprise personnel, who in turn convey the concepts, technologies, and products of "green production" to teachers, students, and parents. For example, the "Children's Day" event was held in collaboration with Sensata Technology Co., Ltd.. On May 31, the company team entered the campus, collaborated with the science research and development group, and set up booths with children to promote "sensor technology and low-carbon products." On June 1, students visited the company in teams and jointly held a "STEM public welfare class." Enterprise lecturers taught new energy and new technologies, engineers collaborated with students to make remote-controlled electric car models and held group competitions. Students presented their research on climate change to company employees.

(2) Promoting "scenario-based" integration from students' perspectives: Schools, families, and communities are the main arenas of students' life. Starting from the actual needs of students in these environments, the integration of multiple themes and scenarios in climate change education through school-enterprise cooperation is enhanced. For instance, during the Book Reading Festival, the school collaborated with Sensata to conduct the "Books for Vegetables" public welfare activity, which spread the concept of circular economy. Additionally, activities like "waste classification" with Evergreen Environmental also involved cooperation with surrounding communities.

(3) Promoting mutual learning among more entities in the context of lifelong education: The project team extends the school's model of "driven by multiple forces, integrated with multiple environments, and empowered by multiple learning" in family-

school-community cooperation to school-enterprise cooperation in climate change education. Efforts focus on integrating entities such as ecological departments, meteorological institutions, and lifelong education systems. Inter-generational and mutual learning among student families and enterprise employee families are encouraged, involving more groups in learning and action against climate change.

Effectiveness & Reflections

1. Improved Content System, but Further Enhancement Needed for Structural Internalization of Climate Change Education

The collaboration between schools and enterprises has expanded the horizon of primary school climate change education beyond conventional understanding and awareness, green planting, and green lifestyle. It now includes the production sector and focuses on human activities as a major factor affecting climate change. This not only enriches the educational content but also helps students perceive the "green productivity" or "new quality productivity" called for by the new era. This is beneficial for planting the seeds of green innovation from a young age and cultivating future talents in green innovation.

However, to release internal motivation through structural adjustment and truly stimulate the innovative vitality of climate change education, it is necessary to systematically sort out the dimensions within the production sector, between production and life, and between production, life, and climate change. This should be done through school-based courses, public classrooms, and other forms to enhance the overall construction of educational content. Additionally, from the perspective of developing educational value, it is crucial to find the zone of proximal development for learners to enhance their intrinsic motivation. Particularly, there should be a long-term vision linking the cultivation of green literacy in schools with the employment needs of future green industries, thus contributing to the mechanism of cultivating innovative talents.

2. Expanded Cooperation System, but Further Activation Needed for the Ecological Empowerment of Climate Change Education

Through school-enterprise cooperation in climate change education, the collaborative education pattern among schools, families, and communities has been extended. Beyond the previous involvement of students' families, career departments, and communities, schools have also established cooperative relationships with ecological environment bureaus at various levels, the Ministry of Ecology and Environment, lifelong education research institutions, economic and technological bureaus serving enterprises, as well as internal corporate foundations and social service organizations. This has effectively enhanced the influence of climate change education.

However, for such a vast cooperation system to truly create a green ecology and mitigate climate change, it cannot stop at a single activity, classroom, or base for research and development. It requires connecting scattered climate change education resources into a "map," transforming effective cooperation experiences into a set of routine action mechanisms, and expanding educational benefits focused on one party to be shared by multiple parties...

3. Innovated Mutual Learning Network, but Further Enhancement Needed for Nationwide Climate Change Education Action

In over ten years of research on family-school-community cooperation, the school has initially established a horizontal mutual learning network among the three main subjects of family, school, and community, and a vertical mutual learning network among three generations of families. In the school-enterprise cooperation on climate change education, it has strengthened mutual learning with government and institutional personnel related to ecological environment and economic development, as well as corporate employees. Special attention is paid to the linkage between students' families and corporate employees' families. Additionally, more "citizens" are involved in mutual learning through corporate WeChat accounts and national climate change education alliance groups.

Based on students' learning insights and achievement evaluations, as well as feedback from corporate and government personnel, the awareness of low-carbon living and green development has been reinforced. However, there is still a long way to go for the awareness of responding to climate change to truly transform into everyday actions of learners. Schools and enterprises need to further explore the "whole school" and "whole enterprise" practice models in their cooperation, refine the implementation paths and specific strategies of the "whole institution" model, thereby driving localized implementationin more units and cells, and encouraging more people to take on climate responsibility and implement climate actions in their work and life.

(Huifen Gu, Principal of Longhutang Experimental Primary School, Xinbei, Changzhou, Jiangsu; Yalan Chen, Vice Principal of Longhutang Experimental Primary School, Xinbei, Changzhou, Jiangsu)

Case Study on Student Collaboration in Green Enterprise Visits at Yellow River Estuary and Yangtze River Estuary

Research Background

In May 2023, under the Shanghai Municipal Institute for Lifelong Education, ECUN, the project "Sustainable Development Education through Linkage between the Yellow River Estuary and the Yangtze River Estuary" was successfully launched. Educators and researchers from both regions, under the context of climate change, innovatively conducted sustainable development education through school, family, and societal collaboration, linking the Yellow River Estuary and the Yangtze River Estuary. This initiative not only facilitated an alternative form of interaction between the Yellow and Yangtze Rivers but also bridged the friendship between Shengli Gudao No.1 Primary School in Dongying City, Shandong Province, and the Second Affiliated Elementary School of Shanghai No.6 Normal School, both located at the estuaries.

Dongying City, as the main production and location area of the Shengli Oilfield, was built and thrived due to oil. Shengli Gudao No.1 Primary School, formerly a children's school of the Shengli Oilfield Gudao Oil Extraction Plant, still has nearly half of its students' parents engaged in professions related to oil.

Mitigating and responding to climate change is a global consensus and a shared responsibility of all humanity. The combustion of fossil fuels such as coal, oil, and natural gas produces large amounts of greenhouse gases, which are the main cause of severe climate conditions. According to Article 2 of the Paris Agreement, the global average temperature rise should be kept below 2°C this century, with efforts to limit it to 1.5°C. This poses requirements for fossil energy enterprises to transition towards green and low-carbon development. On October 21, 2021, the General Secretary visited the Shengli Oilfield located at the mouth of the Yellow River, encouraging everyone to secure the energy bowl, concentrate resources on key core technologies, accelerate clean and efficient development and utilization, and enhance the quality of energy supply, utilization efficiency, and carbon reduction levels.

In December 2023, students from the "LuYaEr" Low-Carbon Environmental Protection Club at Shengli Gudao No.1 Primary School learned about the 28th Conference of the Parties (COP28) to the United Nations Framework Convention on Climate Change held in Dubai, United Arab Emirates, focusing on issues such as developing emission reduction plans, reducing the production of fossil fuels like oil and coal, and increasing the use of renewable energy. They immediately thought of the enterprise where their parents work—the Shengli Oilfield Gudao Oil Extraction Plant. As a fossil energy enterprise, how has the Gudao Oil Extraction Plant transitioned towards green and high-quality development under the "dual carbon" background? With this question, Shengli Gudao No.1 Primary School connected with the Gudao Oil Extraction Plant to jointly organize a study tour practice activity in Dongying and Shanghai, synchronizing online and offline.

Project Design

1. Objectives

Through simultaneous online and offline visits by students from Dongying and Shanghai to the production operations of the Shengli Oilfield Gudao Oil Extraction Plant's Management Zone Seven, the project aims to understand how fossil energy enterprises respond to climate change and promote green and low-carbon transformation and development, and to build a platform for mutual learning among the youth of the two estuary cities. Specific objectives include:

For students: To understand the low-carbon and energy-saving production of oil enterprises in photovoltaic power generation and energy digitalization, focus on renewable energy, feel the innovative spirit of oil workers, and learn to think, explore, and cooperate through project-based learning.

For enterprises: To open up to students, teachers, and parents, provide practical opportunities, introduce low-carbon transformation practices in fossil energy enterprises, popularize new energy knowledge, enhance learners' practical understanding of climate change, and fulfill social responsibility.

For schools: To explore the path of cooperation between schools and enterprises, and to improve the mechanism of collaborative climate change education among schools, families, and society.

2. Process(see Figure 1)

3. Assessment

Based on students' performance in activity participation, learning mindsets, teamwork, achievement presentation, and reflection, conducting star ratings for Students, ultimately awarding "Low-Carbon Environmental Protection Little Talent" and "Low-Carbon Environmental Protection Little Guardian" comprehensive awards, as well as individual awards such as "Climate Change Exploration Award" and "Climate Change Action Award" for the "Learning Thinking" and "Achievement Presentation" sessions.

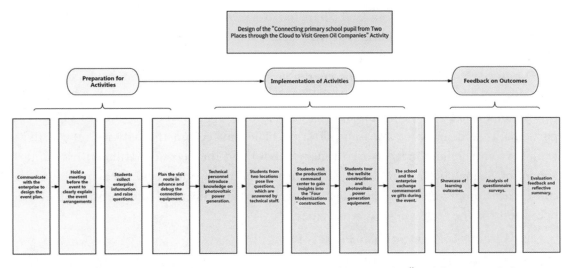

Figure 1. "Two-City Youth Cloud Linkage, Visiting Green Oil Enterprises" activity process design

Project Implementation

1. Activity Preparation Phase

To formulate a detailed activity plan, after the New Year in 2024, the Youth League Committee of Gudao Oil Extraction Plant and Shengli Gudao No.1 Primary School communicated several times to discuss the theme, time, location, and content of the activity, formulating the "*Big Hands Pulling Little Hands, Visiting Green Enterprises*" *Activity Plan*. Later, encouraged and guided by Professor Jiacheng Li, Executive Vice President of the Shanghai Municipal Institute for Lifelong Education, ECNU, the idea of simultaneous cloud visits with the Second Affiliated Elementary School of Shanghai No.6 Normal School was realized, upgrading the plan to *Two-City Youth Cloud Linkage, Visiting Green Oil Enterprises*.

Most of the students visited the oil field for the first time, were curious about oil production but lacking awareness of specific details. Therefore, the school compiled the *Visiting Green Oil Enterprises "Clarity Paper"*, which provided information on the time, content, process, duration, and organizational form of the activity. and held an online meeting with students and parents on the evening of January 12 to clarify activity requirements and conducting relevant training.

Security is crucial during the visit. On January 16 and 19, the school collaborated twice with the Youth League Committee of Gudao Oil Extraction Plant and the responsible comrades of Management Zone Seven to plan the visit route and conduct on-site "stomping", controlling traffic safety to ensure the entire indoor and outdoor activity was safe and orderly.

Learning and thinking complement each other. After the activity plan was issued, teachers from both Dongying and Shanghai organized students to gather information about the visit and raise questions to stimulate students' inquiry enthusiasm and prepare them intellectually and psychologically for the study tour.

Multi-party collaboration is the guarantee for activity implementation. In addition to communicating with the enterprise, the school used WeChat linking and documents sharing to connect with students and parents, ultimately determining the list of activity participants and grouping by transportation. The day before the activity, the activity leaders from Shengli Gudao No.1 Primary School and the Second Affiliated Elementary School of Shanghai No.6 Normal School tested the connection equipment in the online conference room, ensuring seamless network communication and proficient device operation for the following day's activities.

2. Activity Implementation Phase

On the morning of January 20, 2024, at 8:30 a.m., members of the "LuYaEr" Low-Carbon Environmental Protection Club at Shengli Gudao No.1 Primary School arrived at the Seventh Management Zone of Gudao Oil Extraction Plant. Through the cloud connection, they simultaneously conducted the "Big Hands Pulling Little Hands" green enterprise visit activity with students from the Second Affiliated Elementary School of Shanghai No.6 Normal School, one thousand kilometers away(see Figure 2).

Figure 2. Online greeting interaction between teachers and students of Shanghai No. 6 Teachers College Second Affiliated Elementary School and Shengli Remote Island First Primary School in Dongying City, Shandong Province

During the activity, the oilfield technicians first introduced the basic production situation and photovoltaic construction of the unit. Then, students from the Second Affiliated Elementary School of Shanghai No.6 Normal School and Shengli Gudao No.1 Primary School took turns asking questions, which were answered one by one by the project director of Management Zone Seven.

The questions raised by the students included "What technologies are involved in tertiary oil recovery?" "Is oil going to run out??" "Does photovoltaic power generation pollute the environment?" "Will photovoltaic power generation affect climate change?" and others, covering aspects of oil production, photovoltaic new energy, biodiversity, and climate change, and showing breadth and depth, and indicating strong inquiry and ecological environmental awareness among the students.

After the popular science learning and question-and-answer session, everyone visited the Safety Production Command Center of Management Zone Seven, observed the scene of the well site in real time through a large screen, and learning through dialogue that since 2017, Shengli Oilfield has explored and implemented the "Four Modernizations" of standardized design(see Figure 3), modular construction, standardized procurement, and information technology enhancement. The "Four Modernizations" system not only achieves real-time data collection from the production source but also allows for intelligent operations such as remote parameter adjustment and well opening/closing. Previously, it took 30 minutes to patrol one well, but now it takes only 30 minutes to patrol over 300 oil and water wells in the management zone, effectively improving work efficiency.

Figure 3. A staff was introducing the "Four Nodernizations" system

Leaving the Safety Production Command Center, the students continued to the well site of Management Zone Seven. Oilfield technicians provided on-site teaching, led everyone to understand the structure and working principle of the beam pumping unit, emphasized safety precautions, and introduced the large solar photovoltaic panels standing located at the corner of the well site, allowing students to understand how renewable energy generates electricity.

Before leaving Management Zone Seven, the students from Shengli Gudao No.1

Primary School presented their carefully prepared small gifts to the oilfield workers, expressing their gratitude and sincere New Year blessings. The Youth League Committee of Gudao Oil Extraction Plant presented the students with pumping unit assembly models, hoping that they would continuously think and learn through hands-on operation and growing into the pillars of the nation of tomorrow(see Figure 4).

Figure 4. Students from the Shengli GuDao First Elementary School in Dongying City demonstrating the pumping unit model they hold at the well site

This visit activity not only broadened the horizons of the youth from Shanghai and Dongying, familiarized them with new photovoltaic energy, but also allowed them to feel the hard work of oilfield workers. Finally, teachers, students, parents, and oil workers faced the camera together, shouting the activity slogan loudly: "Sustainable development at the Yellow and Yangtze River Estuaries, low-carbon living starts with me!"

3. Achievement Feedback Phase

After the two-city youth cloud linkage and green oil enterprise visit, teachers assigned a task for achievement presentation. Seventeen students from the Second Affiliated Elementary School of Shanghai No.6 Normal School created beautiful learning newsletters, graphically recording the knowledge they learned that day and their most interested content. Sixteen "LuYaEr" Low-Carbon Environmental Protection Club members from Shengli Gudao No.1 Primary School assembled the pumping unit models, painted them, installed batteries, and the small pumping units started "working", which the students loved.

To truly understand the evaluation of schools and enterprises regarding this collaborative climate change education practice activity, Shengli Gudao No.1 Primary

School designed the *Shanghai-Dongying Two-City Linkage*, *"Big Hands Pulling Little Hands" Green Enterprise Visit Activity Feedback Questionnaire*. The questionnaire was divided into student and adult versions, with 8 questions in the student version and 7 in the adult version; both included 4 subjective questions, covering the special aspects of the activity, memorable points, harvest from participating, and suggestions for similar activities. The questionnaire was distributed to teachers, parents, and students from both schools, as well as managers and technicians from the Shengli Oilfield Gudao Oil Extraction Plant. A total of 25 adult questionnaires and 33 student questionnaires were collected, with a 100% recovery rate and 100% validity rate. Additionally, the Second Affiliated Elementary School of Shanghai No. 6 Normal School provided 3 activity feedback forms written by parents. The positive feedback from participants laid the foundation for the scientific analysis of the gains and losses of this activity.

During the recent green enterprise visit event, students demonstrated commendable abilities in various aspects, including thinking, learning, practical skills, and comprehensive problem-solving. Based on their participation level and the submission of learning outcomes, Shengli Gudao No. 1 Primary School selected seven students as "Low-Carbon Environmental Protection Whizzes" and nine as "Low-Carbon Environmental Protection Sentinels." Additionally, based on students' learning, thinking, and questioning, six were awarded the "Exploration Prize for Addressing Climate Change." According to the submission of assembled oil extraction machine models and activity handbooks, five students received the "Action Prize for Addressing Climate Change."

Effectiveness & Reflections

1. Activity Effectiveness

Through the analysis of the *Shanghai-Dongying Two-City Linkage*, *"Big Hands Pulling Little Hands" Green Enterprise Visit Activity Feedback Questionnaire*, the following effectiveness can be reflected:

Firstly, digital learning became the biggest highlight. The activity made full use of digital resources and relied on digital platforms for interactive communication and questionnaire surveys, which improved the learning efficiency. "Without the support of digitalization, such a scenario would be almost unimaginable; it is because of the development of educational digitalization and the participation of more diverse subjects that richer and more complex educational processes are brought about, and educational benefits are multiplied."

When asked, "Have you ever participated in such an activity before?" the students' answers were unanimous: no. In the adult version, only 1 person (4%) had participated before. When asked, "What do you think is the most special about this activity?" more than a third of the respondents mentioned "connection" or "linkage." Some excerpts from the student questionnaire answers are as follows:

Student Wang: Overcoming geographical boundaries, just like a real visit, learning while watching and being able to interact in real time.

Student Xu: The most special thing is learning more knowledge together with students far away in Shanghai.

Some excerpts from the adult questionnaire answers are as follows:

Parent Zhao: Through video calls, children from two places were brought closer, allowing Shanghai children to get firsthand information about green enterprises!

Teacher Lu: The linkage between the two places can deepen the school's involvement with enterprises, promoting comprehensive development of students.

Secondly, new energy in photovoltaics has become a focal point of interest. Question 5 of the student questionnaire asks, "What do you find most special about this green enterprise visit?" Surveys reveal that students are highly interested in leaarning about photovoltaic power generation and oil extraction. The term "photovoltaic" has emerged as a frequent answer to this question. In the learning newsletters of Shanghai students, it is evident that every child has drawn the appearance of photovoltaic power generation, indicating its strong appeal to them.

Thirdly, interactive questioning has left memorable impressions. Surveys show thatstudents who excel in thinking and asking questions during activities tend to stand out from the crowd. Question 6 in the student questionnaire, "Who made the deepest impression on you?" reveals that students like Guo Qiyao from Shengli Gudao Primary School No. 1 and Jing Zimeng from No. 2 Affiliated Primary School of Shanghai No. 6 Normal School made a profound impact due to their proactive, brave, and outgoing personalities. It is evident that students from both locations serve as role models for each other, collectively enhancing their learning experiences.

Fourthly, the visit to the enterprise has sparked interest points. The survey results (see Figure 5) indicate that during the joint visit to the oil enterprise by students, parents, and teachers from both locations, there was significant interest in the enterprise's operational models, remote monitoring and intelligent management, and the application of new energy in photovoltaics. The feedback below illustrates this:

Teacher from Shanghai: The most special aspect was the collaboration between the enterprise and the school, which is relatively rare in Shanghai. Students showed a keen interest in enterprise culture, which they had only previously heard about vaguely from their parents.

The Enterprise Technician: The activity combined theory with practice, extending what was learned in the classroom into various aspects of life.

Parent in Shanghai: The most memorable aspect was the desolate construction site and the tough working environment, which made the children of oil workers very understanding of their parents' hardships.

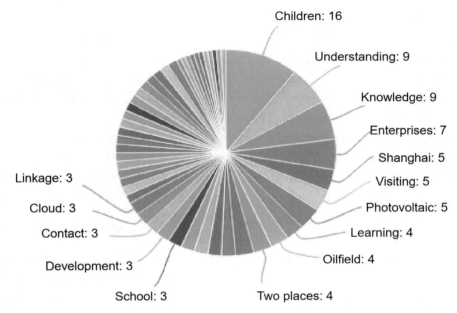

Figure 5. Statistical chart of the results for Question 5 in the "Shanghai-Dongying Joint Visit to Green Enterprises" feedback survey (Adult Edition), "What left the deepest impression?"

Fifthly, the visit stimulated corporate responsibility towards climate change. Oilfield enterprises opened their production command centers and well sites to host visits by teachers, students, and parents, showcasing the transformation of their production methods in response to climate change. This helped students understand governance approaches such as "developing renewable energy and achieving energy conservation and emission reduction through technological innovation."

Enterprise Feedback:

(1) As an enterprise, this was the first time collaborating with a school on such anactivity, allowing children to experience oil and new energy development up close, which also enhanced the enterprise's reputation and social responsibility.

(2) In preparation for the students' visit, the oil extraction plant made thorough preparations, formulated safety plans, and during the visit, showcased the basic production operations of the oilfield to the "little oil kids," enabling them to understand their fathers' working environment and the efforts the oilfield is making in low-carbon environmental protection to adapt to climate change.

2. Reflection and Follow-up

Firstly, it is essential to strengthen the sense of responsibility. Climate change is a global challenge that transcends national boundaries. Climate change education should be directed towards multi-stakeholders and themes, requiring collaboration among schools, families, society, government, and enterprises to enhance the awareness of the "community of shared future for humanity," to recognize the relationship between small efforts and systemic updates, and to envision the impact of "doing" versus "not doing" in the present on the future, thereby strengthening low-carbon environmental measures to achieve the "dual carbon" goals sooner.

Secondly, it is important to highlight inheritance and promotion. The Shengli Oilfield Gudao Extraction Plant actively responds to environmental calls, takes on social responsibility, and achieves sustainable development through measures such as renewable energy power generation, energy efficiency improvement, and energy digitalization promotion. Previously, although the "little oil kids" grew up in an oil city, they rarely had the chance to deeply understand the production process at the oilfield frontline and experience their parents' hard work. This visit, along with the hands-on activity of making oil extraction machine models afterward, allowed the "little oil kids" to understand the process of oil extraction and their parents' working environment, and sparked thoughts on how to inherit and carry forward the oil spirit and enhance a sense of social responsibility.

Thirdly, it is necessary to enhance collective learning and mutual learning. Students from both locations exhibit different learning styles during collective learning. For example, Shanghai students extensively gather information before activities, boldly ask questions during activities, and complete learning newsletters afterward to consolidate what they've learned. Dongying students make gratitude gifts before activities and engage in hands-on model makingfor concrete perception after activities. Both schools indicate that they will promote mutual learning and exchange in multiple dimensions in the future.

In the future, schools will take climate change education as a lever, integrating new energy, ecological environment, national defense technology, agricultural production, low-carbon living, green cities, and green campuses, to deeply engage in multidisciplinary integrated project-based learning, promoting students' integration of climate change knowledge into their daily lives. Schools will also leverage digital media technology to expand activity content, innovate activity forms, and promote the sustainable development of school-family-society collaborative educational practices.

(Pinting Zhao, Vice Principal of Shengli Gudao Primary School No. 1, Dongying, Shandong)

Investigation of the Principles of Green Refrigeration for Home Use and Practical Operation

Project Background

The "Green-Light Year" is a public welfare organization dedicated to the mission of "making the public enjoy the concept of sustainable development." As an institution for sustainable development education, Green-Light Year recruits and trains for college student mentors, collaborates with elementary and secondary schools to conduct research projects, and jointly designs and polishes courses. It is committed to the continuous propagation of sustainable development education to young students.

In March 2022, the news of the Conger Ice Shelf breaking away from the Antarctic continent caught our attention. This massive glacier, equivalent in size to the combined areas of New York City and Rome, is set to disappear entirely. Scientists indicate that a heatwave from Australia swept across the entire Antarctic continent, leading to temperature rises observed at all Antarctic research stations that were a staggering 40℃ higher than the historical average. This event marked the highest temperature increase ever recorded in Antarctica since the beginning of human documentation.

During the summer of 2023, due to a nationwide heatwave, almost every household had to keep their air conditioner on continuously to cool down, leading to the urban heat island effect, which made cities even hotter, thereby requiring more energy for cooling. So, how much electricity would we use to consume to get through this summer? And how can we reduce the use of air conditioning in a way that is lower in carbon and more energy-efficient, or cool our homes in a greener, more carbon-friendly way that also saves money? With these questions in mind, "Green-Light Year" formed a team composed of 13 families from Shanghai, Hangzhou, Suzhou, and Wuxi to embark on a study of the principles of intelligent green cooling at home.

Project Design

The project design(see Figure 1) showcases the following features:

Firstly, the core objective is to solve real social issues. Green cooling has significant implications for slowing climate change and improving the welfare of workers in high temperatures, and it is a reality present in everyone's daily life. Our aim is to tackle the problem of green cooling in household settings, encouraging students to gain a deep understanding and conduct research on top of their knowledge of refrigeration and energy-saving techniques, to discover an economical and energy-efficient method of

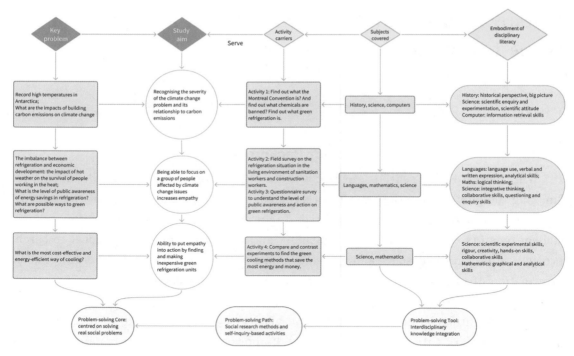

Figure 1. Project design approach

green cooling. In the process, students need to integrate multidisciplinary knowledge, such as physics, chemistry, environmental science, etc., through a series of activities including field investigations, questionnaire interviews, and experimental research, to truly grasp societal needs and enhance their sense of social responsibility.

Secondly, the problem-solving approach is based on social investigation and self-directed inquiry. We encourage students to step out of the classroom and into the real world, to understand the application of green cooling methods in real life through field investigations, observations, interviews, etc., and to collect primary data to provide an empirical basis for subsequent research. At the same time, we encourage students to engage in self-inquiry, such as designing experiments and conducting data analysis, to delve into the principles of refrigeration technology and explore ways to save energy and reduce emissions. This project fully reflects a student-centered approach that empowers students and responding to their diverse needs for knowledge learning.

Thirdly, interdisciplinary knowledge integration serves as the tool for problem solving. During the research process, students are encouraged to integrate knowledge from different disciplines, such as understanding refrigerants from an environmental science perspective, analyzing the cost-effectiveness of cooling solutions from an economic standpoint, and optimizing the design of cooling devices from an engineering perspective. Through such interdisciplinary integration, students can better understand the problems more deeply and find more effective solutions.

Project Implementation

1. Introducing the Driving Question

Before the project kicked off, we consider the characteristics of a driving question, focusing on four main features: relevance to daily life, openness, multidisciplinary nature, and focus. Based on these, we posed a question corresponding to the project research: How to design an economical and practical energy-saving **cooling device**?

However, behind this core driving question, there are many guiding background questions. We adopted the "4W1H questioning method," which includes "What, Why, Where, Who, and How?". This method is also an important one for students to master. For example, the following questions were raised in this research:

What has been your experience with high temperatures in recent years? Why do you think such changes are occurring?

Where have you experienced or seen any extreme weather conditions?

What impact do you think these extreme weather conditions have had on your life? What impact might they have on other groups of people?

Why do you think these extreme weather events occur?

How do you think the occurrence of extreme weather can be reduced? What has your family done to address this?

...

Based on these progressive questions, which lay the groundwork for student reflection, a global issue is brought back to the level of family action. From macro to micro, the real core driving question naturally emerges.

2. Project Development Phases

1) Phase 1: desk research

The project team initially conducted literature research and learning, consulting the official website of the Ministry of Ecology and Environment of the People's Republic of China, the official website of the China National Environmental Monitoring Centre, Baidu, and other websites. They collected and organized information about the content and development process of the Montreal Protocol, the refrigeration principles and methods of various refrigeration equipment, and the development history and characteristics of green refrigerants.

2) Phase 2: questionnaire survey

From July 23 to 29, 2023, an online survey was conducted, and a total of 517 feedback forms on "household refrigeration methods" were collected. At the same time, under the teacher's guidance, the project team members learned to understand and analyze the data represented in the charts and drew the following conclusions(see Table 1):

Firstly, 95.36% of the 517 respondents have air conditioning installed in their homes, with 70% having 3 to 5 units; 37.14% of the respondents have never paid attention to energy consumption, 16.63% believe that refrigeration equipment has a small impact or no impact on the environment, and 17.80% chose not to purchase or were unsure about purchasing more efficient refrigeration equipment for environmental reasons.

Secondly, regarding knowledge of energy saving: A total of 7 questions about energy saving were asked. Only 47.97% of respondents knew that "setting the air conditioner 1℃ higher for cooling or 2℃ lower for heating can save 10% on electricity," which was the least known energy-saving fact among the respondents.

Table 1. Analysis of questionnaire survey results

Question	Analysis
1. Climate Change Observations	The main climate changes noticed by respondents are increasing temperatures every year (about 85%) and more frequent natural disasters (nearly 64%). This indicates that abnormal temperature changes are attracting more attention, highlighting the urgency of the climate change issue.
2. Primary Cooling Methods in Homes	The main cooling methods used in respondents' homes are air conditioning and fans, with air conditioning having a high prevalence of 95% and fans at 69%. This suggests a very high penetration rate of air conditioners, which also implies a greater impact on ozone depletion, necessitating research into green cooling methods.
3. Number of Cooling Devices at Home	Over 50% of households have 3 to 4 cooling devices, with only 2% of households having none.
4. Energy Efficiency Rating of the First Air Conditioner at Home	32% of respondents have a first-class energy efficiency rated air conditioner, which is the international advanced level and the most energy-saving. However, 37% of respondents have never paid attention to this. Advocating for manufacturers to sell lower energy consumption air conditioners could be helpful.
5. Energy Efficiency Rating of the Second and Third Air Conditioners at Home	Over 29% of respondents have a first-class energy efficiency rated air conditioner for their second and third units, with about 36% not paying attention to the energy consumption of air conditioners. Questions 5 and 6 highlight the necessity of spreading knowledge about energy consumption.
6. Is it Correct to Close Windows Before Turning on the Air Conditioner?	64% of respondents believe it is correct to close windows before turning on the air conditioner. Some believe that it is better to open the windows first, turn on the air conditioner for 5 to 10 minutes to ensure that the hot and polluted air is expelled before closing the windows.

(Continued)

Question	Analysis
7. Duration of Air Conditioner Use	26% of respondents use the air conditioner all day, while 74% adjust the usage appropriately.
8. Do You Know: Setting the air conditioner 1 degree higher for cooling or 2 degrees lower for heating can save 10% on electricity?	52% of respondents did not know this energy-saving tip, indicating that knowledge about energy saving is not well disseminated.
9. Do You Know: When starting the air conditioner, setting it to high cool/high heat to quickly reach the desired temperature, then switching to medium or low fan speed, can reduce energy consumption and noise?	30% of respondents were not aware of this energy-saving tip.
10. Do You Know: It is best to use thick curtains in air-conditioned rooms to reduce air loss?	20% of respondents were not aware of this energy-saving tip.
11. Do You Know: When using a refrigerator, do not pack items too tightly, leave gaps for cold air circulation to reduce compressor run times?	22% of respondents were not aware of this energy-saving tip.
12. Do You Know: Using the dehumidifying function in conjunction with the air conditioner's cooling feature not only saves electricity but also reduces the harm to the human body from air conditioning?	Nearly 35% of respondents were not aware of this energy-saving tip.
13. Do You Know: Before opening the refrigerator, you should think about what you need to take out first?	12% of respondents were not aware of this energy-saving tip.
14. Do You Know: When the temperature is not too high, you can use a fan instead of an air conditioner?	Nearly 9% of respondents were not aware of this energy-saving tip.
15. Do You Know Any New Energy Cooling Equipment?	50% of respondents have not heard of new energy cooling equipment.
16. What Are Your Expectations for the Development of Indoor Cooling Methods?	Environmental protection, energy saving, green energy, low energy consumption, physical cooling...

3) Phase 3: field interviews

To understand the human perception of climate change, the project team conducted interviews with sanitation workers, construction workers, cleaning staff, and food delivery personnel(see Figure 2). The interview results showed that the respondents had limited knowledge about green cooling equipment. However, both outdoor and indoor workers have a relatively keen awareness of climate change. A food delivery person said, "We work outside every day, exposed to the elements, and even slight changes in the weather can greatly affect our work." A cleaning lady said, "Nowadays, the summers are hotter than in previous years."

During the interviews, team members promoted the use of green cooling technologies to the interviewees, who indicated that there should be more widespread dissemination of these devices in the future.

Figure 2. Junior researchers interviewing outdoor workers

4) Phase 4: experimental research

After conducting desk research, the project team identified nine common coolingmethods used in daily life. Team members randomly combined curtains, water, ice cubes, ice crystals, and bottle air conditioners to compare indoor and outdoor temperatures to study the optimal cooling method(see Table 2, Figure 3).

Table 2. Research results of the junior researchers' comparison of nine cooling methods

Cooling method	Maximum temperature difference between indoor and outdoor/℃	Temperature difference between indoor and outdoor/℃
no curtains	4.5	0.5
only with curtains	4.9	1.3
curtains & water	7.0	1.2
curtains & ice	7.1	1.8
curtains & ice crystals	7.1	3.2
curtains & bottle air conditioner	7.8	2.2
curtains & bottle air conditioner & ice	6.7	1.8
curtains & bottle air conditioner & ice crystals	4.8	1.8
curtains & bottle air conditioner & ice & ice crystals	8.4	2.5

Figure 3. Junior researchers conducting refrigeration experiments

Junior researchers' analysis of the reasons for the experimental results

First, curtains block out more heat than no curtains, resulting in a greater temperature drop.

Second, "Bottle Air Conditioning" + Water: This method uses the evaporative effect of the fan to reduce indoor temperature. The effectiveness of this method depends on the indoor humidity and the temperature of the water. It can provide a slight cooling effect indoors. However, as the water evaporates, the cooling effect will gradually weaken. Also, the evaporation of water releases moisture into the indoor air, which may lead to an increase in relative humidity.

Third, "Bottle Air Conditioning" + Ice: This can have a short-term cooling effect. The low temperature of the ice can lower the surrounding air temperature in a short period. However, as the ice melts, the cooling effect diminishes. The melting of the ice also leads to an increase in indoor humidity.

> Fourth, "Bottle Air Conditioning" + Ice Crystals: Ice crystals are a reusable cooling material, and the air conditioning fan with ice crystals can provide a sustained cooling effect over a longer period of time. Similar to ice, the melting of ice crystals will also lead to increased humidity.
>
> Fifth, "Bottle Air Conditioning": The cooling effect of this method is not very significant. If the area of the bottle air conditioner used is large enough and the external wind is stronger, the cooling effect might be better.
>
> Sixth, if the highest indoor temperature is lower than the highest outdoor temperature, and the lowest indoor temperature is higher than the lowest outdoor temperature, the reason is due to the lack of ventilation.

The research found that the cooling effect is influenced by various factors such as ambient temperature, relative humidity, space size, and the quality of the cooling medium. However, generally speaking, if there is a significant temperature difference indoor and outdoor, it indicates that the cooling method is relatively efficient and can quickly reduce the indoor temperature. Conversely, if there is only a small difference between the highest and lowest temperatures indoor and outdoor, it may be due to the inefficiency of the cooling method or poor indoor insulation performance.

The change in the temperature difference between indoor and outdoor also reflects the insulating properties of materials and their ability to maintain temperature stability. Under the same conditions, air conditioners with ice crystals currently seem to have the strongest cooling effect, while also having relatively low power consumption and electricity usage. Therefore, the project team recommends that the public use air conditioners with ice crystals in their daily lives.

5) Phase 5: model conception

Based on the practical results, the project team members used their imagination to design three cooling technologies: a heat pump technology using an air compressor for cooling, roof insulation technology, and the introduction of natural cooling sources.

First, heat pump technology(Figure 4). The principle of this technology is to use a compressor to compress a green refrigerant into a high-pressure state, after which the heat is dissipated through a heat exchanger, turning it into a low-pressure state. Next, a fan draws external air into the machine, which turns into condensate water after passing through the water vapor in the collection tank, and exchanges heat with the refrigerant through an insulating panel, cooling the air. Finally, the cold air is blown back into the room through the fan via the air outlet. Throughout this process, the green refrigerant is continuously circulating without causing environmental pollution. At the same time, the collection tank reduces the waste of tap water resources by recycling. Therefore, this model is a forward-looking, low-carbon, environmentally friendly, and highly energy-efficient green cooling model of the future(see Figure 4).

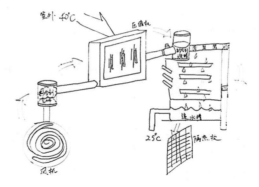

Figure 4. Schematic diagram of heat pump technology

Second, roof insulation technology. As shown in Figure 5, the principle of this technology is to construct a swimming pool on the roof. When the wind blows over the pool, its temperature decreases, and then it continues to flow to every room of the building on each floor. Each floor has windows and curtains to ensure that the wind can enter the room. Inside the room, there is a bottle air conditioner, an air cooler fan with ice crystals, an ice box, and two windows, thus achieving a cooling effect. This model is more environmentally friendly than traditional refrigeration equipment because it can utilize natural resources to create a cool sensation. At the same time, the junior researchers also conceived ideas for indoor green cooling by arranging an air cooler fan and a regular fan in the room, using both at the same time to achieve the purpose of rapid cooling while saving power (see Figure 6).

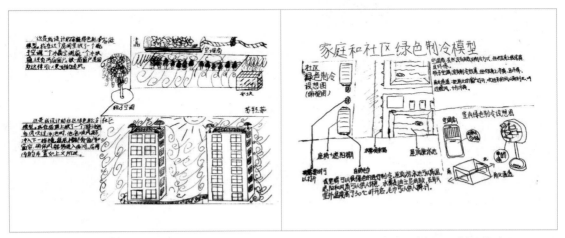

Figure 5. Schematic diagram of the roof insulation system

Figure 6. Left-top view of the roof insulation system, right-conceptual diagram of indoor green cooling system

Third, natural cooling sources. A cooling source, also known as a refrigerant, is a working substance that continuously circulates in a refrigeration system and achieves cooling through its own state changes. Common refrigerants include chemical substances such as propane, ethylene, and ammonia, which can cause substantial carbon emissions and exacerbate the greenhouse effect. In contrast, natural cooling sources are more environmentally friendly. Based on this, the junior researcher devised a plan(see Figure 7). Firstly, select a natural cooling source from the community or nearby (such as water from a pond), place the water pump into the cooling source connected to the water pipes, and use the pump to pressurize the water up into the pipes. Then, use a booster pump to increase the system pressure to overcome the pressure loss of the air conditioner's heat exchanger. Next, connect the inlet and outlet of the air conditioner main unit with water pipes. Inside the air conditioner, there is a heat exchanger with dense thin iron fins to increase the heat exchange area and improve heat exchange efficiency. Finally, the water discharged after heat exchange in the heat exchanger will be released into a specially designated pool for the cold water air conditioning system (the cold water will automatically sink and stratify with the warmer water).

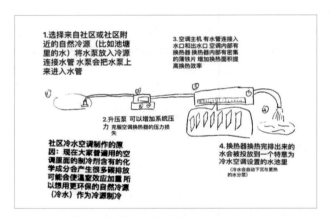

Figure 7. Natural cooling source system

Effectiveness & Reflections

1. Primary Results

Firstly, the project has innovated the path of climate change education. Climate change education is not only about guiding students to pay attention to energy transformation but also about focusing on those groups of people who are the first to be affected by climate change. This is a necessary yet often overlooked starting point—for the survival and development of all humanity, not just for solving problems for the sake of solving them. After conducting field surveys to understand the living conditions of the workers, the students involved in this project found the scenes they investigated to be unforgettable. Their empathy was aroused, and they became more caring and conscious about the problems caused by climate change.

Secondly, it has stimulated the innovative thinking and practical abilities of young people. The students used new methods, technologies, and theories to explore green cooling methods to increase sustainable applications. During the model conception process, they fully considered the use of natural cooling sources and physical insulation to develop more environmentally friendly cooling methods, thereby further reducing energy consumption. Through this research, the students gained a deeper understanding of climate change issues, such as discussions on carbon emissions and the greenhouse effect. At the same time, they also proposed ideas for the long-term application and promotion of green cooling technology, such as how it can be applied in homes over the long term. Through this process, the students gained a deeper understanding of the causes, effects, and solutions to climate change. Their innovative thinking and practical skills were exercised.

2. Reflection

This project has actively explored innovative paths in climate change education by leading students to conduct research on household cooling equipment and social practice, cultivating their interdisciplinary thinking, practical abilities, and climate literacy. However, due to the specialized nature of some knowledge, which is not easy to learn and disseminate, there is a need to strengthen learning methods and styles. It is necessary to hire professional talents to serve as mentors for exploring the principles of green cooling and to use more interesting methods to communicate the importance of green cooling to the public.

In the future, the project team will continue to refine the project design, broaden the scope of implementation, and promote the concept of green cooling to the public eye, thus continuing to contribute to the global response to climate change.

(Nannan Xie, Deputy Secretary-General of Project Green Light-Year; Han Ni, Project Officer of Green Light-Year)

Experiences of New Energy Companies in Supporting Climate Change Education

Research Background

Climate change is a global issue concerning the destiny of humanity and a severe challenge to sustainable development in the 21st century. Addressing global climate change, protecting the ecological environment, and achieving carbon neutrality are the common missions and responsibilities of all mankind, as well as the inherent requirements for high-quality development across various industries. As a typical representative of clean energy, photovoltaics is a crucial field in responding to climate change, promoting green development, and leading a low-carbon future. Distributed photovoltaics, in particular, play a key role in developing green towns and promoting harmony between humans and nature, contributing significantly to addressing climate change and becoming the backbone of the green and low-carbon energy transition.

Founded in 1997, Trina Solar Co., Ltd. is one of the earliest companies to enter the photovoltaic field in China and has now developed into a global leading provider of photovoltaic smart energy solutions. "Let the sun benefit thousands of households" is Trina Solar's unchanging pursuit. For more than 20 years, Trina Solar has adhered to the green development concept, using clean energy to protect green mountains and clear waters, and has always been an active participant and firm supporter of ecological protection, with sustainable development engraved in the company's genes. The company has also set sustainable development goals for 2020 to 2025, including carbon emission management, energy management, water resource management, and waste management, each with specific targets.

Trina Solar has been focusing on "climate change" for over ten years. Chairman Gao Jifan believes that new energy companies have the responsibility to grasp the key to addressing climate change and commit to making carbon reduction in energy the main force for zero carbon. Therefore, on the one hand, the company seizes the focus of the "14th Five-Year Plan" on "Photovoltaic+", strives to explore new models and new businesses of "Photovoltaic+", promotes the integrated and innovative development of "Photovoltaic+" application scenarios, and uses photovoltaic power to respond to climate change and build a beautiful new zero-carbon world. On the other hand, recognizing that children are the future of the country and the hope of the Chinese nation, the company established the "Kelin Environmental Protection Association", launched dual-carbon education public welfare projects, and leveraged education to exert more "Trina Power", so that the ecological protection of the youth like a prairie fire can continue to help the mitigation and response of climate change.

Project Design

Based on Trina Solar as a new energy company and combining the characteristics of climate change education, the project fully utilizes resources such as the "Trina Solar Dream and Innovation Exhibition Center" and "Trina Ecological Farm", leveraging thepower of the "Kelin Environmental Protection Association", and collaborating with schools and society. The project adopts two methods: "Going In" (sending company lectures to schools) and "Bringing Out" (students going to company bases for research and practice), continuously developing the "Trina Cup" dual-carbon education series of public welfare activities, such as the "Youth Silicon Valley" project for primary and secondary students and the "New Energy Explorers" project for college students.

1. Project Objectives

Through the implementation of the project, the power of new energy companies will be used to popularize "Photovoltaic+" new energy knowledge, enhance the promotion and dissemination of low-carbon living concepts, and provide more opportunities for students of all kinds to experience research and social practice. This will stimulate their interest in new energy research in real situations, and promote the vast youth group to become the backbone of developing China's green and low-carbon cause and addressing climate change.

2. Project Implementation Plan

The project is implemented in three phases: The first phase focuses on scattered and flexible environmental education activities. The second phase focuses on fixed and continuous dual-carbon education cooperation projects. The final phase focuses on school-enterprise cooperation in climate change education.

Project Implementation

1. Phase 1: Scattered and Flexible Environmental Education Activities

Trina Solar and its subsidiary "Trina Family" have been actively fulfilling their corporate ESG responsibilities. In 2011, the current chairman of Trina Family, Gao Haichun, together with several outstanding young environmentalists, initiated and founded the non-profit environmental protection organization—Changzhou Xinbei District Kelin Environmental Protection Public Welfare Association, referred to as "Kelin" (Co-link)."Kelin" represents the love of environmental protection into the small trees, let the small trees into a forest, let pieces of forest to protect the earth. Today, Kelin has also established branches in the United States and Germany, with more than 2,000 members, growing its power to protect the ecology.

Since its establishment, "Kelin" has not only relied on the power of Trina Solar but also established close cooperation with companies such as Youze Group and Junhe Company, providing sustainable development consulting services to more than 20 companies, including advice on carbon emissions, carbon footprint, sewage treatment, and lean production. It has also consciously linked up with schools, starting with pilot projects at schools such as Jiangsu Changzhou High School, Changzhou No. 1 Middle School, Changzhou Zhengheng Middle School, and Changzhou Xinhua Experimental Primary School. Through public welfare activities such as environmental publicity and environmental experiences, "Kelin" promotes healthy, green, and low-carbon lifestyles.

In the early days, "Kelin" launched flexible environmental education activities in cooperation with various non-profit organizations, government agencies, and schools, including environmental art exhibitions, ecological love practices, tree planting with visiting American scholars, public welfare environmental cycling from Yunnan to Tibet, etc. Although these activities were point-based and one-off, they showed a strong vitality inenvironmental education. For example, the "Public Welfare Environmental Cycling from Yunnan to Tibet" (see Figure 1) involved over ten middle school students over 33 days, crossing nine high-altitude mountains, traversing the Three Parallel Rivers, following the Niyang River, revisiting the Tea Horse Road, and holding 15 environmental advocacy activities along the way. This activity was recognized and praised by the public and reported and reprinted by 76 media outlets, effectively extending the influence of environmental education actions.

Figure 1. Commemorative items from the early environmental education activity "Yunnan-Tibet Cycling" organized by Trina Solar in cooperation with "Kelin"

Emerging from Trina Solar, "Kelin" shoulders the responsibility of green environmental protection and carries the gene of sustainable development. In the years of implementing environmental education activities, it has acted as a link and bridge of love among schools, companies, and social organizations, lighting up the starfire of environmental education with its determination to "protect the environment", lighting up the green growth path of young people with its heart-to-heart connection and tree-to-forest transformation.

2. Phase 2: Fixed and Continuous Dual-carbon Education Cooperation Projects

"Kelin's" attempts in environmental education activities have laid a solid foundation for Trina Solar to support climate change education. In 2023, as climate change education becomes a global focus, Trina Solar, as a new energy company in a "billion-dollar GDP city" and "new energy capital", realizes the importance of revitalizing the nation through science and education and strengthening the country with talent. Therefore, Trina Solar, leveraging "Kelin", launched the "Dual-Carbon Education into Campus" project, signing cooperation agreements with a number of schools. The project, divided into "Walking with Children" and "Entering Universities", implements long-term series of school-enterprise cooperation in dual-carbon education.

1) Dual-carbon education project in universities: "New Energy Explorers"

To promote the realization of the Party and the country's dual-carbon goals and allow college youth to correctly understand and grasp the theoretical knowledge of carbon peak and carbon neutrality, establishing a new era of green and low-carbon concepts, Trina Solar launched the "New Energy Explorers—2023 Dual-Carbon Education into Universities Series Activities" on November 12, 2023, at the new campus of Hohai University in Changzhou. This project adopts an innovative form of "charity + education + dual-carbonpopularization", effectively integrating university and various social resources, uniting school and enterprise forces. Through activities such as research and practice in new energy companies, integration into zero-carbon life exhibitions, new energy public welfare classes, the first low-carbon life short video contest, and the first wind and photovoltaic power model building contest, the project provides college students with a rich platform for employment innovation practice, attracting more students to participate as promotional ambassadors and action stars, enhancing their awareness and ability of innovation and entrepreneurship in the new energy field, and contributing to the realization of dual-carbon goals and talent team building.

On the afternoon of March 15, 2024, the "New Energy Explorers—Dual-Carbon Education into Universities Series Activities" summary meeting was held at the Jiangning Campus of Hohai University (see Figure 2), along with the award ceremonies for the "First Wind and Photovoltaic Power Model Building Contest" and the "First Low-Carbon Life Short Video Contest". The representative of the first prize-winning team, Lu Wen, expressed their gratitude: "As a student of Hohai Broadcasting and Television, I will continue to use the lens to record life and convey environmental protection concepts. I know that one person's power is small, but if everyone works together, the power will be great. I also hope we can all take action and enjoy a low-carbon life together." Trina Solar also recognized and encouraged the participation of college students in the project, driving more people to step out of the campus and into companies, improving their ability to solve practical problems and innovate in the school-enterprise cooperation platform, and promoting the completion of the stage and the spiral rise of the project.

Figure 2. "New Energy Explorers" dual-carbon education activity in universities organized by Trina Solar in cooperation with "Kelin"

The successful implementation of the "New Energy Explorers" project at Hohai University allowed college students to see the positive role of new energy in addressing global climate change and to gain a deeper understanding and recognition of new energy and the new future.

2) Dual-carbon education project with children: "Youth Silicon Valley"

To strengthen cooperation between primary and secondary schools and off-campus practice bases, and to support climate change education through student development, activity networking, and resource sharing, the "Trina Cup" Dual-Carbon Education Project with Children was launched on September 25, 2023, at the second experimental primary school of Longhutang, Xinbei District, Changzhou (see Figure 3). The Trina Charity Foundation, Trina Family Energy Co., Ltd., and Longhutang No. 2 Primary School jointly promoted the "Youth Silicon Valley" project construction, aiming to organize students to experience new energy through research-based learning regularly during after-school services and holidays, and fostering their interest in new energy research.

Figure 3. "Youth Silicon Valley" dual-carbon education activity for children organized by Trina Solar in cooperation with "Kelin"

Since promoting the "Youth Silicon Valley" project, Trina Solar has collaborated with schools to focus on new energy and climate change education, carrying out a series of "going in" and "bringing out" activities: researchers going in, and public welfare environmental classes starting—Trina Solar's new product and technology research and development staff lectured on the theme of "Responding to Climate Change, Seeking a Better Life with Carbon" for students, helping them understand different types of clean energy such as photovoltaic power, wind power, hydropower, and their roles, and guiding green and low-carbon lifestyles, promoting green development in "big hands holding small hands"; children playmates bringing out, environmental creativity moving towards the future-students formed playmate groups, carrying research topics such as "principle of photovoltaic panels" and "how photovoltaics are applied in farms", visiting Trina Solar bases such as the "Trina Solar Dream and Innovation Exhibition Center" and "Trina Ecological Farm" in batches for off-campus practice experiences, researching new energy facilities like photovoltaic panels, enhancing their understanding of new energy knowledge, and genuinely stimulating their research interest.

The effective promotion of the "Youth Silicon Valley" project at the second experimental primary school of Longhutang, Xinbei District, Changzhou revealed the curiosity and exploratory power of primary school students towards new energy, opening new ideas for school-enterprise cooperation.

3. Phase 3: Focusing on School-enterprise Cooperation in Climate Change Education

On December 22, 2023, Professor Jiacheng Li, a researcher at the Institute of Basic Education Reform and Development of East China Normal University and Executive Vice President of the Shanghai Municipal Institute for Lifelong Education, and Principal Gu Huifen of Longhutang Experimental Primary School in Xinbei District, Changzhou, visited the Trina Solars' exhibition hall. They discussed sustainable development education and the role of parent committees in school-enterprise cooperation with relevant personnel from Trina Family, and reached a consensus on jointly promoting climate change education.

During the 2024 winter vacation, a national science and innovation painting contest themed "Entering Clean Energy, Creating a Green Future", sponsored by the "Longhuang Family Cup", was held in school-enterprise cooperation. The event encouraged studentsto visit nearby clean energy companies and bases with their parents and partners to look, visit, understand, research, and then use brushes and imagination to outline the future world in their hearts. The contest received more than 70 student works from across the country, with awards presented at the second "Climate Change Education through a cooperative mechanism made up of schools, families and communities" seminar(see Figure 4). The event not only promoted students' focus on clean energy and resistance to global climate warming but also provided Trina Solar with children's perspectives and different ideas for product development, fostering the continuous development of climate change education cooperation actions between Trina Solar and Longhutang Experimental Primary School.

At the beginning of 2024, in a school-enterprise cooperation consultation meeting (see Figure 5), four types of cooperation were further discussed and approved: ecological campus construction—introducing climate and weather observation equipment, creating an ecological secret realm, completing the photovoltaic power construction on the roofs of the library and the campus, and creating aerial green courtyards; featured curriculum development—adding school-based courses integrating new energy and modeling, popularizing clean energy knowledge, organizing new energy model design contests, etc.; featured activities—organizing "green ecological playmate groups", involving families, schools, enterprises, and society in regular cooperation to carry out climate change education-themed practice activities; promoting results dissemination—using climate change education seminars, academic achievements publications, and other channels to spread cooperation results.

The exploration of school-enterprise cooperation focusing on climate change education is still in its early stages. It is believed that in the process of continuously enriching school climate change education resources and creating new, immersive, high-quality educational experiences for all students, school-enterprise cooperation will evolve into unity, generating enormous educational and driving forces for climate change education.

Figure 4. Award ceremony of the "Long Dada Family Cup" National Science Innovation Painting Contest

Figure 5. School-enterprise cooperation meeting between Trina Solar and Longhutang Experimental Primary School

Effectiveness & Reflections

New energy inherently has the ability to improve global climate warming. As a new energy company, Trina Solar, with the vision of creating a zero-carbon new energy world, has the responsibility to promote global sustainable green development with photovoltaic intelligence. From focusing on "climate change" to supporting "climate change education", Trina Solar has formed some experiences and visualized achievements from the perspective of school-enterprise cooperation.

From the practice of the project, the most significant achievement is that Trina Solar has kept pace with the times in its cooperation model, making the project more systematic, deep, and sustainable. From the initial point-based, flexible environmental education activities to fixed and continuous dual-carbon education cooperation projects, and then to school-enterprise cooperation focusing on climate change education, Trina Solar has continuously optimized its model of participating in climate change education, signing a number of cooperation schools (units) along the way, such as Hohai University and Longhutang Experimental Primary School in Xinbei District, Changzhou. It has also developed a series of projects that effectively promote climate change education. The scope and number of participants have been expanded continuously, and each activity has received attention from all sectors of society and has been reported and reprinted by different platforms, thereby driving and radiating secondary influence, continuously enhancing the positive role of climate change education. Trina Solar remains passionate about low-carbon environmental protection, concerned about global climate change issues, and persistently and systematically intervenes in climate change education, co-creating a future of high-quality development for enterprises and society.

Of course, looking back at the entire process of project development, we also see greater space for development and optimization. In 2024, Trina Solar will continue to explore the campus model based on the "Trina Cup Dual-Carbon Education Project", launch new cooperation projects, integrate industry and education resources, formulate teaching evaluation standards, develop a series of core public welfare courses and creative practice projects related to dual-carbon into campuses, and cooperate with schools to build low-carbon campuses. The promotion and dissemination of the low-carbon lifestyle concept will be intensified, providing more social practice opportunities for students and offering excellent sample cases for the nationwide replication and transfer of our newenergy projects. At the same time, uniting the forces of schools and enterprises, we will strive to create "dual-main bodies" of school-enterprise education, "dual identities" of students and apprentices, conducting in-depth exploration in talent training, technological innovation, employment and entrepreneurship, social services, cultural inheritance, and climate change education, contributing to the high-quality development of the region and co-creating a better future with improved nature.

(Yuhan Wu, Brand Manager of Trina Family Energy Co., Ltd., Co-Founder and Deputy Director, Changzhou Kelin Environmental Protection Public Welfare Association; Hong Pan, Director of the Student Development Center, Longhutang Experimental Primary School, Xinbei, Changzhou)

Learning in the Nature: A Practical Study of Climate Change Education in Primary Schools Under the Concept of Common Worlds Pedagogies

Research Background

The ecological civilization thought from President Xi Jinping has put forward a relatively complete ecological civilization thought system, and formed four core psychological concepts for green development. The contemporary educational reform based on the new view of nature needs to face the new social transformation, move toward the reality of education, and create a new stage of the internal connection between education and nature. *The Guidelines for Climate Change Education (Trial)* states that "Education is a core initiative to tackle climate change, and helps learners form the literacy needed to tackle climate change by influencing values, ways of thinking, behaviors and life choices."

Class 3 Grade 2, Xincheng Campus from Hushan Primary School, Wuyi County, Zhejiang Province, guided by "Common Worlds Pedagogies",① emphasizes the importance and necessity of "symbiosis", opposes the separation of nature and school, and focus on the real growth of life. Integrate regional resources and multidisciplinary forces commit to opening up the four major education fields of family, school, community and nature, creating a good education ecology with the practice mode of "multi-dimensional curriculum + ecological experience + climate public benefit activities", so that every student can learn to live and grow together with nature.

Project Design

1. Objectives

Through the design and implementation of the "Learning in the Nature" project, contribute to the integration of local climate education resources, contribute to the synergy of home school and community forces, construct a complete and continuous learning time and space, expand more subjects to participate in the value system and cognitive pattern for teachers and students, from the individual extended to the the earth community, thuspromoting the effective improvement of ecological civilization

---

① Common Worlds Pedagogies integrates the ontology and cosmology with ecological significance from various cultural resources, and it is an educational model based on ecological coordination and restoration. Its main teaching method is for teachers and children to walk together in their living environments such as fields, hills, streets, urban gardens, and garbage dumps, encounter everything, and then promote the transformation of thinking mode from "subject-object dichotomy" and "human-centered" to "interconnectedness of all things" and "co-generation".

accomplishment.

2. Project Planning

The "Learning in the Nature" project is based on the wholeness of life, integrating the resources of school, community, family and nature. The target system, content system and link system are determined and a series of activities are constructed (see Figure 1). In this process, we should pay attention to the value development of specific activities, the connection of the activity time, content and requirements, and give full consideration to the possibility of the activities can go smoothly.

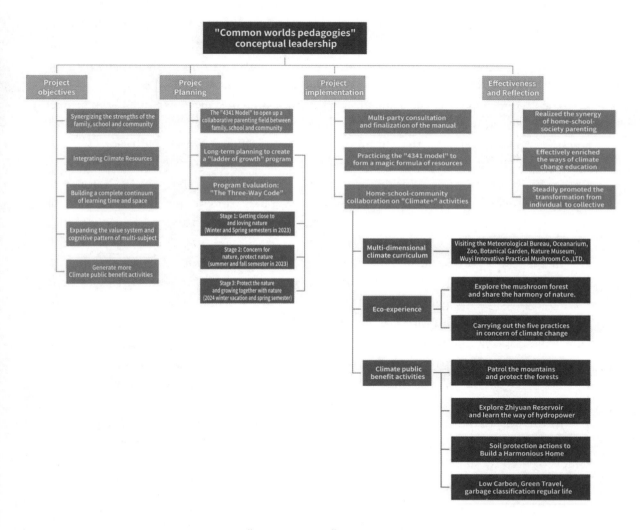

Figure 1. "Learn in Nature" project path diagram

1) With the "4341 Model" to open up a collaborative parenting field between family, school and community

UNESCO has called for a "school-wide" approach to climate change education, and building community partnerships to support school teaching. Climate resource development is not limited to nature, is integrated into campus life, family life, community life, so we use the "4341 model", namely the "4" adherence to the student subject, the whole life, the five different dimensions for education and the collaborative education; the "3" namely linkage between school and community, school and enterprise, school and family; the "4" namely open up the four field of family, school, community and nature, ultimately strive to achieve the "1" goal of "I grow up with nature".

2) Long-term planning to create a "ladder of growth" program

The first stage is in the winter vacation and spring semester of 2023. With the theme of the project "Magic Mushroom", it's to synergize the strengths of family school and community to open up the four natural education fields, so that students can feel the wonder of the mushroom world in the real ecological experience.

The second stage is in the summer vacation and autumn semester of 2023, the project team focus on the theme practice of "travel with the mushrooms" and jointly plan the practice of "Traveling with Mushrooms", jointly plan "Water Nutrition", "Air Nutrition", "Food Nutrition", "Medicine Nutrition", and "Body Nutrition", make a manual of "Exploring the World of Mushrooms and Creating Harmonious Nature" to encourage students to go into the nature, carry out "Climate +" practices such us "Ecological Experience" "Climate Public Benefit Activities" and promote effective ecological civilization literacy.

In the third stage, during the winter and spring semesters of 2024, taking the growing environment of mushrooms as an entry point, we linked up with Wuyi Yingxiang community, Wuyi Wanggu village and Xinxin farm of the school to formulated "Building a better life together with the Dragon", jointly developed a series of activities such as soil knowledge, spring planting soil protection, action, and drawing of spring blueprints so as to explore the mysteries of soil, form an awareness of soil care, expand our value system and cognitive pattern, and achieve from the individual extended to the the earth community(see Figure 2).

In this way, from being close to nature, loving nature, protecting nature, symbiosis and growing with nature, and finally carrying out ecological benefits, our project has achieved "step by step growth".

3. Assessment

The project is guided by "three-way codes", that is organizes the developmental evaluation of the whole subject, the whole space and the whole body; organically combines teachers, students and parents to build a"growth community", then carry out multiple evaluations from the emotional attitude, ability, behavior quality, specialty and so on. The evaluation is carried out in three periods: immediacy, process and summation.

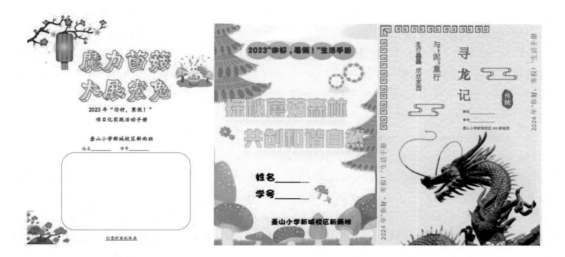

Figure 2. Project manual cover of "Learn in Nature"

Project Implementation

1. Multi-stakeholder Consultation and Finalization of the Manual

In each project, we collect and organize golden ideas in the form of "teacher-student interaction, family-school and community communication, questionnaire survey and negotiation project manual". On the basis of respecting the students' willingness, the class teachers and subject teachers, student representatives, Tashan community staff, the village head of Daheyuan, family committee representatives sit together to hold a "project discussion". It mainly discusses the exploitation of resource, activity personnel, the way, time and place to carry out activities, the formulation of activity manuals, the significance of activities, etc. So that the whole activity can integration children's fun, teachers' wisdom, parents' power, community support, and attract children's active participation, parents' active input, and community's active cooperation. On the day of program registration, the number of applicants in each group was more than 20 people. The heat of each activity remained high, even in the class group formed a "grab" boom.

2. Practice the "4341 Model" for Resource Sharing

The project team practiced the "4341 model" and carried out three major actions of "venue education", "pioneer forum" and "community construction", forming an "whole members, whole-process and all-round" educational pattern driven by multiple forces and integrated by multiple integration. That is with the parents 'forum as the activity place and the "Learning in the Nature" project as the incision, together with Tashan community, Wuyi Meteorological Bureau, Innovative Edible Fungi Co., LTD., Wuyi Fanbei forestry station and other educational venues, enterprises. All the projects have been formed a multi-dimensional resource library, giving full play to the educational value of each subject and each place. Such as Suchen's father is a weather station worker, he led the students to visit the weather station; Tangbao's mother is a teacher, she led the students to make ecological bottles, Huihui's grandfather is a forest ranger, he led the students into the Fanbei forestry station, etc.

3. School-family-community Collaboration on "Climate+" Activities

1) Multi-dimensional climate curriculum

In addition to the use of school classroom teaching resources, the project team also fully tapped off campus climate education resources, linked the strength of families schools and communities, lead students to visit the Meteorological Bureau, Aquarium, Zoo, Botanical Garden, Natural History Museum, Research Base, Wuyi Innovative Practical Fungus Co., LTD. and other places to develop a multi-dimensional climate curriculum with class characteristics, so that the students acquire climate knowledge and enhance ecological literacy in a richer, more authentic, more open and more contextual environment. Such as under the guidance of Mumu's mother, the "weather team" go into the weather bureau (see Figure 3). In the explanation of the weather bureau staff, the team members took a close look at meteorological equipment such as weighing rain gauges, lightning locators, and evaporation sensors, also understand the function and principle of meteorological equipment, know the outdoor observation of wind direction, ground temperature, air temperature, rainfall and other operating methods. We explored the mystery of meteorology and understood the role and significance of meteorological data in the weather forecasting. For another example, after visiting the Natural History Museum, the students exclaimed that "Every species on earth deserves to be loved!"

Figure 3. The "weather team" walked into the Wuyi Meteorological Bureau

2) Ecological experience

Climate education emphasizes the learning method of "participation" or "experience" and respects the dominant status of students. Therefore, representatives from various parties worked hard to explore natural and community resources and jointly designed a series of ecological experience activities.

Sub-project 1: Exploring the mushroom forest and share harmonious nature.

Starting from the mushroom forest, carry out activities such as understanding mushrooms and related natural environment and climate; understanding the morphological structure of mushrooms, learning to identify edible mushrooms and poisonous mushrooms; exploring the microscopic world and understanding the growing environment of mushrooms; making microscopic ecological bottles and understanding the natural ecosystem.

Sub-project 2: In concern of climate change and carry out the five practices.

The 2022 *Annual Report on China's Policies and Actions on Climate Change* proposes to carry out the creation of demonstration counties of "Lucid waters and lush mountains are invaluable assets". Wuyi, Zhejiang Province has put this concept into practice. While the green economy is booming, it has also performed outstandingly in environmental protection and ecological advantages. For example, Tantou Village, Lvtan Town, Wuyi County won the "Examples of Global Habitation Environment Village" award. The "Learning in the Nature" project uses "environment, climate, care, protection, and practice" as keywords, and based on Wuyi County's "five kinds of regimen" cultural and tourism characteristics, develops a series of water, medicine, food, air and sport activities, pays attention to ecological protection, and shares green life(see Table 1).

Table 1. "Five aspects" resource exploration

Five aspects	Climate change education resources
Airculture	Surrounded by mountains on three sides, Wuyi is full of greenery, with a forest coverage rate of 74%. It is a "global green city" and "China's natural oxygen bar". There is a provincial demonstration base, the bamboo garden, near the Wuyi-Lishui Line, with lush bamboo forests and fresh air that makes people feel relaxed and refreshed. Xiayeshan Leisure Service Station explores the local "bamboo" industry, adopts imitation bamboo corridors, bamboo pavilions and other forms, and highlights the theme of "air culture".
Water culture	Hot springs are a precious gift from nature to Wuyi. Wuyi hot springs are large in quantity, high in quality, and of suitable temperature. They are rich in more than 20 minerals and trace elements such as meta silicic acid that are beneficial to the human body. They have the effects of moisturizing the skin, nourishing the face, and curing diseases and maintaining health. They are known as "the best in Zhejiang and the best in East China". In addition, Wuyi also has many reservoirs such as Yuankou Reservoir, Xili Reservoir, and Zhiyuan Reservoir, as well as several power stations such as Shuangyuankou Hydropower Station, Hengshan Hydropower Station, and Zhiyuan Power Station.

(Continued)

Five aspects	Climate change education resources
Sports culture	The way to keep in good health lies in the combination of movement and stillness. The ancient cultural wall with the theme of "Sports Culture" is located at the fork of Dagong Mountain, which is the entrance to the Wuyi large-scale racing track. At the racing track, different from the traditional way of health care, racing enthusiasts can enjoy the speed and passion. Not only a racing track, the beautiful Wuyi-Lishui Line itself is an excellent cycling greenway.
Food culture	Wuyi ranks first in the province in terms of certified organic tea area and output. The three industries of tea, Chinese herbs and edible fungi have been identified as the exemplary full industrial chain of Zhejiang Province, ranking first in the province in terms of number of selected industries. The Wuyi-Lishui Line has been cleverly designed a large-scale relief landscape of "food culture" on the retaining wall of the curve, displaying various alpine vegetables and organically raised poultry and livestock, making it clear at a glance. It is through rural roads that characteristic agricultural products have walked out of the mountains and into thousands of families. Datian Township along the line is rich in blueberries and has held several consecutive blueberry festivals, driving the development of agricultural and tourism integration.
Medicine culture	Shouxian Valley Organic Chinese Medicine Base is the "Pharmaceutical Science Demonstration Base for Chinese Medicine" and "National Youth Agricultural Science Demonstration Base" of the Chinese Medicine Association, and the Zhejiang Province Traditional Chinese Medicine Culture Health Tourism Demonstration Base. In order to restore the pure and pollution-free environment in the wild, the Ganoderma lucidum in Shouxian Valley Organic Chinese Medicine Health Garden is planted in an intelligent greenhouse that simulates wild organic cultivation. The temperature and humidity are strictly monitored, allowing the Ganoderma microsporum to grow in the most suitable environment. In addition, Wanpu Village has a 100-mile medicinal material base that focuses on "eight flavors of health and four seasons of flowers" and grows a wide variety of Chinese herbal medicines.

Through the "Five kinds of regimen" practice, students have a better understanding of the collaborative contributions of the people of their hometown in the construction of ecological civilization, and thus actively participate in it. For example, "Air Care Wuyi" is to understand that Wuyi's forest coverage rate is 74%, to know the role of forests, and to go to Fanbei Forest Farm to promote forest fire prevention; "Water Care Wuyi" is to feel that "every kilowatt-hour of electricity is hard-earned", to "clean up garbage to protect the mother river", and to go to the community on World Water Day to promote water conservation; "Sport Care Wuyi" is to hold an orienteering treasure hunt in Hushan and observe animals and plants along the way; "Food Care Wuyi" is to plant, observe, pick, and cook mushrooms, tea, and Chinese herbal medicines; "Medicine Care Wuyi" is to understand the Ganoderma lucidum cultivation technology and to develop an ecological civilization awareness of "the coexistence of science and technology and tradition, and the symbiosis of nature and society".

3) Climate public benefit activities

As the project progressed, the students put the concept of "harmonious coexistence between man and nature" into practice, began to pay attention to the climate issues facing mankind, and carried out "Climate public benefit activities" practices such as Patrol the mountains and protect the forests, exploring hydropower stations, and soil protection actions to Build a Harmonious Home.

Patrol the mountains and protect the forests: In the process of exploring the world of mushrooms, students Muhui Shenand Yanxi Hong learned that fungi mushrooms play an important role in the decomposition of forest residues. With the help of their mother, they organized a "Patrolling Mountains and Forests to Protect Green Water and Green Mountain" activities. On the way into the mountain, the forest ranger guided the team members to discover Ganoderma lucidum and mushrooms, and introduced the role of forests in conserving water sources and regulating climate. The team members intuitively felt the growth environment that mushrooms like, and also understood the truth that nature and our survival are interdependent. They said that they would definitely keep in mind and practice "caring for every tree and grass, protecting green waters and green mountains". After going down the mountain, the team members and the forest rangers hung tree protection signs and promoted forest fire prevention knowledge (see Figure 4).

Figure 4. Hanging tree protection signs and promote forest fire prevention knowledge

"Explore Zhiyuan Reservoir and learn about hydropower": Under the guidance of Zhixing's Dad, the team members went up the stream and found the source of the clear spring; followed the stream down and saw the water storage reservoir. Under the leadership of the staff of "Zhiyuan Power Station", the team members visited the power station and learned about hydropower generation.

Student Ningning said: "I learned the principle of power generation. So this is how we get our electricity. We have to save water and electricity."

Student Zhixing said: "I will start with the small things around me and refuse the garbage fall to the ground. I hope everyone can gradually become a little environmentalist who loves the earth."

Student Ruohan said: "We live together on the earth, and we must protect nature."

"Soil protection actions to build a harmonious home": As the activity deepens, the project team uses the mushroom growth environment as an entry point to explore the secrets of the soil. The school, family and community collaborate to carry out a series of "Walking with Soil" activities on "The source of mud, the use of mud, the language of mud, mud journey, the beauty of mud" (see Figure 5). For example, Su Chen's father led the team members to understand the function of soil, Zihan's mother led the team members to observe the organisms in the soil, Jiayi's mother led the team members to recite soil poetry, Lening's mother led the team members to make ceramic art in the Wuzhou Kiln, Xin Ni's mother led the team members to carry out the spring planting action... Under the guidance of various forces, the team members discovered, analyzed and solved problems through observation, comparison, operation, experiment and other methods, calling on everyone to protect the soil and work together to build a sustainable ecological environment, make a contribution.

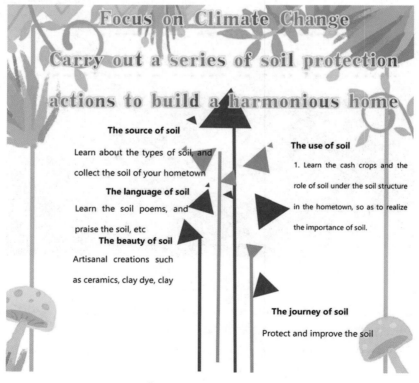

Figure 5. "Walking with Soil" activity design

"**Low carbon, green travel, garbage classification in regular life**": Class head teacher Miss Gong has always encouraged students to start from small things and practice green culture. Through the "Climate +" series of practices, more and more students walk to school, and they use practical actions to promote the concept of "low-carbon, eco-friendly, green travel". In addition, students will pick up garbage and sort it on their way home, becoming responsible little environmental protection guardians inside and outside the school.

Effectiveness & Reflections

The effectiveness of the "Learning in the Nature" project is mainly reflected in the following three aspects.

The first is to realize the coordination of school, family and social education. The project widely carried out in the "4341 Model" of "Multi-dimensional Climate Curriculum", "Hello, Winter (summer) Vacation" and other practices. Each activity has the guidance of teachers, parents' participation, and community support, so as to achieve the synergy of family, school, community education. In the progress of the project, we have gained 100% support from parents both in spirit and action (see Figure 6). Each Parent is a social worker, with their own social network, each network contains rich resources. These families in different activities also formed a high-quality relationship and the three-dimensional resources cube which can shareable. In this new pattern of collaborative education, teachers keep pace with the times and broaden their perspective; parents are consciously engaged and learn from each other; community's active participation, collaborative development; children grow up to be fit for the future.

Figure 6. Parental support questionnaire feedback

Secondly, effectively enrich the ways of climate change education. Climate change education should not be confined to books, also not a stiff lecture. Under the leadership of the "Common Worlds Pedagogy", students, teachers, parents, community workers, enterprise staff gather together in fields, forests, hydropower stations, weather stations, museums, communities, night markets, and mushroom greenhouses etc., Through "listening, reading, checking, asking, doing, tasting, speaking, writing, drawing, singing, playing" and other colorful and diversified ways to carry out a series of experiential and interactive "Climate +" activities, so that they can learn and grow in a real, complete and continuous life experience. At this moment, the children's heart, brain and hands are blend together, and they present their learning results in the form of photos, videos, calligraphy, painting and handwork. This greatly enriched their climate knowledge, and then promoted the change of thinking mode from "subject-object dichotomy" and "human-centered" to "all things are connected" and "co-generation", helped them better cope with the challenges brought by climate change, and truly realized teaching for the future.

Thirdly, the transformation from individual to collective has been continuously promoted. The students have had new experiences, discoveries and creations by interacting with each other across subjects. Through reflection and dialogue after the activities, with the help of feedback from parents, teachers and peers, students have formed a clearer, more holistic and more objective understanding of self-growth. They are interested in nature from the beginning, to in-depth understanding of nature, and then initiate the idea of "harmonious coexistence between man and nature". It improves their awareness and attention to climate change, so as to promote students to carry out a series of "climate benefit action" practices. This is the value enhancement and perspective expansion from individual to collective, which realized self-awakening and development, and also promoted the joint construction of the earth community.

On reflection, we also found the direction of further efforts: the development of resources for climate change education should not only be broad but also deep; participation in the group should be from the point to surface, multi-party involvement. In the future, the project team will carry out in-depth development on the basis of comprehensive integration of resources, form multiple project practices, and create an atmosphere of public participation.

(Shuqing Gong, Teacher of Xincheng Campus, Hushan Primary School, Wuyi, Zhejiang; Jun Ma, Teacher of Wuyi Experimental Primary School, Wuyi, Zhejiang)

Yangtze River Ship Explorations

Research Background

As one of the longest rivers in Asia, the Yangtze River has carried China's civilization and development for thousands of years. It not only provides China with abundant water resources and supports the livelihood of a large population and industrial production of cities, but also gives rise to numerous cultural cities and historical monuments, and is a source of pride and a symbol of cultural confidence for the Chinese people.

However, in the context of increasing global climate change, the Yangtze River areas face multiple challenges posed by climate change. According to the objective of the *Climate Change Education Guidelines* issued by the Shanghai Municipal Institute for Lifelong Education in February 2024, which is to "enhance the learners' awareness of the community of human destiny and the community of life on earth between human beings and nature," it is therefore urgent to carry out climate change education.

Shanghai is located at the mouth of the Yangtze River, in the alluvial plains of the Yangtze River Delta, located in the golden waterway, with unique geographical advantages. The No. 2 Primary School Attached to Shanghai No. 6 Normal School is located near the Huangpu River, a tributary of the Yangtze River, at a distance of only one kilometer or so, and is a major transportation hub on both sides of the river. Focusing on ships on the Yangtze River, this activity is a parent-child activity that guides students to discover the connection between ships and climate change.

At the same time, with the cross-regional and multi-modal linkage, collaboration with primary and technical secondary schools in the areas where are in the mouth of the Yellow River, China's second largest river, this activity is to carry out sustainable education, and realize the same lesson through the way of borrowing each other's resources and doing digital collaborative teaching and research.

Project Design

This Yangtze River Exploration chose ships as the entry point for students' interest, borrowed social resources and the power of many parties, and led students to explore the close connection between climate change and ships, to improve students' knowledge of climate change, to cultivate students' developmental knowledge and action ability, to guide students to understand the impact of ship transportation on the environment, and to be able to penetrate the knowledge of climate change into their daily lives. At the same time, strengthen the cooperation between school, family and community to form a good educational linkage. Through promotion and publicity, the significance and importance of the activity will be publicized to more parents and students through school newsletters, parent groups and other channels.

It is hoped that through this design concept, a long-term and stable cooperation mechanism between home, school and community will be established, laying the foundation for similar educational activities on climate change in the future. We are willing to learn from the experience of other successful cases to continuously improve and enhance the quality and impact of the educational activities.

Project Implementation

1. Design of the Entry Activities

During the winter vacation of 2024, parents and children from some of our school's second grade classes were invited to the kick-off meeting for the program through Tencent meetings(see Figure 1). Under the lively atmosphere, the teacher introduced the relevant basic geography knowledge to the students while the students held a simple version of the map of China, accompanied by their parents, and together embarked on the first journey of ship exploration.

Figure 1. Online launch of Ship Exploration Camp

2. Journey of Discovery—Exploring Ship History and Culture

The China Maritime Museum (No. 197, Shengang Avenue, Pudong New Area) is close to Dishui Lake and has rich historical and cultural values that show the development of navigation and the richness of marine culture, and therefore can provide students with the opportunity to gain an in-depth understanding of marine knowledge and navigation technology. In addition, it is an ideal place for interdisciplinary learning, involving a variety of disciplines such as geography, history, science, etc., which can promote students' comprehensive quality and interdisciplinary study. Then, it reflects the spirit of human exploration of the sea and respect for the natural environment, which helps to cultivate students' humanistic sentiments and senseof social responsibility. This one-day study is led by students who actively participate in the activity.

1) Brainstorming before the study tour

The study tour is different from the traditional one where teachers make plans and lead students to carry out activities. This ship exploration journey is student-centered. On January 27, a dozen groups of parents and children got together in the campus conference room to brainstorm. The transportation, goals and itineraries were all decided by students through group discussions, with parents helping them search information online (see Figure 2). After discussion, a unanimous of "green commuting" was made. Although it was a long way to travel, the students were well prepared and planned their tour.

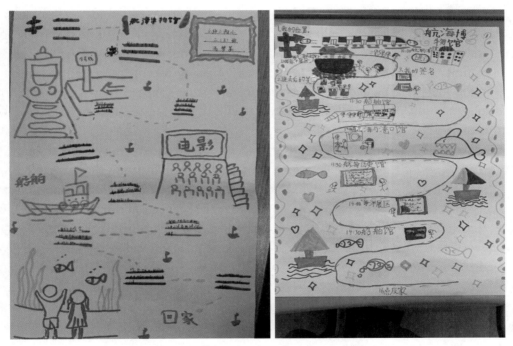

Figure 2. Offline planning of student activities

2) Interaction in the study tour

On January 28, the one-day study route of the Maritime Museum was discussed and planned by group members. During the one-day tour, they still worked in groups. Even if the opinions could not be unified, the group members chose to think about the big picture, and no one wondered away from the team and acted alone. All students well performed their duties of "behavior monitor", "group leader", "little assistant" and other pre-selected jobs. The activities of the day were well organized, and students helped each other, stimulating the spirit of teamwork among members. At the end of the day, many parents and students expressed their affirmation and love for this activity:

Sitong Zhu: "When I came to the Maritime Museum today, what impressed me most was the big Chinese junk ship in the hall. There are two boats in the junk, and I also saw four beds in the cabin. I know these beds are for people sailing away for the night."

Menghan Tang: "After visiting the Maritime Museum, I was most impressed by the Universal Map of Great Ming, which is one of the treasures of the museum. It used to be labeled in Chinese, but in the Qing Dynasty, it became labeled in Manchu, and this is theUniversal Map of Great Ming we see today."

Keixuan Li: "I went to the Maritime Museum today and I learned about icebreakers with different bows which can break ice because of their pointy bows."

Yichen Zhou: "I learned a lot today at the Maritime Museum, and one of my favorites was the sailboat. Whenever the wind blows, the sails are raised, and another function of the sails is to change direction."

Sitong Zhu's mother: "Today is also the first time for me to come to the Maritime Museum, I am very happy to follow my children to this very beautiful place to see some objects related to navigation, including the Chinese junk ship, icebreakers, and ship model exhibit, all of which took my breath away. Usually it is the mom and dad who take the children to visit somewhere, but today we are very honored that it is the children who take us to a different place and follow the children to see the world, thank you very much to the school and Ms. Fan."

Yichen Zhou's mother: "Thanks to the school for organizing the children from Class 2 and Class 3, grade 2 with their parents to come to the Maritime Museum today, the children have gained a lot today, and the adults have learned a lot of knowledge by following the children. We are very delighted that the school can organize such activities during the winter holidays. First, this is an exchange of learning for children who can meet new friends from different classes. Second, it expands the children's extracurricular knowledge and opens their eyes to the knowledge of maritime history. The third is that in the Maritime Museum we come together, learn together, communicate together, the atmosphere is really great. Here I also want to thank the school and the teachers, I hope that there will be more activities like this in the future!"

3) Reflection after study tour

The one-day study tour brought the children and parents full of harvests. On March 13th, a summing-up meeting was held within the grade 2 of the school, where two representatives of parent-child groups shared with the students what they had seen and heard at the Maritime Museum during the winter holiday, and introduced the concept of climate change education. Many students also gained a new understanding of climate change education after the activity, and went on a study tour with their parents along the Yangtze River Basin to investigate and visit the neighboring provinces and cities to explore the impact of climate change on shipping.

3. Words of Wisdom — Experts' Sharing and Interpretation

Principal Gu of Yangjing Community School told students stories about ships in the mouth of the Yangtze River and ships on the Huangpu River to enrich their knowledge. Each lecture chooses a hot topic related to climate change, environmental protection, and the ecology of the Yangtze River, such as the scientific principles of climate change, the characteristics of the ecosystem of the Yangtze River Basin, and the impact of climate change on the water resources of the Yangtze River Basin. In the lectures, the experts introduce the basic concepts, the latest research progress and the practical experience of the related topics in depth. The experts have set up a rich interactive session where students can communicate and discuss with the experts on issues they are interested in. In addition to passing on the conceptual knowledge, the experts also provide practical guidance and advice on how to cope with the changes and protect the ecologicalenvironment of the Yangtze River, so as to promote environmental protection and sustainable development.

4. Creative Joy — Boat Model Handicraft

In this activity, boats made of different materials representing some historical periods, such as: wooden dragon boats, block junks, plastic lifeboats, etc., were selected to complete the construction of boat models under the guidance of the teacher, aiming at cultivating students' hands-on ability and creativity.

Students got together and used simple materials to make boat models, including hulls, masts, sails and other parts under the teacher's guidance and demonstration, and they also painted and decorated the appearance of the boat models according to their own ideas and creativity. In the process, not only improved the students' hands-on ability, but also promoted the spirit of mutual help. Although the making process was a little difficult for some students, the moment they saw the final works, even the greatest difficulties became worthless in front of the sense of achievement.

The creative activity was over, but the enthusiasm of the students did not fade. After returning home, the students chose special ships to make according to their own preferences, including ancient ships with a long history, modern great naval ships, and even modern new energy-powered ships. There were so many different kinds of ships that in order to show the students' achievements, a small school ship model exhibition was held to share the students' achievements and knowledge of ships with other students(see Figure 3).

Figure 3. The school ship model exhibition

This activity was not only an interesting handcraft experience, but also allowed students to learn more about the importance of the shipping industry through hands-on practice, enhanced their interest in the field of shipping, and raised their awareness of climate change and the challenges it may bring to the maritime industry, thus increasing their concern and sense of responsibility towards environmental issues.

5. Beauty of Art — Inspirational Painting Competition

After understanding the close connection between ships and climate change in the early stages, students have gained a new understanding of the structure and materials of ship hulls, the selection of ship types, the optimization of ship power systems, and the adjustment of ship sailing routes, and each student has new expectations and fantasies about the ideal ships in their minds. This activity aims to let students design their own future ships, which can be imaginative. Students were asked to draw a picture of the future ship using the following prompts:

What kind of ideal ship do you have in mind? What is its form of energy power? Is it big or small? What would be special about it that would win the hearts of the crew and passengers? And the answers to those questions have created the ships you see today. Sailboats have sailed on the seas, submarines have dived under water, and expedition ships have traveled to exotic places, all of which could have been changed by a thought that crossed people's minds.

6. Cooperative Linkage — Cross-regional and Multi-modal Learning Experience

After the one-day study tour at the Maritime Museum during the winter vacation, students' knowledge of ships was enhanced, the interests of parents and children were activated, and the focus was enlarged from Shanghai to the whole country, which madestudents think of the Yellow River, the second greatest river in China, and driven by curiosity, the students asked many questions they wanted to know. In the WeChat group, the teacher also collected and sorted out the students' questions. After that, the ship exploration trip will soon organize an on-line and off-line interconnection activity, focusing more on inter-regional learning with the students in Dongying at the mouth of the Yellow River.

Effectiveness & Reflections

The project has shown initial results. Through this project learning, co-learning and co-growth has vigorously emerged between children and parents, resulting in a closer parent-child relationship and a deeper understanding of climate change education.

1. Project Effectiveness

1) Uphold green and low carbon traveling principles

During the winter vacation, we led parent-child groups to carry out a one-day study tour at the China Maritime Museum, focusing on three ideas. The first is that the commuting is green. The school is nearly 150 kilometers away from the Maritime Museum, which is a long way and inconvenient to transfer. The commuting time is nearly 5 hours including walking, taking the subway and bus. Secondly, the whole process is student-oriented. Whether it is the design of the visiting route, the determination of the precautions, the arrangement of the content of the activities, are the children on the spot to discuss, and their parents helped search online. The third is the mutual learning and assistance. Whether from the pre-planning to the implementation of the project, and then to the summing-up, students, parents, teachers have been in a good state of mutual learning, mutual encouragement and mutual achievement, there is no teacher beforehand to provide detailed plans for the activity and travel. Everything is generated in the discussion and slight adjusted in the action. Thanks to the tacit understanding between the parents and the teacher, and the mutual learning among students, a lot of good outcomes has been yielded.

2) Focusing on ships to enhance climate change awareness

The ship study tour focuses on the Yangtze River Estuary, using ships as an entry point to draw students' attention to the connection between ship transportation, energy, emissions and climate change. Climate change education has planted a seed in the hearts of students, and we have already seen the active participation of children and the enthusiastic commitment of parents. In the future, under the leadership of teachers, we will continue to teach students how to take personal and feasible actions to respond to climate change, so as to form a more positive awareness of environmental protection and the power of action, which will contribute to the construction of a green and low-carbon society in the future.

3) Promoting the growth of parent-child relationship and developing sustainable education

The school not only provides a project theme for parent-child families, but also builds a bridge for parents and children to explore and grow together. The process, in which parents and children help and learn from each other, is a valuable growth experience. Children will show their growth and progress to parents through their own active learning and exploration, while parents can provide children with their experience and intellectualsupport.

During the activities, parents can see another side of their children, which is different than when they have textbook learning. The wisdom, courage and hard work shown by the children in planning offline activities as well as during trips not only gratify parents, but also make them more confident in their children's abilities and understand their children comprehensively, so that they can better guide and support their growth.

Although the school-family-community cooperation on the Yangtze River ship exploration of Climate Change Education has achieved certain results, there are still some areas that need to be reflected upon and improved.

2. Project Reflection

1) Strengthen multi-party participation and community actions

The establishment of a solid school-home-community cooperation mechanism is the key to ensuring the smooth implementation of educational activities. Communication and collaboration between home and school can be strengthened by expanding parent committees and parent volunteer teams; organizing regular parent meetings or symposiums to let parents know about the school's educational philosophy and activity arrangements; and making use of the resources of the Yangjing community, inviting community organizations and professional institutions to participate in educational activities to enrich the educational content.

2) Expand the linkage of educational resources

Our school is located in the Yangjing neighborhood along the Huangpu River and is a member of thePrimary School Attached to Shanghai No.6 Normal School educational group. Yangjing has rich humanistic educational resources, surrounded by environmental schools, ship research institutes, community schools, etc., which are all about one kilometer away from our school. Zooming out to the whole country, our school and the primary school at the mouth of the Yangtze River in Dongying, Shandong Province, already established online connections. Due to the unique geographic locations of the two places—at the mouth of the Yangtze River and the mouth of the Yellow River, respectively—there is a need to boost the partnership in the subsequent project promotion to form a linkage of educational resources from multiple sources and to expand the impact of climate change education.

3) The key role of publicity

The school will make better use of new media platforms, such as the WeChat public account, website and social media, to disseminate environmental protection knowledge and information on related activities, expand the coverage and influence of education, and attract more students and parents to participate in climate change education to form a virtuous education cycle.

Through continuous reflection and improvement, the effectiveness of this project will be further enhanced, laying a solid foundation for cultivating more future citizens who are concerned about climate change and have a sense of responsibility.

(Yiyi Fan, Teacher of No.2 Primary School Attached to Shanghai No.6 Normal School)

School-Home-Community Collaboration Creates a New "Climate and Carbon" Education Paradigm

Research Background

Primary school students are future citizens. It is the responsibility and responsibility of education to enhance their awareness and ability to cope with the challenge of climate change and try to formulate certain coping strategies for the issue of climate change. In primary school textbooks, various subjects have already included or permeated the content related to climate change. However, we can also clearly see the problems of climate change education in primary schools: First, the knowledge is scattered, and each discipline has a little, which is not systematic. Second, the relationship is fragmented, the knowledge learned does not create value, which is mainly manifested as: each discipline is in its own array, the discipline is separated from each other, the lack of integration; classroom knowledge and social life are separated from each other and lack of correlation; semester learning and vacation life are separated and lack of integration; knowledge learning and problem solving are separated from each other and lack of fusion and innovation.

Knowledge that is meaningful to the life of the learner is likely to have long-term vitality. Therefore, students' action to "cope with climate change" is a link to form a "community of common destiny" of the whole society, a carrier for education to create life value, a magic stone to promote "comprehensive integration" literacy, and an accelerator to cultivate comprehensive talents. Taking school as the leading role and working with families and society to cope with climate change can not only mitigate the impact of climate change with an overall effort, but also help cultivate students' comprehensive qualities such as pro-social ability, leadership ability and learning ability.

Project Design

Climate change is a fait accompli that human beings must facenow, but climate change education is not only the responsibility of schools. Schools, families and society can cooperate to find scientific solutions. On the one hand, we integrate climate change-related content into campus life, and adopt task-driven, project-oriented and interdisciplinary learning methods to guide students to explore from a disciplinary perspective. On the other hand, according to the characteristics of primary school students, we have sorted out the content related to "climate change" in the six years of primary school, formulated the annual goals, content system and evaluation mechanism of "climate change education", and carried out various interdisciplinary activities relying on family and social forces. At the same time, with the help of the school's holiday parent-child playgroup, which is characterized by "mutual learning and playing, connecting life and learning together", to stimulate creativity and carry out a series of long-term and diversified holiday research activities(see Figure 1).

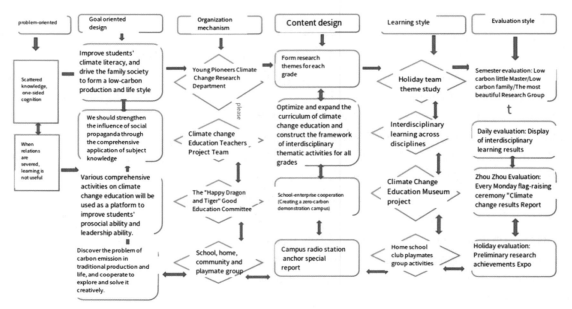

Figure 1. Flow chart of "school-home-community cooperation to carry out climate change education" in Longhutang Primary School

In short, our "climate change education" adheres to the goal-orientation, problem-orientation and practice-orientation, and insists on the multi-subject "co-education and mutual education, co-learning and co-growth", seeking to achieve success in the comprehensive and integrated learning process, and becoming adults in embodied cognition and practice.

Project Implementation

1. Formulate the Coordinated Action Target of "Catching the Weather"

In order to have a more comprehensive understanding of the actual situation of "global climate change", the school joined the National Climate Change Education Research Alliance. The members of the project team carefully studied the Paris Agreement, the National Climate Change Adaptation Strategy 2035, and the 2023 Annual Report on China's Policies and Actions to address climate change. And made it clear that "mitigation and adaptation are the two major strategies to deal with climate change, and neither is indispensable." Focusing on the requirements of "green life", "green consumption" and "green production", according to the age characteristics of students, and through interdisciplinary learning, we have formulated different goals for different grades, focusing on cultivating students' social responsibility(see Table 1).

Table 1. Targets of "Climate Change Education" for each year of Longhutang Experimental Primary School

Lower elementany grades	Middle elementany grades	Upper elementany grades
Permeate the consciousness of "being close to nature andguarding the diversity of plants and animals".	Focus on the study of "green consumption, resource conservation, low-carbon life".	Focus on understanding the "new energy enterprises" of "newenergy capital".
Parent-child play partners, intergenerational learning and other ways to inspire children to be willing to go into nature, participate in "grandparent planting", "claiming green trees" and "green travel".	Through the independent recruitment of partners, understand the impact of "climate change" on human beings, and use the surrounding venues to experience and publicize.	Strengthen the knowledge and understanding of new energy and new quality productivity, widely publicize and actively use.
Cultivate students' ability of group, expression and hands-on practice.	Cultivate students' organizational ability and creativity to find problems, organize activities, cooperate in solving problems, and form results.	Cultivate students' ability to explore projects, try small inventions and small modifications.

2. Build a "Crowd-power" School-home-community Cooperation Organization

Efficient organizations have the effect of pooling and amplifying forces, which can improve efficiency. We do this by building effective organizations at the neighborhood, school, and student levels.

Regional "Climate Change Education" Committees: In order to promote the formation of a micro-system of climate change education, enhance social education, and pool resources, our school has established a "Climate Change Education Committee" at the street level, which includes students, parents, teachers, community personnel, urban leaders, leaders of the Ecological Environment Bureau, public welfare organizations, representatives of new energy enterprises, and other representatives. It has a clear work goal, organizational structure and responsibilities. Relying on the Wechat group and regular forums, the committee is able to operate on a daily basis, which has greatly improved the effectiveness of interaction between the school, family and community in climate change education.

School "Climate Change Education" ProjectTeam: There are contents related to "climate change and environmental protection" in all disciplines of the national curriculum. In order to effectively implement the relevant contents of the national curriculum and extend, expand and transform them into practice, we have recruited teachers of all grades and disciplines who are interested in exploring in this aspect, and formed core members of the project team to communicate and discuss regularly, leading the in-depth and continuous research of this discipline. All the class teachers are the main members of the project team, and lead the tutor group of their respective classes to carry out various class-based climate education activities.

Red Scarf "Climate Change" Research Department: Using the Young pioneers position, divided into "knowledge propaganda research team", "green electricity research team", "family carbon reduction research team", "green travel research team", "early adopter research team" five sub-teams, responsible for guiding and organizing the teamcommittee, the team committee to carry out related activities.

"School, home and community mixed play group": In order to make "action against climate change" become an important part of students' holiday life, the school recruits a group of "climate change research leaders" through independent registration in each grade, with the help of their power, and then choose their favorite theme, recruit like-minded partners to carry out research activities. "A single spark can start a prairie fire", led by a small head, carried out a variety of "low-carbon" learning and practice activities.

3. Design of "Grounding" Interdisciplinary Theme Courses

There are many subjects in the national curriculum for primary and secondary schools. On the premise of not increasing the extra burden on students, we carry out "interdisciplinary" activities based on the implementation of the national curriculum, and develop class-based courses focusing on "climate change" by relying on "class team meetings" and "comprehensive practical activities".

1) Explore national curriculum resources and implement "interdisciplinary" activities

We have made full use of the requirement in thenew curriculum standards that "10% of class hours in all disciplines should be devoted to interdisciplinary thematic activities", screened out suitable content, and launched "climate change education" activities. Taking Chinese as an example, the third grade text "Flower Clock" introduced various plants in different seasons and the exercise "Making Business Cards for Plants". Teachers and students jointly designed the interdisciplinary theme activity "Colorful Flowers": After class, they went to the campus library and urban library to look for "Plant Encyclopedia" and other books about plants to read; in the art class, observe the plants in the campus and the community, select 1-2 kinds of "portraits" for them, and then add a paragraph of introduction, and hold a campus "exhibition"; labor class, with the students to plant flowers, do "small gardener"; weekends, and grandparents together to create a "one-meter balcony colorful world", and with the help of parents, and grandparents in the campus planted a "Chinese rose field", increase carbon sink.

2）Link to school-level characteristic activities and develop personalized "class-based courses"

Campus activities are rich and colorful, and if the design of many activities can be scientifically integrated into "climate change education", it will not simply do addition and increase the burden, but can play a "1+1" greater than 2 effect. For example, the characteristic culture of the fourth grade group is "bookish reading", when the Earth Day "meet" campus reading festival activities, the class teachers design "our gas change action" course: read "gas change", say "gas change", see action. Launch students to the library to find books related to "climate change", make a good book recommendation card after reading, and recommend to the whole school; I took the team as a unit to find the strategies to deal with climate change, and made a team report after trying the practice. In this process, students visited and experienced the "Meteorological Bureau", "Natural Disaster Experience Museum" and other venues, and wrote proposals with the knowledge they had mastered to publicize the "climate change" initiative, "low-carbon travel" and "green consumption" to the public.

4. Build a "Sustainable" Holiday Playmate Group Activity

School "Playmate group activities" need parents and teachers based on the position of children, driven by students' interests and hobbies, with the help of social resources and give full play to the joint education of parents and teachers to guarantee and guide the activities. In the activities, parents and children adopt exploratory learning, master learning and other ways to achieve mutual learning. Exploratory learning is to explore the unknown, while master learning is to use the known. Students are more enthusiastic about mastering learning and making what they have already learned become second nature. This is more conducive to the integration of knowledge and action, the comprehensive use of knowledge, problem solving, exercise ability.

1）Implementing top-level planning

In order to make students' actions on "climate change" targeted and tangible, the school focuses on the five major actions of "Climate change I publicize", "Double carbon Plan I practice", "Energy saving and sustainable daily life", "Increasing carbon sink and green" and "School-enterprise cooperation exploring new capabilities", and has formulated the theme of holiday playmate group study for each grade (see Table 2), according to which students can discover relevant problems in life. To find out the solutions.

Table 2. Topics of "Climate Change Education" playmate research activities for each grade of Longhutang Experimental Primary School

Grade	Holiday Playmates research topics	Grade	Holiday Playmates research topics
1	We are good at turning waste into treasure	4	We are a small model for saving resources

(Continued)

Grade	Holiday Playmates research topics	Grade	Holiday Playmates research topics
2	We are the little gardener of knowledge and action	5	We are the small owner of ecological garden
3	We are a small pioneer of green travel	6	We are "New new energy" small makers

2) Constructing activity model

Althoughthe students of each grade focus on the content of exploration under the guidance of corresponding courses, in order to facilitate students to effectively carry out activities under the guidance of parents, the school designs the activity process by task, and proposes the "six ones" to be completed jointly in the process: planing a program, issuing a recruitment order, taking a set of activity photos, designing an evaluation form, making a materialized result (experience, works, creativity, etc.), and obtaining a set of chapters (commemorative and accomplishment chapters).

3) Building resource think tanks

The learning field of climate change should not only be in the classroom, but also in the scenes of family and social life. In order to allow students to learn in the vast natural world and social space, we have formed a preliminary learning resource base with the help of the "Climate Change Education Committee"(see Table 3).

Table 3. Climate change research resource library (some venue resources)

Library	Enterprise	Visiting experience category		Parks	Pastoral
Changzhou Library	Trina Solar	Changzhou Museum	Changzhou Meteorological Museum	Bauhinia Park	One Meter Sunny Garden
Autumn White Book Garden	Sensata	Aneng Natural Disaster Experience Pavilion	Chang Jin Rice Industry Exhibition Hall	Dongpo Park	Gannon Probing
Community Library	Yongqi Bike	The Little Dragon Man Safety Experience Hall	Garbage in Changzhou Treatment Plant	Pier Park	Xixia Villa Art Rice Field
Dazhong Book Company	Winconor & Co.	Sanjiangkou Park Civilization Practice Base	Earthquake Prevention and Disasterreduction Science Education Base	Forest Park	Changzhou Jiuhong Ecological Park

5. Enriching the "Wisdom Communication" Achievement Exhibition and Evaluation Mechanism

1) Visualization: stimulating Student Enthusiasm for Activity Participation through Diverse Assessments of Commemorative and Literacy Badges

One is the "Small Leaders on Climate Change" commemorative badge that can be obtained by participating in the event. The other is the literacy badge based on activity performance. In order to ensure the quality of the activity and highlight the value of parent-child learning, the school collaborates with parents, students and the community to develop activity evaluation standards for each grade(see Table 4). Before the end of each activity, multiple evaluations are carried out: peer evaluation and parent-child evaluation. Such two-way evaluation further stimulates the democratic atmosphere of parent-child co-learning and family co-learning.

Table 4. Standard of "Climate Change" research activity literacy chapter contention

First-level index	Second-level index
Mutual learning	1. Take the initiative to learn knowledge related to climate change, feel the joy of learning new knowledge, and form a strong learning ability. 2. Respect and understand others in exploration activities, and be good at discovering and learning from others' strengths. 3. Be good at using knowledge, using resources, and working with parents and playmates to effectively solve life and study problems.
Dare to explore	1. Be curious and interested in climate change related phenomena in your environment. 2. Be able to take the initiative to explore with peers according to the content of questions and form common research results. 3. Confident and optimistic, with the indomitable will and ability to overcome difficulties.
Dare to take responsibility	1. Abide by the rules and develop a preliminary sense of citizenship and a sense of community with a shared future. 2. Take the initiative to participate in climate change exploration and publicity activities, and initially form pro-social emotions and abilities. 3. I am willing to assume the posts and responsibilities in the exploration activities of the playmates group, and enhance my team cooperation ability.

2) Re-creatable: building Display Stages in Large-Scale Events to Expand Their Impact

First, the playmate group activities in the annual winter and summer vacation will be displayed and evaluated on the spot at the school's initial fair and sales after re-processing, and the Playmate group activities in the daily weekends and small holidays will also be displayed and interactive in the flag-raising ceremony by the class as a unit, and then radiate out through the school public account. This way not only allows more students to participate in the experience, but also promotes creativity, cooperation and stimulates a sense of achievement. Second, we have created an exclusive platform like using campus radio to set up a special "Longwa talk about climate" column, and opened a "climate change column" on campus public account to publicize students' research achievements.

3) Gatherability: building a Climate Change Museum that Combines Exhibition and Educational Functions

The climate change museum includes two parks, one corridor and one center. The two gardens, namely the half-acre flower field of the school and the "Sky garden" built in cooperation with Trina Solar, can not only generate electricity from the solar roof, but also be planted by students. The first corridor, a secret ecological environment, has functions such as meteorological monitoring, climate learning, and exploring the 24 Solar Terms. The first center is the Playmate Center, which displays the achievements of the students' playmate group's climate change research process and can interact with carbon emission calculation.

Effectiveness & Reflections

1. Project Effectiveness

First, it has created a good environment for multiple actors to pay attention to climate change, and formed a practical force for a "community of shared future". This winter vacation, the "climate change" playmate group activities carried out by our school accounted for 59% of all the playmate group activities, and the cooperation between teachers of various disciplines has become frequent, promoting the effective integration of educational forces(see Figure 2、Figure 3). The school, the family and the community have established a friendly and cooperative relationship, and the awareness and feelings of "community of destiny" of parents, community, class teacher, subject teacher and students are developing. Through the creation of a good ecology featuring students' autonomy, teachers' guidance, parents' protection and social support, the pro-social force of multiple subjects has been enhanced.

Figure 2. Students present their research achievements at the Playmate Center

Figure 3. Students present their research achievements at the preliminary trade fair

Second, it promotes the integrated learning ecology of applying knowledge to practice, and improves the creativity of extracurricular life. To reconstruct the comprehensive target system, focusing on "climate change"; To drive diversified activities and bring together team efforts; to realize the mutual benefit of resources sharing, enhance the joint efforts of family, school and community education as the path, through family and school cooperation, school-enterprise cooperation, parent-child interaction and other ways, multi-subject and multi-interaction to carry out playmates group activities, so that students feel the sense of value and achievement brought by the application of learning, make extracurricular life more autonomous and rich, become the extension, extension and supplement of semester life, but also let families effectively integrate into society.

The third is cultivating the growth thinking of the school, the family and the community, and enhance the core quality of the students. In the process of initiating, planning, recruiting, organizing, evaluating, summarizing and reflecting on research activities, students make progress in trying, introspection in communication, confidence in challenges, and mutual discipline in cooperation. Parents who actively participate in the project activities nourish the courage and wisdom to accompany their children's common growth, enhance their own awareness of the role of family educators, and more harmonious parent-child relationship. At the same time, the teachers participating in the project understand that "learning must consider practice as a whole, and take into account the diversity of relationships—both within the community and with the whole world", breaking through the constraints of disciplines and moving towards interdisciplinary and extra-disciplinary life education.

Fourth, it builds a diversified platform for the project's research and expand its social influence. Since the implementation of the project, more than 70 related interdisciplinary theme activities have been carried out, more than 300 related holiday playmate group activities have been carried out many National Climate Change Education Alliance live display activities (see Table 5), which have been publicized by *China Education News*, Jiangsu Young Pioneers, *Changzhou Evening News*, Changzhou TV station and other platforms for more than 20 times, and teachers have shared more than 10 lectures in the National Climate Change Education Alliance seminar. With an audience of about 50,000 people. The research results were also brought to the "3rd China-Japan-ROK Environmental Teachers Exchange Meeting and China-Japan-ROK Environmental Education and Public Environmental Awareness Seminar" by Principal Gu Huifen and were highly appreciated... In the implementation of school, family and community cooperation in education and the construction of ecological civilization under the background of the new era, a new educational ecology of all-domain co-education has been formed.

Table 5. Examples of activities carried out by the "Climate Change Playmates Group" (Part)

Category of activities	Activity content	Number of classes	Number of participants (around)
Save energy and get the most out of everything	Visit a sewage treatment plant	12	300
	Explore the mysterious water world	5	80
	Visit a waste treatment plant	15	260
	Don't let the dinosaur today, become the human tomorrow	30	1000
	Garbage sorting—our special action	20	1200
Increase carbon sinks, Action for all	Grow meaty together	12	200
	Conservation of plant diversity	12	600
	I make business cards for plants	12	500
	Step into Eco-Farm	5	50
	Strive to be small farmers	24	1200
	Green the campus	36	200
	I plant rose with my fathers	12	100
There is no limit to creativity in turning waste into treasure	Waste for green public fair	5	200
	Barter Love Bazaar	20	500
	"Bottle" empty recycling	5	200
	Carton turn 'Big player'	12	1000
	Cloth "bags" are king	5	50
School-enterprise cooperation, wisdom science and innovation	Into Trina Solar	13	90
	Exploring "new energy vehicles"	2	90
	"Sensata" learn from each other	13	50
	Work together to create a solar-powered sky garden	12	500
	Long Dada Cup Clean Energy Science and Technology Creative Painting Competition	15	100
	Eco-friendly Fireworks Show	3	100

2. Deficiencies and Improvements

First, we should strengthen the accumulation of visual results in climate change education practice and promote the popularization of experience. Although the school has a top-level design of climate change education, with rich activities and diverse forms, and profound participation experience of students, parents and community members, the visual results are still relatively few. First, the project team has a weak awareness of collecting results, and has not timely collected, classified, stored and displayed them; Second, there is little text experience, such as the climate change bibliography, school-based playmate group activity manual, climate change publicity copy (such as poetry, slogans, allegos, etc.), which is suitable for all grades. There is still a large space for experience radiation.

Second, it is necessary to improve the integration and validity of climate change education and school brand projects, and promote characteristic practices. The school has carried out research on intergenerational learning projects for many years, and the subject of climate change education is precisely the whole society and all mankind. Therefore, we have tried to bring grandparents and relatives into the climate change education classroom, and set up a multi-time and multi-field climate change education college inside and outside the school, so that multiple generations can learn from each other to live a green life. In the future, we will further strengthen the intergenerational cooperation to carry out climate change education and other integrated school characteristics.

Third, it is necessary to enhance the task-driven nature of climate change education and social collaborative education to promote sustainable mechanisms. With the help of top-level design, on the one hand, we rely on the content of climate change education in the national curriculum to design activities such as "happiness homework" and strengthen normal implementation. On the other hand, we stimulate the initiative and creativity of students and parents to carry out after-school playmate group activities to enhance dynamic generation. At the same time, relying on the strength of the community and cooperation between schools and enterprises, in the way of "orienteering", climate change action is launched on the National Ecological Day and the primary time node of each month, and new learning and dissemination of results are constantly promoted.

In short, under the background of "double reduction", climate change education projects have effectively improved the quality of family life and the quality of parents' accompanying their children, changed the way of learning, promoted the "three-in-one" of multiple co-learning, internalization and practical action, and created a powerful learning field with children as the center and relying on the joint efforts of family and school communities. It has opened an optimal path for all people to pursue a better green and low-carbon life.

(Yalan Chen, Vice Principal of Longhutang Experimental Primary School, Xinbei, Changzhou; Huifen Gu, Principal of Longhutang Experimental Primary School, Xinbei, Changzhou)

Expert Review

Climate change is a common challenge faced by all humankind in today's world and is a central concern within the United Nations Sustainable Development Goals. Climate change education is a fundamental measure for the current and next generations to understand, and respond to climate change, and it also extends the scope of human education to a wider range of contents, a longer period of time and a wider space between human beings and nature.

Addressing this critical and complex issue, this book encapsulates the proactive thinking and solid practice of Chinese educators in climate change education through a compilation of case studies covering various important aspects of climate change education, such as theoretical research, policy backgrounds, program design, implementation processes, andreflection on the effectiveness of climate change education.

In this chapter, "climate change education through a cooperative mechanism made up of schools, families and communities" is the central theme, which provides readers with a comprehensive perspective on climate change education from the perspective of the main implementers of climate change education, and encompasses diversified educational approaches such as school-enterprise cooperation and community participation. The compilation of cases demonstrates how families, schools and society can work together to promote climate change education.

The cases in this chapter emphasize not only the learning of theoretical knowledge, but also the application of what has been learnt to practical problem solving. The chapter provides detailed steps and strategies for the practical operation of climate change education through specific project implementation cases, demonstrating high practicality and educational value. The innovative nature of its project design also demonstrates the practical effectiveness of educational activities in promoting social participation and raising public awareness. This in-depth case study provides readers with valuable first-hand information on every aspect of a climate change education program, from conception to implementation. The case "Climate Change Classroom Targeted the Elderly and Children" expands the audience of climate change education from students to the old and young. Through the grandparent-grandson-together learning model, it enhances the younger generation's understanding of climate change and their ability to cope with it, while at the same time updating the older generation's knowledge of environmental protection, thus realizing a two-way transfer of knowledge. The case "Phoenix Station" combines climate change education with bicycle culture through school-enterprise cooperation, which not only raises students' environmental awareness, but also enhances their practical ability and innovative thinking. The case "Learning in the Nature" program, builds a complete learning space and time, allowing students to learn and grow in a real natural environment, and strengthening the social and practical nature of education through a cooperative mechanism made up of schools, families and communities.

The "dual-teacher" teaching mode in project design, namely, the cooperation betweenschool teachers and enterprise experts, offers students the chance to combine theory and practice. In addition, the strategy of constructing an all-round and hierarchical climate education system is put forward in this chapter, which not only focuses on the transmission of knowledge, but also pays more attention to the cultivation of interdisciplinary thinking and problem-solving ability of students, which embodies the concept of educational innovation.

The compilation of cases in this chapter offers profound insights on the process of project implementation, in addition to demonstrating its effectiveness. The reflection on the educational activities in the case studies goes beyond the superficial and goes into the evaluation of the methodology, content and effectiveness of the programs. Such reflection is not only for the improvement of individual projects, but also for the development of a replicable educational model and a "sustainable" educational paradigm. Such Reflection help to optimize the design of the program and provide key guidelines for teachers to follow and improve.

(Yong Zhang, Professor and Director of Department of Environmental Science, East China Normal University; Qianhua Xu, Graduate Student of Department of Environmental Science, East China Normal University)

Appendix

Climate Change Education Guidelines(Draft)

February 2024

Education is an important force in responding to climate change. Carrying out climate change education is a significant way to raise public awareness of climate change and improve their ability to cope with it, reinforce climate action, and realize sustainable development.

Based on the combination of existing results at home and abroad, especially the experience of a series of studies promoted by the Shanghai Municipal Institute for Lifelong Education of the East China Normal University together with various educational institutions, and several academic conferences and seminars, this guideline is hereby formed.

1. Concept and Characteristics of Climate Change Education

1.1 Concept

Climate change is one of the major challenges of our times, which is related to human survival and development of humankind. Addressing to climate change is an overall strategy of the United Nations.

China attaches great importance to address climate change issues, and actively promotes fair, rational, cooperative, and win-win global climate governance. China has incorporated climate change adaptation into its national development strategy, as well as economic and social development planning and reaction.

Education is a core initiative to address climate change and helps learners develop the skills they need to cope with climate change by influencing values, ways of thinking, behavior, and life choices.

1.2 Characteristics

Climate change is a global challenge that transcends national borders. It ranges from concrete, perceptible climate change to understanding climate change on a global scale and through historical processes. The complexity of climate change itself requires multi-faceted, interdisciplinary thinking to conduct climate change education that is comprehensive, practical, and multifaceted.

(1) Comprehensiveness: From the learning contents, climate change education includes both the natural and human social sciences. From the learning process, climate change education aims to help learners comprehensively understand eco-environment systems from multiple perspectives and to master the close links and interactions between the social environment and the ecological environment and its internal elements. From the learning methods, climate change education can be integrated into the teaching of the various disciplines through interdisciplinary methods; it can also be set up separately in the climate change-themed education lessons; and it can be carried out through integrated practical activities or relying on the family, enterprise, and community to form a new pattern of school, family, and society to work together to conduct climate change education.

(2) Practicality: Climate Change Education emphasizes the discovery and creation of knowledge by learners through personal experiences; developing innovative and critical capacity in solving real climate change problems; enhancing communication and understanding in participation to form the right ecological environmental values; forming healthy lifestyle habits in harmony with the natural environment in practice; and enhancing the sense of active participation in action for sustainable development. Therefore, it is required to avoid single-disciplinary education and split educational methods, especially to avoid the separation of knowledge teaching and productive life practice.

(3) Multifacet: The subject, object, time, space, carrier, and method of climate changeeducation are diverse, with both general principles and new strategies that are constantly generated. It is necessary to continuously summarize the experience that has been gained and deal with the new problems. In particular, when it is aimed at all people, highlighting lifelong learning and paying attention to different educational subjects and educational institutions, climate change education is more infinite, multifaceted and creative.

2. Objectives and Contents of Climate Change Education

2.1 Objective

Climate change education needs to renew and develop the values and thinking of learners, develop the corresponding knowledge and capabilities, focus on life education and capacity-building, and contribute to the transformation of production, daily life and participatory governance.

2.1.1 Development of values and thinking

Enhance learners' awareness of the community of human destiny and develop the values of respecting, collaborating, and contributing to human civilization and human development; develop a way of thinking that rationally understands the relationship between the individual and humankind.

Enhance the orientation of unity of knowledge and action, leadership innovation, and form a rational understanding between bit-by-bit efforts and systematic renewal.

Strengthen the value orientation of the present and the initiative, and form a rational understanding of the current and future.

2.1.2 Development of knowledge and capacity

Develop natural science knowledge related to climate change.

Develop human and social science related to climate change.

Develop learners' leadership, practicality, comprehensive problem-solving and lifelong learning skills.

2.1.3 New habits in production, living and participatory governance

Integrate and reflect climate change knowledge and competencies in daily life.

Apply the climate change knowledge and competencies in production.

Apply the climate change knowledge and competencies in social and global governance.

2.2 Contents

The content of climate change education is diverse. To facilitate a comprehensive understanding, concern, and response to the complexity of climate change, it can be selected and systematized according to a specific learning population to serve the goal. The following can be used as the basic content structure:

2.2.1 Status and urgency of climate change

Concept and characteristics of climate change: Understand the definition in the United Nations Framework Convention on Climate Change (UNFCCC): "Climate change" means a change of climate which is attributed directly or indirectly to human activity that alters the composition of the global atmosphere and which is in addition to natural climate variability observed over comparable time periods.

Status and problems of climate change: Climate change has far-reaching implications for the geographical environment and human activities, including temperatures rising, extreme weather events, sea levels rising, and ecosystem damage.

2.2.2 Multidimensional impacts and responses to climate change

Science Dimension: Explain the scientific principles and mechanisms of climate change and promote scientific knowledge.

Cultural Dimension: Explore the perception and response of different cultures to climate change and promote cross-cultural understanding.

Ethics Dimension: Emphasize the ethical responsibility of climatechange and foster global citizenship and responsibility.

Economic Dimension: Analyze the impact of climate change on the economy and guide the economic concept of sustainable development.

Political Dimension: Review the impact of political decision-making on climate policy, keep a high level of political acuity and promote political participation.

2.2.3 Climate change governance practices

Interpretation of the content and implementation mechanisms of the United Nations Framework Convention on Climate Change and the Paris Agreement: Analyze the mechanism of global cooperation, as well as cases of local innovation and action, emphasis on the relevance of emission reduction and adaptation actions and strengthen monitoring assessments.

Climate change response and action: Emphasize climate change governance measures, including greenhouse gas emissions-reducing, renewable energy development, a circular economy, and forest protection and restoration.

2.2.4　Transformation of the mode of production

Energy Efficiency: Study the technologies and applications of renewable energy sources to improve energy efficiency and promote sustainable energy development.

Greenhouse gas emission management: Focus on control and reduction strategies for methane and other non-carbon dioxide greenhouse gases.

Circular economy and resource utilization: Promote the transition of production to a circular economy and improve resource efficiency.

Biodiversity protection: Introduce the importance and methods of protecting biodiversity, emphasizing the relationship between biodiversity protection and climate change.

2.2.5　Lifestyle change

Environmental Awareness: Encourage proactive awareness of the environment and promote sustainable lifestyles.

Consumer habits update: Guide consumers to choose environmentally friendly products and reduce their carbon footprint.

Sustainable choices for clothing, food, housing, and transportation: Introduce concrete sustainable options and practices in daily life.

2.2.6　Cultural heritage and innovation

Heritage: Explore the impact of climate change on traditional culture and emphasize the importance of cultural heritage conservation.

Innovation: Encourage social involvement in tackling climate change and promote the integration of emerging cultures with sustainable development practices.

2.2.7　The driving role of technology and the use of digital tools

Discover the role of technological tools such as climate science, meteorological technology and data modeling in monitoring, predicting and understanding climate change.

Explore innovative applications of emerging technologies such as artificial intelligence, big data, etc. in climate research and solutions.

Use digital tools like climate simulation software, and data visualization tools, to analyze and demonstrate climate data.

Explore the driving role of technological innovation in emission reduction technologies, clean energy and sustainable development technologies.

2.2.8 Multilateral participation and community action

Community action awareness: Guide learners to understand the climate challenges in communities and the impact of climate change on local communities. Explore collaboration with community organizations, local governments, etc., to promote community-level climate response programs and develop community sensitivity to theenvironment.

Sustainable Development Concept: Encourage learners to reflect on community sustainability concepts and define community development goals in the context of climate change. Introduction to community sustainability assessment tools to help learners understand the relationship between community sustainable development and climate change.

Climate education at the community level: Conduct climate education activities at thecommunity level, including school-to-community collaboration, a series of educational activities at community schools, climate change thematic events, lectures, research tours, etc.

2.2.9 International cooperation and global perspective

Understand climate change challenges and response strategies in different countries and regions. Provide practical cases and practices for learners to broaden their global perspectives.

Develop international cooperation to promote transnational cooperative learning among learners and to collaborate in promoting climate change education.

2.2.10 The key role of education and publicity

Emphasize the key role of education in raising social awareness and promoting climate action.

Explore the concrete role of media and social media in publicity.

2.3 Education objectives and content structure diagram for all people

Table 1. Education objectives and content structure diagram for all people

Objectives		Reference Points							
Primary	Secondary	Kinder-garten	Primary School	Middle School	High School	Vocational School	University	Comm-unity School	Elderly School
1. Form the right values and rational ways of thinking	1. Enhance learners' awareness of the community of human destiny and develop the values of respecting, collaborating, and contributing to human civilization and human development; develop a way of thinking that rationally understands the relationship between the individual and humankind; and possess the value of realizing the unity of the community of life between human beings and nature and the community of human destiny.								
	2. Enhance the orientation of unity of knowledge and action, leadership innovation, and form a rational understanding between bit-by-bit efforts and systematic renewal.								
	3. Strengthenthe value orientation of the present and the initiative to form a rational understanding of the current and future.								

(Continued)

Objectives		Reference Points							
Primary	Secondary	Kinder-garten	Primary School	Middle School	High School	Vocational School	University	Community School	Elderly School
2. Develop knowledge and capacity to address climate change	1. Develop natural science knowledge related to climate change.								
	2. Develop human and social science related to climate change.								
	3. Develop learners' leadership, practicality, comprehensive problem-solving and lifelong learning skills.								
3. Cultivate ways and habits of producing, living and participating in governance for sustainable development	1. Integrate and reflect climate change knowledge and competencies in daily life.								
	2. Apply the climate change knowledge and competencies in production.								
	3. Apply the climate change knowledge and competencies in social and global governance.								

3. Climate Change Education Approaches and Methods

3.1 Approaches to climate change education

(1) Interdisciplinary integration and project-based Learning: Educators can promote

deep, shared learning among learners, parents and community members by integrating the teaching and learning of various disciplines within educational institutions, developing school-based curricula and organizing project-based learning, developing a school culture, and guiding and facilitating home- and community-based learning activities.

(2) Parent-child, intergenerational learning and family community learning: Parents and grandparents can promote the joint participation of people with different backgrounds and expertise in climate change education activities by supporting home-based learning, supporting learners' school and community learning, developing parent-child and intergenerational learning, and conducting family and community learning.

(3) Openness and cooperation in social resources: Encourage open educational resources, such as government, enterprise, private sector and social organizations, to work with families and schools to provide practical opportunities and expertise support for staff, students, parents, teachers, community residents, etc. Explore collaborative projects between enterprises and schools such as field trips and workshops for enterprises to enhance the practical understanding of climate change by learners.

(4) Digital learning: Pay attention to digital resources, guide digital learning, and use digital platforms to disseminate learning outcomes, and interactive exchange, and improve the effectiveness of learning.

3.2 Methods of climate change education

Carry out climate change education for all people, advocate individual learning, group learning, co-learning and mutual learning, advocate a variety of approaches such as project-based learning, observation learning, experiential learning, inquiry-based learning and classroom learning, and emphasize multiple types of learning resources such as the real world of production and daily life, books, videos, simulation software, and so on.

The educational process should emphasize targeted education for the characteristics of learners. For example, for adolescents, particular emphasis should be placed on the awareness and understanding of this comprehensive and complex issue through formal learning, supplementing informal learning, acquisition of systematic knowledge and skills, leadership exercise in project-based learning, and motivating their parents and community. For adults, it is important to emphasize the high integration with production, daily life and governance and the integration of the family, school interaction, and parent-child learning, and encourage participation in climate change education and advocacy activities. For the elderly, the emphasis should be placed on human heritage, encourage older practitioners to strengthen scientific learning and technology application, encourage intergenerational learning, and stress service learning.

The above are general recommendations that educators and learners can specifically reference, adopt, update and develop in climate change education.

4. Educational Evaluation

(1) Emphasize the formation of learners' literacy. Evaluation results are formed through the changes in literacy before and after the learners' participation in learning.

(2) Emphasize the creation and application of circumstances. Evaluate the effectiveness of climate change education in specific urban and rural development, daily life, production and governance.

(3) Emphasize data collection and analysis. Attach great importance to feedback from learners, educators and participants, accumulate multiple types of data, and emphasize data-supported conclusions.

(4) Emphasize the subject and method of evaluation. Give full play to multi-subject evaluation, such as self-evaluation, peer evaluation, teacher evaluation, expert evaluation, social evaluation, etc., and be able to jointly promote and practice feasible programs to address climate change.

5. Educational Support

(1) Support educators, parents and people in society to learn the expertise of climate change themes, to continuously develop green skills, to accumulate educational experience and to enhance the wisdom of sustainable development education.

(2) Establish a curriculum framework for pre-service and in-service professional development training of teachers. Develop monitoring tools and institutionalize teacher training for climate change education.

(3) Promote the updating of educational policies, especially encouraging the participation of adults and the elderly in climate change education programs.

(4) Promote the formation of research alliances, encourage educators, climate professionals, parents and the public to establish climate change education alliances, form cross-border and cross-regional research and practice, and innovate collaborative and linkage mechanisms. Not only to carry out research in key areas based on regional characteristics and make effective breakthroughs but also to enhance the quality of climate change education as a whole by complementing each other and learning from each other and conducting joint studies and research.

(5) Establish special projects to guarantee project funding, encourage research and promote the production, dissemination and transformation of educationalknowledge.

(6) Encourage international exchanges, establish or participate in the construction of an online resource library as a platform for exchanges to share China's excellent cases and practices in climate change education, and contribute China's strength to the realization of the global sustainable development goals.

Written by:
Jiacheng Li, East China Normal University
Danlei Zhu, East China Normal University

Huifen Gu, Longhutang Experimental Primary School, Xinbei, Changzhou, Jiangsu

Yining Yang, East China Normal University

Caiying Chen, Bao'an Middle School (Group) Experimental School, Shenzhen, Guangdong

Jie Huang, The Fifth Affiliated School of East China Normal University, Shanghai

Yifan Shi, Xinchang Experimental Middle School, Pudong, Shanghai

Chunfei Guo, Chongming District Education College, Shanghai

Qiuju Li, Nanhui New Town Community (Elderly) School, Shanghai

Pinting Zhao, First Primary School of Shengli Gudao, Dongying, Shandong

Yaning Wang, Dongying Vocational College, Dongying, Shandong

Translated by:

Fuli Wang, The Fifth Affiliated School of East China Normal University, Shanghai, China